Measurement and Instrumentation in Engineering

MECHANICAL ENGINEERING

A Series of Textbooks and Reference Books

Editor: L.L. FAULKNER Columbus Division, Battelle Memorial Institute, and Department of Mechanical Engineering, The Ohio State University, Columbus, Ohio

Associate Editor: S.B. MENKES Department of Mechanical Engineering, The City College of the City University of New York, New York

Measurement and Instrumentation in Engineering

PRINCIPLES AND BASIC LABORATORY EXPERIMENTS

Francis S. Tse
Ivan E. Morse
University of Cincinnati
Cincinnati, Ohio

MARCEL DEKKER, INC.   New York and Basel

Library of Congress Cataloging-in-Publication Data

Tse, Francis S., [date]
   Measurement and instrumentation in engineering: principles and basic laboratory experiments / Francis S. Tse, Ivan E. Morse.
       p.   cm.   —   (Mechanical engineering: 67)
   Bibliography: p.
   ISBN 0-8247-8086-8
   1. Engineering instruments.   2. Engineering instruments—
Experiments.   I. Morse, Ivan E.   II. Title.   III. Series:
Mechanical engineering (Marcel Dekker, Inc.); 67.
TA165.T75 1989   89-7884
620'.0044—dc20   CIP

This book is printed on acid-free paper.

MARCEL DEKKER, INC.
270 Madison Avenue, New York, New York 10016

Current printing (last digit):
10 9 8 7 6 5 4 3 2 1

PRINTED IN THE UNITED STATES OF AMERICA

In Memory of
Karl Hans Moltrecht

Preface

Experimental data must be valid to the desired degree of accuracy, consistent with the purpose of the experiment. Our objective here is to present a mathematical basis for obtaining valid data and basic concepts in measurement and instrumentation. This differs from some other excellent books on the subject, which tend to either describe specific instruments in detail or emphasize the design of experiments and statistical methods. The data must be valid, however, before statistical analysis is applicable.

The sources of noise and loading are identified and documented in order to obtain valid data in a measurement. To this end, a measuring system is examined in terms of (1) physical laws describing the interaction between a transducer and its environment, (2) modeling, basic methods of measurement, errors and uncertainty, (3) arrangement of components in a system and the interaction between components, and (4) system dynamics. These related attributes are presented in the first four chapters. The subsequent chapters are oriented toward current engineering practice and applications.

The conventional functional stages of a measuring system are described in the next two chapters. The transducer/detector stage for non-self-generating transducers is presented in Chapter 5. It takes the viewpoint that if a physical law describing a transducer relates several variables, then it can be used to measure each of the variables if only one of the variables is desired, and all others are undesired and are sources of noise. The intermediate stage for signal conditioning and the terminal stage for data utilization are treated in Chapter 6.

Current engineering practice for measurements of motion, flow, and temperature are described in greater detail in Chapters 7, 8, and 9. Some topics of interest, such as acoustics, are omitted in order to limit the size of the book. There is no unique method to classify transducers. Evidently, many phenomena or physical laws can be used to measure a variable, such as a temperature. The other side of the coin is that the same phenomenon can also be used to measure many variables. For example, a strain gage is susceptible to both strain and temperature. Thus, a resistance wire can be a strain gage or a temperature sensor. The deduction is that any object that has a property influenced by a variable is potentially a transducer for that variable. By viewing measurement from physical laws and effects, the current practice in measurement can be extrapolated to future needs. For example, if a physical law relates flow, temperature, tubulence, and drag force in an equation, then heat transfer, turbulence, or drag force can be utilized for the construction of a flow transducer. Again, the equation also points at the possible sources of noise in a flow measurement.

The laboratory experiments in Chapter 10 complement the text material. Basic, general-purpose laboratory instruments are the equipment required, and the capital investment for the equipment is nominal for an engineering laboratory. The experiments are given in sufficient detail to minimize the instructional time. Some topics are used as demonstrations in order to cover the scope of a broad study, but many other suitable experiments, such as an air gage, a transistor amplifier, or digital techniques, are not included in order to limit the size of the book.

This book is suitable for a concurrent lecture and laboratory course at the junior level in engineering. The prerequisite is basic electrical network. The concurrent laboratory may be the first laboratory course in the engineering department. It is difficult to synchronize the lectures and the laboratory at times, and we wish to ask the instructor and students for their indulgence. When the experiments are used for an independent laboratory course, the laboratory session is preceded by a lecture, and the other chapters are used as references. The book is suitable for a senior or beginning graduate level course in measurement and instrumentation if students have prior laboratory experience.

The book includes a wide scope of material, and the selection of topics is fairly flexible. Chapters 1 and 2 and parts of Chapter 3 give the framework for a course in measurement. The instructor may wish to select topics from other chapters to utilize the background of students. Our experience is that it is difficult to teach a meaningful lecture course on the subject to students with no prior laboratory experience, or without an associated laboratory course.

We wish to acknowledge our indebtedness to friends and students for their suggestions, to the authors listed in the references, and to those who have contributed to this area. Particular thanks are due to Professors Kurt S. Lion and Peter K. Stein, who pioneered in the field of measurement.

Francis S. Tse
Ivan E. Morse

Contents

Measurement and Instrumentation in Engineering

1
The Place of Measurement, Instrumentation, and Laboratory

1-1. INTRODUCTION

Innumerable measurements are made everyday with countless instruments. The field of measurement, instrumentation, and laboratory means different things to different people in our pluralistic society. The subject is more difficult to teach than a "traditional" laboratory course, because students tend to associate the subject with specific instruments instead of with a more general approach to the study. Hence the roles of measurement, instrumentation, and laboratory are discussed in this chapter rather than in the preface. The discussion is in greater detail than in the usual textbook introduction. Hopefully, the digression will bring into sharper focus the organization of the topics to be presented.

The book is intended as an undergraduate text and/or a supplement for a laboratory course. The objectives are (1) to examine the methodology of measurement to judge the validity of test data, (2) to present principles by which the hardware can be understood, (3) to follow the principles with applications, and (4) to suggest a sequence of laboratory exercises aimed at acquiring a working familiarity with basic laboratory instruments.

The methodology for obtaining valid measurements is discussed in the early chapters. Although numerous measurements are made, the same methodology is used regardless of the type of measurement or the instrumentation.

A measurement must be valid. When an engineer has experimental data, the first question should be: Are the data worth analyzing? Or, phrased another way: Are the data valid? What accuracy can be expected? How much of your reputation are you willing to bet? These are difficult questions, but they must be answered satisfactorily before the data analysis.

The discussions in the early chapters should enable the reader to understand the general principles of the hardware. The engineer should know what an instrument can and cannot do in an application. Few of us need to be convinced of the importance of instrumentation or reminded of proliferation of electronic instruments in recent years. The field is literally overflowing with hardware. These devices cannot be mastered by one person in a lifetime. Yet there is the uneasy feeling that the information is important and that one's own knowledge can become obsolete. By discussing principles first, we will not be overwhelmed by details. Furthermore, we will find a means to organize the knowledge, which may otherwise appear as pieces of unrelated information.

Subsequent chapters deal with more specific measurements, such as pressure and temperature. Principles are emphasized rather than instrumentation technology. As an introductory course, the perspective presented will be more useful to readers than will the detailed coverage of specific topics.

A group of laboratory exercises is included in Chapter 10. The methodology and principles provide the necessary background for the exercises. There is no substitute for hands-on experience. Furthermore, familiarity with equipment is a prerequisite for meaningful measurements. Only simple and basic exercises are suggested. Since skill is transferable, the reader should be able to advance from the basic to more complex problems.

This book is intended as a teaching text at the junior or senior level. It is neither a compendium of transducers nor an operating manual for specific instruments. Excellent information in this regard can be found in the literature or obtained from vendors of instruments. Statistics is treated only superficially, since the text focuses on the "front end" of experimentation. Statistics can be applied after it has been ascertained that data are worth analyzing. There are excellent books available on statistical analysis and the design of experiments.

It is assumed that readers are somewhat conversant with basic subjects such as electrical networks. Alternatively, the instructor can give a brief introduction to the subjects as needed. The background of measurement is drawn from a broad range of subjects, which are separate studies in themselves. In reality, engineers must be conversant with the fields in which they intend to make measurements. For example, it is futile to measure temperature without an understanding of the heat transfer process. If the error due to the heat transfer is unknown, the temperature measured must be questioned.

With the advent of computers, digital processing of experimental data is fast becoming an important field. This is a more advanced subject but will not invalidate the basic information presented in this book.

1-2. SIGNIFICANCE OF MEASUREMENT AND INSTRUMENTATION

Measurement is the common thread that runs through the fabric of all science and engineering. In this section we briefly relate measurement and experimentation to the development of science, research, engineering design, and the manufacture of goods. Measurement and experimentation can also be justified on their own merits.

The development of science can be traced through experimentation, observation, the generalization of facts, and subsequent formulation of hypotheses and theories. Experiments supply the facts. It is easy enough to collect a massive amount of data. An accumulation of facts, however, is no more a science than a pile of bricks is a house. The facts must be simplified and stripped of "nonessentials" such that each item of information fits into the framework of a physical law. A law is observable experimentally, and a theory seeks to explain the law. A measurement must be valid; invalid data will lead to a faulty theory and then be verified by other erroneous experiments. After all, a theory only represents knowledge at a given time. Theories, as devised by human beings, impose on science. They do not impose on nature, for nature dictates the laws.

Measurement is a vital link in the chain of events in research and development. The chain begins with a definition of the problem and objectives, and ends with the utilization of information. The events can be involved, but the chain is only as strong as its weakest link. Valid measurement is no less important than any other phase of the work.

Experimentation is an integral phase of engineering design. Theories predict the performance of a design. Experimentation measures what actually happens. Theoretical analysis and experimental verification are complementary, and together they form the basis of our design method.

Measurement is necessary for proper operation, maintenance, and control of equipment and processes in manufacturing. Proper maintenance and control are the prerequisites for efficient production, which is the ultimate payoff in manufacturing. Without the means for measurement, automation would not be feasible. In fact, the industrial revolution was made possible through the introduction of mechanical power, the success of which was hinged on the flyball governor for measuring and controlling the speed of the steam engine.

Experimentation can be justified on its own merit as a source of information. Alexander Bell, Louis Pasteur, the Wright Brothers, and Thomas Edison were experimentalists who left their marks in history. An experiment can furnish specific or general information. For example, the measurement of the tensile strength of steel is specific. Data from experiments are also used to formulate empirical relations. Alternatively, phenomenological relations are studied with the hope of developing a general theory.

Engineers often resort to measurements even for familiar objects. An automobile proving ground is an example. A complex design, such as the control system of an aircraft, is often investigated experimentally. It is expedient to obtain the composite operating relations by measurements, even when the characteristics of the individual components of the system are known. Other experiments may consist of some hardware interfaced with a computer, when certain hardware is not yet avaialble or is very expensive.

1-3. MEASURING SYSTEMS

Measurements are generally made with instruments. Human senses can be very keen but are lacking in latitude. Our ability to sense hot and cold is only relative, and only over a very limited temperature range. Measuring a high electric voltage with human senses can be hazardous. An overview of measuring systems, the configuration, and the process of sensing are described in this section. For a more realistic discussion, the thermocouple and the resistance strain gage are used as examples of typical sensors. These are described at the end of this section. Conforming to common usage, the terms *sensors, transducers, instruments,* and *systems* are used interchangeably in the discussions.

A. The Transducer and Its Environment: An Overview

Let us examine the process of *sensing.* A transducer G is placed in an environment E for a measurement shown in Fig. 1-1a. In this section we examine the interaction between G and E.

The environment may not be uniform or at a steady state. There must be interaction between G and E, or else a measurement cannot be made. To this end, G is subjected to all the influences of E, and vice versa. These influences, denoted as inputs and outputs, are shown as double arrows in the figure to indicate the multivalues. Hence a transducer is potentially a multi-input, multioutput device. Not all the influences are of equal magnitude. Some influences may be large and others insignificant, but they always exist.

Only one of the many inputs is the *measurand.* This is the desired input and the object of the measurement. All other inputs are undesired. The extraneous inputs can contribute to corruption of the signal at the output. One example concerns malfunctioning in airborne electronic equipment.

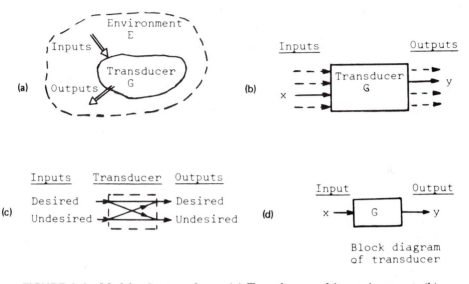

FIGURE 1-1   Models of a transducer. (a) Transducer and its environment. (b) Input/output quantities of a transducer. (c) Coupling in transducers. (d) Block diagram of transducer.

The equipment was calibrated on the ground, but repeated malfunctioning occurred when airborne. The engineer finally traced the trouble to mechanical vibration. The vibration was the undesired and extraneous input.

A *noise* is any unwanted signal mixed in with the desired signal at the output. It has the same form as the signal and the two may or may not possess distinct characteristics. For example, if the output of an instrument is a voltage, the signal and noise are both voltages. Methods of noise treatment are discussed in Sec. 3-7.

Consider a metallic resistance strain gage (see Sec. 1-3E for a description). Let the gage be subjected to both strain and temperature inputs. Although both inputs will produce a resistance change in the gage, only the strain is the measurand. The temperature in a strain measurement is an undesired input. The corresponding resistance change will give a noise output. On the other hand, the resistance of a metallic wire will change with temperature, and a strain gage could be used as a temperature sensor. In fact, a strain gage is furnished with a temperature calibration. The discussion shows that an undesired input can well be the desired one in another application. Once a horticulturist was asked to define a weed. His reply was, "In a tulip bed, a rose is a weed."

Only one of the outputs is the *signal*. This is the output corresponding to the measurand. All others are undesired and can compound the difficulties of a measurement. Consider a resistance wire sensor. Depending on the application, many effects can be produced by passing an electric current of various magnitudes through the wire. It is feasible to use each of the detectable effects for a different measurement. Hence each detectable effect is an output. For example, when the heating due to the current is negligible, the wire is a strain gage or a resistance thermal detector (RTD). The current can also produce heating in the wire. An airflow will cool the heated wire, and the wire becomes a flow meter as in a hot-wire anemometer. There are many applications for a heated wire. In the extreme case, the wire becomes incandescent as in an electric light bulb. The highly heated wire can be compared with the temperature of steel in a heat-treating furnace. This is a basis for optical pyrometry. The discussion does not imply that the instruments enumerated have the same physical form. The common denominator is that many effects can be produced by an electric current through a resistance wire. It is feasible to utilize each of the effects for measurement.

The discussions above show that the input/output quantities of a sensor are coupled as shown schematically in Fig. 1-1c. Cross-sensitivity, coupling, or crosstalk is common in instruments. Consider an accelerometer for measuring the vibration along its z axis. Normally, the accelerometer is relatively insensitive to vibrations along its x or y axis. The output due to vibrations along the x or y axis is due to the cross-sensitivity and is a part of the specification. Coupling can also cause severe interference between the signal channels in a multichannel instrument. There was a tool dynamometer (see Fig. 1-3 for a description) borrowed from a research laboratory for a thesis project. The dynamometer was for measuring the cutting forces in the x and y directions in a milling machine. Under static conditions, the output voltage V_x of the dynamometer would correspond to the force F_x in the x direction. Similarly, V_y would correspond to F_y. Probably due to resonances under dynamic conditions, V_y was greater than V_x when the force F_y was zero. This was perhaps an extreme case of dynamic coupling in an instrument. The matter was brought to the attention of the research laboratory. The reply was: "We never thought of it that way."

Now consider the consequence of a measurement on the environment E in Fig. 1-1a. When a transducer G is placed in E, the interaction between G and E will alter the state of E in the neighborhood of the transducer. This is loosely called a *loading effect,* which is a consequence of the act of the measurement itself. Loading is due to (1) an energy transfer between the transducer G and the environment E, and (2) a possible structural change in E and/or G.

First, a measurement is a *transfer of information* between the transducer and its outside world. The information transfer must be accompanied by an *energy transfer* [1]. For example, placing a large thermometer in a cup of hot tea could cool the tea sufficiently to yield an erroneous reading. Alternatively, placing a thermometer at room temperature in a glass of cold water could warm up the water adjacent to the thermometer. The difference between the true and the indicated temperature is an error attributed to loading due to the energy transfer. This is examined by means of modeling in Sec. 2-3 and in terms of input/output impedance in Sec. 3-3B.

Second, the local condition or the *structure* of a medium may be altered by the presence of the probe for the measurement. For example, a probe in a flow stream can disturb the flow pattern, which is the object of the measurement. A strain gage cemented to a piece of thin plastic can stiffen the plastic when the strain in the plastic is the object of the investigation.

The loading effect can be large or insignificant, but it always exists in a measurement. It can be disregarded when insignificant, but it should not be dismissed without some consideration. Since loading is unavoidable, "perfect" measurement is not possible. A perfect measurement is obtainable if (1) "the transducer is not there," and (2) the response of the transducer is instantaneous.

A transducer G is often modeled by means of a simple single-input, single-output block diagram, as shown in Fig. 1-1d. The input is the measurand, the output is the corresponding desired signal from the transducer, and G relates the output to the input. Loading and other ramifications are ignored in this simplified representation.

B. The Nature of Measurement: A Design Problem

Measurement can be viewed as the reconstruction of the input from the observed output of an instrument. The input is the measurand and the output is the signal from the instrument. If a thermometer gives a reading, what is the true input temperature? If a thermocouple (see Sec. 1-3E for description) gives a voltage, what is the temperature in the furance? Although this viewpoint is appropriate, it is not expedient for applications. Inherently, measurement is a design problem. We shall discuss these two views of measurement.

Let an instrument G be represented by the simplified model in Fig. 1-1d. The input x causes a change in G, and the subsequent effect observed is the output y. The input x is commonly called the *excitation* and the output y, the *response*. This *causal relation* is denoted by the *transfer function* G, describing the input/output relation of the system. Hence G is a mathematical model representing the characteristics of the system, including the system dynamics:

$$\text{output } y = (\text{transfer function } G) (\text{input } x) \tag{1-1}$$

or

$$\frac{\text{output } y}{\text{input } x} = \text{transfer function } G \tag{1-2}$$

The transfer function G in the equations above is an operator. It may be a simple constant or a mathematical expression. G is used in Chapter 4 to describe the dynamics of systems.

The relation between the input x, the output y, and the system transfer function G is divided into three classes of problems, as shown in Table 1-1.

1. If x and G are given in order to find the output y, the problem is one of *analysis*. We are familiar with such typical textbook problems.
2. If x and y are given in order to find G, we have a *synthesis* problem. G must be designed to meet the stipulated input/output relations.
3. If y and G are known and it is required to find the input x, it becomes a *measurement* problem. For example, if the characteristics of an instrument is known and a temperature is recorded, what is the true temperature in the jet engine?

It is evident that measurement differs from an analysis or a design problem. The reconstruction of the input information from the output signal is based on Eq. 1-1 or 1-2. The equations seem simple enough, but the implementation will introduce errors of its own. The reconstruction procedure is used only for exceptional cases.

In practice, measurement is a design problem. Suppose that the experiment is to measure the exhaust temperature of a gas turbine. The environment E and the transducer G must be considered together. If the flow in the exhaust of a turbine is not uniform, locating G in the exhaust for obtaining a representative temperature is a design as well as a

TABLE 1-1 Types of Problems

Problem type	Input x	System G	Output y
Analysis	Known	Known	Find
Synthesis	Known	Find	Known
Measurement	Find	Known	Known

measurement problem. Furthermore, for locating the temperature sensor at one location, the type of data, the accuracy desired, and whether the data obtained will satisfy the problem must be considered prior to the experiment.

A measurement is designed to meet the specification for accuracy, since a perfect measurement is not possible. If the problem calls for a 5% accuracy, it behooves the engineer to meet this specification. An accuracy of 20% will be inadequate, but an accuracy of 0.5% could be a waste of resources. An instrument more sophisticated than needed can hinder an experiment. We often hear: "Gee. We pay so much for the instrument, it's got to be good." The reply is: "Good for what?"

Furthermore, a temperature sensor is designed for a given environment to yield an accuracy consistent with the intended purpose of the experiment. It is axiomatic that the sensor can indicate only its own temperature; and it cannot indicate what it does not sense. Thus the sensor must be exposed only to a properly designed environment and be exposed only to a properly designed environment and be safeguarded against all other influences. This viewpoint of measurement seems artificial. It tacitly assumed that a great deal is known about the source of information. There is no alternative, because G and E must be considered together to obtain a true measurement. This also implies that the calibration of an instrument under idealized conditions is only a reference. A realistic calibration must be done under in-service conditions.

When a sensor is designed for a specific measurement, it does not give all the information of the source. For example, a common voltmeter gives only the nominal root-mean-square (rms) value of an alternating-current (ac) voltage. It does not indicate the true rms value, the waveform, the voltage regulation, or even the line frequency. It seems redundant to state that an instrument is normally designed for measuring a specific item, but readers may feel that a common voltmeter will give all that is to know about an electrical outlet, or an instrument will yield all the information of a source. In reality, it is neither possible nor necessary to have "all" the information of a source in an experiment.

C. Functional Stages of Measuring Systems

The block diagram of the *functional stages* of a measuring system is shown in Fig. 1-2. The grouping of components into functional stages is a convenient way to describe a system. This representation is shared by all measuring systems, regardless of the variable being measured or the hardware deployed. Loading and other effects are ignored in this representation.

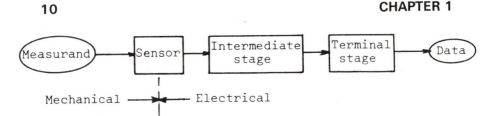

FIGURE 1-2 Functional states of measuring systems.

The first stage is the *sensor* or detector. Its function is to detect the measurand and to produce a corresponding output signal. The function of the intermediate stage is *signal conditioning* and *transmission*. This may include amplification and many other necessary modifications of the signal. The terminal stage prepares the signal for *utilization*, such as signal monitoring, indication, or data storage for processing. It is understood that the software and hardware interfacing the stages must be compatible.

Although the measurand in our discussion is "mechanical," the signal output from the sensor is often electrical, as shown in Fig. 1-2. One cannot help but notice the "black boxes" in a research or a test laboratory. A pure mechanical system has certain disadvantages, such as friction, tolerance, and the inherent mass of the parts. These drawbacks can be overcome to some degree by miniaturization and the use of precision parts, if one is willing to pay the price. The main disadvantage of mechanical devices is the inherent inability for power amplification. It is not possible to obtain an output at a reasonable power level without seriously loading the source of information, since the output energy must be from the source. On the other hand, the amplification of a weak electrical signal to a high power level, such as in a hi-fi amplifier, can readily be done.

Let us illustrate a measuring system with a tool dynamometer for measuring the vertical cutting force in a lathe. The functional stages are illustrated in Fig. 1-3a. The first stage is the sensor, the second stage for signal conditioning consists of the amplifier and filter, and the terminal stage is an oscilloscope for monitoring the cutting force.

The sensor, as shown in Fig. 1-3b, consists of a cutting tool, a metal block with a slot, and a strain gage. Assuming that the tool is rigid, the applied force F causes the block to deform, thereby narrowing the open end of the slot by a small amount. The metal above the slot is strained accordingly. The strain is sensed by means of a strain gage. The output of the gage is an electrical voltage analogous to F. Hence the primary sensor in the dynamometer is a heavy spring in the form of a metal block. The secondary sensor is the strain gage, which is electromechanical. The com-

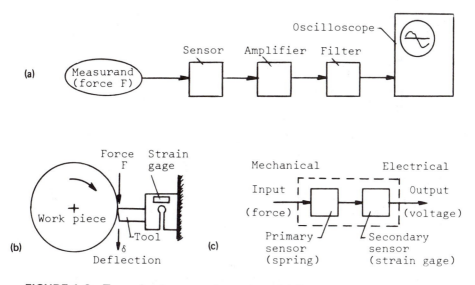

FIGURE 1-3 Example of a measuring system. (a) Components of a measuring system. (b) Tool dynanometer. (c) Force transducer.

bination converts a mechanical signal into an electrical output, and is often called a transducer, as shown in Fig. 1-3c.

A *transducer* is an energy conversion device. The sensor is called a transducer because it transforms the energy at its input into a more convenient form at the output. In this illustration, the input energy is due to the force F and its associated displacement δ. The metal block converts the input energy into strain energy at the location of the strain gage. The gage transforms its own strain energy into an electrical output. An external electrical input is required for driving the gage, and this will be described later. Since an information transfer is accompanied by an energy transfer, all sensors are energy conversion devices and are therefore transducers. Similarly, all components in a measuring system transmit information and convert energy from one form to another. Only sensors are commonly called transducers.

The signal from the tool dynamometer in Fig. 1-3 is amplified and filtered and then monitored by means of an oscilloscope. Here the dynamometer is the sensor. The amplifier converts a weak signal from the strain gage into a stronger one for transmission. The filter eliminates the noise from the signal. If the amplifier and the filter are subminiaturized and incorporated into the dynamometer, the combined package is called a force transducer. The grouping of components to form functional stages is a convenience. There is no rigid rule for defining the stages, but the general functions described above are shared by all measuring systems.

D. The Sensing Process

The sensing process is governed by physical laws. At the same time, a transducer is potentially a multi-input, multi-output device, as illustrated in Fig. 1-1. These concepts are combined in this section. It will be shown that (1) different phenomena or physical laws can be used for sensing the same measurand, and (2) the same phenomenon can be used for sensing different measurands. In fact, the views are but two sides of the same coin.

Table 1-2a shows different phenomena for sensing the same measurand. Let us enumerate several temperature sensors to illustrate the table. A mercury-in-glass thermometer utilizes the differential expansion of two materials, mercury and glass. When the mercury expands more than the glass in the bulb, some mercury rises in the stem to show a temperature change. A thermistor is made of semiconducting material that has the property of resistance change with temperature. Thermistor temperature probes are commonly used. The resistance of a metallic wire changes with temperature. Resistance thermal detectors (RTDs) are used for precision temperature measurements. The pressure of a gas under constant volume increases with temperature according to Charles' law. Hence temperatures can be measured by means of pressure gages. A thermocouple utilizes the Seebeck effect to convert a temperature into an electrical signal. The melting points of gold and silver are standard fixed points for tmeperature calibration. Gold and silver are therefore "thermometers." It seems that the types of thermometers or the phenomena that can be utilized for temperature sensing are almost without limit.

TABLE 1-2 Causal Relations Between Measurands and Phenomena for Sensing

(a) Phenomena for sensing a measurand		(b) Effect of input variables on a phenomenon for sensing	
Measurands (inputs)	Phenomena for sensing	Input variables (measurands)	Phenomena for sensing
Flow	Expansion	Flow	Expansion
Temperature	Resistance	Temperature	Resistance
Pressure	Inductance	Pressure	Inductance
Motion	Capacitance	Motion	Capacitance
- - - - -	- - - - -	- - - - -	- - - - -

If a temperature causes a detectable change in an object, that object is potentially a temperature sensor. This is simply a statement of the causal relation in Eq. 1-1. The open-endedness of the list of phenomena in Table 1-2a shows that a listing of possible temperature sensors cannot be comprehensive in a transducer compendium or in a textbook.

The same statement can be made for other measurements. To measure the distance between two points, must the instrument be a yardstick? Depending on the application, a distance can be measured by electricity as in radar, by sound as in sonar, or by light as in laser.

The generalization from Table 1-2a is that if a measurand causes a detectable change in a device, that device is potentially a sensor for the measurand. The phenomena for sensing can be linear or nonlinear and reversible or irreversible. This generalization gives the bases for organizing the later chapters, such as measurements of flow or temperature. Furthermore, it shows that we should not be taken by surprise if there is a new sensor on the market using a new phenomena. The generalization also enables us to search systematically for a new type of transducer when one is not available. Be cautioned, however, not to attempt to design or build a transducer when one is available in the market. The development cost of the instrument has already been paid by the manufacturer and spread over a large number of units.

The other side of the coin is that the same phenomenon can be used for sensing many variables. Consider the types of sensors that can be constructed from the phenomenon of variable resistance, as shown in Table 1-2b. It is feasible to use a resistance wire as a strain gage, a thermometer, a hot-wire anemometer, or an optical pyrometer, as enumerated in Sec. 1-3A. If a wire is subjected to high hydrostatic pressure, it will undergo an elastic size change and thereby a resistance change. Thus it is feasible to use the wire for a high-pressure transducer, as in a Bridgman-type gage. This shows that the phenomenon of resistance change in a wire has numerous applications. If a measurand causes a detectable resistance change in the wire, the resistance change can be utilized to indicate the measurand. We do not imply that the details of the various instruments are of secondary interest. The overriding fact is that the transducers enumerated can be modeled as a metallic resistance wire.

The model study will enable us to generalize, to discuss physical laws, and to investigate sources of noise in measurements in Chap. 2. If a phenomenon such as a resistance change in a wire is susceptible to many inputs, and if only one of which is desired, all others are undesired and must be sources of noise in the measurement. For example, if strain is the measurand, the strain gage must be guarded against the effects of temperature, thermal gradient, hydrostatic pressure, electrical pickups, and other influences. Again, the list of noise sources must be open ended. Some inputs will produce greater effects on the sensor than others. This reinforces the statement that measurement is a design problem. The transducer must be exposed only to the correctly designed environment to yield the correct answer.

E. Examples of Typical Sensors

Two typical sensors, the thermocouple and the metallic strain gage, are described here to supplement the background for the discussions. The thermocouple is representative of the self-generating transducer and the strain gage that of the non-self-generating type. The modeling of the two classes of transducers is discussed in Chapter 2.

1. Thermocouples

A *thermocouple* for temperature measurement is formed by two wires A and B of different metals, as shown in Fig. 1-4. Let the junctions of the wires be at the temperatures T_1 and T_2. The voltage V developed in the circuit in Fig. 1-4a is due to the thermoelectric or *Seebeck effect,* and V is a function of $(T_2 - T_1)$. Actually, V is due to two effects. The voltage due to the contact of the metals is the *Peltier effect,* and that due to the thermal gradient along the wire is the *Thomson effect.* Only the overall effect is of interest for normal temperature measurements. Alternatively, the current I can be used to indicate $(T_2 - T_1)$, as shown in Fig. 1-4b. Voltage measurement with a voltmeter or a potentiometer (see Fig. 3-1) is used almost exclusively, since this avoids the effect of resistance in the circuit and loading in the measurement.

The voltage developed in the circuit is not linear with the differential temperature $(T_2 - T_1)$ for the normal range of the thermocouple. Tables [2] are used to convert the output voltage of the couple to a temperature with respect to a reference temperature. In the laboratory, one of the

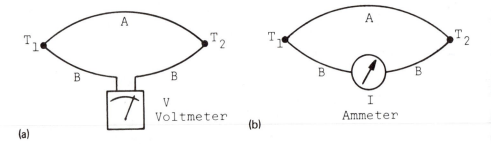

FIGURE 1-4   Basic thermocouple circuits. (a) Voltage output. (b) Current output.

junctions is at the reference of 0°C, as shown in Fig. 1-5. An automatic reference junction compensation is often found in instruments.

Metals commonly used for thermocouples are iron–constantan, copper–constantan, and chromel–alumel. The sensitivity of these couples is of the order of 50 μV/°C. Platinum/platinum–rhodium couples can be used up to 1500°C and the sensitivity is about 6 μV/°C. Temperature measurements with thermocouples are discussed further in Chapter 9.

Thermocouple circuits can be analyzed by the *thermoelectric laws*. Assuming that the metals are homogeneous, the laws are briefly states as follows:

1. The thermal electromotive force (EMF) is a function of temperatures T_1 and T_2 at the junctions, and is independent of the other temperatures in the circuit. For example, the thermal EMF of the couple in Fig. 1-5 is dictated by $(T_2 - T_1)$ and is independent of intermediate temperatures T_3 and T_4. It is independent of T_5 and T_6 of the copper leads, provided that $T_7 = T_8$ at the meter.

2. If a thermocouple produces an EMF V_1 when its junctions are at T_1 and T_2, and V_2 when at T_2 and T_3, it will produce an EMF $V_3 = V_1 + V_2$ when the junctions are at T_1 and T_3. This allows the use of standard tables (reference at 0°C) when the reference for the actual measurement is not at 0°C. For example, assume an iron–constantan couple, $T_1 = 0°C$, $T_2 = 20°C$, T_3 is the unknown, and T_3 is referenced to T_2. If the voltage due to $T_2 - T_1 = (20 - 0)°C$ is $V_1 = 1.019$ mV, and that due to $T_3 - T_2 = (T_3 - 20)°C$ is $V_2 = 3.167$ mV, then

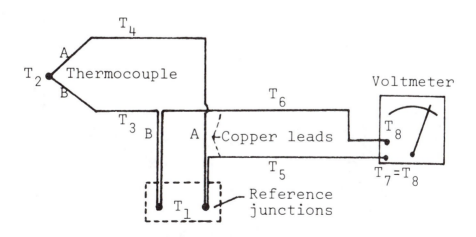

FIGURE 1-5   Thermocouple circuit with reference junction.

$V_3 = V_1 + V_2 = 1.019 + 3.167 = 4.186$ mV. This is due to $(T_3 - 0)°$C. The corresponding temperature from a standard table is $T_3 = 80°$C.

3. If the thermal EMF of metals A and B is V_1 and that of B and C is V_2, the thermal EMF of A and C is $V_3 = V_1 + V_2$. This allows us to pair thermocouples of different metals if each is calibrated with respect to a reference material. For example, the sensitivity of iron–platinum is 18 μV/°C and that of constantan–platinum is 35 μV/°C. The sensitivity of an iron–constantan couple is $(18 + 35) = 53$ μV/°C.

2. Resistance Strain Gages

The common metallic strain gage consists of a fine wire or an etched foil (cross-sectional area $\simeq 10^{-6}$ in^2) bonded to a paper or plastic backing as shown in Fig. 1-6. The assembly is cemented to the surface of a base material. Once cemented, the gage effectively undergoes the same strain as the surface of the base material. Strain as small as 10^{-7} or as large as 10^{-3} can be measured. The resistance wire of some gages are not bonded. These are called unbonded gages and are generally for special applications.

When a gage wire is elongated, its diameter decreases due to the Poisson's effect. Consider a gage of effective length L and cross-sectional area $A = CD^2$, where D is a characteristic dimension. For a circular wire, D = diameter and $C = \pi/4$. The resistance R of the gage is defined by the functional relation

$$R = f(\rho, L, A) = \frac{\rho L}{CD^2} \tag{1-3}$$

where ρ is the resistivity of the material. Differentiating the equation gives

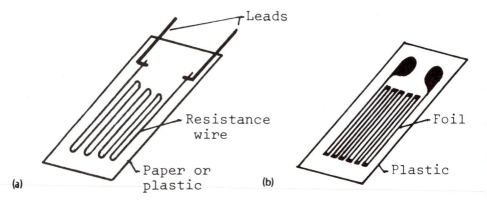

FIGURE 1-6   Resistance strain gages. (a) Wire gage. (b) Foil gage.

$$dR = \frac{1}{CD^2}\left(\rho\, dL - 2\rho L\, \frac{dD}{D} + L\, d\rho\right)$$

Dividing this equation by R from Eq. 1-3 yields

$$\frac{dR}{R} = \frac{dL}{L} - 2\frac{dD}{D} + \frac{d\rho}{\rho} \qquad (1\text{-}4)$$

Since the axial strain is $\varepsilon = dL/L$, the lateral strain is dD/D, and Poisson's ratio is $\nu = -(dD/D)/\varepsilon$, Eq. 1-4 can be simplified to give

$$\frac{dR/R}{\varepsilon} = 1 + 2\nu + \frac{d\rho/\rho}{\varepsilon} \qquad (1\text{-}5)$$

This is the basic equation for resistance strain gages. The gage factor G_f is defined as

$$G_f = \frac{dR/R}{\varepsilon} \qquad (1\text{-}6)$$

G_f is a measure of the sensitivity of the strain gage: The larger the gage factor, the more sensitive the gage. G_f and R are part of the specifications. The value of R ranges from 120 to 500 Ω and G_f from 2 to 3.5 for common gages. Note that the resistivity ρ is strain sensitive and $(d\rho/\rho)/\varepsilon$ in Eq. 1-5 cannot be zero. Let's say, for example, that the value of ν is about 0.3. If ρ is independent of the strain ε, then $(d\rho/\rho)/\varepsilon = 0$ and G_f in Eq. 1-6 would be approximately $1 + 2 \times 0.3 = 1.6$. Since $G_f > 2$ for commonly used gages, the quantities ρ, L, and D cannot be independent.

1-4. THE MEASUREMENT PROBLEM

The measurement problem is described here in terms of valid measurements and the considerations necessary for valid measurements. Although it is convenient to describe a measuring system by means of functional stages as in Sec. 1-3C, this description does not prescribe conditions for valid measurements. Hence valid measurements and measuring systems are reexamined. The discussion also furnishes the rationale for organizing the topics in the next three chapters.

A. Valid Measurements

A measurement must be valid. With invalid data, no amount of sophisticated analysis will give the correct answer. The usefulness of statistics and other techniques is never questioned. Statistics, however, would be misapplied if the data were analyzed before its validity is assured. Pushing the argument to the extreme, if a measurement does not have to be valid, anyone can make any measurement on any process with any instrument. It would be nonsensical to wave a red pencil in front of a vibrating table and say: "Hocus-pocus, the vibration is red." What difference does it make if the statement does not have to be valid?

The mere repeatability of results could imply a valid measurement or simply the repeatability of errors. In ancient China, children beat on pots and pans during a lunar eclipse to save the moon. The sound would scare away the celestial frog that tried to swallow the moon. The event was repeatable and it never failed.

The data are valid if they give the desired information, which may be only a part of the information of an event under observation. Alternatively, the data may have been modified by means of signal conditioning, to get the desired information. It was stated in Sec. 1-3B that an instrument generally measures only a specific item. It is neither possible nor necessary to have "all" the information of a source. A digital voltmeter indicates only the magnitude of the input voltage. A frequency spectrum is a plot of the amplitude of the harmonic components of a periodic signal versus frequency. Some spectrum analyzers on the market do not give the phase angle of the harmonic components. Yet the phase information is pertinent for some related analysis.

A measurement is valid when the data reflect the value of the measurand with the required accuracy. Due to the inherent noise and loading in measurements, perfect measurement is not possible. A measurement can only be made to a degree of accuracy. It is incumbent on the investigator to design the experiment and to verify beyond a reasonable doubt that the data are valid and that the accuracy meets the specification. The most expensive equipment that money can buy will not automatically yield valid data. The same instruments are available to all. A master mechanic and his 10-year-old son use the same screwdriver. It is not the fault of the screwdriver when it is being misused. This prompted a student to remark: "There are no bad instruments, only bad guys."

For most cases, the data are valid if there is a time lag in the measurement but the waveform of the analog data is preserved. The time lag is due to the inherent dynamics of the measuring system. Data with a time lag must be interpreted carefully, especially when an experiment measures several variables simultaneously with a multichannel instrument.

B. Measuring Systems

The concepts of functional stages, loading, noise, and sensing process were presented in the preceding sections. Considerations for valid meas-

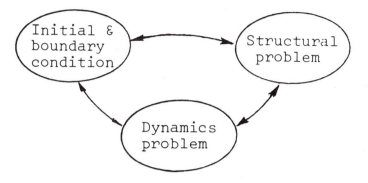

FIGURE 1-7   Interrelated aspects of measuring systems.

urements are introduced in this section. These are (1) properties of sensors, (2) the structure of measuring systems, and (3) the system dynamics, as shown in Fig. 1-7. These aspects are interrelated, since the performance of a measuring system must be viewed in its entirety. For purpose of discussion, they are examined separately.

1.   The Sensor

The process of sensing is governed by physical laws. This is described in Chapter 2. For the time being, we examine the initial and boundary conditions in sensing.

Initial conditions describe the state of a sensor at an operating point or at time zero. A state is defined in terms of properties, as in the study of thermodynamics. Hence initial conditions are synonymous with the properties of a sensor at time zero. The performance of a sensor must be influenced by its properties, about which the initial conditions are defined.

The problem is complicated by the fact that the state of a sensor is dependent on its past history. For example, when a strain gage in a transducer is replaced by a new gage from the same batch of gages, one can no longer expect the same performance. The old gage could have gone through a large number of strain cycles or peak loads. The calibration of a thermocouple may be changed by mechanical straining during installation. An accelerometer could be damaged by dropping it through a small distance. The effects of past history may be irreversible. When high precision is required, the materials in a transducer may not be completely stable with time and oeprating conditions.

Due to the cummulative effect of its past history, each transducer has its own individuality. Two transducers of the same make cannot be assumed identical in every respect. Theory deals with averages, whereas a measurement is made on a particular event with a given instrument. A medical doctor treats a patient as an individual; an underwriter uses mortality tables. Owing to the individuality of instruments, a basic requirement in experimentation is to identify all instruments completely, including the model and serial numbers. It is often debated whether an experiment is truly reproducible, because of the individuality of events in the real world.

The sensing process is a boundary problem, since the interaction between the sensor and its environment must occur at an interface. The corollary is that the environment for the sensor must be designed for the sensor to detect the measurand correctly. The environment is that which affects the sensor, and it may not be the general environment in which the sensor is placed. An instrument never lies and will always indicate what it senses. A thermometer can, however, only indicate its own temperature. Whether the indication is the correct value is a measurement problem.

Calibration serves to reveal the individuality of an instrument. It is often said that all instruments must be calibrated by comparison with a reference standard, and the standard must be traceable to a standards laboratory. Only one variable at a time is changed during calibration. It would be utopian if an instrument were employed under the calibrating conditions of a standards laboratory.

A calibration is generally conducted under idealized conditions; that is, loading and noise due to extraneous inputs are controlled. It may not account for the boundary conditions or the initial conditions of the instrument in applications. For example, a thermometer is calibrated in a constant-temperature bath, the thermal capacity of which is infinite compared with that of the thermometer. The thermometer is immersed to the top of the mercury thread if it is specified for "full immersion." This avoids cooling of the exposed mercury thread and therefore heat loss to the environment. It would be impossible to duplicate the calibrating conditions when the same thermometer is used to measure the temperature of a cup of tea, since the cup has limited size and the tea has limited thermal capacity. Due to the differences in operating conditions, when an instrument works well for one application, there is no assurance that it will perform adequately for another, despite its proven performance and certificate of calibration.

Calibrations are commonly conducted under static conditions. This would not reveal the dynamics of the instrument, such as the coupling in the tool dynamometer described in Sec. 1-3A. Since an information transfer requires an energy transfer, the energy transfer under dynamic conditions must be investigated to evaluate the dynamic loading effect (see Chapter 3).

2. Structure of Measuring Systems

Structure describes the configuration or the arrangement of components in measuring systems. The term will be used to denote the electrical, mechanical, or any configuration. The subject is treated in Chapter 3.

Structure could include a large number of topics, such as feedback. It is evident that the interaction between the components and the performance of a measuring system are influenced by the arrangements of components. Expressing this colloquially, there is a difference between putting the horse before the cart and putting the cart before the horse.

3. System Dynamics

It takes time for an instrument to respond to an input, because information transfer requires energy transfer and it is not possible to transfer energy with zero time. System dynamics in the time and frequency domains are discussed in Chapter 4.

Theoretically, the response of an instrument cannot be instantaneous. The time response of a real instrument will lag the input. The time lag may be large or insignificant. The lag is insignificant when a "fast" instrument is used for measuring a "slowly" varying event. The fastness and slowness are only relative. A barometer is sufficiently fast for measuring changes in atmospheric pressure. The response of the barometer is inadequate for measuring the pulsating pressure at the discharge of an air compressor.

Time, frequency, and transient response can be used to examine the dynamics of systems. *Time response* is a study of system dynamics by means of differential equations for arbitrary inputs. The solutions are time functions, which may indicate a time lag and/or oscillations at the output. Both imply distortions of the input signal at the output. *Frequency response* is a study by means of the sinusoidal steady-state method. *Transient response* is a study by means of Fourier methods.

Since the same system is being examined, the methods are merely different views and descriptions of the same problem. The system behavior must be independent of the method of study, and there must be correlation between the methods. Moreover, the input function to a measuring system is never known beforehand, or else a measurement is superfluous. Knowledge of the methods is necessary because each brings out certain characteristics of the measuring system. The method used for an application depends on convenience and equipment available, although the trend in recent years has been toward transient testing methods.

1-5. THE INSTRUMENTATION PROBLEM

Since this book is written for non-electrical engineering students, the instrumentation problem is viewed as the proper and efficient use of instruments. This entails familiarization with basic laboratory instruments, some hands-on experience, and a knowledge of the applicability and limitations of an instrument. This does not have to involve every type of instrument. The objective can be accomplished with reasonable time and expense.

Except for the sensor, most components of a measuring system are electrical, as illustrated in Fig. 1-2. The field of instrumentation is flooded with electronics and can be perplexing. This has led some people erroneously to equate measurement with electronics or to feel that an experimenter must be knowledgeable in electronics.

The knowledge required of an experimenter is the proper and efficient use of instruments, not electronics per se. One should understand the general principles, the functions performed, how the functions are performed, and the operating characteristics of instruments. Knowledge of the analysis and design of electronics belongs to another field. Except when it is absolutely necessary, experimenters should not attempt to repair a commercial electronic device, even when they are well qualified. Their time is valuable; and their performance is judged by the results of experiments, not by prowess in repairing instruments.

There is no substitute for hands-on experience in the laboratory. To begin with, students should perform some exercises on simple transducers and electronic devices, including an electronic amplifier. It is paradoxical that with the advent of microelectronics, it takes relatively little experience to construct a device from "chips" and readily available circuit diagrams. In the very early days of radio, three control knobs on a receiver were manipulated to tune to an AM station. Now a color TV set is tuned by remote control by the push of a single button. Although electronics has greatly simplified instrumentation, an experimenter must have a feeling for the controls on the front pannel. To use instruments properly and efficiently, the experimenter must have an understanding of the do's and don'ts.

1-6. THE LABORATORY PROBLEM

The laboratory in engineering education is a broad subject and somewhat controversial regarding both content and approach. There is always room for difference in our pluralistic society. The laboratory problem will be viewed in terms of (1) experimentation, (2) the goals of the laboratory and means of implementation, and (3) the types of laboratory exercises. Hopefully, this presentation will give readers a perspective for the laboratory experience.

A. Experimentation

Other than routine testing, experimentation is diverse and elusive, let alone teaching it in a first engineering laboratory. The crux of experimentation is (1) to know what and how to measure, (2) to predict whether the test results will solve the problem, and (3) to obtain the information efficiently for the time, effort, and expense.

To know what to measure and whether the results will solve the problem are the diagnostic aspects of experimentation. In medicine, diagnosis is a skill and an art. It is a fundamental step toward the treatment of a patient. In engineering, the investigator must be conversant with the subject on which a measurement is made. Other than that, the skill of diagnosis is developed largely through experience and reinforced by case studies.

The discussions in previous sections centered on how to measure. These are by no means trivial, since an experimenter must know both what and how to measure. In one story about a neurosurgeon operating on a patient, electrodes covered the patient and instruments were recording furiously. The surgeon knew which vital signs of the patient to monitor but did not have a working knowledge of the instruments. The engineering consultant took a little black box from his hip pocket and found that the patient was being cut up to the tune of a local radio station. The veracity of this tale has yet to be confirmed. In this case, to know what to measure without knowing how is chaos. Conversely, to know how without knowing what to measure is an exercise in futility.

Experimental procedure to get information efficiently is more tangible than the skill of diagnosis. The necessary steps in the course of an investigation are outlined systematically in an excellent book by Wilson [3]. Some textbooks on experimental procedure stress the detailed planning of an experiment—from the initial conception and a careful statement of objectives, to a defintiion of when the experiment will be considered final. Others emphasize student projects as a means to teach experimentation.

B. Goals and Implementation

The goals [4] of the engineering laboratory can be defined, but the types of laboratories for their implementation remain controversial and a topic of discussion among engineers and educators. In this section we consider the goals and to comment on the types of laboratories.

The goals of the engineering laboratory can be classified as lecture related and laboratory related. The laboratory is subservient to the *lecture-related objectives*, which are:

1. To reinforce theory
2. To demonstrate principles
3. To enhance knowledge of system behavior

The *laboratory-related objectives* are:

1. To use experimentation as a design procedure
2. To develop the ability to design experiments
3. To acquire a working familiarity with measurement methods, instrumentation, and equipment
4. To enhance the ability of data analysis and evaluation
5. To cultivate report writing as a communication skill

Although students should be familiar with the goals above, it should be understood that the goals cannot all be implemented in a laboratory course.

Divergent types of laboratories for implementing the goals are probably the result of pedagogical philosophies and local constraints. The value of laboratory experience for students is recognized, but the place of the laboratory in a curriculum has not been defined. There can be many types of well-justified laboratories. Remember the Cheshire Cat in Alice in Wonderland? "Would you tell me, please, which way I ought to walk from here?" asked Alice. "That depends a good deal on where you want to get to," answered the Cat.

It would be difficult to find a "typical" laboratory. At one extreme, a laboratory may be highly structured. The experiments are built on test codes, properties of materials, and performance tests. On the other hand, some laboratories are loosely structured and are variously called *personalized, self-paced, innovative, open-shop,* and *projects.* Some heavily stress computer simulation; others stress electronic instrumentation. Although most engineering laboratories cannot be stereotyped, we shall comment on "types" of laboratories.

A *traditional laboratory* tends to be structured, consisting of codes and performance tests. Such tests have served the profession well and will continue to serve a large segment of industries in the future. Some students enjoy the routine performance test of an internal combustion engine. Furthermore, some traditional experiments can be reincarnated. An engine test can become an investigation of pollution. A steam turbine test can be updated to become an experiment in digital control. It would be a mistake to discard the old indiscrimantely, as it is wrong to ignore the new that is of lasting value. Since experiments for the traditional laboratory are found in the literature [5], they will not be repeated in this book. Note that a test is not routine for students if being encountered for the first time.

An *innovative laboratory* means different things to different people. Its main purpose is to enhance the initiative, creativity, and motivation of students. However, projects can be time consuming for the student and the instructor. Despite careful planning, there can be delays in the delivery of components and equipment can be tied up for long periods. Projects should not be the first laboratory course, because students have yet to learn the fundamentals and the proper use of instruments. Furthermore, it is difficult to maintain continuity in an innovative laboratory over the years.

A *computer simulation laboratory* has many advantages. First, a model study is concise. Many problems have the same model and the same formal solution. Second, the ramifications of a problem can be investigated and the solution optimized. A model study also has certain disadvantages:

1. A model can be deceptively simple. A model relates the variables in a problem as dictated by the theory, but it is not an explanation of the physical law. Under different conditions, the same device could exhibit different behavior and would result in a different model.
2. It takes experience and a "feel" for the subject to deduce an appropriate model or to extrapolate from a model to a real problem.
3. Some students tend to forget that a theory cannot be verified by means of a computer solution when the basis of the computer program is the theory to be verified. This would be like a dog chasing its own tail.

C. Laboratory Exercises

The exercises suggested in this book are for a first laboratory course. They focus on measurement and instrumentation to complement the types of laboratories enumerated above. Although most of the exercises are electrical, our interest is on the proper use of instruments, not on the electronics. Students have no qualm in using instruments when they have some hands-on experience: The pocket calculator is not envisioned as being electronic.

Three guideposts are used for selecting the exercises.

1. The exercises should be fundamental and the equipment simple, such that the basic principles and their possible adaptations for new problems are not obscured.

2.   If skill is transferable, it is not necessary to be drilled in the details of every piece of hardware.
3.   On the other hand, skill is transferable only when one has mastered a subject, including the concepts of the fundamentals, a working knowledge of the equipment, and a feel for the quantities measured and their order of magnitude.

A basic laboratory course is justified because engineers should have some feeling for the data. When using outside data, it is incumbent on the experimenter to know that the data are valid. Consider the tool dynamometer with extreme cross-sensitivity described in Sec. 1-3A. The vendor could refund the price of the dynamometer. In the meantime, the dynamometer might have been used in the research laboratory for collecting cutting data for years. Furthermore, the data generated might be used by others. Engineers directly involved in experimentation must be competent with the instruments in their laboratories, even when the data logging is entrusted to others. There is a story about a research project in which an engineer could not get the results to agree with the predictions. While the engineer was at lunch, a technician manipulated the controls of the instruments arbitrarily to get what was expected.

Finally, from experience in laboratory teaching, (1) all equipment should be in good working condition, and (2) experiments should not be rushed in the name of coverage. It is often said that students will learn more from a bad experiment than from a good one. This can only be interpreted to mean that students should have both a working familiarity with the equipment and the type of data so that they can spot malfunctions and can make critical evaluations. Only the proper use of equipment and critical evaluation of reasonable results belong in a beginning laboratory. Poorly maintained equipment discourages students, some of whom may be struggling simply to learn to use the equipment, let alone spot a malfunction. Equipment repair logially belongs to the instrument room staff.

1-7. REPORTS

An investigation is not complete until it is reported. The importance and format of technical reports are discussed briefly in this section.

The importance of a report cannot be overstated. After a research project is done and the hardware abandoned, the report is often the sole evidence of the effort. The sponsor of the research gets a report for the thousands of dollars invested. An engineer should be as meticulous with the report as with the investigation. The report is a means of communicating to the outside world, making the investigation known, gaining recognition for the ideas, and selling the fruit of the endeavor to the supervisor or sponsor of the research. Yet report writing has been treated at best as a necessary evil.

Reports take many forms. A routine report from a test laboratory, the writeup of a design project, a research report, a company memo, a news release on a product, and a student thesis are all reports of one form or another. They are all necessary. A report is written to serve a specific purpose, and the writer must keep in mind the party for whom the report is intended. The report is a sales tool, not a display of the writer's prowess. A business owner once said: "When you sell baby shoes, advertise to the mothers."

The style of a report is optional, except that it is in good taste to write formal reports in the third person. Whatever the style, there is no substitute for clarity of thought. Readers cannot be expected to follow the ideas in a report when the writer is unclear. An engineer once said that he owed his report-writing style to an army training manual. "First, I state what is to be done. Second, I do exactly what was described. Finally, I tell them what has been done." There is some truth in this approach.

The format of a report can be organized by sections and subsections. A short introduction summaries the section. There can be only one main idea in each paragraph. The leading sentence of each paragraph summarizes the paragraph. The section should be ended with concluding remarks.

A technical report may consist of (1) the preliminary section, (2) the body of the report, and (3) the concluding section. The preliminary section may contain the following items:

1. Letter of transmittal
2. Title
3. Abstract
4. Objective

These items should not be treated as routine. In addition to the proper salutation, the *letter of transmittal* should, completely and concisely, identify the nature of the report, such as interim or final, and its background, such as contract number. The *title* gives the first impression. It is the basis on which a prospective reader will decide whether to read the report. The *abstract* is an expansion of the title. It is the basis on which the reader will decide whether to go further with the reading. The abstract generally gives a concise description of the nature of the problem, the purpose of the investigation, and the significance and limitations of the results. The *objective* should be clear and complete. It gives the specifications for the entire project, the direction for the experiment, the guideposts for procedures to meet the objectives, and an outline of the report.

The body of the report generally includes the following items:

1. Introduction/theory
2. Apparatus
3. Procedure
4. Results

The *introduction* gives a perspective on the investigation in its field or related fields of study. Some historical background, the nature of th problem, and other pertinent information may be included. The *theory* furnishes the background for the investigation, the equipment, its operation, the test procedure, and a computation of results. The theory is explained in detail only if it is new to the prospective reader. Similarly, the *apparatus* is described in detail only when it is new to the reader. Line drawings are better suited than photographs for the purpose of explanation, although photographs can present a sense of reality and show the relative size of components. The *procedure* describes what and how to measure. The validity of the measurement and the uncertainty anticipated should be stated. The description of the procedure and apparatus must be sufficient for the experiment to be repeated. Only the final *results* are presented in the body of the report. It is preferable to plot the results to show trends. The supporting raw data, the intermediate results, and the units employed are placed in the appendices.

The concluding sections generally contain the following items:

1. Conclusion/recommendations
2. Discussion
3. Acknowledgments
4. Appendices

There is no rigid rule regarding the demarcation between items in the concluding sections. The *conclusion,* as deduced from the experimental results, is an attempt to answer the stated objectives. Depending on the audience, "normal" trends may be omitted from the *discussion,* but unusual trends must be honestly stated. If a conclusion is not directly supported by the results, as is true of a generalization, a conjecture, or an extrapolation, it should be clearly stated as such. The validity of the data, the limitations of the conclusions, the conditions under which the results will apply, and other helpful information may be included in the concluding section. It is courteous to give credit in the *acknowledgments.* The *appendices* should include the original data, all supporting material, and any details that would overburden the text of the report.

1-8. NOMENCLATURE

Information from diverse technical fields is encountered in measurement and instrumentation. Often the same symbol is used to denote different quantities. The use of subscripts and superscripts is not standardized. The symbols used in the text are defined in the context used and conform to common usage whenever possible.

For convenience, a few symbols shown in Table 1-3 will be reserved for particular applications. The beginning letters of the alphabet ($A, B, C, \ldots$ and $a, b, c, \ldots$) are generally used to represent constants and parameters. The final letters (W, X, Y, Z and w, x, y, z) are used for variables. Uppercase letters (W, X, Y, Z) denote the instantaneous total values, constant or dc values, or phasor or rms values. Lowercase letters (w, x, y, z) represent the corresponding incremental values. An incremental value is a deviation from some reference or datum. For example, if X_0 is an initial value and X the total value, then $\Delta X = (X - X_0) = x$, where Δ denotes an increment.

1-9. UNITS AND STANDARDS

Let us first clarify the terms *units* and *standards*. The value of a measurand is the ratio of its magnitude compared with that of a standard.

TABLE 1-3 Symbols Generally Reserved for Particular Applications

Symbol	Description
f	Function of
G	Transfer function of a system (transducer, instrument)
g	Gravitational constant, 32.2 ft/sec^2 or 9.81 m/s^2
I, i	Electric current in amperes, A
ln	Log to the base e
t	Time in seconds, s
V, v	Electric voltage in volts, V
X, x	Variable or input measurand
Y, y	Variable or desired output
Z, z	Variable or auxiliary input
Δ	Small change or increment
ζ	Viscous damping ratio (actual damping/critical damping)
ϕ	Phase angle in radians, rad
ω	Circular frequency, rad/s
ω_n	Undamped natural frequency, rad/s
ω_d	Damped natural frequency, $\omega_d = \sqrt{1 - \zeta^2}\, \omega_n$

For example, the meter is a standard length as well as a unit of measure. If the length of a rod is so many meters, its length is so many units of the standard. Hence the standard defines the unit and the number of units describes the magnitude of a measurand. We shall briefly describe the common SI units and the nature of standards.

A. Units

Until recently, almost every discipline had its own set of units. *SI (Système International d'Unités)* [6] is the modern version of the metric system, adopted by international agreement, including the United States. It is anticipated that other systems of units will eventually be phased out.

The SI system consists of (1) the base units, (2) two supplementary units, and (3) the necessary derived units. The base units and the supplementary units are shown and described briefly in Table 1-4. The base units are the meter, kilogram, and second. Independently defined base

TABLE 1-4   SI Base Units and Supplementary Units

Quantity	Unit, Symbol	Description
	Basic units	
Length	meter, m	1 650 763.73 wavelengths of krypton-86
Mass	kilogram, kg	International kilogram at Sèvres, France
Time	second, s	9 192 631 770 period of a specified transition of cesium 133
	Independently defined units	
Temperature	kelvin, K	1/273.16 of the thermodynamic temperature of the triple point of water
Electric current	ampere, A	Defined in terms of the force between parallel conductors
Amount of substance	mole, mol	Amount of substance containing as many entities as there are in 0.012 kg of ^{12}C
Luminous intensity	candela, cd	Luminance of 1/600000 of a square meter of a blackbody at the temperature of freezing platinum (2045 K)
	Supplementary independent units	
Plane angle	radian, rad	2π rad $= 360°$
Solid angle	steradian, sr	A sphere subtends 4π steradians

units are the kelvin, ampere, mole, and candela. The supplementary units are the radian and steradian. Derived units are formed from the base units according to the algebraic relations linking the corresponding quantities. Derived units given special names and symbols are shown in Table 1-5. The common prefixes for the multiples and submultiples of SI units are shown in Table 1-6.

Examples of multipliers for conversion from English to SI units are shown in Table 1-7. A common error in the conversion is to become ensnared in too many decimal places. The result of a computation cannot have more significant figures than were in the data before the conversion.

The recommendations for uniformity in the use of SI units are as follows:

1. In numbers, a period (dot) is used only to separate the integral part of a number from the decimal part. Numbers are divided into groups of three to facilitate reading. For example, the meter is defined as "equal to 1 650 763.75 wavelengths."
2. The type used for symbols is illustrated in Table 1-5. Lowercase roman type is generally used. If the symbol is derived from a proper name, a capital is used for the first letter, such as J for joules and Pa for pascal. The symbols are not followed by a period.
3. The product of units is denoted by a dot, such as $N \cdot m$. The dot may

TABLE 1-5   SI Derived Units with Special Names

Quantity	Name	Symbol	Expressed in terms of: Other units	Expressed in terms of: SI base units
Frequency	hertz	Hz		s^{-1}
Force	newton	N		$m \cdot kg \cdot s^{-2}$
Pressure, stress	pascal	Pa	N/m^2	$m^{-1} \cdot kg \cdot s^{-2}$
Energy, work, heat	joule	J	$N \cdot m$	$m^2 \cdot kg \cdot s^{-2}$
Power	watt	W	J/s	$m^2 \cdot kg \cdot s^{-3}$
Electric charge	coulomb	C	$A \cdot s$	$s \cdot A$
Electric potential	volt	V	W/A	$m^2 \cdot kg \cdot s^{-3} \cdot A^{-1}$
Capacitance	farad	F	C/V	$m^{-2} \cdot kg^{-1} \cdot s^4 \cdot A^2$
Electric resistance	ohm	Ω	V/A	$m^2 \cdot kg \cdot s^{-3} \cdot A^{-2}$
Electric conductance	siemens	S	A/V	$m^{-2} \cdot kg^{-1} \cdot s^3 \cdot A^2$
Magnetic flux	weber	Wb	$V \cdot s$	$m^2 \cdot kg \cdot s^{-2} \cdot A^{-1}$
Flux density	tesla	T	Wb/m^2	$kg \cdot s^{-2} \cdot A^{-1}$
Inductance	henry	H	Wb/A	$m^2 \cdot kg \cdot s^{-2} \cdot A^{-2}$
Luminous flux	lumen	lm		$cd \cdot sr$
Illuminance	lux			$m^{-2} \cdot cd \cdot sr$

TABLE 1-6 Prefixes for Multiples and Submultiples of SI Units

Multiple	Prefix	Symbol	Submultiple	Prefix	Symbol
10^{12}	tera	T	10^{-1}	deci	d
10^{9}	giga	G	10^{-2}	centi	c
10^{6}	mega	M	10^{-3}	milli	m
10^{3}	kilo	k	10^{-6}	micro	μ
10^{2}	hecto	h	10^{-9}	nano	n
10^{1}	deca	dc	10^{-12}	pico	p
			10^{-15}	fento	f
			10^{-18}	ato	a

be omitted if there is no risk of confusion with another unit symbol, such as m N (but not mN).

4. The division of units may be indicated by a solidus (/), a horizontal line, or a negative power. A velocity can be expressed as m/s or m · s^{-1}. The solidus must not be repeated in the same term unless the am-

TABLE 1-7 Examples of Conversion from English to SI Units

To convert from:	To:	Multiply by:[a]
Barrel (for petroleum, 42 gal)	meter^3 (m^3)	1.589 879 E−01
British thermal unit (mean)	joule (J)	1.055 87 E+03
Btu (thermochemical)/hr · ft^2 · °F	watt/meter^2-kelvin (W/m^2 · K)	5.674 466 E+00
Btu (thermochemical)/lb$_m$ · °F	joule/kilogram-kelvin (J/kg · K)	4.184 000 E+03
Foot	meter (m)	3.048 000 E−01
Foot-pound-force (Ft · lb$_f$)	joule (J)	1.355 818 E+00
Horsepower (550 ft · lb$_f$/sec)	watt (W)	7.456 999 E+02
Inch	meter (m)	2.540 000 E−02
Inch of water (60°F)	pascal (Pa)	2.488 4 E+02
Miles/hour (international)	meter/second (m/s)	4.470 400 E−01
Pound-force (lb$_f$ avoirdupois)	newton (N)	4.448 222 E+00
Pound-force-inch (lb$_f$ · in.)	newton-meter (N · m)	1.129 848 E−01
Pound-force/inch (lb$_f$/in.)	newton/meter (N/m)	1.751 268 E+02
Pound-force/inch^2 (psi)	pascal (Pa)	6.894 757 E+03
Pound-mass (lb$_m$ avoirdupois)	kilogram (kg)	4.535 924 E−01
Pound-mass/inch^3 (lb$_m$/in^3)	kilogram/meter^3 (kg/m^3)	2.767 990 E+04
Slug	kilogram (kg)	1.459 390 E+01
Ton (long, 2240 lb$_m$)	kilogram (kg)	1.016 047 E+03
Ton (metric)	kilogram (kg)	1.000 000 E+03

Source: Metric Practice Guide, E380-74, American Society of Testing and Materials, Philadelphia.
[a]The number followed by E (for exponent) indicates the power of 10. For example, the value in inches is multiplied by 2.54 × 10^{-2} to obtain the value in meters.

biguity is avoided by the use of parentheses. For example, acceleration may be expressed as m/s^2 or $m \cdot s^{-2}$ but not as $m/s/s$.

5. The prefix symbols shown in Table 1-5 are used without spacing between the prefix symbol and the unit symbol, such as mm. Compound prefixes formed by using more than one SI prefix are not used.

B. Standards

To be reproducible in any laboratory, standards were originally defined in terms of objects that could be measured with the greatest accuracy by the instruments "available." As the history of units and measure changes with technology, even the primary standards are not time invariant. Hence standards are steeped in tradition [7]. For example, time was defined in terms of a mean solar day, the average time for one rotation of the earth. The second as a time standard is now defined by means of the atomic resonant frequency of cesium 133. The meter has been redefined since its original conception, as shown in Table 1-4.

The accuracy of a measurement cannot exceed that of the *primary standard*. Unless one is in the field of metrology, the extreme accuracy of the primary standard is neither required nor available to the engineer. Greater tolerances can be expected in *working standards*, although all standards should be traceable to the National Bureau of Standards. A pragmatic view is that standards are defined by general acceptance. This view is justified because (1) extreme accuracy is the exception rather than the rule in engineering, and (2) only interlaboratory standards are generally available to the engineer.

Let us briefly examine the accuracy of standards commonly available. The frequency standard is broadcast by the National Bureau of Standards' radio station WWV with the precision of 1 part in 10^8. Frequency and time standards that have an accuracy of $\pm 7 \times 10^{-12}$ and a long-time drift of 1×10^{-11} are commercially available [8]. Gage blocks (e.g., C. E. Johansson Gage Co., Dearborn, Michigan; DoAll Co., Des Plaines, Illinois) made of dimensionally stable steel can be stacked for dimensional measurement. Their dimensions are accurate to ± 8 μin. for working-grade blocks, ± 4 μin. for reference grade, and ± 2 μin. for the master grade. The International Practical Temperature Scales of 1968 are based on six fixed points [9]. These are the oxygen point ($-182.962°C$), the triple point of water ($+0.01°C$), the steam point ($100°C$), the zinc point ($419.58°C$), the silver point ($961.93°C$), and the gold point ($1064.43°C$). For example, the gold point is the equilibrium temperature between solid gold and liquid gold.

The accuracy of temperature measurement is elusive. There can be precision without accuracy. The problem lies in the uncertainty of fixed points rather than precision in the interpolation between the fixed points. Fixed points are defined by nature as shown above, but they do not come with numbers. Accuracy in temperature measurement is governed by the determination of these numbers. The interpolation between the fixed points is accomplished by means of (1) a platinum resistance thermometer for the range -190 to $660°C$, and (2) a platinum/platinum–10% rhodium thermometer for 660 to $1063°C$. The resistance thermometer is sensitive. High precision can be obtained from interpolation between the fixed points, but the exact value of the temperature is less certain. As another example, the resonant frequency of quartz crystals is temperature sensitive and quartz can be used for temperature sensing. A quartz resonator has been developed to measure temperature from -40 to $230°C$ with resolution up to $\pm0.0001°C$ [10].

Standards for the derived units are defined in terms of the base units shown in Table 1-4. Derived units are related to the base units by physical laws. For example, by Newton's second law, we get

$$\text{force} = (\text{mass})(\text{acceleration}) \tag{1-7}$$

A unit force is expressed in terms of mass, length, and time. Similarly, pressure is force per unit area. Although very accurate instruments are available for the calibration of pressure transducers, these "standards" must ultimately depend on the standards of the base units for their accuracy.

Before leaving the subject, let us reflect briefly on the philosophy of units. SI has many advantages. It is expedient, however, to regard SI as being established by general agreement. Since any system of units will change with the demand of technology, it is anticipated that SI units are not time invariant. SI has been hailed as a coherent system in which all units are related to each other by coefficients of unity. For example, the newton (N) is defined as $1 \text{ N} = (1 \text{ kg})(1 \text{ m} \cdot \text{s}^{-2})$, and the system is coherent. A force, however, is also related to the base units by the law of gravitation:

$$\text{force} = C \frac{(\text{mass})_1 \times (\text{mass})_2}{(\text{distance})^2} \tag{1-8}$$

Since force is defined by Eq. 1-7, it cannot be redefined in Eq. 1-8. In other words, the coefficient C is not unity and is dimensional. It can readily be shown that the dimension of C is $\text{m}^3/(\text{kg} \cdot \text{s}^2)$. Thus when a unit appears in two equations, as in two physical laws, only one of the equations can be coherent. Further discussions of units can be found in the literature [11].

PROBLEMS

1-1. With the aid of a sketch when necessary, briefly discuss each of the following:
    (a)   A valid measurement.
    (b)   Applications of measurement.
    (c)   Loading effects.
    (d)   Noise.
    (e)   Calibration of transducers.
    (f)   Statistical data analysis.
    (g)   Functional stages of measuring systems.
    (h)   Measurement as a design problem.
    (i)   Initial and boundary conditions of transducers.
    (j)   Phenomena for sensing.
    (k)   Examples of structural changes in measurements.

1-2. A strain gage is used in the tensile test of 18–8 stainless steel. The gage resistance is $R = 120 \ \Omega$, the gage factor is $G_f = 2.0$, and the yield strength of the steel is 35×10^3 psi. Find the strain and the ΔR of the gage at yield.

1-3. A ballast circuit for dynamic strain measurement is shown in Fig. P1-1. The static or dc component of the strain is blocked by means of the capacitor C, and the output $v_o = \Delta V_o$ gives the dynamic or ac signal. (a) Assuming that V_i and R_b remain constant, find the sensitivity $\Delta V_o / \varepsilon$ of the circuit. (b) Assuming that V_i is constant but R_b is varied, find R_b for maximum sensitivity in the strain measurement. (c) Plot sensitivity versus R_b / R for the range $0.7 < R_b / R < 1.3$.

1-4. The strain of a rotating steel shaft is measured by means of a ballast circuit as shown in Fig. P1-2 (see also Fig. 3-51). Assume that the gage resistance is $R = 120 \ \Omega$, the gage factor is $G_f = 2.0$, the fluctuating resistance of the brushes at the slip rings is $\Delta R = \pm 2 \ \Omega$, and the input impedance of the voltmeter is infinite. Find the equivalent strain due to the brush noise.

1-5. Strain gages R_1 and R_2 are mounted on a gun barrel for measuring the circumferential stress when the gun is fired. The circuit is as shown in Fig. P1-3. Assume that $R_1 = R_2 = 350 \ \Omega$, the gage factor $G_f = 2.1$, $E = 30 \times 10^6$ psi for the base material, the initial current in the circuit is $I = 20$ mA, the oscilloscope is ac coupled, and the input impedance of the oscilloscope is 1 MΩ. The sensitivity of the oscilloscope is 10 mV/cm and its trace deflects 2 cm when the gun is fired. Find the circumferential stress of the gun.

1-6. Repeat Prob. 1-3 for the setup shown in Fig. P1-4. Let $R_1 = R_2 =$

$R_3 = R_4 = 120\ \Omega$, $G_f = 2.0$, and the deflection of the trace of the oscilloscope is 1.5 cm.

1-7. The output of the potentiometer in Fig. P1-5a is $V_o = xV$, where x is the position of the wiper (contact) and $0 < x < 1$. When V_o is measured with a voltmeter with input impedance R_m, shown in Fig. P1-5b the measured voltage is $V_m < xV$, due to the loading effect. If $R_m = R$, (a) find the error due to loading as "percent of reading" and as "percent of full scale," (b) find the maximum errors, and (c) sketch the error versus x for $0 < x < 1.0$.

1-8. A 50-μA moving coil (D'Arsonval) meter is shown in Fig. P1-6. Assume that $R_m = 1\ \text{k}\Omega$. Find R_1 and R_2 when the meter is used for dc current measurements for the ranges from 0 to 100 mA and 0 to 1 A.

1-9. A multimeter for dc current measurement is shown in Fig. P1-7. The meter itself has a range of 0 to 50 μA with $R_m = 5\ \text{k}\Omega$. Neglecting the contact resistance at the switch, find R_1 to R_4 when the meter is used for the ranges 0 to 15 mA, 0 to 150 mA, 0 to 0.5 A, and 0 to 15 A.

1-10. The dc voltmeter in Fig. P1-8 consists of a microammeter in series with a resistance R_s. If $R_m = 300\ \Omega$ and the meter reads 200 μA full scale, find R for the voltage range 0 to 20 V. Repeat for the voltage range 0 to 100 V.

1-11. The multimeter in Fig. P1-9 is a voltmeter when the switch is in the position shown. It is an ammeter, when the switch is in the down position. The dashed line indicates the mechanical link for the contacts to operate together. Find R_1 for the voltage range 0 to 20 V and R_2 for the current range 0 to 1 A.

1-12. A dc microammeter is used to measure an ac voltage as shown in Fig. P1-10a. A diode in Fig. P1-10b has low resistance to the current flow in one direction and very high resistance for the current flow in the opposite direction. The operation of a diode is analogous to that of a check valve in a hydraulic line. (a) Show the directions of the current flow for one cycle of the ac applied voltage. (b) Find the scale factor for converting the reading of the meter to rms voltages.

1-13. The schematic of an ohmmeter is shown in Fig. P1-11. The meter is calibrated by adjusting R_1 for the meter to read 0 Ω when terminals a and b are shorted. Note that the current through the meter is 50 μA for this initial adjustment. (a) Find the value of R_1. (b) If a 12.5-Ω resistor is connected across a and b and the meter indicates one-third full scale, find the value of R_2. (c) Construct the scale for the dial if the meter is to read from 0 to 1 kΩ.

1-14. Iron–constantan thermocouples are used for measuring the temperatures of an air-conditioning unit. The references temperature for couples is at 30°C. Find the temperatures for the thermal EMF of 4.004 mV, 2.113 mV, −1.787 mV, and −2.531 mV.

	Thermal EMF (absolute mV)—iron–constantan				
°C	−100	(−)0	(+)0	100	200
0	−4.632	0	0	5.268	10.777
10	−5.036	−0.501	0.507	5.812	11.332
20	−5.426	−0.995	1.019	6.359	11.887
30	−5.801	−1.481	1.536	6.907	12.442
40	−6.159	−1.960	2.058	7.457	12.998
50	−6.499	−2.401	2.585	8.008	13.553
60	−6.821	−2.892	3.115	8.560	14.108
70	−7.122	−3.344	3.649	9.113	14.663
80	−7.402	−3.785	4.186	9.668	15.217
90	−7.659	−4.215	4.725	10.222	15.771
100	−7.890	−4.632	5.268	10.777	16.325

1-15. An iron–constantan differential-temperature thermocouple reads 3.115 mV. What is the differential temperature?

1-16. The iron–constantan thermocouples in Fig. P1-12 are electrically insulated. If $T_1 = 150°C$, $T_2 = 40°C$, and $T_3 = 20°C$, and copper leads are used to connect the junctions at T_3 to a millivolt meter, find the indicated mV. Find the indicated mV if the junctions at T_1 and T_2 are not electrically insulated.

1-17. Find the mV indicated by the circuit in Fig. P1-13. Assume that $T_1 = 120°C$, $T_2 = 110°C$, $T_3 = 100°C$, $T_4 = 60°C$, and $T_5 = 20°C$.

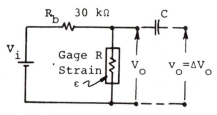

FIGURE P1-1

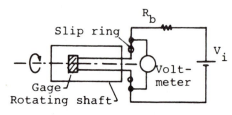

FIGURE P1-2

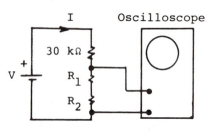

FIGURE P1-3

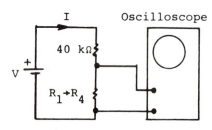

FIGURE P1-4

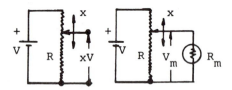

FIGURE P1-5

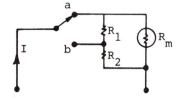

FIGURE P1-6

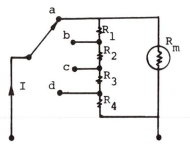

FIGURE P1-7

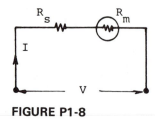

FIGURE P1-8

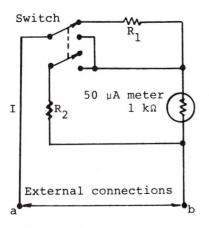

FIGURE P1-9

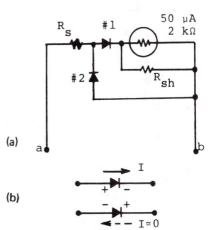

(a)

(b)

FIGURE P1-10a, b

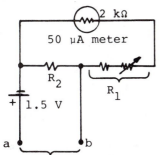

FIGURE P1-11

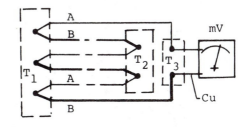

FIGURE P1-12

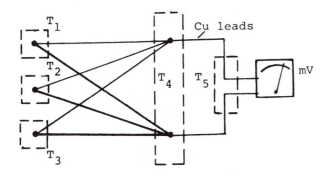

FIGURE P1-13

REFERENCES

1. Tribus, M., and E. McIrvine, Energy and Information, *Sci. Am.*, (1974), pp. 179–188.
2. Powell, R. L., et al., *Thermocouple Reference Tables Based on the ITPS-68,* NBS Monograph 125, Omega Press, Omega Engineering, Inc., Stamford, Conn., 1975.
3. Wilson, E. B., Jr., *An Introduction to Scientific Research,* McGraw-Hill Book Company, New York, 1952.
4. Rice, S., Objectives of Engineering Laboratory, *Eng. Educ.* (Jan. 1975), pp. 285–288.
5. Shoop, C. F., and G. L. Tuve, *Mechanical Engineering Practice,* 5th ed., McGraw-Hill Book Company, New York, 1956. Ambrosius, E. E., R. D. Fellows, and A. D. Brickman, *Mechanical Measurement and Instrumentation,* The Ronald Press Company, New York, 1966.
6. Page, C. H., and P. Vigoureus (eds.), *The International Systems of Units (SI),* Special Publication 330, U.S. National Bureau of Standards, Washington, D.C., revised 1974. *Some References on Metric Information,* Special Publication 389, U.S. National Bureau of Standards, Washington, D.C., revised 1974. *ASME Orientation and Guide of SI (Metric) Units,* ASME Guide SI-1, ASME, United Engineering Center, New York. ASEE Metric (SI) Resource Kit Project, American Society for Engineering Education, Washington, D.C.
7. Cohen, E. R., Fundamental Units and Constants, in *Methods of Experimental Physics,* Vol. I, *Classical Methods,* I. Estermann (ed.), Academic Press Inc., New York, 1959, Chap. 2.
8. *Electronic Instruments and Systems,* Hewlett-Packard, Palo Alto, Calif., 1979, pp. 274–285.
9. Benedict, R. P., *Fundamentals of Temperature, Pressure, and Flow Measurements,* 2nd ed., John Wiley & Sons, Inc., New York, 1977, Chap. 4.
10. The Linear Quartz Thermometer—A New Tool for Measuring Absolute and Differential Temperatures, *Hewlett-Packard J.,* Vol. 16, No. 7 (Mar. 1965).
11. Chertov, A. G., *Units of Measurement of Physical Quantities,* Scripta Technica, Inc. (trans.) Hayden Book Company, Inc., Hasbrouck Heights, N.J., 1964. Whitaker, R. O., Philosophy of Units and Their Realization, *M & D Home Study Course* No. 6, Measurement & Data Corporation, Pittsburgh, Pa., 1967.

2

Transducers

The aspects for valid measurement are the transducer, the structure of the measuring system, and the system dynamics, as shown in Fig. 1-7. The interaction between a transducer and its environment is examined in this chapter. We shall describe (1) physical laws in sensing, (2) static characteristics, (3) modeling of transducers, and (4) errors and uncertainties in measurements. Loading is discussed in Chapter 3.

2-1. PHYSICAL LAWS

In this section we examine physical laws in sensing [1]. A transducer is susceptible to many inputs, as shown in Table 1-2a. The desired input D is the measurand. All other inputs U are undesired and are potential sources of noise. Let a physical law relating the output Y and the inputs D and U be expressed by a general functional relation

$$Y = f(D, U) \qquad (2\text{-}1)$$

where D and U are independent. At the initial state (0), we have

$$Y_0 = f(D_0, U_0) \qquad (2\text{-}2)$$

A perturbation of the variables D and U in Eq. 2-1 gives

$$Y = Y_0 + \Delta Y = f(D_0 + \Delta D, U_0 + \Delta U) \qquad (2\text{-}3)$$

where Δ denotes an increment of the variables. The incremental values about the initial state can be positive or negative. The initial state is at equilibrium. Generally, Y_0 is adjusted to zero before a measurement is made. Expanding Eq. 2-3 in a Taylor series about (0) and subtracting Eq. 2-2 from the series expansion, the output ΔY is

$$\Delta Y = \frac{\partial f}{\partial D}\bigg|_0 \Delta D + \frac{\partial f}{\partial U}\bigg|_0 \Delta U + \frac{1}{2}\frac{\partial}{\partial D}\left(\frac{\partial f}{\partial D}\right)\bigg|_0 (\Delta D)^2$$

$$+ \frac{1}{2} \frac{\partial}{\partial U} \left(\frac{\partial f}{\partial U} \right)\Big|_0 (\Delta U)^2 + \frac{1}{2} \frac{\partial}{\partial U} \left(\frac{\partial f}{\partial D} \right)\Big|_0 \Delta D \, \Delta U$$

$$+ \frac{1}{2} \frac{\partial}{\partial D} \left(\frac{\partial f}{\partial U} \right)\Big|_0 \Delta U \, \Delta D + \cdots \tag{2-4}$$

Results from the interaction between a transducer G and its environment E are as shown on the right-hand side of Eq. 2-4. The partial derivatives are evaluated about (0). Each term in the series indicates the effect of an input on the transducer and can be interpreted physically [2]. For small changes in the input variables, the third- and higher-order terms of the series expansion are neglected.

Note that both Eqs. 1-1 and 2-4 relate the transducer output to the inputs.

1. Only the desired input D is considered in Eq. 1-1, since it describes a single-input, single-output system. The transfer function, however, includes the dynamics of the system.
2. All inputs are included in Eq. 2-4, but only the static characteristics are considered (see Sec. 2-2). It will be shown in Chapter 3 that static characteristics are also the basis for dynamic analysis.

Loadings are neglected in both cases, because the Taylor series describes the interaction between G and E for an existing condition.

A. First-Order Effects: Signal/Noise Ratio

The first-order approximation in Eq. 2-4 gives the *signal/noise ratio* for linear operations:

$$\Delta Y = \underbrace{\frac{\partial f}{\partial D}\Big|_0 \Delta D}_{\text{signal}} + \underbrace{\frac{\partial f}{\partial U}\Big|_0 \Delta U}_{\text{noise}} \tag{2-5}$$

output

The output ΔY has two parts. The part desired is the signal due to the measurand D. The associated partial derivative is the *signal sensitivity*, which is a constant of proportionality relating ΔY and ΔD. The undesired part is the noise output due to all other inputs, and each undesired input U has its associated *noise sensitivity*.

A high signal/noise ratio requires that

$$\frac{\partial f}{\partial D}\Big|_0 \Delta D \gg \frac{\partial f}{\partial U}\Big|_0 \Delta U \tag{2-6}$$

This is achieved when

(1) $\quad \dfrac{\partial f}{\partial D}\Big|_0 \gg \dfrac{\partial f}{\partial U}\Big|_0$

(2) $\quad \dfrac{\partial f}{\partial U}\Big|_0 = 0$ $\hspace{4cm}$ (2-7)

(3) $\quad \Delta U = 0$

(4) $\quad \Delta D \gg \Delta U$

1. Referring to Eq. 2-7, a transducer should be sensitive to the measurand and relatively insensitive to the noise inputs. Since the sensitivities $\partial f/\partial D|_0$ and $\partial f/\partial U|_0$ are evaluated about an *operating point* (0), the signal-to-noise ratio can be altered through a judicial choice of the partial derivatives at an operating point (0). Many techniques can be used. First, this can be achieved by any physical and/or chemical changes in the properties of the sensor. For example, by using certain alloys, some strain gages are less temperature sensitive than others. Cold work, heat treat, aging, and prestressing will change the properties of a sensor. Second, the environment in which the sensor operates can be changed, such as by placing the sensor in a strain-free environment or in a constant-temperature oven. Furthermore, the operating characteristics of a measuring system, denoted by the partial derivatives, can be changed by means of signal modulation, feedback, system compensation, or other structural changes.

2. The transducer is insensitive to noise when $\partial f/\partial U|_0 = 0$. Again, this is achieved by a judicial choice of (0). The requirement is that the transducer be insensitive to the noise inputs only for its range of operation about (0). This is satisfied by any of the four conditions shown in Fig. 2-1.

3. The condition $\Delta U = 0$ is achieved by methods of noise reduction described in Chapter 3.

4. When $\Delta D \gg \Delta U$, the magnitude of the measurand is greater than that of the noise inputs. The signal can only be increased to a degree for the non-self-generating transducers, such as a strain gage, by increasing the power level of the drive. It cannot be assumed that the signal input is inherently greater than the noise inputs in a measurement.

Example 2-1.   Assume that the metallic wire strain gage in Fig. 1-6 is subjected to a strain ε and a temperature T input. Find the signal/noise ratio when the gage is used for strain measurements.

Solution: Let the resistance R of the strain gage be a function of ε and T. Following Eqs. 2-1 to 2-5, we have

$\quad R = f(\varepsilon, T) \qquad \text{and} \qquad R_0 = f(\varepsilon_0, T_0)$

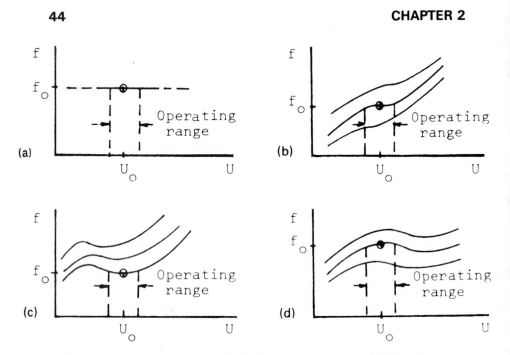

FIGURE 2-1 Methods to obtain $\partial f/\partial U|_0 = 0$. (a) f constant. (b) Inflection point. (c) Minimum point. (d) Maximum point.

$$R = R_0 + \Delta R \qquad \text{and} \qquad T = T_0 + \Delta T$$

$$R_0 = f(\varepsilon_0 + \Delta\varepsilon, T_0 + \Delta T)$$

$$\frac{\Delta R}{R_0} = \underbrace{\frac{\partial R/R_0}{\partial \varepsilon} \Delta\varepsilon}_{\text{signal}} + \underbrace{\frac{\partial R/R_0}{\partial T} \Delta T}_{\text{noise}}$$

Thus the signal/noise ratio is obtained. The equation is normalized by dividing by R_0. The normalized strain sensitivity is the gage factor G_f in Eq. 1-6. The noise output is due to the temperature ΔT. A resistance change due to temperature will be interpreted as an equivalent strain in the measurement.

Example 2-2. A capacitor essentially consists of two metal plates separated by a dielectric, as shown in Fig. 2-2a. The equation for the capacitance C is

$$C = f(k, A, d) = \frac{kA}{d}$$

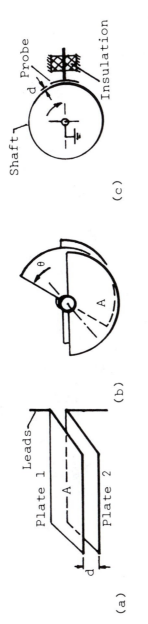

FIGURE 2-2 Variable capacitance transducers. (a) Capacitor. (b) Capacitor in radio. (c) Micrometer.

where k is the dielectric constant, A the area of the plates, and d their separating distance. Deduce the types of transducers that can be constructed from this physical law.

Solution: The output of the transducer is the change in capacitance ΔC. From Eq. 2-5, we get

$$\Delta C = \underbrace{\frac{\partial C}{\partial k}\bigg|_0 \Delta k}_{\text{type 1}} + \underbrace{\frac{\partial C}{\partial A}\bigg|_0 \Delta A}_{\text{type 2}} + \underbrace{\frac{\partial C}{\partial d}\bigg|_0 \Delta d}_{\text{type 3}}$$

Since the inputs are independent, three basic types of transducers can be constructed. These are by changing the dielectric constant k, the common area A, or the separating distance d. Each of them gives a basic type of transducers. For a given application, however, only one of the variables is the measurand. The other two are sources of noise.

Referring to the type 1 transducer in Example 2-2, the dielectric k is usually constant. If the dielectric is hydroscopic, the capacity could be a humidity gage [3]. Another example is a mass flow meter, in which k is a function of the density of the fluid between the capacitor plates (see Sec. 8-8). For the type 2 transducer, the common area A between the plates of a capacitor can be changed as shown in Fig. 2-2b. This device is used for tuning a radio or measuring an angular displacement θ. Furthermore, if the plates are of particular shape, such that $A = f(\theta)$, the device is a function generator. For the type 3 transducer, the distance d separating the plates is changed if one plate is fixed and the other movable. A capacitance pressure transducer has a fixed plate and a movable plate in the form of a diaphragm. When pressure is applied to the diaphragm, its motion is a measure of the applied pressure. Similarly, a capacitance micrometer is constructed as illustrated in Fig. 2-2c. The movable plate is a rotating shaft and the fixed plate is the probe. The device is used for measuring precisely the runout of a rotating shaft.

It should be reiterated that the variables k, A, and d in Example 2-2 must be independent for the three possible types of transducers. For the strain gage described in Sec. 1-3E, the variables ρ, L, and D are dependent. Hence only one type of transducer, namely, a strain gage, can be construction from the functional relation $R = f(\rho, L, D)$.

In summary, if the functional relation in Eq. 2-1 has several independent variables, each variable can be utilized for the construction of a different type of transducer. Each type can have different forms and for different applications. Some may be more practical and economical than others. Nonetheless, they are all feasible. If an input causes a detectable change in a device, the device is potentially a sensor for that input.

B. Second-Order and Irreversible Effects

The signal/noise ratio was deduced from the first-order terms of Eq. 2-4. In this section we examine the second-order terms and irrevresible effects from the same equation. Regrouping the series expansion in Eq. 2-4 and considering only the signal sensitivity, we get

$$\Delta Y = \underbrace{\frac{\partial f}{\partial D}\bigg|_0 \Delta D}_{\substack{\text{desired} \\ \text{output}}} + \underbrace{\frac{1}{2}\frac{\partial}{\partial D}\left(\frac{\partial f}{\partial D}\right)\bigg|_0 (\Delta D)^2}_{\substack{\text{nonlinearity due to} \\ \text{desired input}}} + \underbrace{\frac{1}{2}\frac{\partial}{\partial U}\left(\frac{\partial f}{\partial D}\right)\bigg|_0 \Delta D\,\Delta U}_{\substack{\text{sensitivity drift due} \\ \text{to other inputs}}}$$

$$(2\text{-}8)$$

where ΔY is the output, ΔD the desired input, and ΔU all the undesired inputs. Only the first- and second-order terms involving ΔD are shown in the equation. All the other terms in the series are lumped together and regarded as source of noise in the output.

Let us examine the effect of the nonlinear second-order terms on the *signal sensitivity* $\partial f/\partial D|_0$. The second-order terms represent changes in the signal sensitivity. The second term in Eq. 2-8 shows the change due to the desired input ΔD itself. A transducer has a finite linear range and becomes nonlinear beyond this range. For example, Hooke's law is not obeyed when a coil spring is elongated excessively. A sensor is inherently nonlinear, because $\partial Y/\partial D|_0$ is evaluated about an initial state (0). Any interaction with the environment will alter the state of the transducer and therefore the sensitivity. Fortunately, most transducers are linear over a reasonable range. Some, such as thermocouples, are nonlinear over their normal ranges. Conversion tables are used to relate the input/output quantities.

The third term in Eq. 2-8 shows the change in $\partial Y/\partial D|_0$ due to the undesired inputs ΔU. This is normally called a *drift,* such as drift in an electronic amplifier due to changes in ambient temperature. Generally, this is not a serious problem, because an instrument can be stabilized by placing it on standby for a longer period. All instruments will drift to a degree. The long-time drift of an instrument is a part of its specifications. In the extreme case, the undesired inputs can have considerable effect on a transducer, such as a measurement under hostile conditions [4], where each case is a special study in itself.

An effect is *irreversible* if the sensor does not return to its original state upon the removal of the inputs. This demonstrates the effect of past history on a transducer, such as large peak amplitudes or the type of duty cycles. For example, the second-order term may denote the presetting of a spring in manufacturing. A spring may also be heat treated to change its operating characteristics. Hence not all irreversible changes are undesired.

Irreversible effects are commonly used for measuring the peak or cumulative values of a measurand. For example, a radiation dosimeter is based on the radiation "damage" to a photographic film. A Brinell hardness test is measured by means of an irreversible indentation in a material. If a coated surface changes color permanently at a rated temperature, the coating is a temperature recorder. Shock can be measured by the permanent deformation of a soft copper ball placed between an inertia weight and an anvil [5]. If a measurand causes a detectable change in an object, the object is potentially a sensor for that measurand. The detectable change can be first- or second-order and reversible or irreversible. An unwanted irreversible change is a damage, but a desired one is a signal. It is the case in which the optimist sees the doughnut and the pessimist sees the hole.

2-2. STATIC CHARACTERISTICS

Properties of a device are generally presented graphically as a single curve or a family of curves. We shall illustrate a procedure for obtaining a static characteristic curve, and discuss common linear and nonlinear characteristics and methods of linearization.

A. Obtaining a Static Characteristic Curve

The calibration curve in Fig. 2-3a is determined under static conditions and is called the *static characteristic* of a transducer. This is a straight line when the transducer is linear and a curve when nonlinear. The small circles in the figure show the scatter of the calibration points. During calibration, all inputs except one are kept constant.

An example for obtaining a static characteristic is the calibration of a pressure gage with a dead-weight tester as shown in Fig. 2-3b. The assembly is essentially a hydraulic balance. A reference pressure is generated by balancing the calibrating weights on a precision plunger of known cross-sectional area. A hand-operated piston controls the pressure of the oil such that the plunger always floats on the oil. As a weight X is added to the platform, the pressure in the oil increases. The pressure is transmitted to the gage to give a reading Y. During the calibration, the platform is rotated slowly to eliminate friction. Similarly, the gage is tapped lightly with a finger before each reading. The procedure is repeated to obtain pairs of values of (X, Y). Since X is the only variable, the characteristic of the gage is described by

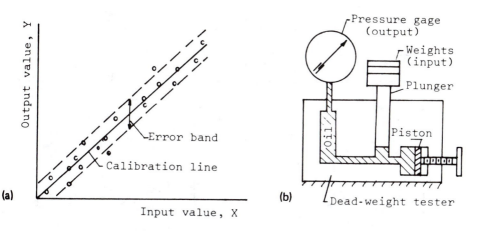

(a)

(b)

FIGURE 2-3 Example of static calibration. (a) Calibration curve. (b) Pressure gage calibration.

$$Y = f(X) \tag{2-9}$$

A similar procedure is used for obtaining the characteristics of other devices involving several variables. For example, if Y is a function of two independent variables X_1 and X_2, we get

$$Y = f(X_1, X_2) \tag{2-10}$$

A family of curves is plotted for Y versus X_1 with X_2 as a parameter. For example, the flow Y through a control valve is a function of the valve opening X_1 and the supplied pressure X_2. Calibration of the valve gives a family of curves.

The calibrating line in Fig. 2-3a is determined by the method of least squares [6]. The equation of a straight line for approximating the static characteristic is

$$Y = mX + c \tag{2-11}$$

where m is the slope of the line and c the intercept on the Y axis. The unknown parameters in the equation are m and c. For the typical data point (X_i, Y_i) in Fig. 2-4, the method minimizes a *sum,* defined as

$$\text{sum} = \sum_{i=1}^{n} (\text{deviation})^2 = \sum_{i=1}^{n} [Y_i - (mX_i + c)]^2 \tag{2-12}$$

where X_i is the independent variable and *deviation* is the vertical distance between a value Y_i and the calibration line.

The sum in Eq. 2-12 is minimized with respect to m and c:

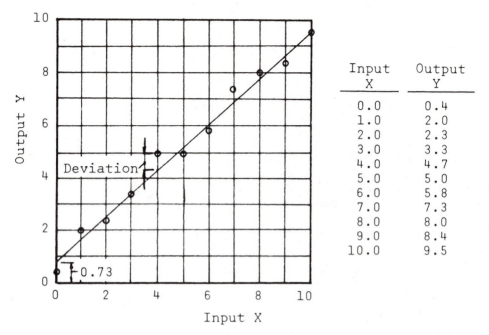

Input X	Output Y
0.0	0.4
1.0	2.0
2.0	2.3
3.0	3.3
4.0	4.7
5.0	5.0
6.0	5.8
7.0	7.3
8.0	8.0
9.0	8.4
10.0	9.5

FIGURE 2-4 Example of static calibration.

$$\frac{\partial(\text{sum})}{\partial m} = 0 \quad \text{and} \quad \frac{\partial(\text{sum})}{\partial c} = 0$$

Thus

$$c \sum X_i + m \sum X_i^2 = \sum X_i Y_i$$
$$cn + m \sum X_i = \sum Y_i$$

(2-13)

The simultaneous solution of the equations for m and c yields

$$m = \frac{n \sum X_i Y_i - (\sum X_i)(\sum Y_i)}{n \sum X_i^2 - (\sum X_i)^2}$$

$$c = \frac{(\sum X_i^2)(\sum Y_i) - (\sum X_i)(\sum X_i Y_i)}{n \sum X_i^2 - (\sum X_i)^2}$$

(2-14)

The values of m and c in Eq. 2-11 are found by substituting the calibrating data (X_i, Y_i) in the equations above. For example, the calibration line for the data points in Fig. 2-4 is

$$Y = 0.88X + 0.73$$

An *error band* is drawn about the calibration line to show the *scatter* of the data. The scatter is expressed statistically in terms of an *estimated standard deviation* s_y of Y:

$$s_y^2 = \frac{1}{n} \sum_{i=1}^{n} [Y_i - (mX_i + c)]^2 \qquad (2\text{-}15)$$

where (X_i, Y_i) are the data points from calibration. Since s_y is a measure of the deviation of Y from the calibration line, s_y is also a measure of the scatter of the data. In other words, if a value of X_i is fixed and the test repeated, Y_i will have scattered values about a mean value of Y, according to a gaussian distribution. The mean value of Y will be on the calibration line.

If an error band is drawn about the calibration line, it can be shown that

68% of the points lie within $\pm 1s_y$ of the band

95% of the points lie within $\pm 2s_y$ of the band $\qquad (2\text{-}16)$

99.7% of the points lie within $\pm 3s_y$ of the band

Note that Eq. 2-15 should be slightly modified when there are fewer than 20 data points.

The calibration line is generally used for obtaining the "true" value of the input X from an observed value of Y. Hence Y_i becomes the independent variable, and the scatter of X_i about the mean value of X is desired. From $Y = mX + c$, we get

$$X_i = \frac{1}{m} (Y_i - c) \qquad (2\text{-}17)$$

The estimated standard deviation s_x of X is

$$s_x^2 = \frac{1}{n} \sum_{i=1}^{n} \left[\frac{1}{m} (Y_i - c) - X_i \right]^2 = \frac{s_y^2}{m} \qquad (2\text{-}18)$$

It can be shown from the data in Fig. 2.4 that $s_x = 0.31$ and $3s_x = 0.94$. For example, if an observed value is $Y = 7.50$, the corresponding value of X from Eq. 2-16 is

$$X = \frac{1}{0.88} (7.50 - 0.73) \pm 3s_x$$

$$= 7.69 \pm 0.94 \ (3s \text{ limits})$$

The interpretation is that 99.7% of the time the value of the input X is between $7.69 - 0.94 = 6.75$ and $7.69 + 0.94 = 8.63$.

B. Linear Characteristics

An instrument is *linear* when the input/output relation is a straight line, that is, the output is proportional to the input. As shown in Fig. 2-4, the slope of the line is the sensitivity of the instrument and the constant of proportionality. In this section we discuss several definitions of linearity, the implications, and some related topics.

Independent linearity specifies that the maximum deviation of the output Y from the calibration line is independent of the scale reading, as shown in Fig. 2-5a. For example, the accuracy of a voltmeter may be specified as 2%. This means that the error is $\pm 2\%$ of the "full scale" of the meter, independent of the actual reading. If the full scale of the meter is 100 units, the deviation from the calibration is defined by a band of ± 2

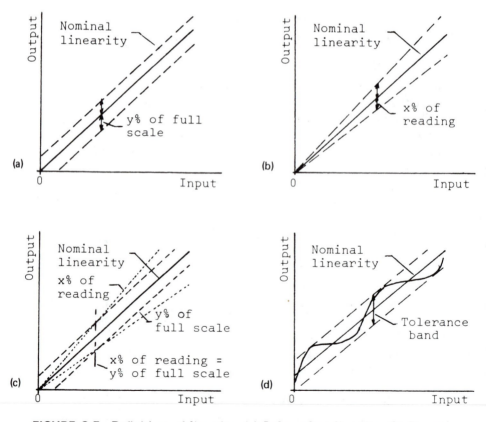

FIGURE 2-5 Definitions of linearity. (a) Independent linearity. (b) Proportional linearity. (c) Linearity specification. (d) The "best" straight line.

units. When the meter reads 10 units, however, the error is ± 2 units or $\pm 20\%$. Hence the accuracy at the lower ranges may be questionable.

Proportional linearity, as shown in Fig. 2-5b, specifies that the deviation of the output from its nominal value is a constant percentage of the actual reading for all ranges of the instrument. If the linearity is 2% and a meter reads 100 units, the error is ± 2 units. If the meter reads 10 units, the error is ± 0.2 unit. This specification is overly stringent for the lower range of the instrument.

A realistic combination of the two definitions of linearity is shown in Fig. 2-5c. An independent linearity is specified for the lower range. This recognizes the overly stringent requirement. A proportional linearity is specified for the upper range. This conveniently states the error as a percentage of the actual reading. The combined specification stipulates that the error is "$\pm x\%$ of full scale" or "$\pm y\%$ of reading," whichever is greater. These errors are equal at some intermediate point, as shown in the figure.

The scatter of the calibration data must also be stated. A 2% (1s limit) specification is not identical to a 2% (3s limit) specification, as shown in Eq. 2-16. If the scatter is unknown, the accuracy specification is questionable.

The method of least squares is generally used for obtaining the calibration line, but this is not always the case. For example, the linearity of a resistance wire potentiometer, as illustrated in Fig. 2-5d, is specified by the "best" straight line that can be "eyeballed" through the calibration points. This specification is satisfactory if only the relative positions along the potentiometer is important in applications.

Let us examine the implications of linearity in instruments.

1. Linearity is desirable. It eliminates the bother of referring to a calibration chart or a conversion table, although this can be mechanized, such as by means of a microprocessor. A system with a nonlinear component is a nonlinear system, and the mathematical analysis is more difficult than that for linear systems. Fortunately, most systems can be linearized and their performances reasonably predictable by linear theory [7].

2. Linearity simply means that the output is nominally proportional to the input. It does not imply better accuracy, higher precision, or greater sensitivity. For example, a household thermometer is nominally linear. It would be absurd to read the thermometer with a microscope and proclaim a $\pm 0.01\,^\circ$C accuracy. In fact, the linear characteristic in Fig. 2-5d does not even pass through the origin of the coordinates.

3. A nonlinear instrument can be accurate and even desired in some cases. For example, a high-quality voltmeter may have a logarithmic scale, which is nonlinear. Moreover, a log scale is linear in decibels (see

Fig. 4-23), a unit used in sound-level measurements. Incidentally, a meter has proportional linearity if the uncertainty in the dial reading is a constant amount, say ±1 mm, and the dial has a logarithmic scale. The paradox is that proportional linearity is achieved by means of a nonlinear scale.

Let us clarify some terms that are often used with instruments. The *scale readability* of an instrument is half of a "least count" or a scale division, unless the scale is provided with a vernier by the manufacturer. A reputable manufacturer will provide a suitable scale such that the accuracy of the instrument is no more than half of a division. If this is not the case, a pressure gage can be furnished with a large or a small dial. If the larger dial has more scale divisions, better accuracy or precision can be claimed for the same gage with the larger dial.

The range of the output that an instrument is designed to measure is called the *span* or the *linear operating range*. This implies that the instrument will become increasingly nonlinear beyond its linear range. A *dynamic range* is the ratio of the largest to the smallest dynamic input that the instrument will measure. For example, if a transducer reads from 0.1 to 100 units, its dynamic range is 1000:1 or 60 dB, where "dB" represents decibels (see Sec. 4-6).

Drifts in instruments are illustrated in Fig. 2-6. A *zero drift* is a vertical translation of the calibration line. A *sensitivity drift* changes the slope of the line. Both types of drifts can occur simultaneously in an instrument. The drifts are due to changes in initial conditions, boundary conditions, or irreversible effects. Common causes of drifting in electronic instruments are changes in operating temperature and the aging of components.

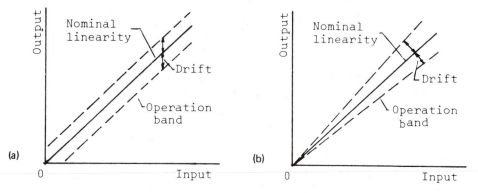

FIGURE 2-6 Drifts in instruments. (a) Zero drift. (b) Sensitivity drift.

C. Common Nonlinearities

An instrument is *nonlinear* when its output is not proportional to the input. Nonlinearities include a multitude of characteristics, and nonlinear components, such as a rectifier, are often used for the manipulation of waveforms. Nonlinearity in transducers, however, will produce distortions and is generally undesired. Some common nonlinearities are described in this section.

Saturation occurs when the magnitude of the input exceeds the linear range of an instrument. The sensitivity of the instrument decreases with a further increase in the magnitude of the input, as shown in Fig. 2-7a. *Limiting* in Fig. 2-7b is an extreme case of saturation.

The *backlash* characteristic in Fig. 2-7c often occurs in gears. Assume that X is the pinion and Y the driven gear. With X increasing, the output Y follows the input X along path 1. When $X = a$ and starts to decrease, X traverses the backlash and goes from b to c until the gears are again in mesh. Then Y follows X along path 2. Similar descriptions can be given for X increasing or decreasing at other positions. Antibacklash gears are commonly used in instruments.

Hysteresis is due to the internal friction in a material. Without internal friction, the stress-strain relation of an elastic material follows Hooke's law. An example of hysteresis is shown in Fig. 2-7d. The output follows the input along one path for increasing input and follows another path for decreasing input. The characteristic shown is multivalued. The area enclosed by the cyclic *hysteresis loop* gives the energy dissipation per cycle in the material. The blowout of an underinflated automotive tire is an example of hysteresis heat buildup in rubber. The blowout usually occurs at the side wall, where the greatest flexing occurs. Hysteresis may be important for high-precision instruments. A common pressure gage exhibits some hysteretic effect. An extremely precise pressure gage can be made of quartz (e.g., Model 156 Pressure Test Sets, Texas Instruments, Inc., Industrial Products Division, Houston, Texas) to minimize this effect. Magnetic devices, such as the head of a tape recorder, are strongly hysteretic. The hysteresis characteristic can be utilized for control in some instruments.

The *threshold effect,* shown in Fig. 2-7e, denotes the minimum input to produce a detectable output near the zero reading. This is due to Coulomb friction or the free play in an instrument. *Dead space, dead zone,* and *dead band* are also used to describe the threshold effect. *Resolution* in Fig. 2-7f describes the minimum input to produce a detectable output for any range of the instrument. For example, a wire-wound potentiometer consists of a resistance wire wound on an insulating material. The contact

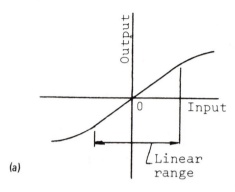

(a)

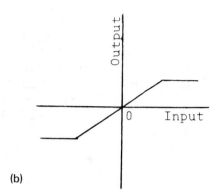

(b)

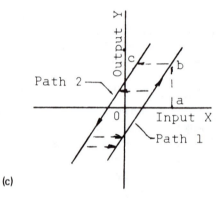

(c)

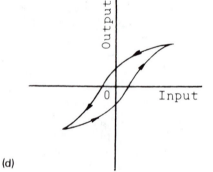

(d)

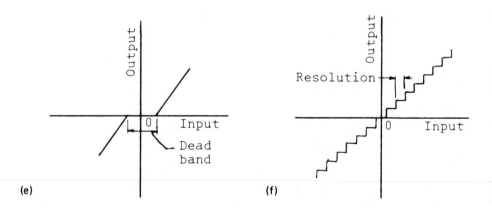

(e)   (f)

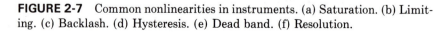

FIGURE 2-7   Common nonlinearities in instruments. (a) Saturation. (b) Limiting. (c) Backlash. (d) Hysteresis. (e) Dead band. (f) Resolution.

at the potentiometer moves from coil to coil. Thus the change in the output voltage of the potentiometer is stepwise. If a potentiometer has 1000 turns, its resolution is 1/1000 or $\pm 0.1\%$. *Infinite resolution* simply means that the input/output relation of the instrument is not stepwise.

D. Effects of Nonlinearities

The effects of nonlinearity in an instrument are amplitude distortion and phase distortion in measurements. Any distortion is equivalent to extraneous signals or noise at the output. We first show examples of distortions and then generalize the results.

An example of distortions due to a limiting characteristic is illustrated in Fig. 2-8. Assume that the input X is sinusoidal. Within the linear range of the instrument, the output Y is proportional to X. Hence both X and Y are sinusoidal, and there is no distortion in waveform. When X exceeds the linear range, the waveform of Y is distorted, as shown in Fig. 2-8a. The construction of Y from X is shown in Fig. 2-8b. At a time $t = t_1$, we have $X = X_1$. The projection from X_1 to obtain a corresponding Y_1 is shown as dashed lines in the figure. Similarly, Y_2 is obtained from X_2 at $t = t_2$.

Let us compare the *Fourier series expansion* of the input X and the output Y to show the distortions. Since X is sinusoidal, its series expansion has only one component:

input $= X \sin \omega t$

where ω is the circular frequency in rad/s and X the amplitude. The output Y is periodic. Its series expansion is

$$\text{output } Y = \frac{a_0}{2} + \sum_{n=1}^{\infty} (a_n \cos n\omega t + b_n \sin n\omega t)$$

where ω rad/s is the frequency of the input. Only the component ($a_n \sin n\omega t$) with $n = 1$ is proportional to the input. All other terms in the series are extraneous. They are generated by the instrument and not from the input. For example, the term $a_0/2$ is the average value of Y. It is a constant and cannot be from a sinusoidal input.

Example 2-3.   Assume that a voltage $X \sin \omega t$ with $X = 1$ is applied to a single-pole, double-throw relay G. (a) Find the output waveform of Y when the characteristic of G is as shown in Fig. 2-9a. (b) Repeat for the characteristic Fig. 2-9b.
Solution: (a) using the procedure in Fig. 2-8, it is evident that the waveform of the output Y is as shown in Fig. 2-9c. The Fourier series expansion of the square wave is

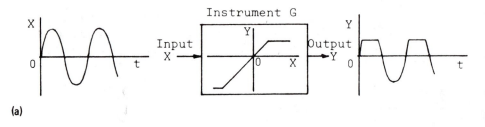

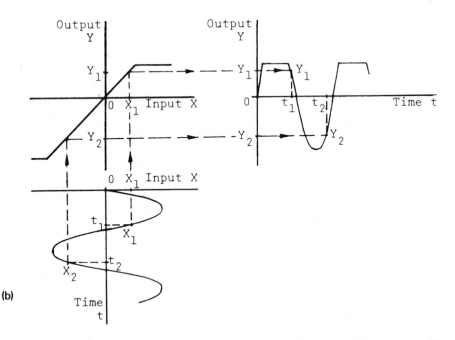

FIGURE 2-8 Distortion due to nonlinearity. (a) Schematic of instrument with limiting characteristic. (b) Construction of output Y from input X.

$$\text{output } Y = \frac{4}{\pi} \sum_{n=1}^{\infty} \frac{1}{n} \sin n\omega t \qquad \text{for} \quad n = 1, 3, 5, \ldots$$

$$= 1.27 \sin \omega t + 0.42 \sin 3\omega t + 0.25 \sin 5\omega t + \cdots$$

The output signal proportional to the input is $1.27 \sin \omega t$. All other terms in the equation are extraneous.

(b) The characteristics in Fig. 2-9b are those of an imperfect relay, that is, a relay with hysteresis. Using the procedure in Fig. 2-8, it can be shown that the output Y is a square wave as before, but the wave is translated

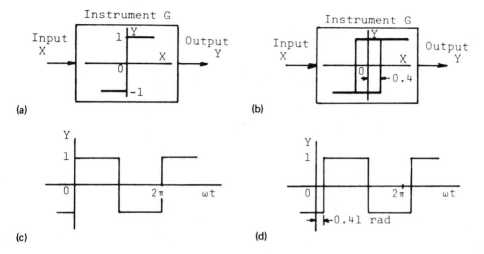

FIGURE 2-9 Distortions due to nonlinearity in relay characteristics. (a) Relay characteristic. (b) Relay with hysteresis. (c), (d) Output waveforms.

along the time axis, as shown in Fig. 2-9d. A shift of the signal along the time axis gives a phase angle. The Fourier series expansion is

$$\text{output } Y = \frac{4}{\pi} \sum_{n=1}^{\infty} \frac{1}{n} \sin\left(n\omega t - n\phi\right) \qquad \text{for} \quad n = 1, 3, 5, \ldots$$

$$= 1.27 \sin\left(\omega t - 0.41\right) + 0.42 \sin\left(3\omega t - 1.23\right) + \cdots$$

where $\phi = 0.41$ rad is the phase angle due to the hysteresis effect.

Let us generalize the foregoing observations: (1) amplitude distortion occurs when the characteristic of G is not linear; (2) a constant term is introduced in the output when G is not symmetric about the origin 0; and (3) a phase shift occurs when G does not pass through 0. A nonlinearity may include all the above.

For example, the saturation characteristic in Fig. 2-10a is symmetric, that is, the curves in the first and third quadrants are identical. The nonlinearity will produce amplitude distortions. It is equivalent to the introduction of harmonic signals, which are not in the input. The characteristic in Fig. 2-10b is linear but does not pass through 0. The output will have a constant term, although the waveform of the input is preserved. An example of phase distortion is shown in Fig. 2-9c, in which the characteristic of the transducer is multivalued. The characteristic in Fig. 2-10c is nonsymmetric but passes through 0. It will introduce additional harmonics at the output as well as a dc component. There will be no phase

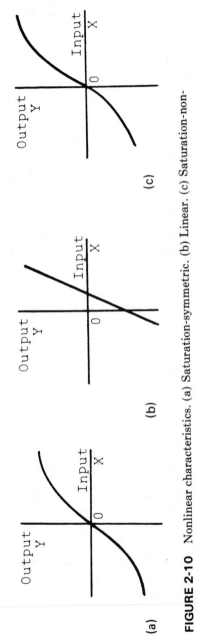

FIGURE 2-10 Nonlinear characteristics. (a) Saturation-symmetric. (b) Linear. (c) Saturation-non-symmetric.

shift, since G passes through 0. Description of common nonlinearities are found in the literature [8].

E. Linearization

Linearization is the process of finding a straight line about an operating point (0) to approximate the nonlinear characteristic of an instrument. In other words, the nonlinear system is represented by means of a linear model for a limited range of operation about (0). The incremental model and the linear model will be discussed in this section.

Let the output Y of a transducer be related to an input X by

$$Y = f(X) \tag{2-19}$$

as shown in Fig. 2-11a. Expanding this in a Taylor series about the given operating point (0), we get

$$Y = Y_0 + \Delta Y = f(X_0) + \frac{df}{dX}\Big|_0 \Delta X + \cdots \tag{2-20}$$

Since $Y_0 = f(X_0)$, retaining only the first-order term of the series expansion, we obtain

$$\Delta Y \simeq \frac{df}{dX}\Big|_0 \Delta X \qquad \text{or} \qquad y \simeq Cx \tag{2-21}$$

where $x = \Delta X$, $y = \Delta Y$ are incremental values, and the slope at (0) is $C = df/dX|_0$.

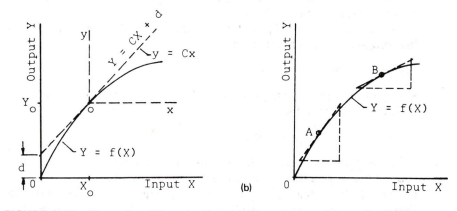

FIGURE 2-11 Examples of linearization. (a) Linearized model $y = Cx$. (b) Effect of initial conditions.

The *incremental model* is described by Eq. 2-21. Steps for finding the model are as follows: (1) establish the operating point (0) at (X_0, Y_0), which becomes the origin of the incremental model (see Sec. 3-3D); and (2) approximate the nonlinear characteristic by its slope at (X_0, Y_0). This gives Eq. 2-21.

Note that selecting a different initial condition will result in a different model for the same system, as shown in Fig. 2-11b for points A and B. The average slope instead of the tangential slope should be used when a wider range of operation is anticipated. Thus Eq. 2-21 becomes

$$\Delta Y \simeq \frac{\Delta f}{\Delta X}\bigg|_0 \Delta X \qquad \text{or} \qquad y \simeq Cx \qquad (2\text{-}22)$$

where $C = \Delta f/\Delta X|_0$ is the average slope. In other words, the range must be estimated in addition to establishing (X_0, Y_0).

Example 2-4.   A transducer has the input/output characteristic

$$Y = 0.5X|X|$$

Find the incremental model for (a) $(X_0, Y_0) = (1, 0.5)$ and $\Delta X = \pm 0.1$, and (b) $(X_0, Y_0) = (0, 0)$ and $\Delta X = \pm 0.5$.
Solution: The characteristic $Y = 0.5\ X|X|$ is plotted in Fig. 2-12. The slope of the curve is $dY/dX = X$.

(a) For $(X_0, Y_0) = (1, 0.5)$ and $\Delta X = \pm 0.1$, the tangential slope is $dY/dX = 1.0$. The incremental model is

$$y = 1.0x \qquad \text{for} \quad x = \pm 0.1$$

An inspection of the plot shows the model is a reasonable approximation of the system characteristic.

(b) For $(X_0, Y_0) = (0, 0)$, the tangential slope is $dY/dx = 0$. It would be inappropriate to represent the system by

$$y = 0 \qquad \text{for} \quad x = \pm 0.5$$

Using an estimated average slope $\Delta Y/\Delta X = 0.2$ as shown in the figure, the incremental model is

$$y = 0.2x \qquad \text{for} \quad x = \pm 0.5$$

A *linear model* uses a straight line to approximate the nonlinear characteristic as before, but the origin of the coordinates remains at $(X, Y) = (0, 0)$ instead of at the operating point (0). For example, the linear model shown in Fig. 2-11a is

$$Y = CX + d$$

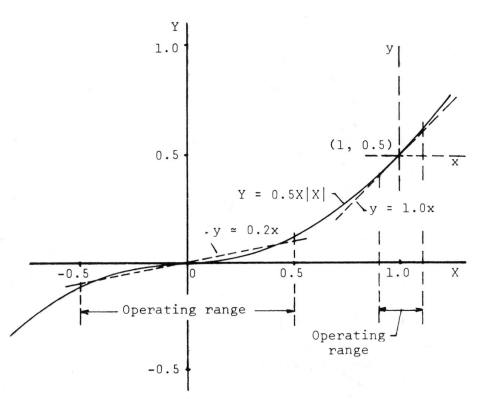

FIGURE 2-12 Methods of linearization.

where C is the slope and d the intercept on the Y axis. Note that an incremental model is often called a linear model in the literature, and the terminology is not standardized.

The linearization of Y as a function of two independent variables X_1 and X_2 is illustrated in Fig. 2-13. Let

$$Y = f(X_1, X_2) \tag{2-23}$$

Expanding this in a Taylor series about (0) and retaining only the first-order terms, we get

$$Y = Y_0 + \Delta Y \simeq f(X_1, X_2)|_0 + \frac{\Delta f}{\Delta X_1}\bigg|_0 \Delta X_1 + \frac{\Delta f}{\Delta X_2}\bigg|_0 \Delta X_2 \tag{2-24}$$

Since $Y_0 = f(X_1, X_2)_0$, the incremental model about (0) is

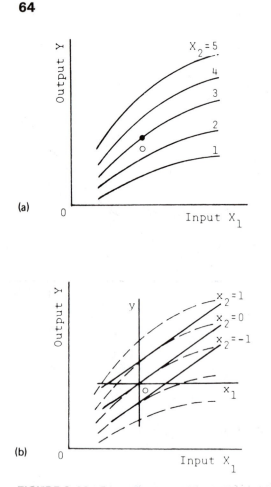

FIGURE 2-13 Linearization of $Y = f(X_1, X_2)$. (a) Nonlinear characteristics. (b) Linearized characteristics.

$$\Delta Y = \frac{\Delta f}{\Delta X_1}\bigg|_0 \Delta X_1 + \frac{\Delta f}{\Delta X_2}\bigg|_0 \Delta X_2 \quad \text{or} \quad y = C_1 x_1 + C_2 x_2 \quad (2\text{-}25)$$

where $x_1 = \Delta X_1$, $x_2 = \Delta X_2$, $y = \Delta Y$, $C_1 = \Delta f/\Delta X_1$, and $C_2 = \Delta f/\Delta X_2$.

The steps for finding the incremental model are as before. (1) Establish the operating point (0) and estimate the ranges of X_1 and X_2. (2) Draw a family of straight lines about (0) to approximate the nonlinear curves, as shown in Fig. 2-13b. The linearization gives a family of equally spaced straight lines with equal slopes. The lines are equally spaced because when x_1 is constant, y is proportional to integer values of x_2. For example, when $x_1 = 0$ in Eq. 2-25, we get $y = C_2 x_2$. The lines have the same slope because, when x_2 is constant, y is proportional to x_1.

Example 2-5.   The experimental test data of an internal combustion engine in Fig. 2-14 are described by the relation $Q = f(N, T)$, where Q is the fuel consumption in lb_m/hr, N is the engine speed in rpm, and T is the engine torque in ft-lb$_f$. Find the incremental model about the given operating point (0) and for the range shown in the figure.
Solution: From Eqs. 2-23 and 2-25 we get

$$Q = f(N, T) \quad \text{and} \quad \Delta Q = \frac{\Delta f}{\Delta N}\bigg|_0 \Delta N + \frac{\Delta f}{\Delta T}\bigg|_0 \Delta T$$

For the given operating range, we obtain from the figure

$$\frac{\Delta f}{\Delta N}\bigg|_0 \simeq \frac{30 - 20}{2300 - 1700} = 0.0167 \quad \text{and} \quad \frac{\Delta f}{\Delta T}\bigg|_0 \simeq \frac{34 - 18}{160 - 80} = 0.20$$

The linearized characteristics about (0) are shown as dashed lines in the figure. Defining $q = \Delta Q$, $n = \Delta N$, and $t = \Delta T$, the corresponding incremental model is

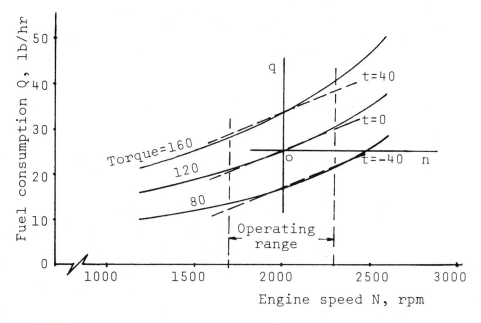

FIGURE 2-14   Linearization of performance characteristics of an engine.

$q = 0.0167n + 0.20t$

2-3. TRANSDUCER TYPES AND MODELING

Transducers are classified as self-generating or non-self-generating. These are examined in this section by means of the information model, energy model, incremental model, and generalized model. Although the terms *self-* and *non-self-generating* are descriptive, they are commonly called *passive* and *active,* respectively. The terminology is not standardized.

Note that a passive electrical network is one that does not contain an energy or power source. An active network contains a power source. In dynamic analysis, however, a constant power source, such as a battery, is omitted from consideration. On the other hand, a photocell is considered an active network, because of the variable external energy supplied for its operation. Thus a self-generating transducer such as a thermocouple should be called active instead of passive. We choose to avoid the controversy in semantics.

A. Information Models

A transducer is potentially a mutli-input, multi-output device. For the *information model,* only the signals (information) at the input and output are considered. Noise and loading in the sensing process are ignored.

A transducer is *self-generating* when the energy at its output is entirely from the input. Its model in Fig. 2-15a has a signal input X from the measurand and a signal output Y. The thermocouple is representative of a self-generating transducer. The input/output relation is

$$Y = f(X) \tag{2-26}$$

Since the energy at the output is entirely from the measurand, a self-generating transducer cannot deliver appreciable power without seriously loading the source of information (see Sec. 3-3A).

A transducer is *non-self-generating* when the energy at its output is mainly from an auxiliary source, that is, the energy from the measurand is minimal. The model in Fig. 2-15b has a signal input X from the measurand, an auxiliary input Z, and an output Y. The input/output relation is

$$Y = f(X, Z) \tag{2-27}$$

Let us examine the non-self-generating transducer by considering the operation of a resistance strain gage. The measurand X is the strain ap-

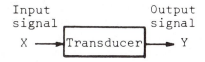

(a)

Self-generating
transducer, Y = f(X)

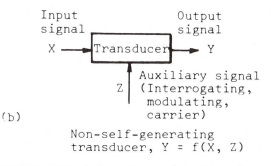

(b)

Non-self-generating
transducer, Y = f(X, Z)

FIGURE 2-15 Information models of transducers. (a) Self-generating transducer, $Y = f(X)$. (b) Non-self-generating transducer, $Y = f(X, Z)$.

plied to the gage. The energy for straining the fine resistance element is minimal. The effect of the strain is a resistance change in the gage, which can only be detected by means of a measurement with an electrical input Z. The auxiliary input Z is necessary for the operation. Since the gage is a resistor, or an impedance, it is deduced that all impedance-based transducers are non-self-generating and will require an auxiliary input Z. The electrical energy at the output Y of the strain gage is supplied by the input Z.

The input Z is also a carrier. The measurand X changes only the impedance of the transducer. The carrier Z conveys this information to the output Y. Another example is a voice spoken into a microphone in a broadcasting station. The energy for broadcasting can hardly be from the voice spoken into the microphone. The voice is conveyed or carried by means of the carrier at a radio frequency for transmission.

Furthermore, the carrier modifies the characteristics of the input signal for transmission. The output Y is the modulated signal. We are familiar with amplitude modulation AM and frequency modulation FM in radio. Had it not been for signal modulation, there would be no television TV. Since AM and FM have different characteristics, it is evident that the nature of the output Y is dependent on the characteristics of the carrier Z. In other words, a desired characteristic in Y can be obtained by proper selection of the type of carrier Z. This important advantage of non-self-generating transducers is not shared by self-generating transducers.

The generalized models [9] for both the self- and non-self-generating transducers are shown in Fig. 2-16. A strain gage can be modeled as non-self-generating, as shown in Fig. 2-16a. The gage resistance R is a parameter. The strain from the environment changes R of the gage. The carrier Z for measuring R is viewed as an interrogating signal. It "interrogates" R and conveys the information on the state of R to the output Y. The Z input must be controlled or regulated, much as the radio frequency of a broadcasting station is regulated.

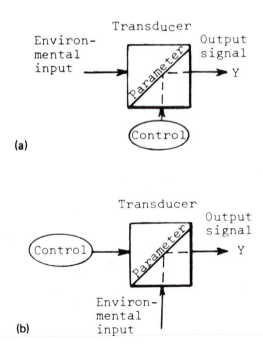

FIGURE 2-16 Generalized information models. (a) Non-self-generating. (b) Self-generating.

A spring scale can be modeled as self-generating, shown in Fig. 2-16b. The spring constant is a parameter, which is a property of the scale. It is essentially constant or controlled during the operation. When the scale is used for a measurement, the environmental input "interrogates" the scale and supplies energy to the output. The tool dynamometer in Fig. 1-3 is another example of a self-generating transducer. The model of the primary sensor is a heavy spring. The applied force from the environment interrogates the dynamometer to produce an output proportional to the applied force.

The generalized model in Fig. 2-17a is deduced from the similarity of the self- and non-self-generating models. It represents both models,

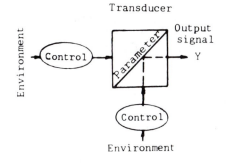

(a)

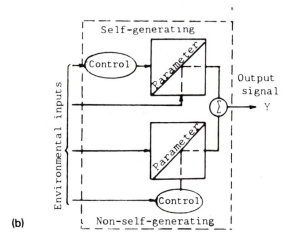

(b)

FIGURE 2-17 Composite information models of transducers. (a) Composite information model. (b) Composite model showing self- and non-self-generating effects.

depending on where the controls and the environmental inputs are applied. In reality, the controls cannot be absolute and are themselves influenced by the external environment, as shown in Fig. 2-17b. Thus a transducer can be both self- and non-self-generating at the same time. For example, a strain gage is normally non-self-generating. It can be self-generating when exposed to a changing magnetic field. A strain gage can also be a thermocouple, since there are two junctions connecting the gage to the copper leads. A thermal emf exists when there is a thermal gradient between the junctions. A thermal gradient occurs even under steady-state conditions, such as the thermal gradient through a wall of a house in winter, or a thermal gradient across a blade of a gas turbine.

B. Energy Models

Since an information transfer requires an energy transfer, let us upgrade the information model by considering the energies in self- and non-self-generating transducers. *Energy* or work is described by two coexisting quantities. Mechanical energy is the product of force and displacement. *Power* is the rate of doing work. Electrical power is the product of voltage and current. Hence an *energy model* has two quantities at each input and output.

The energy model of a self-generating transducer shown in Fig. 2-18a is a *two-port device*. It has one energy "port" for the energy input and one port for the energy output.

First, consider the energy at the input. One of the two quantities is the *primary input* X_p and the other the *secondary input* X_s. The input energy is the product $(X_p X_s)$. The tool dynamometer shown in Fig. 1-3 can be modeled as a cantilever, as shown in Fig. 2-19a, or a heavy spring, as shown in Fig. 2-19b. The input energy is the product of (force F) and

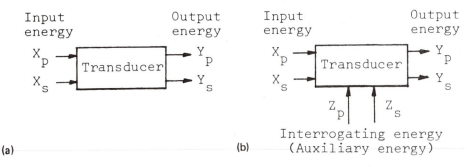

FIGURE 2-18 Energy models of transducers. (a) Self-generating. (b) Non-self-generating.

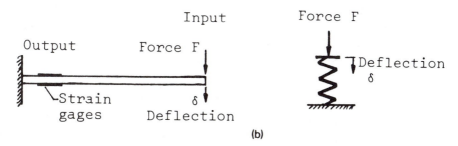

FIGURE 2-19 Model of the tool dynanometer shown in Fig. 1.3b. (a) Cantilever. (b) Spring.

(deflection δ). The cutting force F is of primary interest; therefore, $F = X_p$. The associated deflection δ is only of secondary interest and $\delta = X_s$. Note that δ must exist in a force measurement, or else there will be no signal at the output for the strain gage and the force F cannot be measured. It is paradoxical that δ must be small or else the structure of the cutting process is changed and the force measurement will not be the true force. A perfect measurement is made when δ approaches zero.

The output of the cantilever in Fig. 2-19a is the input to the strain gages or the secondary sensor of the dynometer. The energy is the product of force and displacement at the gages. Strain is the primary output Y_p of the cantilever and input the gages. The associated force to strain the gages is the secondary output Y_s. For normal applications, the force to strain the gage is negligible.

The discussions above show that the designation of primary and secondary quantities depends on the problem. At the input of the tool dynamometer, force is of primary interest and the assocaited deflection secondary. At the output of the same dynamometer, displacement or strain is the primary quantity and the associated force the secondary. In any case, the secondary quantities must approach zero for a true measurement in order that loading due to a structure change in the sensing be minimal.

As an additional example, consider the dial gage for dimensional measurement shown in Fig. 2-10. Again, the dial gage is modeled as a spring. The displacement at the input of the gage is the primary quantity and the associated force the secondary. The force for deflecting the gage must be small for a true measurement. To this end, a dial gage must have a weak spring. There was an experimental thesis on the deflection of a thin shell. Dial gages covered the shell like quills on a porcupine. Yet not

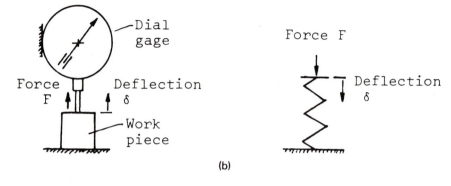

FIGURE 2-20 Model of a dial gage for displacement measurement. (a) Dial gage. (b) Spring.

one word was said in the thesis about the possible loading of the thin shell.

A non-self-generating transducer is a *three-port device*. Its model in Fig. 2-18b has three ports for energy transfer, with two quantities at each of the energy ports. From the discussions on two-port devices, the designation of primary and secondary quantities at each port is dictated by the problem.

The discussions above are summarized as follows:

1. An energy model has two quantities at each energy port, one of which is designated as primary and the other secondary. The primary quantity gives the desired information, and the associated quantity is of secondary interest in a measurement. The product of the two quantities is energy or power.
2. The secondary quantity must exist, else information cannot be obtained. The secondary must not cause a loading in the problem.
3. When the secondary quantities are omitted, the energy model degenerates to the information model.

C. Incremental Models

An *incremental model* is a linearized energy model. Energy models were described in the preceding section. Incremental models of the self- and non-self-generating transducers are derived in this section.

First, consider the self-generating transducer in Fig. 2-18a. The quantities at the input are (X_p, X_s) and those at the output are (Y_p, Y_s), where the subscript (p) denotes the primary quantity and (s) the secondary. There are four variables in a self-generating transducer. The model is derived by (1) finding the function relation between the four variables (X_p, X_s, Y_p, Y_s), and (2) linearizing these relations to obtain the incremental model.

The linearization procedure is similar to that for the static characteristics described in Sec. 2-2E. It can be shown that only two of the four quantities are independent. The selection of independent variables generally depends on the ease with which the system parameters can be measured. For the tool dynamometer in Fig. 2-19, X_p is the input force and is considered independent, while the associated deflection X_s is dependent. It is desired to measure the strain Y_p as a function of the applied force, so Y_p is dependent. With (X_p, Y_s) as the independent variables, we obtain the functional relations

$$Y_p = f_1(X_p, Y_s)$$
$$X_s = f_2(X_p, Y_s) \tag{2-28}$$

It can be shown that there are six sets of such equations, because there are six combinations of the four variables taken two at a time (see Sec. 3-3C).

The equations above are linearized about an operating point (0) to obtain the incremental model. The procedure follows directly from that described in the preceding section.

$$Y_p|_0 = f_1(X_p, Y_s)|_0$$
$$X_s|_0 = f_2(X_p, Y_s)|_0 \tag{2-29}$$

The incremental model is obtained by expanding Eq. 2-28 in a Taylor series. Subtracting Eq. 2-29 from the series expansion and retaining only the first-order terms, we get

$$\Delta Y_p = \frac{\partial f_1}{\partial X_p}\Big|_0 \Delta X_p + \frac{\partial f_1}{\partial Y_s}\Big|_0 \Delta Y_s \qquad \Delta X_s = \frac{\partial f_2}{\partial X_p}\Big|_0 X_p + \frac{\partial f_2}{\partial Y_s}\Big|_0 Y_s \tag{2-30}$$

or

$$y_p = a_{11}x_p + a_{12}y_s \qquad \begin{bmatrix} y_p \\ x_s \end{bmatrix} = \begin{bmatrix} a_{11} & a_{12} \\ a_{21} & a_{22} \end{bmatrix} \begin{bmatrix} x_p \\ y_s \end{bmatrix}$$
$$x_s = a_{21}x_p + a_{22}y_s \tag{2-31}$$

where the a's denote the partial derivatives and the lowercase letters (x, y) the incremental values about the operating point.

The incremental model in Eq. 2-31 reduces to the information model when all the secondary quantities are neglected. Consider the tool dynamometer in Fig. 2-19. If the force required to strain the gages is negligible, then $y_s = 0$. Thus $a_{12}y_s$ and $a_{22}y_s$ are negligible. The parameter a_{21}

defines the compliance (1/stiffness) of the dynamometer. If the dynamometer is extremely stiff, a_{21} approaches zero. Hence the remaining term in Eq. 2-31 gives the information model.

$$y_p = a_{11}x_p \tag{2-32}$$

This is the information model shown in Fig. 2-15a. For dynamic measurements, a_{11} is the transfer function of the system for zero loading conditions.

The incremental model for the non-self-generating transducer in Fig. 2-18b is derived in a like manner. There are six variables for a three-port device. It can be shown that only three of these can be independent. For example, if (X_p, Y_s, Z_p) are independent, the functional relations are

$$
\begin{aligned}
Y_p &= f_1(X_p, Y_s, Z_p) \\
X_s &= f_2(X_p, Y_s, Z_p) \\
Z_s &= f_3(X_p, Y_s, Z_p)
\end{aligned}
\tag{2-33}
$$

The incremental model is obtained by means of linearization.

$$
\begin{bmatrix} y_p \\ x_s \\ z_s \end{bmatrix} =
\begin{bmatrix} a_{11} & a_{12} & a_{13} \\ a_{21} & a_{22} & a_{23} \\ a_{31} & a_{32} & a_{33} \end{bmatrix}
\begin{bmatrix} x_p \\ y_s \\ z_p \end{bmatrix}
\tag{2-34}
$$

where the a's denote the partial derivatives evaluated about the operating point, and the lowercase letters (x, y, z) the incremental values.

Under zero loading conditions, the incremental model above also reduces to the information model as shown in Fig. 2-15b. Furthermore, when the Z input is ideal and is not influenced by the measurement, we get $\Delta Z = z = 0$. Hence the Z input does not have to be considered. The model is further reduced to the information model for the self-generating transducer in Fig. 2-15a.

2-4. CALIBRATION

Calibration of instruments was discussed briefly in Sec. 1-4B. The generalized model in Fig. 2-21 will be used to reexamine calibration. A transducer should be calibrated for static and dynamic operations as well as for in-service conditions. The thrust of the discussion in this section is to caution against the misuse of calibration.

Calibration serves to reveal the individuality of an instrument. An instrument is "traceable" if the standards for calibration can be traced to the National Bureau of Standards. The certificate of calibration, however, is only a reference and should not be accepted as legal tender for the

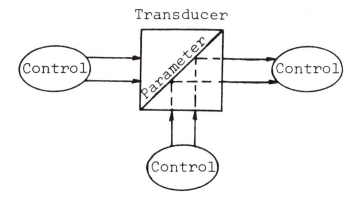

FIGURE 2-21 Conditions for calibration.

instrument to measure accurately. Due to different in-service conditions, when an instrument works well for one application, there is no assurance that it will perform adequately for another, despite the proven performance and certificate of calibration.

The generalized model for calibration is shown in Fig. 2-21. A calibration is normally conducted under idealized conditions, because it is not possible for a standards laboratory to calibrate instruments for all imaginable in-service conditions. Thus all quantities at the energy ports are controlled. Loading due to energy transfer and structural changes are minimal, and only one variable is changed at a time. This means that all the undesired inputs shown in Eq. 2-4 are suppressed.

During calibration, the loading *due to energy transfer* is negligible. For example, when a thermometer is calibrated in a constant-temperature bath, the thermometer must not change the temperature of the bath. Similarly, the energy transfer at the output is controlled. For example, when a "full-immersion" thermometer is being calibrated and a part of the mercury in the stem is not immersed, a correction for energy transfer must be made for the exposed stem. In other wrods, the output of an instrument must be controlled during a calibration because the output is also susceptible to loading.

Loading *due to a structural change* is more subtle. Assume that a pitot tube (see Fig. 3-6) for a flow measurement is calibrated in a large duct. If the pilot tube is subsequently used in a smaller duct, the flow patterns in the two ducts could be different. The discrepancy due to the structural effect cannot be eliminated by repeated calibration.

Instruments are commonly calibrated under *static conditions*. This cannot be extended automatically for dynamic applications. For example, the frequency response (see Sec. 4-6) of a strip-chart recorder is shown in Fig. 2-22. It is desired to have a "flat response," that is, the response of the recorder is independent of the input amplitude and frequency. Assume a sinusoidal input with amplitude of about 12 mm. The response is flat up to about 110 Hz, then it attenuates or "rolls off" with a further increase in frequency, as shown in the figure. This shows that the response is frequency dependent. Furthermore, the range of the flat response decreases with increasing input amplitude, as shown in the figure. Hence the low-frequency response cannot be extrapolated for applications at higher frequencies, and the response of the recorder is both amplitude and frequency dependent.

The guidepost for in-service calibration is that a known input will give a known output. The corollary is that a zero input will produce a zero output. The output for zero input is from the inherent noise in the instrument and it gives a datum for the measurement. The known input, however, must be carefully defined for the application. For example, a strain gage type of force transducer (see Fig. 3-33) cannot always be calibrated directly by applying a known force. An indirect calibration may be necessary. Hence the "known" input or its equivalent for the calibration should be examined.

Calibration for in-service conditions is not always possible. Assume

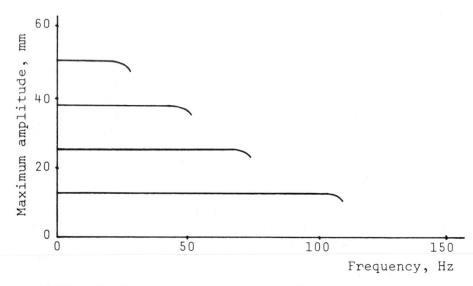

FIGURE 2-22 Characteristics of a typical strip-chart recorder.

that a small accelerometer is used to measure the natural freqeuncy of a structure of thin metal. Despite its small size, the accelerometer adds mass to the structure to cause a mass loading. In fact, a small accelerometer can lower the natural frequency of the "dishpan" of an electric typewriter as much as 30%. The solution is to use a different method of measurement, such as a noncontact or proximity transducer.

2-5. ERRORS IN MEASUREMENTS

Errors in measurement can be classified as static or dynamic and systematic or random. Mistakes are often called illegitmate errors. Static errors are discussed in this section. Dynamic errors are more elusive and are treated in Chapter 4. Let us first classify some terms commonly used in error analysis.

A. Accuracy, Precision, Uncertainty, Mistakes

Accuracy is the deviation of the output from the calibrating input or the *true value*. If the accuracy of a voltmeter is 2% full scale as described in the preceding section, the maximum deviation is ±2 units for all readings.

Precision denotes the repeatability of readings on successive observations, that is, how small a difference can be detected on repeated readings. If the difference is random in nature, the readings will differ by a small amount for each replication of the observations. The precision of an answer is usually expressed as so many significant figures.

Uncertainty is generally stated as a number, indicating the tolerance from the true value of the measurand. The tolerance is only estimated. It represents the confidence level of the investigator in the results, since the true value of the measurement is unknown.

Precision and *accuracy* are not synonymous. Accuracy implies closeness to the true value with reasonable precision. For example, it would be precise to state that the numerical value of π is 2.14159, while a more accurate value is 3.14. When a precision instrument is off calibration, its indications may still be precise. Probably the readings will be off by a constant amount from the calibrated value. Moreover, there can be high precision without a corresponding high accuracy, as described in Sec. 1-9B for temperature measurements. Another example is in a game of American football. The ball carrier ran out of bounds. A referee 20 yards away rushed over to mark the location. The chain was brought out for a measurement, and it was shy of a first-down by inches. The inches describe the precision, but the accuracy was anyone's guess.

It is difficult to define *mistakes* to cover all contigencies. Some mistakes are obvious, whereas others are subtle. Some mistakes can be discovered by a clear definition of the problem and objectives. Following the discussions on aspects of measurement in Fig. 1-7, a method for investigating experimental mistakes is to reexamine the effects of loading, the initial and boundary conditions, method of calibration, the structure of the measuring system, and the dynamics of the system.

An obvious mistake can be an oversight or due to an extreme disturbance affecting the experiment. The lack of control of the test conditions is an oversight. Vibration of the surroundings may be sufficient to render a test meaningless. It is senseless to continue an experiment unless such mistakes are corrected. A mistake is obvious "after" it is discovered.

A mistake can be subtle. Consider the measurement of the diameter of a bar with three lobes shown in Fig. 2-23. The "diameter" is constant if it is measured with a micrometer. The same is true for bars with any odd number of lobes. Such "round" bars could be from a centerless grinder. The mistake could be avoided by clearly defining the problem. "Is the bar round?" A logical follow-up to avoid further mistakes is to ask whether the method of measurement and the results obtained will fulfill the objectives.

An experiment cannot by itself reveal its own errors by means of repetition. Consider the example of a mistake in boundary conditions. Students were to measure the temperature of steam in a drum as shown in Fig. 2-24. The instructor purposely removed the oil from the thermometer well. Clearly, the lack of oil in the thermometer well resulted in poor thermal contact. The experiment was conducted with great care. After repeated readings, the temperature of steam at 140 kPa ($\simeq$ 5 psig) was reported as $98.8 \pm 0.1°C$ ($3s$ limits), where s is the estimated standard deviation. The report received a bad grade, and students blamed the instructor for the "monkey wrench" in the experiment. The reply was: "The minute you use an instrument, it is your baby."

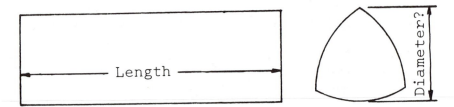

FIGURE 2-23   "Round" bar.

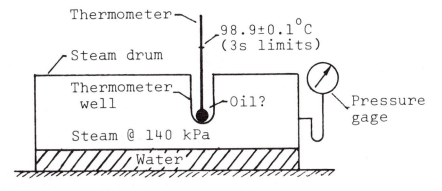

FIGURE 2-24 Measurement of steam temperature.

B. Systematic and Random Errors

Errors in measurement are broadly classified as systematic and random [10]. A systematic error will bias the data by a consistent amount, while a random error shows the scatter of the data about a mean value. The calibration line in Fig. 2-4 illustrates the two types of error. The intercept on the Y axis represents a systematic error. The scatter of the data points about the calibration line shows the random error.

A *systematic error* is a fixed and consistent quantity, which remains the same for the observations in an experiment. The error in the measurement of steam temperature in Fig. 2-24 is an example of systematic error due to faulty boundary conditions. A systematic error can be due to (1) the design, alignment, calibration, or the condition of the apparatus; (2) the technique and procedure of the experiment; or (3) subconscious human bias. When the cause of the error is found, the error can be corrected.

A *random error* is due to a large number of independent small effects that cannot be identified or controlled. It is a statistical quantity. As such, it will vary for each replication of the observations. If a large number of readings is observed for the same quantity, the scatter of the data about a mean value can be evaluated. The scatter generally follows a guassian distribution about a mean value, which is assumed to be the true value.

Not all scatter of data in an experiment should be called random without due consideration. The scatter of data in a poorly controlled experiment could be due to a nubmer of small systematic errors. In other words, statistics applies only for random errors. Statistical analysis is appropriate only after the validity of the results is assured and systematic errors eliminated.

C. Engineering Data

Uncertainty, instead of accuracy, is commonly used to describe engineering data. The nature of engineering data is discussed in this section to justify the terminology.

Single-sample data are those from a succession of readings in an experiment, taken under identical conditions except for time. A systematic error cannot be revealed by the data from a single-sample experiment. This was illustrated in the measurement of the temperature of steam in Fig. 2-24. Neither can the error be revealed by the application of statistics. Hence the true value in a single-sample test is not known. The scatter of data about a mean value can be determined, but the mean may not be the true value. Hence the analysis of random error before the elimination of systematic errors is not meaningful. Two wrong readings repeated precisely will not improve the confidence in the data.

Engineering data are generally from single-sample tests, in which the same principle, equipment, procedure, and personnel are employed. Since neither the systematic nor the random error is certain, one can only speak of the uncertainty or the confidence level of the results. Data can be expressed in terms of a mean value and an uncertainty [11] after (1) a careful analysis of the principle, equipment, procedure, test conditions, and possible sources of error, and (2) a due consideration of past experience and the judgment of the investigator. For example, a temperature may be reported as

$$\underbrace{\text{temperature}=}_{\text{variable}} \quad \underbrace{157.4}_{\text{mean value}} \pm \underbrace{1.6\,^{\circ}\text{C}}_{\text{uncertainty}} \underbrace{(20:1)}_{\text{confidence level}} \qquad (2\text{-}35)$$

The expression states that the odds are 20:1 (2s limits) that the correct temperature is between 155.8 and 159.0°C, to the best judgment of the investigator. Many scientists double the estimated error provided by others before using the data [10].

Multisample data are those from repeated measurements of a given quantity, employing alternative test conditions, such as different principles, instruments, procedures, and observers. The reliability of the results can be assured only when different experiments are performed for the same quantity. This is asking a great deal for most engineering experiments.

Engineering data can be considered multisampled when the range of an experiment partially overlaps those by other investigators. If the data in the overlapping area are in agreement, it is reasonable to assume that the results of the experiments under different conditions are correct.

Finally, the decision to reject an "error" or an unusual data point can be difficult. The search for a justification to reject a result after the fact can be trying and subject to emotional rationalization. There is always the philosophical question of whether an observation can truly be repeated. Sometimes "the exceptions prove the rule." Unusual results can lead to new discoveries. Lord Rayleigh discovered argon because of unusual data [12].

2-6. UNCERTAINTY ANALYSIS

Errors involving a single variable, such as in the calibration of a pressure gage, were described previously. In this section we examine the compounding of errors in the results of an experiment involving several independent variables. The terms *error* and *uncertainty* are used interchangeably since most errors in engineering data can be regarded as uncertainties.

Uncertainty analysis serves three purpose: (1) to find the overall uncertainty, when the uncertainties of the individual variables are known; (2) to estimate the allowable tolerance for each of the variables in a design, when the overall error is specified; and (3) to select alternative methods for testing.

A. Overall Uncertainty

The overall result Y of an experiment is a function of X number of independent variables, the test conditions, and time.

$$Y = f(X\text{'s, test conditions, time}) \tag{2-36}$$

The effect of the variables X's on Y will be examined. Test conditions and time of the experiment cover a broad spectrum and are discussed only briefly.

Test conditions describe the framework from which the data are obtained. When the test conditions remain unchanged, the test is essentially a single-sample experiment. In other words, errors due to test conditions cannot be revealed by mere repetitions of the experiment. A technique to estimate the effect of test conditions is by means of extrapolation. For example, the effect of the wire size of thermocouples for a temperature measurement can be evaluated by using thermocouples of different wire sizes under identical conditions. The "true" temperature is estimated by plotting measured temperatures versus wire sizes and extrapolating the temperatures to obtain the value for a thermocouple of zero size. The technique is cumbersome and is seldom used, but it is applicable for many problems.

Experiments conducted at different times can have different initial and environmental conditions, particularly when the conditions are not completely controlled. In most cases, the effects are due to changes in ambient temperature and humidity. This may be important, such as in a heat transfer experiment. In some cases, the effects can be due to changes in properties, such as the aging of rubber for tests at elevated temperatures.

For simplicity, we assumed that the overall result Y is a function of X number of independent variables:

$$Y = f(X_1, X_2, \ldots, X_n) \qquad (2\text{-}37)$$

The uncertainty of Y due to the individual uncertainties in the X's about a given operating point (0) is found by expanding Eq. 2-37 in a Taylor series. Following Eq. 2-4 and assuming small variations in the X's, we get

$$\Delta Y \simeq \frac{\partial f}{\partial X_1}\bigg|_0 \Delta X_1 + \frac{\partial f}{\partial X_2}\bigg|_0 \Delta X_2 + \cdots + \frac{\partial f}{\partial X_n}\bigg|_0 \Delta X_n \qquad (2\text{-}38)$$

where the partial derivatives about (0) are constants.

The assumptions for the subsequent derivations are as follows:

1. Each variable X_i is independent and follows a gaussian distribution.
2. Each X_i is described by the format shown in Eq. 2-35, that is, $X_i = \bar{X}_i \pm \Delta X_i$ (20:1), where $\bar{X}_i$ is a mean (nominal) value of X_i and (20:1) the odds of the estimated uncertainty for X_i. Note that (20:1) is equivalent to ($2s$ limits).
3. The same odds are used for each X_i.

The *maximum uncertainty* in Y occurs when all the variations in the X's err in the same direction.

$$\Delta Y_{max} = \bigg|\frac{\partial f}{\partial X_1} \Delta X_1\bigg| + \bigg|\frac{\partial f}{\partial X_2} \Delta X_2\bigg| + \cdots + \bigg|\frac{\partial f}{\partial X_n} \Delta X_n\bigg| \qquad (2\text{-}39)$$

$$\Delta Y_{max} = \sum_{i=1}^{n} \bigg|\frac{\partial f}{\partial X_i} \Delta X_i\bigg| \qquad (2\text{-}40)$$

It is unlikely that ΔY_{max} will occur, but this gives the bound for the problem.

The *root-sum-square* (rss) [13] *criterion* for combining the separate uncertainties in X_i is more realistic.

$$\Delta Y_{\text{rss}} = \left[\left(\frac{\partial f}{\partial X_1} \Delta X_1 \right)^2 + \left(\frac{\partial f}{\partial X_2} \Delta X_2 \right)^2 + \cdots + \left(\frac{\partial f}{\partial X_n} \Delta X_n \right)^2 \right]^{1/2}$$

(2-41)

$$\Delta Y_{\text{rss}} = \left[\sum_{i=1}^{n} \left(\frac{\partial f}{\partial X_i} \Delta X_i \right)^2 \right]^{1/2}$$

(2-42)

If the same odds are used for each ΔX_i, the odds are preserved in the overall uncertainty ΔY_{rss}. For examples, if $X_i = \bar{X}_i \pm \Delta X_i$ (2s limits), then $Y = \bar{Y} \pm \Delta Y_{\text{rss}}$ (2s limits), where $\bar{X}_i$ and $\bar{Y}$ are their nominal or mean values.

The calculations for the partial derivatives above can be very tedious. If the expression for Y in Eq. 2-37 is in a "factored form" as shown below, the computation is simplified by using relative values. For example, let

$$Y = f(X_1, X_2, \ldots, X_n) = (X_1^a)(X_2^b) \cdots (X_n^c)$$

(2-43)

It can readily be shown that

$$\frac{1}{Y} \frac{\partial f}{\partial X_1} \Delta X_1 = a \frac{\Delta X_1}{X_1}$$

Considering all the X's in Eq. 2-43, we get

$$\frac{1}{Y} \Delta Y = \frac{1}{Y} \left(\frac{\partial f}{\partial X_1} \Delta X_1 + \frac{\partial f}{\partial X_2} \Delta X_2 + \cdots + \frac{\partial f}{\partial X_n} \Delta X_n \right)$$

$$\frac{\Delta Y}{Y} = \left(a \frac{\Delta X_1}{X_1} + b \frac{\Delta X_2}{X_2} + \cdots + c \frac{\Delta X_n}{X_n} \right)$$

(2-44)

If the expression for Y is in a factored form, the corresponding expressions for Eqs. 2-40 and 2-42 become

$$\frac{\Delta Y_{\text{max}}}{Y} = \sum_{i=1}^{n} \left| a_i \frac{\Delta X_i}{X_i} \right|$$

(2-45)

$$\frac{\Delta Y_{\text{rss}}}{Y} = \left[\sum_{i=1}^{n} \left(a_i \frac{\Delta X_i}{X_i} \right)^2 \right]^{1/2}$$

(2-46)

Note that the $\Delta X_i / X_i$ in the expressions above represent a fraction of actual reading and not the commonly used percent of full scale.

The rss criterion in Eqs. 2-42 and 2-46 amplifies the effect of large errors. If the ratio of the large to the smaller errors is about 5:1, contributions of the smaller errors to the overall rss error are negligible. For example, let

$$\frac{\Delta Y_{\text{rss}}}{Y} = [(0.05)^2 + (0.05)^2 + (0.01)^2 + (0.01)^2 + (0.005)^2]^{1/2} = 0.073$$

$$\frac{\Delta Y_{\text{rss}}}{Y} \simeq [(0.05)^2 + (0.05)^2]^{1/2} = 0.071$$

Example 2-6. The discharge coefficient C_d of an orifice of area A is found by collecting the water flowing under a constant head h for a time interval t. C_d is expressed as

$$C_d = \frac{M}{\rho t A (2gh)^{1/2}} \quad \text{where} \quad \begin{aligned} M &= 270.0 \pm 0.3 \text{ kg} \\ t &= 278 \pm 2 \text{ s} \\ d &= 15.00 \pm 0.02 \text{ mm} \\ h &= 4.00 \pm 0.01 \text{ m} \end{aligned} \quad \begin{aligned} \rho &= 1000 \text{ kg/m}^3 \pm 0.1\% \\ A &= \pi d^2/4 \\ g &= 9.81 \text{ m/s}^2 \pm 0.1\% \end{aligned}$$

(a) Using Eqs. 2-40 and 2-42, find the maximum and rss error for the measurement of C_d. (b) Repeat, by using Eqs. 2-45 and 2-46.
Solution: (a) The calculations are as follows:

$$C_d = \frac{270}{(1000)(278)(\pi/4)(15/1000)^2[2(9.81)(4)]^{1/2}} = 0.620$$

$$\frac{\partial f}{\partial M} \Delta M = \frac{\Delta M}{\rho t (\pi/4)(d^2)(2gh)^{1/2}}$$

$$= (2.298 \times 10^{-3})(0.3) = 0.689 \times 10^{-3}$$

$$\frac{\partial f}{\partial \rho} \Delta \rho = (-0.620 \times 10^{-3})\left(\frac{0.1}{100} \times 1000\right) = -0.620 \times 10^{-3}$$

$$\frac{\partial f}{\partial t} \Delta t = (-2.232 \times 10^{-3})(2) = -4.463 \times 10^{-3}$$

$$\frac{\partial f}{\partial d} \Delta d = (-82.72 \times 10^{-3})\left(\frac{0.02}{1000}\right) = -1.654 \times 10^{-3}$$

$$\frac{\partial f}{\partial g} \Delta g = (-31.62 \times 10^{-3})\left(\frac{0.1}{100} \times 9.81\right) = -0.310 \times 10^{-3}$$

$$\frac{\partial f}{\partial h} \Delta h = (-77.55 \times 10^{-3})(0.1) = -0.775 \times 10^{-3}$$

Substituting the numerical values in Eqs. 2-40 and 2-42 gives

$$C_{d|\text{max}} = 10^{-3}(0.689 + 0.620 + 4.463 + 1.654 + 0.310 + 0.775)$$

$$= 8.51 \times 10^{-3}$$

$$C_{d|\text{rss}} = 10^{-3}(0.689^2 + 0.620^2 + 4.463^2 + 1.564^2 + 0.310^2 + 0.775^2)^{1/2}$$

$$= 4.492 \times 10^{-3}$$

Note that the dominating term for the individual error is 4.463×10^{-3}, which is due to the error in Δt. The rss error can be estimated from this quantity alone, since the other errors are small. In other words, the accuracy in the measurement of the coefficient C_d can be greatly improved by changing the procedure for the timing.

(b) From Eq. 2-44 we get

$$\frac{\Delta Y}{Y} = \frac{\Delta M}{M} - \frac{\Delta \rho}{\rho} - \frac{\Delta t}{t} - \frac{2\Delta d}{d} - \frac{1}{2}\left(\frac{\Delta g}{g}\right) - \frac{1}{2}\left(\frac{\Delta h}{h}\right)$$

Thus the maximum fractional error and the rss error are

$$\frac{\Delta C_d|_{max}}{C_d} = \frac{0.3}{270} + \frac{0.1}{100} + \frac{2}{278} + \frac{2 \times 0.02}{15} + \frac{1}{2}\left(\frac{0.1}{100}\right) + \frac{1}{2}\left(\frac{0.1}{4}\right) = 0.0137$$

$$\Delta C_d|_{max} = 0.0137 C_d = (0.0137)(0.620) = 0.0085$$

$$\frac{\Delta C_d|_{rss}}{C_d} = \left[\left(\frac{0.3}{270}\right)^2 + \left(\frac{0.1}{100}\right)^2 + \left(\frac{2}{270}\right)^2 + \left(\frac{2 \times 0.02}{15}\right)^2 + \left(\frac{0.1}{200}\right)^2 \right.$$
$$\left. + \left(\frac{0.01}{2 \times 4}\right)^2\right]$$

$$= 0.0079$$

$$\Delta C_d|_{rss} = 0.0079 C_d = (0.0079)(0.620) = 0.0049$$

B. Estimation for Design

If the overall error in the design of an apparatus is specified, what are the allowable tolerances for the components? This is a design problem and the converse of the error analysis described in the preceding section.

Let us recapitulate the background from Eqs. 2-37 and 2-38 for convenience. The overall result Y in the operation of an apparatus is a function of X number of independent variables. The overall error ΔY is the contribution of the individual errors, ΔX's. By means of a Taylor series expansion, we get

$$\Delta Y = \frac{\partial f}{\partial X_1}\bigg|_0 \Delta X_1 + \frac{\partial f}{\partial X_2}\bigg|_0 \Delta X_2 + \cdots + \frac{\partial f}{\partial X_n}\bigg|_0 \Delta X_n \qquad (2\text{-}47)$$

Many combinations of the individual errors ΔX's will give the stipulated overall error ΔY. This is an open-ended design problem. The designer must judge the relevance of each of the contributing terms. If the requirement for a particular ΔX_i is overly stringent to be practical or economical, its requirement can be relaxed and closer tolerance relegated to the other variables.

As a first approximation, assume "equal contributions" for each of the terms in Eq. 2-47; that is, the terms are of equal magnitude. Thus for the maximum error criterion we have

$$\Delta Y_{max} = n \left| \frac{\partial f}{\partial X_i} \Delta X_i \right| \quad \text{or} \quad \Delta X_i = \frac{1}{n} \frac{\Delta Y_{max}}{|\partial f/\partial X_i|} \tag{2-48}$$

Similarly, the root-sum-square criterion yields

$$\Delta Y_{rss} = \left[n \left(\frac{\partial f}{\partial X_i} \Delta X_i \right)^2 \right]^{1/2} \quad \text{or} \quad \Delta X_i = \frac{1}{\sqrt{n}} \frac{\Delta Y_{rss}}{|\partial f/\partial X_i|} \tag{2-49}$$

When Y has a factored form as shown in Eq. 2-43, the corresponding equations are

$$\frac{\Delta Y_{max}}{Y} = n \left| a_i \frac{\Delta X_i}{X_i} \right| \quad \text{or} \quad \Delta X_i = \frac{1}{n} \frac{X_i}{a_i} \frac{\Delta Y_{max}}{Y} \tag{2-50}$$

$$\frac{\Delta Y_{rss}}{Y} = \left[n \left(a_i \frac{\Delta X_i}{X_i} \right)^2 \right]^{1/2} \quad \text{or} \quad \Delta X_i = \frac{1}{\sqrt{n}} \frac{X_i}{a_i} \frac{\Delta Y_{rss}}{Y} \tag{2-51}$$

C. Selection of Alternative Test Methods

There are usually several methods for conducting an investigation, such as by means of a transient or a steady-state test. The method that gives the least error for the range of the test should be adopted. The selection of alternative methods is illustrated with examples.

Example 2-7. The power dissipation P in a 2% resistor R can be measured by three methods, as shown in Fig. 2-25. The accuracy of the meters is assumed to be 3% of the reading. Compare the overall uncertainties in the measurement of power by the three methods.
Solution: From Fig. 2-25a the power dissipation P and the uncertainty $\Delta P_{rss}/P$ are

$$P = I^2 R$$

$$\frac{\Delta R_{rss}}{P} = \left[\left(\frac{2\Delta I}{I} \right)^2 + \left(\frac{\Delta R}{R} \right)^2 \right]^{1/2} = [(2 \times 0.03)^2 + (0.02)^2]^{1/2}$$

$$= 0.063 \quad \text{or} \quad 6.3\%$$

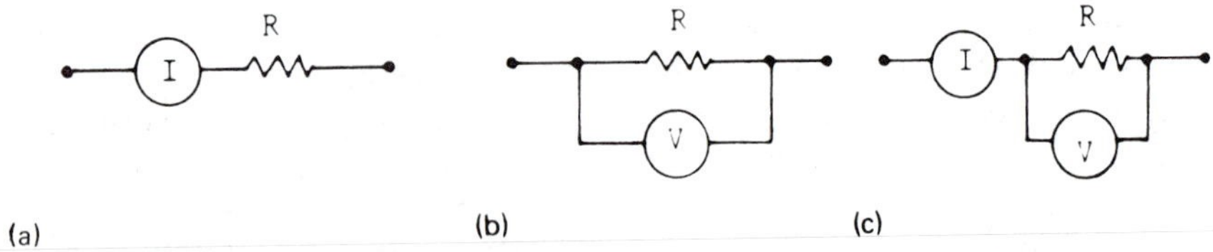

(a)　　　　　　　　　　(b)　　　　　　　　　　(c)

FIGURE 2-25 Comparison of methods to measure power dissipation in resistor R. (a) Ammeter. (b) Voltmeter. (c) Ammeter and voltmeter.

From Fig. 2-25b we have $P = V^2/R$. The corresponding uncertainty is also 6.3%. From Fig. 2-25c we get

$$P = VI$$

$$\frac{\Delta P_{rss}}{P} = \left[\left(\frac{\Delta V}{V} \right)^2 + \left(\frac{\Delta I}{I} \right)^2 \right]^{1/2} = [(0.03)^2 + (0.03)^2]^{1/2}$$

$$= 0.042 \quad \text{or} \quad 4.2\%$$

Evidently, the last method has the least error.

Example 2-8. A damper consists of a sliding piston and a cylinder filled with lubricating oil. Two methods described below are proposed for measuring the viscous damping coefficient c for $100 < c < 2000$ N · s/m. (a) The first method installs the damper in a variable-speed eccentric-weight exciter. The amplitude X at resonance is

$$X = \frac{me\omega}{c} \quad \text{or} \quad c = \frac{me\omega}{X} \tag{2-52}$$

where the unbalance of the exciter is $me = 0.1$ kg · m $\pm 1\%$, the resonance frequency is $\omega = 100$ rad/s $\pm 1\%$, and the estimated $\Delta X = 0.5 \times 10^{-3}$ m. (b) For the second method, the cylinder is vertical. A mass is placed on the piston. The assembly is allowed to slip down freely in the cylinder. The equation for the system is

$$m_1 \ddot{x} + c\dot{x} = m_1 g$$

where m_1 is the mass of the piston and the added mass. It can be shown that m_1 will reach a steady-state velocity $\dot{x}_{ss}$ readily. Then m_1 is allowed to slip through a distance $X = 0.05$ m $\pm 1\%$. The equation for $\dot{x}_{ss}$ is

$$\dot{x}_{ss} = \frac{m_1 g}{c} \quad \text{or} \quad c = \frac{m_1 g t}{X} \tag{2-53}$$

where $m_1 g = 100$ N $\pm 1\%$ is the force due to gravitation, and $\Delta t = \pm 0.05$ s is the uncertainty in the time t.

Solution: (a) From Eqs. 2-44 and 2-52 we have

$$\frac{\Delta c_{rss}}{c} = \left[\left(\frac{\Delta me}{me} \right)^2 + \left(\frac{\Delta \omega}{\omega} \right)^2 + \left(\frac{\Delta X}{X} \right)^2 \right]^{1/2}$$

This is an implicit equation, since X is dependent on c. Substituting $X = me\omega/c$ from Eq. 2-52, we obtain

$$\frac{\Delta c_{rss}}{c} = \left[\left(\frac{\Delta me}{me} \right)^2 + \left(\frac{\Delta \omega}{\omega} \right)^2 + \left(\frac{c\Delta X}{me\omega} \right)^2 \right]^{1/2}$$

$$= [(0.01)^2 + (0.01)^2 + (0.05 \times 10^{-3}c)^2]^{1/2}$$

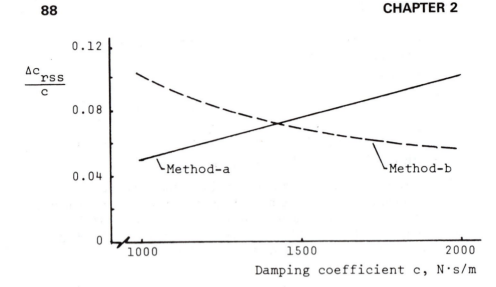

FIGURE 2-26 Comparison of experimental methods.

The equation is plotted as a solid line in Fig. 2-26.
  (b) From Eqs. 2-44 and 2-52 we have

$$\frac{\Delta c_{rss}}{c} = \left[\left(\frac{\Delta m_1 g}{m_1 g}\right)^2 + \left(\frac{\Delta X}{X}\right)^2 + \left(\frac{\Delta t}{t}\right)^2\right]^{1/2}$$

Again, substituting $\Delta t = cX/(m_1 g)$ from Eq. 2-53 yields

$$\frac{\Delta c_{rss}}{c} = \left[\left(\frac{\Delta m_1 g}{m_1 g}\right)^2 + \left(\frac{\Delta X}{X}\right)^2 + \left(\frac{m_1 g \Delta t}{cX}\right)^2\right]^{1/2}$$
$$= [(0.01)^2 + (0.01)^2 + (100/c)^2]^{1/2}$$

This equation is plotted as a dashed line in Fig. 2-26 to compare the relative advantages of the two methods for the given range of testing.

PROBLEMS

2-1. With the aid of a sketch when necessary, briefly *discuss* each of the following:
  (a) Physical laws for measurement.
  (b) Sensitivity and first-order effects.
  (c) Second-order effects in measurement.
  (d) Linear instruments.
  (e) Effects of nonlinearities.

(f) Linearization of $Y = f(X_1, X_2)$
(g) Calibration certification.
(h) Modeling of sensors.
(i) Engineering data.
(j) Standard deviation.
(k) Uncertainty analysis.

2-2. A constantan strain gage with $G_f = 2$ and $R = 120\ \Omega$ and copper leads is used as the sensing element in a transducer. (a) Find the resistance–strain sensitivity and the resistance–temperature sensitivity of the gage. (b) Estimate the equivalent strain ε_{eq} due to a temperature input of $10°C$. (c) Can the sensor act as a thermocouple? Estimate the ε_{eq} due to a temperature differential of $10°C$ at the junctions.

2-3. A fine manganin wire (84% Cu, 12% Mn, 4% Ni) is used as the sensor in a high-pressure transducer. (a) Derive an expression for the resistance–pressure sensitivity. (b) Assume that $(\Delta R/R)/(\Delta P) = 1.7 \times 10^{-7}$ and $(\Delta R/R)/(\Delta T) = 1.5 \times 10^{-5}$. If $\Delta P = 10^4$ psi and $\Delta T = 10°C$, find the error in the pressure measurement. (c) Suggest a method for eliminating the noise due to temperature.

2-4. A sinusoidal input $x = X \sin \omega t$ is applied to the transducer G shown in Fig. P2-1. Let $X = 1$. (a) Sketch the output waveforms for the transducer with characteristics shown in Fig. P2-4b to e. (b) Find an analytical expression for each of the output signals.

2-5. Repeat Prob. 2-1 for the transducers with characteristics shown in Fig. P2-2. Show graphical solutions only.

2-6. The characteristic of a transducer is

$$Y = X^2 + 3X - 4 \qquad \text{for} \quad 0 < X < 3$$

(a) Find the incremental model and the linear model at $X = 0.5$.
(b) Repeat for $X = 1.5$.

2-7. The static characteristics $Y = f(X_1, X_2)$ of a transducer and the operating point Q are shown in Fig. P2-3. (a) Find the incremental model $y = C_1 x_1 + C_2 x_2$, where the x's are incremental values of the X's and C's are constants. (b) Repeat for $x_1 = C_3 y + C_4 x_2$. (c) Repeat for $x_2 = C_5 x_1 + C_6 y$. (d) Find the relations between the C's.

2-8. Data from the calibration of a spring scale:

Input X	0	1	2	3	4	5	6	7	8	9	10	11
Output Y	0	0.58	1.11	0.72	1.28	1.44	1.41	1.81	1.60	1.78	1.92	2.55

(a) Find the calibration curve $Y = mX + c$ by the method of the least square. (b) If $Y = 1.50$, find the input X (3s limits). (c) Plot the calibration curve by "eyeballing."

2-9. Data from the calibration of an instrument:

Input X	0	10	20	30	40	50	60	70	80	90	100
Output Y (X increasing)	−112.	2.1	11.8	20.9	33.3	45.0	52.6	65.9	77.3	86.8	98.0
Output Y (X decreasing)	−6.9	4.2	16.5	24.8	36.2	47.1	58.7	68.9	79.2	91.0	102

Find X for $Y = 80$ from the calibration curve.

2-10. Given:

$$\text{brake hp} = \frac{2\pi}{33,000} \frac{FLR}{t}$$

$$F = 10.20 \pm 0.04 \text{ lb}_f$$
$$L = 15.62 \pm 0.05 \text{ in.}$$
$$R = 1500 \pm 1 \text{ rev}$$
$$t = 60 \pm 0.5 \text{ sec}$$

(a) Find ΔBhp_{max} and ΔBhp_{rss}. (b) Find the allowable ΔF, ΔR, ΔL, and Δt if the allowable rss error of the dynanometer is <1.5%.

2-11. Given:

$$Y = \frac{A(B + C)^2}{\sqrt{D}}$$

$$A = 100 \pm 1\%$$
$$B = 20 \pm 0.3$$
$$C = 50 \pm 0.4$$
$$D = 900 \pm 3.0$$

(a) Find $\Delta Y/Y_{rss}$. (b) If the allowable $\Delta Y/Y_{rss} < 0.05$, estimate the allowable ΔA, ΔB, ΔC, and ΔD.

2-12. Given 1% precision with $R_1 = 100 \,\Omega$ and $R_2 = 500 \,\Omega$. (a) Calculate the uncertainty for the R's in series. (b) Repeat for the R's in parallel.

2-13. Repeat Prob. 2-12 for $R_1 = 100 \,\Omega \pm 1\%$ and $R_2 = 500 \,\Omega \pm 10\%$.

2-14. A nozzle for measuring low-velocity airflow is shown in Fig. P2-4. Estimate the uncertainty in the mass flow rate $\dot{m}$.

$$\dot{m} = CA \left[\frac{2gP_1}{RT_1} (P_1 - P_2) \right]^{1/2}$$

Discharge coefficient, $C = 0.92 \pm 0.01$
R = gas constant
Area, $A = 1.00 \pm 0.001 \text{ in}^2$
$P_1 = 25 \pm 0.5 \text{ psia}$, $P_2 = 24 \pm 0.5 \text{ psia}$
$T_1 = 70 \pm 2°F$

2-15. Estimate the uncertainty in the mass flow rate in Prob. 2-14 if the differential pressure $P_{12} = P_1 - P_2$ is measured directly. Assume that $P_{12} = 1 \pm 0.05$ psi.

2-16. Let the mass flow rate $\dot{m}$ in Prob. 2-14 be expressed as

$$\dot{m} = (\text{constant})[P_1(P_1 - P_2)]^{1/2}$$

P_1 and P_2 are measured independently. The accuracies in measurement are $\Delta P_1/P_1$ and $\Delta P_2/P_2$. (a) Find the effect of the accuracies for $0.55 < P_1/P_2 < 1$ when $\Delta P_1/P_1 = \Delta P_2/P_2$. (b) Repeat for $\Delta P_1/P_1 \neq \Delta P_2/P_2$. (c) Is it correct to use the variable $P_{12} = P_1 - P_2$ for the error analysis?

2-17. The equation $Y = (X_2 - X_1)/(X_3 - X_1)$ is used in an experiment where the X's are measured independently. (a) Find $\Delta Y/Y_{rss}$. (b) Which of the X's is critical in the experiment for $0.1 < Y < 1.0$? (c) It is proposed to use $Y = X_{21}/X_{31}$ to simplify the calculations, where $X_{21} = X_2 - X_1$ and $X_{31} = X_3 - X_1$. Does this give the correct answer? Give sufficient details to justify your answer.

2-18. Two methods, shown in Fig. P2-5, are available for measuring the heat transfer film coefficient h of a heated rod. (a) For the transient method, the rod is heated to a constant temperature, the electric heat is cut off, and the rod is allowed to cool. The coefficient h is calculated from the equation

$$\tau = \frac{Mc}{\pi h D L} \qquad$$
Mass, $\Delta M/M = \pm 0.001$
Specific heat, $\Delta c/c = \pm 0.01$
Diameter, $\Delta D/D = \pm 0.001$
Length, $\Delta L/L = \pm 0.02$
Time constant, $\Delta\tau/\tau = \pm 0.05$

(b) For the steady-state method, we have

$$h = \frac{10^3 \times W}{\pi D L \Delta T} \qquad$$
For $h = 0.001$: $W = 10 \pm 0.5$
$\Delta T = 30 \pm 0.25$
For $h = 1.00$: $W = 5000 \pm 0.5$
$\Delta T = 60 + 0.25$

where W is the electrical power input, T the temperature of the rod, and $\Delta T = (T - T_a)$. Assume that the units are consistent. Recommend a test method for $0.001 < h < 1.00$.

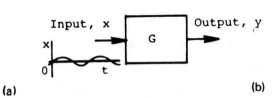

(a)

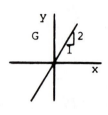

(b)

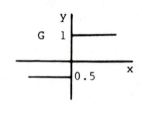

(c)

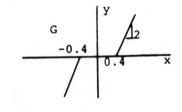

(d)

(e)

FIGURE P2-1 a-e

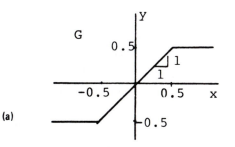

(a)

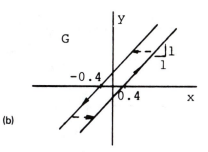

(b)

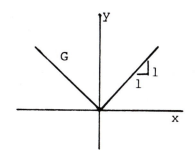

(c)

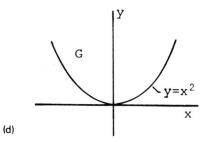

(d)

FIGURE P2-2 a-d

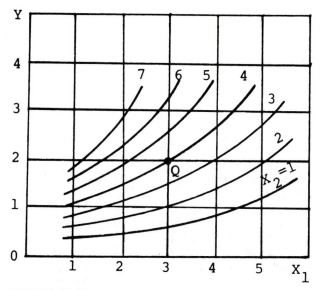

FIGURE P2-3

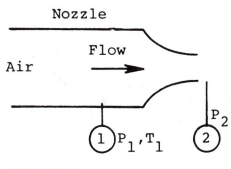

FIGURE P2-4

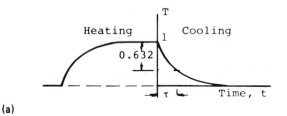

(a)

(b)

FIGURE P2-5 a, b

REFERENCES

1. Hix, C. F., Jr., and R. P. Alley, *Physical Laws and Effects,* John Wiley & Sons, Inc., New York, 1958.
2. Stein, P. K., A New Concept and Mathematical Transducer Model Application to Impedance-Based Transducers Such As Strain Gages, *VDI Ber. (Ver. Dtsch. Ing.),* Vol. 176 (1972), pp. 221–236.
    Stein, P. K., A New Concept Model for Components in Measurement and Control Systems—Practical Applications to Thermocouples, *Proceedings, Fifth Symposium on Temperature,* Instrument Society of America, Pittsburgh, Pa., 1973.

3. Lion, K. S., *Instrumentation in Scientific Research,* McGraw-Hill Book Company, New York, 1959, pp. 146–150.

4. Dean, J. W., and T. M. Flynn, Temperature Effect on Pressure Transducers, *ISA Trans.,* Vol. 5, No. 3 (July 1966), pp. 223–232.

5. Stathopoulos, G., and C. E. Fridinger, Mechanical Instruments for Measuring Shock and Vibration, in *Shock and Vibration Handbook,* 2nd ed., C. M. Harris and C. E. Crede (eds.), McGraw-Hill Book Company, New York, 1976, Chap. 13.

6. Young, D. H., *Statistical Treatment of Experimental Data,* McGraw-Hill Book Company, New York, 1962, p. 121.

7. Tse, F. S., I. E. Morse, and R. Hinkle, *Mechanical Vibrations,* 2nd ed., Allyn and Bacon, Inc., Boston, 1978, Chap. 8.

8. Sridhar, R., A General Method for Deriving the Describing Functions for a Certain Class of Nonlinearities, *IRE Trans. Autom. Control,* Vol. AC-5 (1960), pp. 135–141.

9. Stein, P. K., A Unified Approach to Handling of Noise in Measuring Systems, *Lecture Series No. 50,* Flight Test Instru. Adv. Group for Aerospace R & D, NATO, Baden, West Germany, Oct. 1972.

10. Wilson, E. B., Jr., *An Introduction to Scientific Research,* McGraw-Hill Book Company, New York, 1952, p. 232.

11. Kline, S. J., and F. A. McClintock, Describing Uncertainties in Single Sample Experiments, *Mech. Eng.,* Vol. 75 (1953), p. 3.
Thrasher, L. W., and R. C. Binder, A Practical Application of Uncertainty Calculations to Measured Data, *Trans. ASME,* (Feb. 1957), p. 373.

12. Wilson, Ref. 10, p. 255.

13. Scarborough, J. B., *Numerical Mathematical Analysis,* 3rd ed., The Johns Hopkins Press, Baltimore, 1955, p. 429.

SUGGESTED READINGS

Beckwith, T. G., N. L. Buck, and R. D. Marangoni, *Mechanical Measurements,* 3rd ed., Addison-Wesley Publishing Company, Inc., Reading, Mass., 1981.

Doebelin, E. O., *Measurement Systems,* 3rd ed., McGraw-Hill Book Company, New York, 1983.

3

Structure of Measuring Systems

The aspects for valid measurement are the transducer, the structure, and system dynamics, as shown in Fig. 1-7. The transducer and process of sensing were examined in Chapter 2. In this chapter we present the structural aspect or the arrangement of components in a measuring system. Topics included are methods of measurement, interaction between components, transducer circuits. System dynamics are examined in Chapter 4.

3-1. METHODS OF MEASUREMENT

Methods of measurement are the null balance and the unbalance method. Other methods can be viewed as variations of these two. Models showing principles of instruments based on these methods are described. A comparison of the characteristics of the models is presented in the next section.

The process of sensing and methods of measurement are summarized in Table 3-1. The first two columns of the table are similar to those in Table 1-2. The first column is an open-ended list of measurands and the second is an open-ended list of phenomena or physical laws for sensing. Numerous instruments utilizing combinations of these items are commonly used, yet there are essentially only two basic methods of measurement, as shown in the third column.

A. Null-Balance Method

The *null-balance method* compares the unknown value of a measurand with that of a reference standard. The system is nulled or balanced prior to a measurement. The unknown and the standard are then applied to the system. The standard is adjusted until the system is again nulled. Thus the value of the measurand is equal to that of the standard. The technique

TABLE 3-1   Relation of Measurands, Sensors, and Methods of Measurement

Measurand	Phenomena for sensing	Method of measurement
Flow	Pressure drop	Null balance method
Temperature →	Resistance change	Unbalance method
Pressure	Inductance change	
Motion	Capacitance change	

uses the effect of the standard on the system to oppose that of the measurand until there is no detectable output.

For example, the chemical balance shown in Fig. 3-1 is initially nulled before an unknown mass is applied. The mass added will upset the balance. The standard mass is introduced and adjusted to restore the balance to its original null position. Hence the value of the unknown mass is equal to that of the standard. The method essentially compares the effect of the unknown mass on the balance with that of the standard. The output of the balance is zero prior to a measurement and is also zero at the end of the measurement.

The thermocouple potentiometer for temperature measurement, shown in Fig. 3-1b, is another example. It compares the thermal EMF from a thermocouple with that of a standard voltage. The instrument consists of a working battery V in series with an adjustable resistor R_{adj}, a uniform slide wire of resistance R, and a precision resistor R_s. The slide wire R can be calibrated to become a variable standard voltage source.

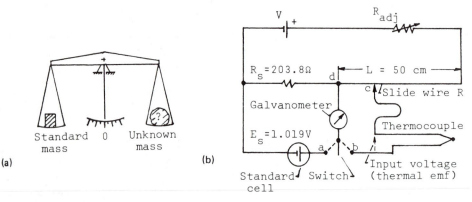

FIGURE 3-1   Examples of null-balance instruments. (a) Chemical balance. (b) Thermocouple potentiometer.

First, the instrument is calibrated by means of a standard cell V_s. The resistor R_{adj} is adjusted and the galvanometer is switched momentarily to position a. The galvanometer will show no deflection if the voltage across R_s due to the battery V is equal to that of the standard cell. Thus the voltage across R_s is standardized to 1.019 V. Since R and R_s are in the same loop, the slide wire R is also calibrated. Let $R = 10 \, \Omega$, $R_s = 203.8 \, \Omega$, and the standard voltage $V_s = 1.019$ V. Hence the voltage across R is $(1.019)(R/R_s) = 1.019(10/203.8) = 0.05$ V. If the length of the slide wire R is 50 cm, it is calibrated to 1.0 mV/cm.

Second, the EMF from the thermocouple is compared with the voltage along R. The contact c is moved along R and the galvanometer is switched momentarily to position b. When it shows no deflection upon switching to position b, the voltage from d to c along R is equal to the thermal EMF of the thermocouple. In other words, the instrument is again nulled at the end of a measurement. When the thermal EMF is equal and opposite to the voltage from d to c of the slide wire R, there is no current flow in the thermocouple circuit. Thus the measurement is independent of the resistance of the thermocouple and the length of the lead wires, and there is no energy transfer between the thermocouple circuit and the potentiometer.

The model of a null-balance instrument is shown in Fig. 3-2. The input X_1 is the measurand and X_2 the reference standard. The transducers G_1 and G_2 are assumed identical. The outputs Y_1 and Y_2 are compared by means of a comparator or a null detector G, which is a high-gain amplifier. Any difference between the inputs X_1 and X_2 is greatly magnified.

The equations describing the system are as follows:

$$Y_1 = G_1 X_1 \qquad Y_2 = G_2 X_2 \tag{3-1}$$

$$X = Y_1 - Y_2 \qquad Y = GX \tag{3-2}$$

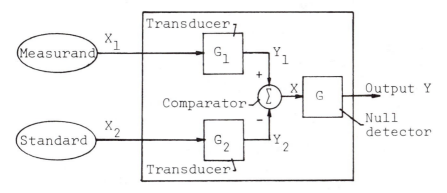

FIGURE 3-2 Models of null-balance instruments.

$$Y = G(G_1 X_1 - G_2 X_2) \tag{3-3}$$

$$Y = G[G_{av}(X_1 - X_2)] \tag{3-4}$$

where $G_{av} = (G_1 + G_2)/2 \simeq G_1 \simeq G_2$. When the instrument is nulled, the output is $Y = 0$, giving $X_1 = X_2$.

The requirements of the null-balance method are that (1) the inputs X_1 and X_2 produce the same effect on the instrument, (2) G is very sensitive near the null psoition, and (3) G is insensitive away from null. When the system is unbalanced, the requirement is that the sensitivity is sufficient only to indicate the direction for restoring the system to null.

A null-balance instrument is capable of high sensitivity and precision by having G in Eq. 3-4 very large for detecting a small difference in $(X_1 - X_2)$. The instrument is also capable of high accuracy, because a reference standard is used for the measurement. Except for the null balancing, the instrument does not have to be calibrated for its range of operation.

Note that the accuracy of the null instrument does not imply equal accuracy for a measurement. For example, the potentiometer itself can be very accurate in comparing the EMF of a thermocouple with that of a reference voltage, but the thermocouple is external to the potentiometer. When the thermal EMF is measured precisely and accurately, it is not automatic that the temperature is also evaluated with equal precision or accuracy.

Finally, only (1) like quantities and (2) quantities of the same order of magnitude can be compared. A force is compared with another force, and a color with another color. The comparison of two unlike quantities is like comparing apples and oranges. Furthermore, the quantities being compared must be of the same order of magnitude. An old Chinese proverb says, "The water in an ocean cannot be measured with a drinking cup." The size of an atom could not be measured until the discovery of x-rays, the wavelength of which is comparable to the dimension of atoms.

B. Unbalance Method

For the *unbalance method,* the effect of the input on the instrument is manifested as an output. The input and output are generally different physical quantities, such as a mechanical input with an electrical output. Hence it is also called the analogous method.

The common bourdon tube pressure gage shown in Fig. 3-3 is an example of an unbalance instrument. The bourdon tube is elliptical in cross section. The pressure forces the tube to become more circular, thereby causing a deflection at the free end of the tube. The deflection is

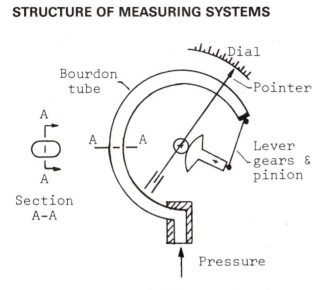

FIGURE 3-3 Bourdon tube pressure gage.

magnified with levers and gears to rotate a pointer for a dial reading. Thus the input is pressure and the output is an angular deflection of the pointer analogous to the input pressure.

The model of an unbalance instrument shown in Fig. 3-4 is similar to that shown in Fig. 3-2 for the null-balance instrument, except that a zero adjustment is substituted for the reference input. The instrument is initially adjusted to zero. For example, the levers and gears of the pressure gage in Fig. 3-3 are initially adjusted to indicate zero gage pressure. The initial conditions of an unbalance instrument are

$$Y = G(G_1X_{10} - G_2X_{20}) = 0$$

where X_{10} is due to some internal unbalance in the instrument and X_{20} is the initial zero adjustment. When the measurand X_1 is applied, we get

$$Y = G[G_1(X_1 + X_{10}) - G_2X_{20}]$$
$$Y = GG_1X_1$$

(3-5)

Terminology for methods of measurement is not standardized. While not intending to "set the record straight," let us comment briefly on the topic. An unbalance method only implies that the input is not balanced by means of a reference standard. Internally, all systems must be in equilibrium, such that the static and dynamic forces must balance. The unbalance method is variously called *indirect, deflection,* or *analogous.* The term *indirect* stems from the fact that the null balance method is

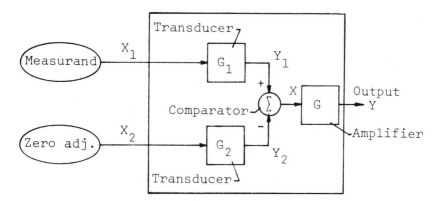

FIGURE 3-4 Model of unbalance instruments.

called the *direct* (comparison) method. Since all measurements are based on physical laws, it is difficult to say that one law is more direct than another. The term *deflection* implies a dial reading, while some instruments are digital. The output of a null balance or an unbalance instrument can be analogous to the input. The terms *null balance* and *unbalance* seem sufficiently descriptive for this book.

C. Differential Method

A differential measures the difference of two like quantities. Every car has a differential. A differential manometer for a flow measurement is shown in Fig. 3-5a. For remote sensing, an electrical differential pressure transducer is substituted, as shown in Fig. 3-5b.

The *differential method* is a variation of the null-balance method. The instrument is nulled prior to a measurement. The inputs X_1 and X_2 are applied, as shown in the model in Fig. 3-2, but X_2 is a second input and not a reference standard. Only the difference $(X_1 - X_2)$ is of interest. From Eq. 3-3 we get

$$Y = G[G_{av}(X_1 - X_2)] \tag{3-6}$$

where $G_{av} = (G_1 + G_2)/2 \simeq G_1 \simeq G_2$. When G_1 and G_2 are not identical, it can be shown by simple algebra that

$$Y = G[\underbrace{G_{av}(X_1 - X_2)}_{\text{signal}} + \underbrace{(G_1 - G_2)X_{av}}_{\text{noise}}] \tag{3-7}$$

output

where $X_{av} = (X_1 + X_2)/2$ is called the *common mode*. The signal and the

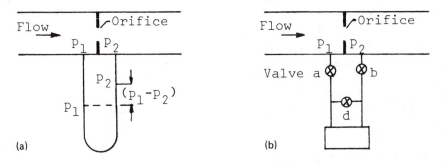

FIGURE 3-5 Differential devices for flow measurement. (a) Differential mano-meter. (b) Differential pressure transducer.

noise terms are identified in Eq. 3-7. When G_1 is not identical to G_2, a noise component is always present in a differential measurement.

A rearrangement of Eq. 3-7 gives the signal/noise ratio.

$$Y = G[G_{av}(X_1 - X_2)]\left[1 + \frac{(G_1 - G_2)X_{av}}{G_{av}(X_1 - X_2)}\right] \tag{3-8}$$

$$Y = \underbrace{(GG_{av})(X_1 - X_2)}_{\text{desired signal}}\left(1 + \frac{\text{noise}}{\text{signal}}\right) \tag{3-9}$$

$G_{av}/(G_1 - G_2)$ is the *common-mode-rejection ratio* [1]. It is a part of the specification and a figure of merit indicating the ability of an instrument to reject the common mode. G_{av} is the gain or amplification of the signal $(X_1 - X_2)$. $(G_1 - G_2)$ is the gain of the common mode X_{av}. The gain of the signal is compared with that of the common mode in Eq. 3-8. The signal/noise ratio is deduced in Eq. 3-9. In electronic instruments, the rejection ratio is dependent on the magnitude and frequency of the signal as well as the sensitivity setting of the instrument. It is intuitive that the larger the common mode, the larger the noise component in the output. In other words, when X_{av} is very large, the instrument attempts to take the difference between two large quantities of almost equal magnitude.

The noise output is due to an imbalance in G_1 and G_2. Consider the differential pressure transducer shown in Fig. 3-5b as an example. Assume that $G_1 = 101$, $G_2 = 99$, and $X_{av}/(X_1 - X_2) = 15/1$. From Eq. 3-8 we obtain

$$Y = G(100)(X_1 - X_2)\left(1 + \frac{101 - 99}{100}\frac{15}{1}\right)$$

$$= G(100)(X_1 - X_2)(1 + 0.30)$$

The example shows that when G_1 and G_2 differ by $\pm 1\%$ and the average pressure to the differential pressure has a ratio of 15:1, the noise can be as high as 30% of the signal.

The noise due to the common mode can be evaluated by applying identical inputs to a differential instrument. For the differential pressure transducer in Fig. 3-5b, the pressure inputs can be equalized by closing valve b and opening valve d. Thus, any output is a noise due to the common mode.

D. Inferential and Comparative Measurements

An *inferential measurement* is one in which the value of a measurand is calculated or inferred from the measurement of other physical quantities. The method is used when a more direct measurement is less convenient or when a transducer cannot be placed at the desired location. Inferential also implies the measurement of quantities other than the measurand.

For example, the measurement of air velocity by means of a pitot tube shown in Fig. 3-6 is inferential. The pitot tube consists of an inner and an outer tube separated by an annular space. The assembly is aligned with the direction of flow. Port a of the outer tube is normal to the flow and it measures only the static pressure p_s. Let p_v be the pressure due to the velocity v. Port b of the inner tube points upstream. It measures the total pressure p_t, which is the sum of p_s and p_v:

$$p_t = p_s + p_v \qquad \text{and} \qquad p_v = \tfrac{1}{2}\rho v^2$$

$$v = \left[\frac{2}{\rho}(p_t - p_s) \right]^{1/2}$$

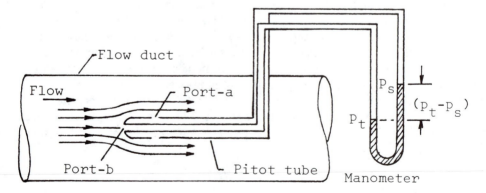

FIGURE 3-6   Pitot tube for fluid velocity measurement.

where ρ is the mass density of the fluid. The velocity v is inferred from the pressure information.

Note that all measurements are inferential in the sense that they are inferred by means of physical laws. For a pressure gage, the pressure is inferred from the elastic deformation of the bourdon tube, according to the stress–strain relation of the material. For a resistance strain gage, the strain is inferred by means of a resistance change. For the pitot tube, flow velocity is inferred from measurements of pressure. It is true that some measurements are made more directly than others. Nonetheless, they are all inferred from physical laws.

A *comparative method* compares the characteristics of an object with the corresponding characteristics of another. The technique is extremely useful, but the measurement is meaningful only on all-things-equal basis. For example, the viscosity of lubricating oils is expressed in Saybolt Universal Seconds [2]. The method determines the viscosity of oil by timing the flow of a certain quantity through a short capiliary tube under specified conditions. The measurement is meaningful only for comparing two oils that are identical in every respect except for viscosity. Pushing the argument to the extreme, a comparison of the viscosity of lubricating oil with that of syrup is meaningless.

Comparative tests are often the bases for specifications and acceptance tests in industries [3]. Executing the tests under standardized conditions is relatively straightforward. Interpretation of the test results is vastly more difficult. For a machine tool, the results from laboratory tests at one cutting speed must not be used for predicting the performance at speeds used in practice. Ideally, a performance test should give information about the behavior of the machine tool for all possible cutting operations [4]. The ultimate performance of a machine must depend on its inherent quality and conditions of application. A comparative test is meaningful only on an all-things-equal basis. This reemphasizes the discussion in Chapter 2 that a static calibration cannot be extrapolated automatically for dynamic applications.

3-2. COMPARISON OF METHODS OF MEASUREMENT

The null-balance and unbalance methods represent two principles of measurement. Their characteristics are examined in this section.

For a null-balance instrument, the measurand X_1 is linear with the reference standard X_2, as shown in Fig. 3-7a. Note that as long as the system can be balanced before and after a measurement, the null balance instrument itself can be nonlinear. The instrument is balanced when $X_1 = X_2$, and the output is

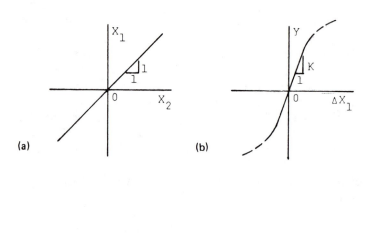

(a)

(b)

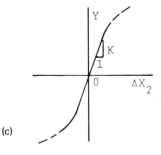

(c)

FIGURE 3-7 Characteristics of null-balance systems. Model shown in Fig. 3-2. (a) X_1 versus X_2. (b) Y versus ΔX_1. (c) Y versus ΔX_2.

$$Y = f_1(X_1 - X_2) = 0 \qquad \text{for} \quad X_1 = X_2 \tag{3-10}$$

A small unbalance exists when

$$Y = f_2(\Delta X) \neq 0 \qquad \text{for} \quad (X_1 - X_2) = \Delta X \tag{3-11}$$

The instrument is made very sensitive for any unbalance ΔX, whether X_1 is larger or smaller than X_2 as shown in Fig. 3-7b and c. The sensitivity is the slope K shown in Fig. 3-7b. K can be large about null when the instrument is almost balanced. For a large unbalance, however, only the direction of the deviation from null is needed for restoring the instrument to its high-sensitivity region. It is neither practical nor necessary to maintain a very high sensitivity over a wide range. Since the sensitivity of a null balance instrument is not constant, the null detector is inherently nonlinear.

For an unbalance instrument, the sensitivity K relating the output Y and the input X_1 is constant for its entire linear range, as shown in Fig. 3-8a. K cannot be excessive in order to maintain a reasonable scale at the output. At the same time, the instrument should be fairly insensitive to the initial zero adjustment X_2. Hence both the input X_1 and the output Y are insensitive to X_2, as shown in Fig. 3-8b and c.

Advantages of the null-balance systems are high sensitivity and high accuracy, and minimal loading due to energy transfer.

1. High sensitivity is possible because the instrument measures only small deviations about a null point, as shown in Fig. 3-7b and c. The potential for high accuracy is by virtue of the high sensitivity and the comparison with a reference standard. An example of the null-balance instrument with high accuracy is a servo-accelerometer [5]. It has a built-in servomechanism to balance the internal forces within the accelerometer. By comparison, a piezoelectric accelerometer [6] is an unbalance transducer and is less accurate.

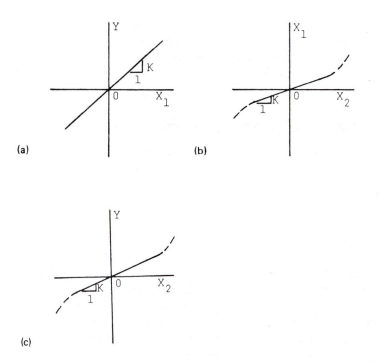

(a)

(b)

(c)

FIGURE 3-8 Characteristics of unbalance systems. Model shown in Fig. 3-4. (a) Y versus X_1 (X_2 constant). (b) X_1 versus X_2. (c) Y versus X_2 (X_1 constant).

2. Loading due to energy transfer in a null balance instrument is minimal. For the thermocouple potentiometer shown in Fig. 3-1b, there is no current flow in the thermocouple circuit at null. Thus there is no energy exchange between the thermocouple and the potentiometer.

Disadvantages of the null-balance systems are the added complexity and cost, and the additional time required for balancing the instrument.

1. The additional complexity is self-evident. As Henry Ford once said: "The gadgets you left off the car cannot cause you trouble."
2. A null balance instrument is inherently slower than an unbalance instrument. It takes time for the balancing, even when it is done automatically. For example, the frequency range of a servo-accelerometer is much lower than that of a piezoelectric type.

Unbalance instruments are generally faster in response and less expensive. They are more commonly used for dynamic measurements. An unbalance system must be calibrated for its operating range, since the measurement is not compared with a standard.

Due to the differences in performance characteristics shown in Figs. 3-7 and 3-8, an instrument designed for the null balance method will not be suited for the unbalance method, and vice versa. A comparison of the performance of the two types of instruments is given in an excellent paper by Stein [7].

3-3. INTERACTION BETWEEN COMPONENTS

Components are interconnected to form a measuring system. The interaction between components is examined in this section by means of input/output impedances. The types of components considered at the one-, two-, and three-port devices.

A. Concept of Impedance

Impedance is that which impedes a flow. For electrical systems, a flow is the current through a component. For mechanical systems, impedance is defined from the generalization of Ohm's law, but the definitions are not standardized [8]. Using the force–voltage analogy, the "flow" in a mechanical system is velocity. Analogies will be used for the discussions to follow, but generalized impedance, the classification of types of energies, and related topics are not presented.

Let us use examples to illustrate the concept of input/output impedance. The schematic of a hi-fi phono system is shown in Fig. 3-9a. In-

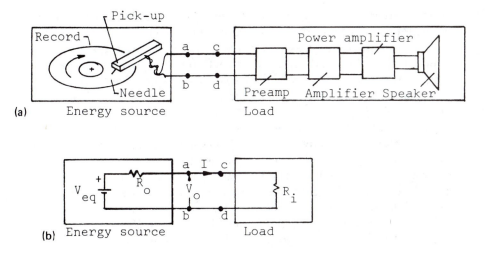

FIGURE 3-9 Example of input/output impedance. (a) Schematic of a hi-fi system. (b) Thévenin's equivalent.

formation is transferred from the energy source to the load. The energy source includes all components upstream of the interface at terminals a and b, including the turntable, the phono record, the needle, and the pickup. Certain quantities, such as the speed of the turntable and the ac power supply, are constants. They are omitted from considerations in dynamic analysis. The energy source is modeled by means of its *Thévenin's equivalent,* consisting of an equivalent voltage source V_{eq} in series with an output impedance R_o, as shown in Fig. 3-9b. The load includes all components downstream from terminals c and d. Since the load does not have an energy source, it is modeled by means of an input impedance R_i, as shown in Fig. 3-9b.

The representation in Fig. 3-9b is a general model, from which the interaction between an energy source and load is derived. At an interface dividing a measuring system, the components can be grouped as upstream and downstream from the interface, and their interaction studied by means of this model.

The *output impedance* R_o limits the power than an energy source can deliver. Let an energy source, such as a D-cell, be represented by its Thévenin's equivalent, as shown in Fig. 3-9b. The D-cell is connected to a resistor R_i. The loop equation is

$$V_{eq} = (R_o + R_i)I \tag{3-12}$$

where V_{eq} is an ideal voltage source, R_o the output impedance of the source R_i the input impedance of the load, and I the loop current. Since V_{eq} and R_o are constant, I increases as R_i decreases. If $R_o = 0$ and R_i approaches zero, the current I would become infinite. It would be possible to get an infinite current from the D-cell. This is not possible because R_o must exist and it is the internal to the energy source. It limits the current that the D-cell can deliver.

Returning to the phono recorder player in Fig. 3-9a, the pickup is an energy source and it delivers "power" downstream. Since its power is derived entirely from the movement of the needle, its capacity for power delivery must be small. Hence the output impedance R_o of the pickup must be high in order to limit the power delivery. If R_o were low, it would be possible to drive the speakers directly from the phono pickup.

The *input impedance* R_i determines the capacity of a load to absorb power. From Fig. 3-9b the power P delivered to the load is

$$P = I^2 R_i = \frac{V_o^2}{R_i} \tag{3-13}$$

where V_o is the actual voltage across R_i. A large value of R_i is necessary to minimize the energy (power) transfer between the energy source and the load. In practice, a phono pickup is connected to a preamplifier (preamp), which has an extremely high input impedance. This minimizes the loading of the pickup. Furthermore, the preamp has a low output impedance in order to deliver sufficient power to the components downstream.

To further elaborate on the concept of impedance, consider the hydraulic pump-and-motor assembly shown in Fig. 3-10a. The pump is properly sized for the motor. A worker uses the same pump to drive a smaller motor, as shown in Fig. 3-10b, and the setup works satisfactorily. To meet a higher power demand, the worker connects the same pump to a much larger hydraulic motor, as shown in Fig. 3-10c. The engineer refuses to authorize the setup, on the ground of impedance mismatch. Although the pump and the motors have the same pressure rating, it takes pressure *and* flow to deliver power.

The hydraulic pump-and-motor assembly can be modeled in Fig. 3-9b. Assume that the pump is represented by its Thévenin's equivalent, with a V_{eq} and R_o in series. The motor is represented by an input impedance R_i. A small hydraulic motor has small passages and therefore a high input impedance R_i. A large motor has large passages and therefore low input impedance. The loop equation for the assembly is

$$V_{eq} = (R_o + R_i)I$$

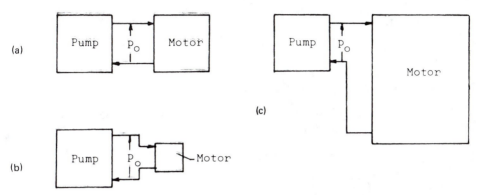

FIGURE 3-10 Hydraulic pump and motor. (a) Pump and hydraulic motor—properly sized. (b) Pump with small motor. (c) Pump with large motor.

When the pump drives the small motor, $R_o \ll R_i$ of the motor. The loop equation becomes

$$V_{eq} = R_i I \qquad \text{for} \quad R_o \ll R_i \tag{3-14}$$

The pump appears as an ideal potential source, since V_{eq} is constant and the flow I depends on R_i of the motor. In other words, the small motor is driven at its rated pressure, and the flow depends on the size of the motor.

When the same pump drives a much larger motor, $R_o \gg R_i$ of the motor. The loop equation becomes

$$V_{eq} = R_o I \qquad \text{for} \quad R_o \gg R_i \tag{3-15}$$

Since V_{eq} and R_o are constant, the flow I is constant. The pump becomes a constant delivery pump, or an ideal flow source. Note that the internal V_{eq} does not changed, but the supplied pressure V_0 at the inlet of the motor cannot be maintained by the pump.

In summary, the output impedance R_o of a power source limits its ability to delivery power at a given potential level, and the input impedance R_i of a load determines its capacity for power absorption, as shown in Eq. 3-13. A power source becomes a constant potential source when $R_o \ll R_i$, and a constant flow source when $R_o \gg R_i$.

Impedance matching describes conditions for minimum loading due to power transfer, or maximum power transfer between adjacent components. Minimum loading is examined in the next section. Maximum power transfer is desired only at the output power stage of instruments. The power P delivered is

$$P = I^2 R_i$$

Substituting $I = V_{eq}/(R_o + R_i)$ from Eq. 3-12 gives

$$P = \left(\frac{V_{eq}}{R_o + R_i}\right)^2 R_i$$

Differentiating the equation with respect to R_o and equating the derivative to zero for maximum, we get

$$R_o = R_i \qquad \text{for maximum power transfer} \tag{3-16}$$

B. One-Port Devices

A *one-port device* has one port for energy transfer. The device is either an energy source or a load, as shown in the model in Fig. 3-9b. The energy source is the source of information and is represented by its Thévenin's equivalent. The load is a transducer or a meter for the measurement. The energy transfer between the source and the load will cause a loading. Loading, such as in a voltage measurement, will be described in this section.

Consider a voltage or a *potential measurement*. The true voltage V_{true} is the open-circuit voltage V_{oc} before the meter is connected, as shown in Fig. 3-11a. Since $I = 0$, $V_{oc} = V_{true} = V_{eq}$. A voltmeter of input impedance R_m for the measurement is shown in Fig. 3-11b. R_m and R_o form a voltage divider for V_{eq} and the measured voltage V_m is

$$\frac{V_m}{V_{eq}} = \frac{R_m}{R_m + R_o} \qquad \text{or} \qquad V_m = \frac{1}{1 + R_o/R_m} V_{true} \tag{3-17}$$

The measured voltage is always less than the true voltage. V_m approaches V_{true} when $R_m \gg R_o$. Hence a voltmeter should have a high input impedance. A rule of thumb is that $R_m/R_o > 10:1$.

This conclusion can be deduced directly from the discussions on energy

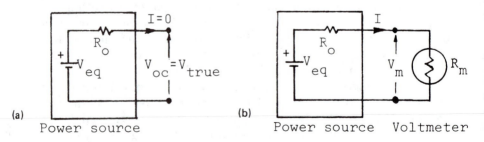

(a) Power source   (b) Power source   Voltmeter

FIGURE 3-11   Voltage measurement. (a) Before. (b) After.

models in Sec. 2-3B. For a voltage measurement, the current is secondary. Hence R_m of the voltmeter must be very high in order to limit the current flow for minimal loading.

The model study is not restricted to electrical systems. For electrical measurements, a voltmeter with extremely high input impedance is readily available, and loading due to energy transfer is usually not a problem. For a complex problem, such as in bioengineering [9], the output impedance of the source of information and the input impedance of the primary sensor must be well considered.

Example 3-1.   Find the measured voltage V_m in Fig. 3-12a. Assume that the input impedance of the voltmeter is 5 kΩ/V$_{ac}$.
Solution: Since an output of 2.5 V is anticipated, the meter is set at the range 0 to 2.5 V. Thus R_m of the meter is

$$R_m = (5 \text{ k}\Omega/\text{V}_{ac})(2.5 \text{ V}) = 12.5 \text{ k}\Omega$$

The equivalent resistance, due to R_m and R_2 in parallel in Fig. 3-12b is $R = (12.5)(100)/(12.5 + 100) = 11.1$ kΩ. R and R_1 form a voltage divider for V_{eq} in Fig. 3-12c. The measured voltage is

$$V_m = \frac{R}{R + R_1} \times 5 \text{ V} = \frac{11.1}{11.1 + 100} \times 5 \text{ V} = 0.5 \text{ V}$$

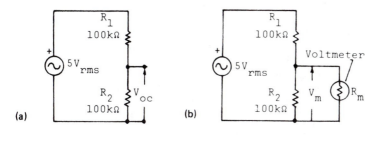

(a)

(b)

(c)

FIGURE 3-12 Loading in voltage measurement. (a) Before. (b) After. (c) Equivalent.

Note that $V_m = 0.5$ V is less than the 2.5 V anticipated. It can be demonstrated that this is the actual measured voltage.

Example 3-2. Find the peak-to-peak voltage V_{p-p} indicated by the oscilloscope in Fig. 3-13a. Assume that the input impedance R_m of the oscilloscope is 1 MΩ and its sensitivity is at 1 V/cm.
Solution: The anticipated measured voltage is 2.5 V_{rms} or $2 \times 2.5\sqrt{2}$ = 7.07 V_{p-p}. The equivalent resistance, due to R_m and R_2 in parallel in Fig. 3-13b, is 0.5 MΩ. From the equivalent circuit in Fig. 3-13c, the measured voltage V_m is

$$V_m = \frac{R}{R + R_1} \times 5 \text{ V}_{rms} = \frac{0.5 \times 5}{0.5 + 1} = 1.67 \text{ V}_{rms}$$

The observed peak-to-peak voltage is $V_{p-p} = 2 \times 1.67\sqrt{2} = 4.71$ V or 4.7 cm, but the anticipated V_{p-p} is 7.1 V or 7.1 cm. An expensive oscilloscope can have a large error of 33% in a simple voltage measurement.

Example 3-3. A force F is applied to a system shown in Fig. 3-14a, where the k's are the stiffness of the springs. A force transducer of stiffness k_m, shown in Fig. 3-14b, is inserted to measure the static force transmitted through k_2. Find the error in the force measurement by means of

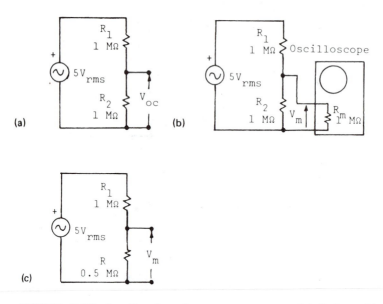

FIGURE 3-13 Loading in voltage measurement. (a) Before. (b) After. (c) Equivalent.

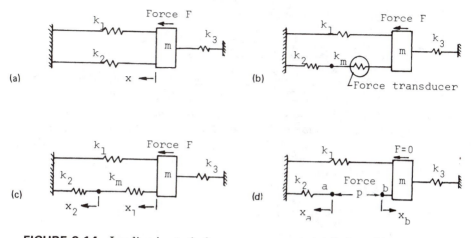

FIGURE 3-14 Loading in static force measurement. (a) Before. (b) After. (c) Direct calculation. (d) Thévenin's theorem.

Thévenin's theorem using the force–voltage analogy. (The problem can be solved directly instead of using Thévenin's theorem, as shown in Fig. 3-14c.)

Solution: The ratio F_m/F_{true} is deduced directly from Eq. 3-17 by analogy, where F_m is analogous to V_m, k_m to R_m, and k_o to R_o.

$$F_m = \frac{1}{1 + k_o/k_m} F_{true} \tag{3-18}$$

The output stiffness k_o is determined by applying a force p at the interface at terminals a and b of the force transducer and noting the corresponding deflections, as shown in Fig. 3-14d. Note that $F = 0$ in applying Thévenin's theorem. The method is analogous to using Ohm's law for finding the resistance of a network by applying a voltage V and noting the current I. Thus

$$k_o = \frac{\text{force}}{\text{deflection}} = \frac{p}{x_a + x_b} \tag{3-19}$$

$$k_o = \frac{p}{p/k_2 + p/(k_1 + k_3)} = \frac{k_2(k_1 + k_3)}{k_1 + k_2 + k_3} \tag{3-20}$$

Substituting k_o from Eq. 3-20 into Eq. 3-18, we get

$$\frac{F_m}{F_{true}} = \frac{k_m(k_1 + k_2 + k_3)}{(k_1 + k_3)(k_2 + k_m) + k_2 k_m}$$

The corresponding fraction error is

$$\text{error} = \frac{F_{\text{true}} - F_m}{F_{\text{true}}} = \frac{k_2(k_1 + k_3)}{(k_1 + k_3)(k_2 + k_m) + k_2 k_m}$$

Error in a *flow measurement* is due to the obstruction introduced by the meter in the flow circuit. The true current I_{true} shown in Fig. 3-15a is before an ammeter is inserted in the circuit. The measured current I_m shown in Fig. 3-15b is after the meter R_m is inserted. The loop equations from the figures are

$$V_{\text{eq}} = (R_o + R_i)I_{\text{true}} \tag{3-21}$$
$$V_{\text{eq}} = [(R_o + R_i) + R_m]I_m \tag{3-22}$$

From the equations above, we get

$$I_m = \frac{1}{1 + R_m/(R_o + R_i)} I_{\text{true}} \tag{3-23}$$

Hence the input impedance R_m of the ammeter must be low for a true flow measurement. In other words, the voltage in a current measurement is secondary, and the voltage drop across the ammeter must be low for minimal loading.

Example 3-4.   A force F is applied at m_2 shown in Fig. 3-16a. The static deflection at m_1 is measured by means of a transducer of stiffness k_m. Find the ratio x_m/x_{true}, where x_m and x_{true} are the displacements at m_1 before and after the transducer is applied.

Solution: The m's and k's in the system are rearranged as shown in Fig. 3-16b for convenience. A force p is introduced to find the output stiffness k_o of the system. Thus

$$k_o = \frac{\text{force}}{\text{deflection}} = \frac{p}{x_a + x_b}$$

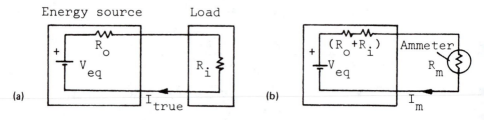

FIGURE 3-15   Current measurement. (a) Before. (b) After.

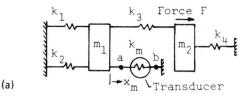

(a)

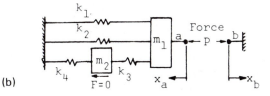

(b)

FIGURE 3-16 Loading in static displacement measurement. (a) System. (b) To find k_0.

$$k_0 = \frac{p}{p/k_{eq} + 0} = k_{eq} = (k_1 + k_2) + \frac{k_3 k_4}{k_3 + k_4}$$

where k_{eq} is due to the series–parallel combination of the k's. By analogy from Eq. 3-23, we have

$$x_m = \frac{1}{1 + k_m/k_o} x_{true}$$

$$x_m = \frac{(k_1 + k_2)(k_3 + k_4) + k_3 k_4}{(k_1 + k_2 + k_m)(k_3 + k_4) + k_3 k_4} x_{true}$$

(3-24)

C. Two-Port Devices

A *two-port device* represents an intermediate component in a measuring system. It has two ports for energy transfer, with two variables at the input port and two at the output port. In this section we discuss the parameters commonly used for describing two-port devices and illustrate their applications under static and sinusoidal steady-state conditions.

The four variables of a two-port in Fig. 3-17 are related by a set of simultaneous equations (see Eq. 2-28), such as

$$y_1 = f_1(x_1, y_2) \quad \text{and} \quad x_2 = f_2(x_1, y_2)$$

(3-25)

where the x's are variables at the input and the y's at the output. Technique for deriving the parameters from the equations for the incremental model was described in Sec. 2-3C. In general, the parameters are con-

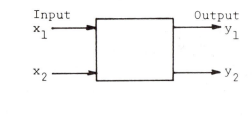

(a)

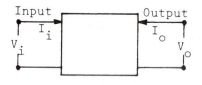

(b)

FIGURE 3-17 Models of two-port devices. (a) Transducer. (b) Network.

stants for static applications and are complex numbers under sinusoidal steady-state conditions.

A two-port device has four variables, but only two of them can be independent. There are six combinations of four variables taken two at a time. Hence there exist six sets of simultaneous equations for describing a two-port device, one of which is shown in Eq. 3-25. The choice of equations depends on the results desired and the ease with which the parameters can be evaluated. Furthermore, these are four parameters in the incremental model for each set of the simultaneous equations as shown in Eq. 2-31. From the six combinations, there exist a total of 24 parameters for describing the same system.

The system will behave in its own way, however, regardless of the equations selected for its description. Thus the parameters from one description must be convertible to those from the others. Readers should not be confused by the large number of parameters in the literature for describing a device, such as a transistor. A familiarity with one set of parameters will lead logically to the interpretation of the others.

For convenience, voltage V and current I will be used to denote potential and flow in the discussions to follow. It is understood that the discussions are equally applicable for nonelectrical systems. The sign convention for network analysis in Fig. 3-17b will be used unless otherwise stated. The parameters commonly used for describing a two-port are presented below.

1. Z Parameters

When the currents in Fig. 3-17b are the independent variables, the parameters are *impedances* or *z parameters*. The functional relations are

$$V_i = f_1(I_i, I_o) \quad \text{and} \quad V_o = f_2(I_i, I_o) \tag{3-26}$$

where the subscript i denotes the input and o the output variables. Following the derivations from Eqs. 2-28 to 2-31, we get

$$\begin{matrix} V_i = z_i I_i + z_r I_o \\ V_o = z_f I_i + z_o I_o \end{matrix} \qquad \begin{bmatrix} V_i \\ V_o \end{bmatrix} = \begin{bmatrix} z_i & z_r \\ z_f & z_o \end{bmatrix} \begin{bmatrix} I_i \\ I_o \end{bmatrix} \tag{3-27}$$

where

$$z_i = \left. \frac{V_i}{I_i} \right|_{I_o = 0} \qquad z_r = \left. \frac{V_i}{I_o} \right|_{I_i = 0}$$

$$z_f = \left. \frac{V_o}{I_i} \right|_{I_o = 0} \qquad z_o = \left. \frac{V_o}{I_o} \right|_{I_i = 0} \tag{3-28}$$

The parameters are identified by their subscripts, where i is for input, r for reverse, f for forward, and o for output. The same subscript notation will be used to identify the nature of the parameters for all descriptions to follow.

The z parameters are measured under open-circuit conditions with either $I_i = 0$ or $I_o = 0$. The z_i and z_o are called the open-circuit driving-point impedances at the input and at the output, respectively. The parameter z_f is the forward transfer impedance. It gives the open-circuit voltage V_o at the output under no-load conditions. Similarly, z_r is the reverse transfer impedance. It indicates the open-circuit voltage V_i at the input due to I_o at the output; that is, the manner the output will influence the input.

Example 3-5.   Identify the z parameters for the transformer shown in Fig. 3-18.
Solution: The induced voltages v_1 and v_2 of a transformer due to the currents i_1 in the primary coil and i_2 in the secondary are

$$v_2 = M \frac{di_1}{dt} \quad \text{and} \quad v_1 = M \frac{di_2}{dt}$$

where M is the mutual inductance between the coils. The loop equations for the primary and the secondary coils are

$$v_1 = R_1 i_i + L_1 \frac{di_i}{dt} + M \frac{di_o}{dt}$$

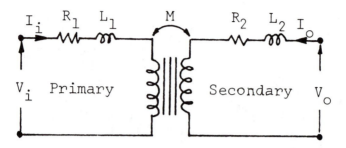

FIGURE 3-18 Transformer: an example of z parameters.

$$v_o = M \frac{di_i}{dt} + R_2 i_o + L_2 \frac{di_o}{dt}$$

For ac operation, $j\omega$ is substituted for the time derivatives (see Sec. A-6, App. A). Thus

$$V_i = (R_1 + j\omega L_1)I_i + (j\omega M)I_o$$
$$V_o = (j\omega M)I_i + (R_2 + j\omega L_2)I_o$$

The parameters in the equations above can be compared with the z's in Eq. 3-27. The z's are complex numbers. Correspondingly, the V's and I's are phasors.

2. Y Parameters

When the voltages are selected as independent variables, the parameters are admittances or y parameters. From the equations

$$I_i = f_1(V_i, V_o) \qquad \text{and} \qquad I_o = f_2(V_i, V_o)$$

we obtain

$$\begin{aligned} I_i &= y_i V_i + y_r V_o \\ I_o &= y_f V_i + y_o V_o \end{aligned} \qquad \begin{bmatrix} I_i \\ I_o \end{bmatrix} = \begin{bmatrix} y_i & y_r \\ y_f & y_o \end{bmatrix} \begin{bmatrix} V_i \\ V_o \end{bmatrix} \qquad (3\text{-}29)$$

where

$$y_i = \left. \frac{I_i}{V_i} \right|_{V_o=0} \qquad y_r = \left. \frac{I_i}{V_o} \right|_{V_i=0}$$

$$y_f = \left. \frac{I_o}{V_i} \right|_{V_o=0} \qquad y_o = \left. \frac{I_o}{V_o} \right|_{V_i=0} \qquad (3\text{-}30)$$

The y parameters are measured under short-circuit conditions; that is, either $V_i = 0$ or $V_o = 0$. Their interpretation follows that of the z parameters. Admittance is the reciprocal of impedance, and the compliance of a spring is the reciprocal of its stiffness. This does not mean that the y's in Eq. 3-30 are the reciprocals of the corresponding z's in Eq. 3-28. The 2×2 matrix for the admittances, however, is the inverse of the 2×2 matrix for the impedances.

Example 3-6.   A cantilever as a force transducer is shown in Fig. 3-19a. The displacement at the end of the cantilever is sensed by means of a dial gage, the model of which is a weak spring (see Fig. 2-20). (a) Derive the equations for the system, assuming zero loading. (b) Repeat by including loading due to the dial gage.

Solution: (a) The cantilever is modeled as a two-port device as shown in Fig. 3-19b. Since a displacement is the desired output, the appropriate equations are expressed in terms of compliance or influence coefficients [10].

$$x_i = \delta_i F_i + \delta_r F_o$$
$$x_o = \delta_f F_i + \delta_o F_o \tag{3-31}$$

where the x's are displacements and F's the forces. The compliances δ's are defined as

$$\delta_i = \frac{x_i}{F_i}\bigg|_{F_o=0} \qquad \delta_r = \frac{x_i}{F_o}\bigg|_{F_i=0}$$

$$\delta_f = \frac{x_o}{F_i}\bigg|_{F_o=0} \qquad \delta_o = \frac{x_o}{F_o}\bigg|_{F_i=0}$$

The equations above are analogous to Eqs. 3-29 and 3-30, where the displacement x is analogous to the flow I, and the force F to the potential V. The dial gage is modeled as a one-port. Since the cantilever and the dial gage are in contact, we have $x_i' = -x_o$ and $F_i' = F_o$, as shown in Fig. 3-19b.

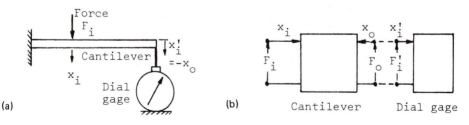

(a)   (b)

FIGURE 3-19   Force measurement: an example of y parameter. (a) System. (b) Model.

For zero loading, the force F_i' to deflect the gage becomes zero, and the cantilever is very stiff. The first condition requires $\delta_r F_o = 0$ and $\delta_o F_o = 0$. Hence Eq. 3-31 becomes

$$x_i = \delta_i F_i \qquad \text{and} \qquad x_o = \delta_f F_i$$

The second condition requires the compliance $\delta_i = 0$. Thus the equations reduce further to that for the information model.

$$x_o = \delta_f F_i \tag{3-32}$$

(b) Now consider the loading due to the force for deflecting the dial gage. The equations for the gage are

$$\delta_m = \frac{x_i'}{F_i'} = \frac{-x_o}{F_o} \qquad \text{or} \qquad F_o = -\frac{x_o}{\delta_m}$$

where δ_m is the compliance of the spring in the gage. Substituting F_o into the second equation in Eq. 3-31 gives

$$x_o = \delta_f F_i - \frac{\delta_o}{\delta_m} x_o \qquad \text{or} \qquad x_o = \frac{1}{1 + \delta_o/\delta_m} (\delta_f F_i) \tag{3-33}$$

Since $x_o = \delta_f F_i$ is the ideal output, the loading is due to the term δ_o/δ_m in Eq. 3-33. Note that Eq. 3-33 is identical to Eq. 3-24 for displacement measurements, where $k_m/k_o = \delta_o/\delta_m$, k_m is th stiffness of the spring in the dial gage, and k_o is the stiffness of the cantilever measured at the location of the dial gage.

3. ABCD Parameters

The analysis of components in cascade can be simplified by means of the *ABCD* parameters using a recurrence formula in matrix algebra. The technique can be used conveniently for complex arrangements of components in a system [11]. The sign of the current at the output in Fig. 3-20 is reversed for convenience.

When the output variables V_o and I_o are assumed independent, the parameters are the *ABCD parameters*.

$$\begin{array}{l} V_i = A V_o + B I_o \\ I_i = C V_o + D I_o \end{array} \qquad \begin{bmatrix} V_i \\ I_i \end{bmatrix} = \begin{bmatrix} A & B \\ C & D \end{bmatrix} \begin{bmatrix} V_o \\ I_o \end{bmatrix} \tag{3-34}$$

The recurrence formula above shows that the output of a preceding component is the input to the following component. For two adjacent components, we have $I_i' = I_o$ and $V_i' = V_o$.

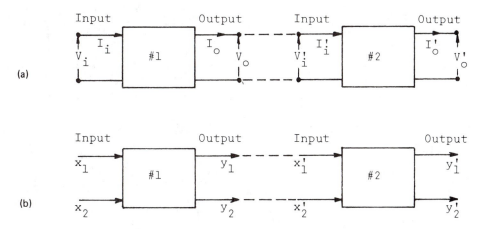

FIGURE 3-20 Examples of *ABCD* and mixed parameters. (a) Circuit components in cascade. (b) Transducers in cascade.

$$\begin{bmatrix} V_i \\ I_i \end{bmatrix} = \begin{bmatrix} A & B \\ C & D \end{bmatrix}\begin{bmatrix} V_o \\ I_o \end{bmatrix} \qquad \begin{bmatrix} V_i' \\ I_i' \end{bmatrix} = \begin{bmatrix} A' & B' \\ C' & D' \end{bmatrix}\begin{bmatrix} V_o' \\ I_o' \end{bmatrix}$$

$$\begin{bmatrix} V_i \\ I_i \end{bmatrix} = \begin{bmatrix} A & B \\ C & D \end{bmatrix}\begin{bmatrix} A' & B' \\ C' & D' \end{bmatrix}\begin{bmatrix} V_o' \\ I_o' \end{bmatrix} = \begin{bmatrix} AA' + BC' & AB' + BD' \\ CA' + DC' & CB' + DD' \end{bmatrix}\begin{bmatrix} V_o' \\ I_o' \end{bmatrix}$$

(3-35)

The *ABCD* parameters can readily be related to the other parameters. For example, when the y parameters are redefined according to the sign convention above, we get

$$I_i = y_i V_i - y_r V_o$$
$$I_o = y_f V_i - y_o V_o$$

(3-36)

It can be shown by simple algebra that

$$A = \frac{y_o}{y_f} \qquad B = \frac{y_i y_o}{y_f} - y_r$$

$$C = \frac{1}{y_f} \qquad D = \frac{y_i}{y_f}$$

(3-37)

Example 3-7. The schematic of a measuring system is shown in Fig. 3-21. Compare the true output voltage V_1 of the transducer with the measured voltage V_2.

Solution: The open-circuit voltage V_{eq} is the true output voltage. The loop equation at the transducer output is

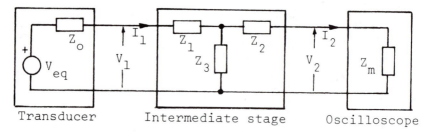

FIGURE 3-21   Components in cascade.

$$V_{eq} = I_1 Z_0 + V_1 \quad \text{or} \quad \begin{bmatrix} V_{eq} \\ I_1 \end{bmatrix} = \begin{bmatrix} 1 & Z_o \\ 0 & 1 \end{bmatrix} \begin{bmatrix} V_1 \\ I_1 \end{bmatrix} \qquad (3\text{-}38)$$

The loop equations for the intermediate stage are

$$\begin{aligned} V_1 &= (Z_1 + Z_3)I_1 - Z_3 I_2 \\ V_2 &= Z_3 I_1 - (Z_2 + Z_3)I_2 \end{aligned} \qquad \begin{bmatrix} V_1 \\ I_1 \end{bmatrix} = \begin{bmatrix} A & B \\ C & D \end{bmatrix} \begin{bmatrix} V_2 \\ I_2 \end{bmatrix} \qquad (3\text{-}39)$$

where

$$A = \frac{Z_1 + Z_2}{Z_3} \qquad B = \frac{Z_1 Z_2 + Z_2 Z_3 + Z_3 Z_1}{Z_3}$$

$$C = \frac{1}{Z_3} \qquad D = \frac{Z_2 + Z_3}{Z_3}$$

The loop equation at the input of the oscilloscope is

$$V_2 = I_2 Z_m \quad \text{or} \quad I_2 = \frac{V_2}{Z_m} \qquad (3\text{-}40)$$

Substituting Eqs. 3-39 and 3-40 into Eq. 3-38, we get

$$\begin{bmatrix} V_{eq} \\ I_1 \end{bmatrix} = \begin{bmatrix} 1 & Z_o \\ 0 & 1 \end{bmatrix} \begin{bmatrix} A & B \\ C & D \end{bmatrix} \begin{bmatrix} V_2 \\ V_2/Z_m \end{bmatrix}$$

Multiplying out the matrices and simplifying, we obtain

$$\frac{V_2}{V_{eq}} = \frac{Z_3 Z_m}{Z_o(Z_2 + Z_3) + (Z_1 Z_2 + Z_2 Z_3 + Z_3 Z_1) + Z_m(Z_0 + Z_1 + Z_2)}$$

4.   Mixed Parameters

Three of the six possible descriptions of two-port devices are presented above. The remaining three are the $A'B'C'D'$, g, and hybrid parameters. These will not be discussed here, since the $A'B'C'D'$ and g parameters are similar to the $ABCD$ and hybrid parameters. Hybrid parameters are examined in detail in the next section. The general case of mixed parameters is illustrated in the example to follow.

Example 3-8. Two transducers in cascade are shown in Fig. 3-20b. The functional relations of the components are

$$
\begin{bmatrix} y_1 \\ x_2 \end{bmatrix} = \begin{bmatrix} M & N \\ P & Q \end{bmatrix} \begin{bmatrix} x_1 \\ y_2 \end{bmatrix} \quad \text{and} \quad \begin{bmatrix} y_1' \\ x_2' \end{bmatrix} = \begin{bmatrix} M' & N' \\ P' & Q' \end{bmatrix} \begin{bmatrix} x_1' \\ y_2' \end{bmatrix}
$$

Find the overall relation between (y_1', x_2) and (x_1, y_2').
Solution: Since $x_1' = y_1$ and $x_2' = y_2$, we get

$$
\begin{aligned}
y_1 - Ny_2 &= Mx_1 \\
x_2 - Qy_2 &= Px_1
\end{aligned}
\quad \text{and} \quad
\begin{aligned}
y_1' - M'y_1 &= N'y_2' \\
y_2 - P'y_1 &= Q'y_2'
\end{aligned}
$$

Expressing the equations in matrices, we have

$$
\begin{bmatrix} 1 & -N & 0 & 0 \\ 0 & -Q & 1 & 0 \\ -M' & 0 & 0 & 1 \\ -P' & 1 & 0 & 0 \end{bmatrix} \begin{bmatrix} y_1 \\ y_2 \\ x_2 \\ y_1' \end{bmatrix} = \begin{bmatrix} M & 0 & 0 & 0 \\ P & 0 & 0 & 0 \\ 0 & 0 & N' & 0 \\ 0 & 0 & Q' & 0 \end{bmatrix} \begin{bmatrix} x_1 \\ 0 \\ y_2' \\ 0 \end{bmatrix}
$$

$$
\begin{bmatrix} y_1 \\ y_2 \\ x_2 \\ y_1' \end{bmatrix} = \begin{bmatrix} 1 & -N & 0 \\ 0 & -Q & 1 \\ -M' & 0 & 0 \\ -P' & 1 & 0 \end{bmatrix}^{-1} \begin{bmatrix} M & 0 & 0 & 0 \\ P & 0 & 0 & 0 \\ 0 & 0 & N' & 0 \\ 0 & 0 & Q' & 0 \end{bmatrix} \begin{bmatrix} x_1 \\ 0 \\ y_2' \\ 0 \end{bmatrix}
$$

It can be shown by simple algebra that the desired overall relation is

$$
\begin{bmatrix} y_1' \\ x_2 \end{bmatrix} = \frac{1}{1 - NP'}
$$
$$
\times \begin{bmatrix} MM' & [N'(1 - NP') + M'NQ'] \\ [MP'Q + P(1 - NP')] & QQ' \end{bmatrix} \begin{bmatrix} x_1 \\ y_2' \end{bmatrix}
$$

D. Three-Port Devices: Amplifiers

A three-port device has three ports for energy transfer, as shown in Figs. 2-15b and 2-18b. It represents a non-self-generating transducer as well as an intermediate component in a measuring system, such as a modulator or an amplifier. Signal modulation is described in Chapter 5. The transistor amplifier is examined in this section, which also provides background for some of the laboratory exercises.

An amplifier is a three-port device. The signal input controls the power supply or an energy source to deliver an output at the desired amplitude or power level. It is expedient to simplify the amplifier as a two-port with controlled sources [12]. The incremental model of a three-port shown in Fig. 2-18b has six variables. Only three of the six variables can be independent. This gives a set of three simultaneous equations, as shown in Eq. 2-33. It can be shown that there exist 20 sets of such simultaneous equations for describing the same system, resulting in a total of $20 \times 9 = 180$ parameters. The problem becomes unwieldy.

The hydraulic amplifier in the form of a car hoist shown in Fig. 3-22 is an example of a controlled-source device. The mechanic regulates the valve at location 1 to control the hydraulic power at location 2 for hoisting the car at location 3. The hoist is a three-port. Since the hydraulic supply is constant, the model of the hoist is a two-port with a controlled source. When the input signal is applied, the output follows the input but at a much higher power level.

A *controlled source device* can be unilateral or bilateral. The household spigot shown in Fig. 3-23 is modeled as unilateral with a controlled flow source. The input is the valve stem position x, the controlled source is a flow generator, and the output is the flow. The flow delivered Cx is controlled by x, where C is a constant. The device is unilateral, because the flow does not influence the valve stem position. Thus the input x is left "dangling" in the model. An engine dynamometer with a dc generator for power absorption is bilateral. The generator can also operate as a dc motor for starting the engine or for finding its frictional horsepower. This is a special of bilateral devices. Normally, the forward and reverse characteristics are not alike.

A simple transistor amplifier will be described as a typical controlled source device. The topics will include static characteristics (see Sec. 2-2A), determining the operating point, and linearization about the operation point for obtaining the dynamic model (see Secs. 2-2E and

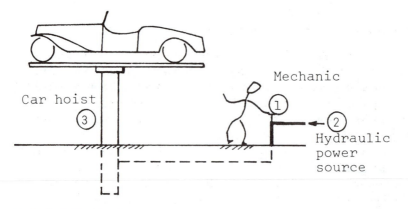

FIGURE 3-22 Hydraulic power amplifier.

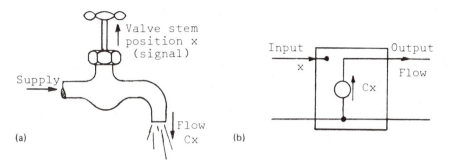

FIGURE 3-23 Example of unilateral controlled source. (a) Spigot. (b) Idealized model with controlled source, Cx.

2-3C). Except for the terminology, most of the principles were presented in previous sections.

1. Static Characteristics of Transistors [13]

A transistor has three terminals. It is either the *NPN* or the *PNP* type. The symbol of a *NPN* transistor is shown in Fig. 3-24. The terminals are labeled as the *base B, collector C,* and *emitter E.* The arrow on the emitter points toward the direction of the flow of positive charges from *P* to *N*. The arrow would point in the opposite direction for a *PNP* transistor. Any one of the three terminals can be used as common lead in a circuit. The common-emitter configuration in the figure is often used for a general-purpose amplifier.

The static characteristics give the functional relations for the variables. The input variables in Fig. 3-24 are the base current I_b, and the input voltage V_{be}, where the first subscript (*b*) denotes the base and the second

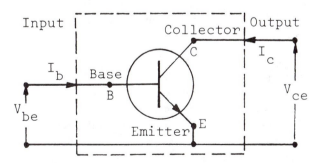

FIGURE 3-24 *NPN* transistor with common emitter.

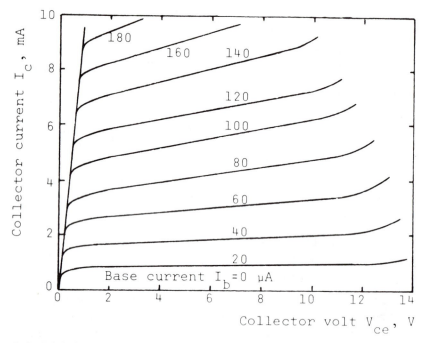

FIGURE 3-25 Typical static characteristics of *NPN* transistor.

subscript (*e*) the common-emitter configuration. The output variables are the collector current I_c, and the output voltage V_{ce}, where *c* denotes the collector. Choosing I_b and V_{ce} as independent, we get

$$I_c = f_1(V_{ce}, I_b)$$
$$V_{be} = f_2(V_{ce}, I_b)$$

(3-41)

The characteristics curves described by the first equation above are illustrated in Fig. 3-25. The second equation can be approximated and the curves are not shown.

2. Operating Point

The operating point *Q* of an ac amplifier is a quiescient point defined by its initial conditions. The ac signals are dynamic quantities, while the initial conditions are dc quantities determined from dc circuit analysis. Hence all ac quantities can be neglected in order to establish *Q*.

The circuit of a common-emitter ac amplifier is shown in Fig. 3-26a. The procedure for finding *Q* is to reduce the circuit to a corresponding dc circuit shown in Fig. 3-26b. First, the ac input and output voltages v_i and

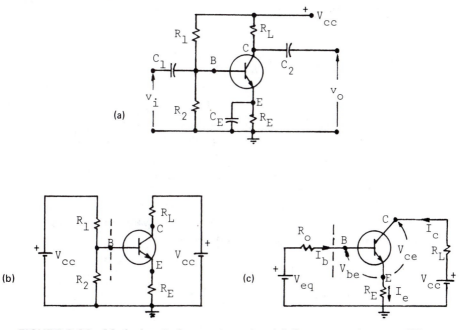

FIGURE 3-26 Method to find operating point. (a) Common-emitter amplifier. (b) Omitting ac components. (c) To find operating point.

v_o are neglected. Second, the capacitors are omitted, since they do not influence the dc analysis. Finally, the battery V_{cc} is shown explicitly by connecting its negative terminal to ground. The dc circuit is further simplified in Fig. 3-26c by reducing the part to the left of B (dashed line) by its Thévenin's equivalent, where

$$R_o = \frac{R_1 R_2}{R_1 + R_2} \quad \text{and} \quad V_{eq} = \frac{R_2}{R_1 + R_2} V_{cc} \tag{3-42}$$

The transistor is a junction, and the node equation for the currents at the transistor in Fig. 3-26c is

$$I_b + I_c - I_e = 0 \quad \text{or} \quad I_c \simeq I_e \tag{3-43}$$

Since I_b is the input signal for controlling I_c, $I_b \ll I_c$. The approximation is also evident from the characteristics shown in Fig. 3-25, where I_c is of the order of milliamperes and I_b is of microamperes.

Since $I_c = I_e$, the equation for the collector loop on the right side of Fig. 3-26c is

$$V_{cc} = (R_L + R_E)I_c + V_{ce} \qquad (3\text{-}44)$$

or

$$I_c = \left(-\frac{1}{R_L + R_E}\right)V_{ce} + \underbrace{\frac{1}{R_L + R_E}V_{cc}}_{} \qquad (3\text{-}45)$$

$$\underbrace{\qquad\qquad}_{\text{slope}} \qquad \underbrace{\qquad\qquad}_{\text{constant}}$$

This gives the *dc load line* [14] shown in Fig. 3-27. Voltages and currents in the collector loop must vary along this line, because they are governed by the loop equation.

The equation for the base loop of Fig. 3-26c is

$$V_{eq} = R_0I_b + V_{be} + R_EI_c \qquad (3\text{-}46)$$

where V_{be} is about 0.2 V for germanium and 0.6 V for silicon transistor. Eliminating I_c between Eqs. 3-44 and 3-46 gives

$$V_{ce} = \underbrace{\left[R_0\left(1 + \frac{R_L}{R_E}\right)\right]I_b}_{\text{slope}} + \underbrace{\left[V_{cc} - (V_{eq} - V_{be})\left(1 + \frac{R_L}{R_E}\right)\right]}_{\text{constant}} \qquad (3\text{-}47)$$

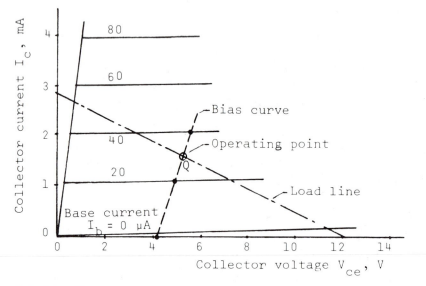

FIGURE 3-27 Locating the operating point Q.

This gives the equation of the *bias curve* shown in Fig. 3-27. The intersection of the dc load line and the bias curve gives the operating point Q.

In summary, the plot in Fig. 3-27 has three variables, I_c, V_{ce}, and I_b. The collector loop gives the load line, which relates I_c and V_{ce}. The base loop gives the bias curve, which relates V_{ce} and I_b. The intersection of the two lines gives the operating point Q.

3. Dynamic Modeling

The dynamic model is obtained from the static characteristics shown in Fig. 3-27 by means of linearization about Q (see Sec. 2-2E). In this section we describe the process of voltage amplification for background information, obtain a controlled-source model for the transistor, reduce the amplifier circuit to its ac equivalent, and deduce the equations for an ac amplifier.

The process of ac voltage amplification is illustrated for the amplifier shown in Fig. 3-26a. Let the ac output voltage v_o be controlled by the ac base current i_b. Assume that i_b is sinusoidal and fluctuating between 20 and 40 μA about Q, as shown in Fig. 3-28. Thus the projections of i_b on the load line gives the corresponding values of i_c and v_{ce} about Q. The ac output voltage is $v_o = v_{ce}$. For the given illustration, the peak-to-peak value of i_b is 20 μA, and the corresponding v_o is approximately $(7.2 - 5.2) = 4$ V_{p-p}. It will be shown that the dc load line should be replaced by an ac load line for the ac amplifier.

The controlled-source model of a common-emitter amplifier is shown in Fig. 3-29. The subscripts i and o denote the input and output quantities. Conforming to standard notations, we define $V_i = V_{be}$, $I_i = I_b$, $V_o = V_{ce}$, and $I_o = I_c$. Let the V's and I's be related by

$$V_{be} = f_1(I_b, V_{ce})$$
$$I_c = f_2(I_b, V_{ce})$$

The incremental model and the controlled sources are deduced from the equations above.

$$\begin{aligned} v_{be} &= h_{ie}i_b + h_{re}v_{ce} \\ i_c &= h_{fe}i_b + h_{oe}v_{ce} \end{aligned} \qquad (3\text{-}48)$$

where

$$h_{ie} = \frac{V_{be}}{I_b}\bigg|_{V_{ce}=0} \qquad h_{re} = \frac{V_{be}}{V_{ce}}\bigg|_{I_b=0}$$

$$h_{fe} = \frac{I_c}{I_b}\bigg|_{V_{ce}=0} \qquad h_{oe} = \frac{I_c}{V_{ce}}\bigg|_{I_b=0} \qquad (3\text{-}49)$$

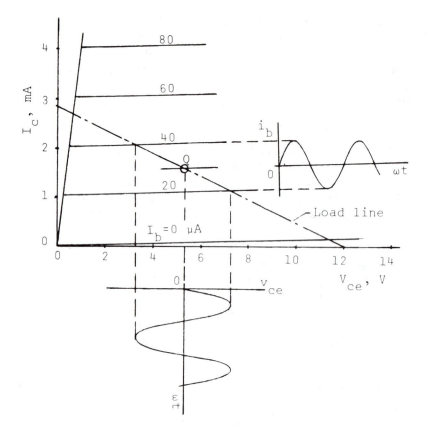

FIGURE 3-28 Amplifying process.

The h's are the *hybrid parameters*. The subscript i is for input, r for reverse, f for forward, and o for output. The subscript e denotes the common emitter. The parameters are mixed dimensionally; h_{ie} has the dimension of impedance, h_{oe} admittance, h_{fe} a current ratio, and h_{re} a voltage ratio. Note that the model has two controlled sources, shown in Fig. 3-29b. The input current i_b controls the current generator $h_{fe}i_b$ at the output, and the output voltage v_{ce} controls the voltage generator $h_{re}v_{ce}$ at the input.

The *ac equivalent circuit* is deduced from the circuit in Fig. 3-26a by omitting all components that do not affect the ac operation. First, all capacitors are replaced with short circuits. The purpose of the capacitors C_1, C_2, and C_E is to block the dc in the circuit. They offer low-impedance paths for the high-frequency ac signals. Second, V_{cc} offers a low-impedance path for the ac signals, and the terminal $+V_{cc}$ is connected

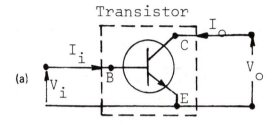

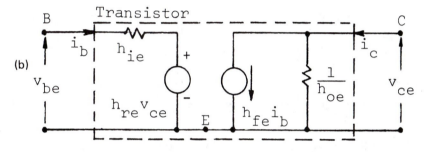

FIGURE 3-29 Hybrid model of transistor. (a) Common-emitter. (b) Hybrid parameters.

directly to ground in the ac circuit. Thus the components remaining in the ac equivalent circuit are the transistor, R_1, R_2, and R_L, as shown in Fig. 3-30a. Finally, substituting the hybrid model from Fig. 3-29b for the transistor, we obtain the ac equivalent circuit in Fig. 3-30b.

Equations for the ac amplifier are deduced from the circuit shown in Fig. 3-30b. The loop equation at the input is

$$v_i = h_{ie}i_b + h_{re}v_{ce} \tag{3-50}$$

Since v_i is applied across B and E, R_1 and R_2 do not affect the equation. The node equation at the output C is

$$i_o + i_c = 0$$

Substituting $i_c = h_{fe}i_b + h_{oe}v_{ce}$ and $i_o = v_o/R_L$, we obtain

$$\frac{v_o}{R_L} + h_{oe}v_o + h_{fe}i_b = 0 \tag{3-51}$$

The open-circuit voltage gain is v_o/v_i. Eliminating i_b between Eqs. 3-50 and 3-51 yields

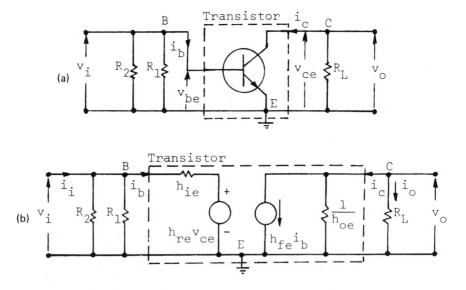

FIGURE 3-30 Ac equivalent circuit of transistor amplifier. (a) Ac circuit— common emitter amplifier. (b) Ac equivalent circuit—common emitter amplifier.

$$\frac{v_o}{v_i} = \frac{-1}{h_{ie}(1 + h_{oe}R_L)/(h_{fe}R_L) - h_{re}} \tag{3-52}$$

The current gain is i_o/i_b. Substituting $v_o = R_L i_o$ into Eq. 3-51 gives

$$\frac{i_o}{i_b} = \frac{-h_{fe}}{1 + h_{oe}R_L} \tag{3-53}$$

Finally, the dc load line is replaced by an *ac load line* for ac operation. The collector loop equation from Fig. 3-30a is

$$v_{ce} + i_c R_L = 0 \qquad \text{or} \qquad i_c = -\frac{1}{R_L} v_{ce} \tag{3-54}$$

The ac load line goes through the operating point Q with the slope of $-1/R_L$, as shown in Fig. 3-31. The process of amplification shown in Fig. 3-28 should be modified by using the ac load line instead of the dc load line.

3-4. BRIDGE CIRCUITS

A bridge denotes a particular arrangement of components in an instrument. It does not pertain to a specific type of hardware. Bridges are used

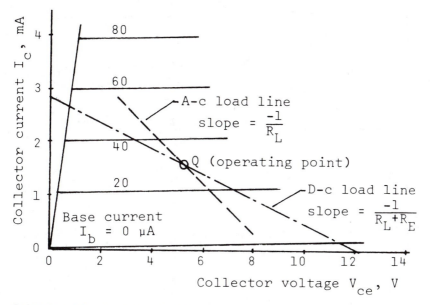

FIGURE 3-31 To find ac load line.

extensively in instrumentation. Bridge circuits for the null balance and unbalance methods of measurement are described in this section.

Examples of bridges are shown in Fig. 3-32. The general configuration in Fig. 3-32a consists of four arms of impedances Z_1 to Z_4, a power supply, and a detector. The Owen bridge [15] shown in Fig. 3-32b is for electrical measurements. It does not appear symmetrical compared with the familiar Wheatstone bridge.

A Wheatstone bridge with strain gages is shown in Fig. 3-32c. The air bridge for measuring the displacement x of the flapper valve, shown in Fig. 3-32d, has orifices R_1 to R_4 instead of strain gages. The air exhausts to the atmosphere to complete the flow circuit. The detector is a sensitive differential pressure transducer or a flow meter (e.g., Hasting-Raydist, Inc., Hampton, Virginia). The bridge is initially balanced for zero differential pressure, or $P_1 - P_2 = 0$. The displacement x unbalances the bridge by changing R_3 and R_4. thus $(P_1 - P_2) \neq 0$, and the value of x is indicated by the detector.

A. Null Balance Systems

A bridge using the null balance method compares an unknown X_1 with a reference standard X_2 (see Fig. 3-2). The bridge is balanced before and

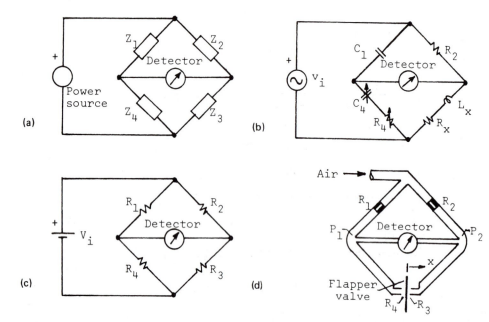

FIGURE 3-32 Examples of bridges. (a) Bridge circuit. (b) Owen bridge. (c) Wheatstone bridge. (d) Air bridge.

after a measurement. The bridge circuit in Fig. 3-33a is rearranged in Fig. 3-33b as a differential for comparing the voltages V_1 and V_2 of two potentiometers. The calibrating resistor R_c will be described later.

The R's in each potentiometer form a voltage divider. The voltages from the circuit are

$$\frac{V_1}{V_i} = \frac{R_4}{R_1 + R_4} \quad \text{and} \quad \frac{V_2}{V_i} = \frac{R_3}{R_2 + R_3} \tag{3-55}$$

$$V_o = V_2 - V_1 = \left(\frac{R_3}{R_2 + R_3} - \frac{R_4}{R_1 + R_4} \right) V_i \tag{3-56}$$

The bridge is balanced when $V_o \doteq (V_2 - V_1) = 0$. Thus

$$\frac{R_3}{R_2 + R_3} = \frac{R_4}{R_1 + R_4} \quad \text{or} \quad R_1 R_3 = R_2 R_4 \tag{3-57}$$

If the components in the bridge are impedances, the corresponding equation is

$$Z_1 Z_3 = Z_2 Z_4 \tag{3-58}$$

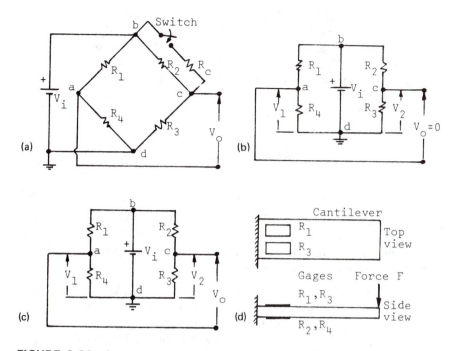

FIGURE 3-33 Strain gage bridge circuits. (a) Strain gage bridge. (b) Null-balance system. (c) Unbalance system. (d) Force transducer.

The equation does not imply that $R_1 = R_2$ or $R_3 = R_4$. It stipulates only the ratio of the R's. For example, let R_1 be the sensor for the bridge in Fig. 3-33b, and the bridge is initially balanced. An input changes R_1 to $R_1 + \Delta R_1$. At the same time, the reference R_2 is adjusted to become $R_2 + \Delta R_2$ to rebalance the bridge. From Eq. 3-57 we get

$$(R_1 + \Delta R_1)R_3 = (R_2 + \Delta R_2)R_4 \qquad \text{or} \qquad \frac{\Delta R_1}{R_1} = \frac{\Delta R_2}{R_2} \qquad (3\text{-}59)$$

Since ΔR_2 is known, ΔR_1 and the value of the input can be evaluated.

B. Unbalance and Differential Systems

For the unbalance method in Fig. 3-33c, one of the potentiometers is for the initial zero adjustment (see Fig. 3-4). Assume that R_1 and R_4 are for this adjustment, and R_2 and R_3 are strain gages in a half bridge with $R_2 = R_3$. when the bridge is unbalanced by an applied strain, R_2 becomes $R_2 + \Delta R_2$ and R_3 becomes $R_3 - \Delta R_3$. The output voltage V_o is obtained from Eq. 3-56 with the new values of R_2 and R_3.

Alternatively, the R's in Fig. 3-33c can be used in a full bridge for a differential measurement. Consider a force transducer in the form of a cantilever, shown in Fig. 3-33d, with gages R_1 and R_3 on the top of the cantilever and R_2 and R_4 at the bottom. When the force F is applied, the gages on the top will be under tension and those at the bottom under compression. R_1 and R_3 become $R_1 + \Delta R_1$ and $R_3 + \Delta R_3$, and R_2 and R_4 become $R_2 - \Delta R_2$ and $R_4 - \Delta R_4$, respectively. The corresponding bridge output from Eq. 3-56 is

$$V_o = V_2 - V_1 = \left(\frac{R_3 + \Delta R_3}{R_2 - \Delta R_2 + R_3 + \Delta R_3} - \frac{R_4 - \Delta R_4}{R_1 + \Delta R_1 + R_4 - \Delta R_4} \right) V_i$$

$$(3\text{-}60)$$

Note that the gages in a bridge are connected such that opposite strains are applied to the adjacent gages. The bridge output is zero if equal strains are applied to adjacent gages. In other words, opposite gages in a bridge must have strains in the same direction.

Example 3-9. The force transducer in Fig. 3-33d consists of a cantilever and a strain gage bridge. Assume that $V_i = 4$ V, the gage factor $G_f = 2.0$, $R_1 = R_2 = R_3 = R_4 = R = 120$ Ω. If the stress at each gage is 140 MPa ($\simeq$ 20,000 psi), find (a) the change in gage resistance, and (b) the output voltage V_o.
Solution: (a) From $G_f = (\Delta R/R)/(\text{strain})$ and stress = (Young's modulus E)(strain), where $E = 200$ GPa for steel, we get

$$\frac{\Delta R}{R} = G_f \times \text{strain} = G_f \frac{\text{stress}}{E} = 2 \times \frac{140 \text{ MPa}}{200 \text{ GPa}} = 1.4 \times 10^{-3}$$

$$\Delta R = 120(1.4 \times 10^{-3}) = 0.17 \text{ Ω}$$

Note that these ΔR for each gage is only 0.14%.
(b) Substituting the values in Eq. 3-60, the bridge output V_o due to four active gages is

$$V_o = \left(\frac{120 + 0.17}{2 \times 120} - \frac{120 - 0.17}{2 \times 120} \right) \times 4 = (1.4 \times 10^{-3}) \times 4$$

$$= 5.6 \times 10^{-3} \text{ V} \quad \text{or} \quad 5.6 \text{ mV}$$

Strain gages can be calibrated directly or indirectly. The force transducer in Fig. 3-33d can be calibrated directly by applying a known weight and observing the output V_o. A direct calibration is generally not possible because once a gage is bonded to a surface, it cannot be transferred to another location for calibration. The *indirect calibration* of a strain gage bridge is shown in Fig. 3-33a. When the gage R_2 and the calibrating resistor R_c are in parallel, the resistance change between the terminals b and c is an *equivalent strain* applied to R_2.

Example 3-10. Use the data in Example 3-9 for the bridge in Fig. 3-33a. (a) Find R_c to give an equivalent strain of 100 μs (microstrain) and the bridge output V_o during calibration. (b) If the bridge has four active gages and $V_o = 2.5$ mV, find the strain indicated by the bridge.

Solution: (a) The resistance for R_2 and R_c in parallel is

$$\frac{R_c R}{R_c + R} = (R - \Delta R) \quad \text{or} \quad \Delta R = \frac{R^2}{R_c + R}$$

From $\Delta R/R = G_f(\text{strain})$, we get

$$\Delta R = RG_f(\text{strain}) = R(2.0)(100 \times 10^{-6})$$

Eliminating ΔR from the equations above gives

$$\frac{R^2}{R_c + R} = R(2.0)(0.0001) \quad \text{or} \quad R_c \simeq 600 \text{ k}\Omega$$

The output voltage V_o is calculated from Eq. 3-60. Since only R_2 is "active" for the calibration, we have

$$\frac{\Delta R}{R} = \frac{R}{R_c + R} \simeq \frac{R}{R_c} = \frac{120 \ \Omega}{600 \text{ k}\Omega} = 0.20 \times 10^{-3}$$

$$\frac{V_o}{V_i} = \frac{R}{R - \Delta R + R} - \frac{R}{R + R} = \frac{1}{2}\left(\frac{1}{1 - \Delta R/2R} - 1\right)$$

$$= \frac{1}{2}\left[1 + \frac{\Delta R}{2R} + \left(\frac{\Delta R}{2R}\right)^2 + \cdots - 1\right]$$

$$\simeq \frac{1}{2}\frac{\Delta R}{2R} \quad \text{or} \quad V_o = \frac{1}{2}\left(\frac{0.2 \times 10^{-3}}{2}\right) \times 4 = 0.2 \text{ mV}$$

(b) The bridge uses one "active" gage for its calibration. If the bridge uses four active gages, V_o is four times that of a single gage. In other words, the actual strain is one-fourth of that indicated by V_o of 2.5 mV. Since 100 microstrain during calibration gives 0.2 mV, we have

$$\text{Strain} = \frac{1}{4}\left(\frac{2.5 \text{ mV}}{0.2 \text{ mV}}\right) 100 \text{ μstrain} = 313 \text{ μstrain}$$

A *bridge constant* C_b is used for finding the actual strain in a bridge when only one gage is used for the calibration.

$$C_b = \frac{\text{actual bridge output}}{\text{output if only one active gage}} \tag{3-61}$$

The bridge constant for Example 3-10 is $C_b = 4$, because the bridge has four active gages and only one gage is used for its calibration.

A high-impedance voltmeter is generally used for measuring V_o at the bridge output, and loading is minimal. Loading should be considered when a low-impedance detector, such as a galvanometer, is used for measuring V_o.

Example 3-11.   The output of the strain gage bridge shown in Fig. 3-34a is measured with a galvanometer of input impedance $R_m = 200\ \Omega$. Using the data from Example 3-9, find (a) the measured voltage V_o at the bridge output, and (b) the deflection of the galvanometer if its sensitivity is 10 μA/cm.

Solution: (a) The bridge circuit is reduced to its Thévenin's equivalent in Fig. 3-34b. The open-circuit voltage in Fig. 3-34c is $V_{oc} = V_{eq} = 5.6$ mV from Example 3-9. The output impedance R_o is the impedance of the bridge with all energy sources removed. The voltage V_i in Fig. 3-34c is removed by replacing with a short circuit, as shown in Fig. 3-34d. Thus R_o is the series combination of R_1 and R_4 in parallel and R_2 and R_3 in parallel.

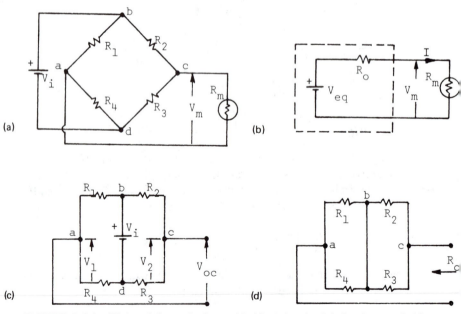

FIGURE 3-34 Thévenin's equivalent of bridge circuit. (a) Strain gage bridge. (b) Thévenin's equivalent. (c) To find V_{eq}. (d) To find R_o.

$$R_o = \frac{R_1 R_4}{R_1 + R_4} + \frac{R_2 R_3}{R_2 + R_3} = \frac{(R + \Delta R)(R - \Delta R)}{R + \Delta R + R - \Delta R}$$

$$+ \frac{(R - \Delta R)(R + \Delta R)}{R - \Delta R + R + \Delta R}$$

$$\simeq R = 120 \ \Omega$$

Since $\Delta R/R$ is small, R_o is approximately 120 Ω. From the Thévenin's equivalent in Fig. 3-34b, R_m and R_o form a voltage divider for V_{eq}. The measured voltage is

$$V_m = \frac{R_m}{R_m + R_o} V_{eq} = \frac{200}{200 + 120} \times 5.6 \text{ mV} = 3.5 \text{ mV}$$

(b) The current I through the galvanometer in Fig. 3-34b is

$$I = \frac{V_m}{R_m} = \frac{3.5 \text{ mV}}{200 \ \Omega} = 17.5 \ \mu A$$

Hence the deflection of the galvanometer is 1.75 cm.

3-5. BASIC TRANSDUCER CIRCUITS

Transducers are self- or non-self-generating. Methods for driving non-self-generating transducers are examined in this section. Self-generating transducers do not require external power sources and are not a part of this discussion.

A resistance strain gage is used to represent typical non-self-generating transducer for this discussion. The gage of resistance R shown in Fig. 3-35a can be placed either in parallel or in series with an ideal power source, as shown in Fig. 3-35b and c. It is assumed that (1) an output voltage due to the applied strain is measured with a voltmeter of high input impedance, that is, the voltmeter is essentially an open circuit, and (2) an output current is measured with an ammeter of low input impedance, that is, the ammeter is essentially a short circuit.

A. Ideal Power Sources

An ideal power source is either a voltage or a current source (see Table A-1 in Appendix A). The strain gage can be placed either in parallel or in series with an ideal power source, as shown in Fig. 3-35b and c, respectively. For both cases the quantities available for measurement are the voltages V_1 and V_2 and the currents I_1 and I_2.

For the *parallel circuit* with an *ideal voltage* source V_i, shown in Fig. 3-35b, an ammeter to measure I_2 will short circuit the gage. Hence I_2 can-

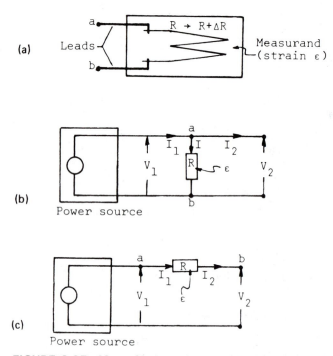

FIGURE 3-35 Non-self-generating transducers with ideal power sources. (a) Non-self-generating transducer. (b) Parallel circuit. (c) Series circuit.

not be used. Since $V_1 = V_2 = V_i$ = constant, the voltages cannot be used. Thus only I_1 can be used for measuring the applied strain in the gage. The equations from the corresponding circuit in Fig. 3-36a are

$$V_i = RI \qquad \text{and} \qquad V_i = (R + \Delta R)(I + \Delta I)$$
$$V_i = RI + I\Delta R + \Delta I(R + \Delta R)$$
$$\Delta I = \frac{-I\Delta R}{R + \Delta R} = -\frac{V_i}{R}\frac{\Delta R/R}{1 + \Delta R/R}$$
$$= -\frac{V_i}{R}\frac{\Delta R}{R}\left[1 - \frac{\Delta R}{R} + \left(\frac{\Delta R}{R}\right)^2 - \cdots\right]$$
$$\simeq -\frac{V_i}{R}\frac{\Delta R}{R} = -\frac{V_i}{R}(\text{gage factor})(\text{strain}) \qquad (3\text{-}62)$$

Note that the output ΔI in Eq. 3-62 for a constant-voltage drive is inherently nonlinear. The degree of nonlinearity depends on the magnitude of $\Delta R/R$. This is small for metallic strain gages, but can be large for semiconductor gages.

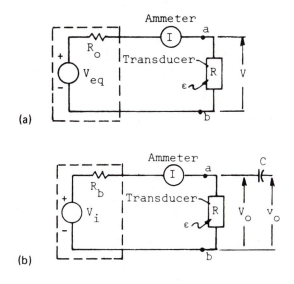

(a)

(b)

FIGURE 3-36 Possible methods of measurement. Non-self-generating transducer with ideal power sources. (a) Constant-voltage drive. (b) Constant-current drive.

For the *parallel circuit* in Fig. 3-35b with an *ideal current source* I_i, I_2 will short circuit the gage. Since $I_1 = I_i = $ constant, only $V_1 = V_2$ can be used for measuring the strain in the gage. The equations from the corresponding circuit in Fig. 3-36b are

$$V = RI_i \quad \text{and} \quad V + \Delta V = (R + \Delta R)I_i = RI_i + \Delta RI_i$$
$$\Delta V = I_i \Delta R \tag{3-63}$$

Since the output ΔV is proportional to ΔR in Eq. 3-63, the constant-current drive is inherently linear. This is an advantage when $\Delta R/R$ is large.

Similarly, ideal power sources can be used for the *series circuit* shown in Fig. 3-35c. With an ideal voltage source, the only appropriate circuit is that shown in Fig. 3-36a. With an ideal current source, the only appropriate circuit is that shown in Fig. 3-36b.

In summary, the two circuits in Fig. 3-36 are the only possible configurations when a non-self-generating transducer is driven by an ideal voltage or current source [16].

B. Nonideal Power Sources

A nonideal power source can be reduced to its Thévenin's equivalent, shown in Fig. 3-37a, where V_{eq} is an ideal voltage source and R_o the output

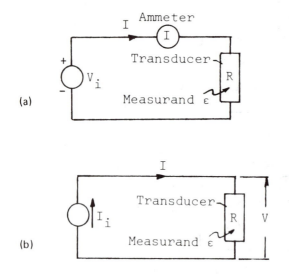

FIGURE 3-37 Non-self-generating transducer with non-ideal voltage source. (a) Thévenin's equivalent. (b) Ballast circuit (from Ref. 17).

impedance. Either the current I or the voltage V can be used for measuring the resistance change of the gage. Note the similarity between this and the ballast circuit shown in Fig. 3-37b.

The output impedance of the source or the external circuitry for the measurement may render an ideal power source nonideal. The loop equation from Fig. 3-37a is

$$V_{eq} = (R_o + R)I$$

The effect of R_o is that (1) when $R_o \ll R$, the source appears as an ideal voltage source, and (2) when $R_o \gg R$, the source appears as an ideal current source. Alternatively, when an ideal power source is used to drive a strain gage bridge and only one of the gages is active, as shown in Fig. 3-38, the remaining gages in the bridge become the R_o of the source. In other words, the external circuitry may render the source nonideal, looking in from the active gage.

Example 3-12. A nonideal power source and a strain gage of resistance R are shown in Fig. 3-37a. Find the output voltage ΔV and the output current ΔI due to the applied strain.
Solution: The equations from the circuit are

$$V_{eq} = (R_o + R)I \qquad \text{and} \qquad V = RI$$

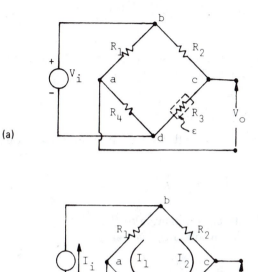

(a)

(b)

FIGURE 3-38 Comparison of voltage and current drive. (a) Voltage source V_i. (b) Current source I_i.

As the strain input causes R to become $(R + \Delta R)$, the equations become

$$V_{eq} = [R_o + (R + \Delta R)](I + \Delta I)$$

and

$$V + \Delta V = (R + \Delta I)(I + \Delta R)$$

The equations can be solved to yield the outputs

$$\Delta V = \frac{R_o}{R_o + R}\left(\frac{\Delta R}{R_o + R + \Delta R}\right)V_{eq} \qquad (3\text{-}64)$$

$$\Delta I = -\frac{1}{R_o + R}\left(\frac{\Delta R}{R_o + R + \Delta R}\right)V_{eq} \qquad (3\text{-}65)$$

Example 3-13.   The ballast circuit shown in Fig. 3-37b consists of an ideal voltage source V_i, a ballast resistor R_b, and a strain gage of resistance R. (a) Find the value of R_b for maximum sensitivity when the output is measured with a voltmeter. (b) What would be the maximum sensitivity when the output is measured with an ammeter?

Solution: Let S_v be the sensitivity for voltage measurement and S_i for current measurement, where

$$S_v = \frac{\Delta V_o / V_i}{\Delta R / R} \qquad S_i = \frac{\Delta I / V_i}{\Delta R / R}$$

(a) Comparing the circuits in Fig. 3-37, we get $V_{eq} = V_i$, $V = V_o$, and $R_o = R_b$. From Eq. 3-64 and the definition of S_v, we get

$$S_v = \frac{R_b}{R_b + R} \left(\frac{R}{R_b + R + \Delta R} \right) \simeq \frac{R_b R}{(R_b + R)^2}$$

S_v is differentiated with respect to R_b and the derivative equated to zero for maximum.

$$\frac{\partial S_v}{\partial R_b} = \frac{R(R - R_b)}{(R - R_b)^3} = 0 \qquad \text{or} \qquad R_b = R \text{ for maximum } S_v$$

It can be shown that S_v is relatively insensitive to R_b.

(b) From Eq. 3-65 and the definition of S_i, we get

$$S_i = \frac{-1}{R_b + R} \left(\frac{R}{R_b + R + R} \right) \simeq \frac{-R}{(R_b + R)^2}$$

The conclusion is that S_i is maximum when R_b approaches zero. This can also be deduced from Eq. 3-62.

Example 3-14. For the bridge circuits in Fig. 3-38, assume that $R_1 = R_2 = R_3 = R_4 = R$ and the bridge has one active gage R_3. (a) Find V_o when the bridge is driven by an ideal voltage source V_i. (b) Repeat for the ideal current source I_i.

Solution: (a) When R_3 becomes $(R + \Delta R)$ due to the applied strain, we obtain from Fig. 3-38a and Eq. 3-56

$$\frac{V_o}{V_i} = \frac{R + \Delta R}{R + R + \Delta R} - \frac{R}{R + R} = \frac{\Delta R / R}{2(2 + \Delta R / R)}$$

$$= \frac{\Delta R}{4R} \left[1 - \frac{\Delta R}{2R} + \left(\frac{\Delta R}{2R} \right)^2 - \cdots \right] \qquad (3\text{-}66)$$

(b) For the circuit in Fig. 3-38b, $I_1 = I_2$ when the bridge is balanced, and the output is $V_o = 0$. The output due to the strain applied to R_3 is

$$V_o = R_3 I_2 - R_4 I_1 = (R + \Delta R) I_2 - R_4 I_1$$

The currents I_1 and I_2 form a current divider for I_i. Thus

$$\frac{I_2}{I_1} = \frac{R_1 + R_4}{R_2 + R_3} = \frac{2R}{2R + \Delta R} \qquad \text{and} \qquad I_1 + I_2 = I_i$$

Eliminating I_1 and I_2 between the three equations above and simplifying, we get

$$\frac{V_o}{I_i R} = \frac{\Delta R}{4R} \frac{1}{1 + \Delta R/4R} = \frac{\Delta R}{4R}\left[1 - \frac{\Delta R}{4R} + \left(\frac{\Delta R}{4R}\right)^2 - \cdots\right] \quad (3\text{-}67)$$

Note that the output V_o for the constant-current drive in Fig. 3-38b is no longer linear with $\Delta R/R$, as it was in Eq. 3-63. The nonlinearity in Eq. 3-66 for the constant-voltage drive is due to the term $\Delta R/2R$, and that in Eq. 3-67 for the constant-current drive is due to $\Delta R/4R$. Hence the constant-current drive is slightly more linear.

3-6. SYSTEMS WITH FEEDBACK

Servomechanism and regulators are used extensively in instrumentation. The power steering of a car is a servo. The home thermostat is a regulator. The guidance and control of missiles and automation in industrial plants are examples of feedback. In view of the diverse applications, feedback is a general principle rather than the description hardware.

Feedback examines the structure of the arrangement of components in an instrument. The information model of single-input, single-output components will be used to describe the effect of feedback on the characteristics of components and the overall system performance. Stability analysis and design of feedback systems belong to another course.

A. System Description

A system is either open loop or closed loop. An *open-loop system* is without feedback, as shown in Fig. 3-39a. The system equation is

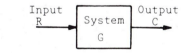

(a)

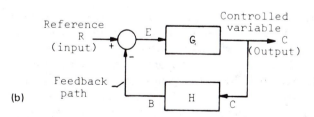

(b)

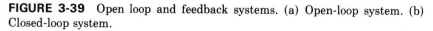

FIGURE 3-39 Open loop and feedback systems. (a) Open-loop system. (b) Closed-loop system.

$$C = GR \tag{3-68}$$

where R is the input, C the output, and G the system transfer function.

The *closed-loop* system shown in Fig. 3-39b has a feedback path to complete the loop. The output C (controlled variable) is the input to the feedback element H. The feedback signal B from H is compared with the input R (reference) in a summer. The actuating signal $E = (R - B)$ is applied to the feedforward element G (process) to produce the output C in order to complete the loop. The descriptions in equation form are

$$C = GE \qquad B = HC \qquad E = R - B \tag{3-69}$$

Substituting the last two equations into the first and simplifying, we obtain the basic equation for feedback theory:

$$\frac{\text{output}}{\text{input}} = \frac{C}{R} = \frac{G}{1 + GH} \tag{3-70}$$

Since $E = R - B$, the feedback is negative or degenerative. Only negative feedback systems are examined. Positive or regenerative feedback is seldom used except in complex systems.

The block diagram in Fig. 3-39b has only two types of symbols interconnected by signal flow paths. The symbols are the transfer function blocks G and H, and the summer that gives $E = R - B$. The summer is also called a differential or a comparator.

As an example, let the block diagram in Fig. 3-39b represent the position controller of a milling machine. C is the position of the turret of the machine, G the electric motor that turns the turret, and H a potentiometer that converts C to a voltage B. The turret is initially stationary and $C = 0$. A step input voltage R directs the turret to a reference position. The instantaneous voltage applied to the motor is $E = R - B = R - HC = R - 0 = R$. As the motor starts to turn, C deviates from zero and $E = R - HC < R$, that is, less voltage is applied to the motor. The motor stops when the turret reaches its final position as commanded by R, that is, when C is at the reference position, $E = R - B = 0$.

Example 3-15. The hydraulic position controller in Fig. 3-40 consists of an actuator (cylinder and piston), a spool valve, and a lever. The arrows indicate the positive directions for the displacements x, y, and z. If the piston is stationary and x is a step input, z opens the valve to admit high-pressure fluid to the cylinder. The piston then moves in the y direction, which is fedback through the lever to close the valve. The valve is closed when y is at the position commanded by x. Derive the system equation, assuming that the velocity of the piston is proportional to the valve opening.

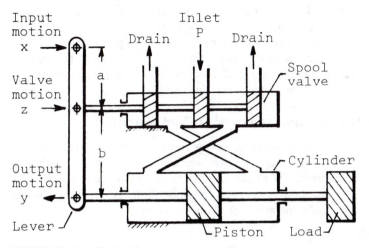

FIGURE 3-40 Hydraulic position controller.

Solution: The equation for the motion of the lever is

$$Y = f(X, Z)$$

$$\Delta Y = \left.\frac{\partial f}{\partial X}\right|_{z=0} \Delta X + \left.\frac{\partial f}{\partial Z}\right|_{x=0} \Delta Z \qquad \text{or} \qquad y = \frac{b}{a}x - \frac{a+b}{a}z$$

where X, Y, and Z are the total displacements, and x, y, and z their incremental values. The motion of the piston is given as

$$\frac{dy}{dt} = Cz$$

where C is constant. Eliminating z from the equations gives

$$\frac{a+b}{aC}\frac{dy}{dt} + y = \frac{b}{a}x \qquad \text{or} \qquad \tau\frac{dy}{dt} + y = Kx \qquad (3\text{-}71)$$

where τ is the time constant and K the sensitivity of the controller. The equation shows that the steady-state response is $y = Kx$; that is, the position y is proportional to the command x.

Feedback has many advantages. As a null balance system, it is capable of high accuracy. The feedback enables the system to perform a task automatically and to be self-correcting. For the example above, if the output C exceeds the position commanded by R, the feedback returns C to the desired position in order to achieve $E = 0$. It will be shown that feedback can be used to modify the characteristics of components as well as to enhance the dynamic performance of a system. Finally, a computer can be a part of the control loop in a complex system. The system can be adaptive to the environment to yield an optimum performance.

Feedback is not without disadvantages. It adds complexity, and the system can be unstable. As a null balance transducer, it is inherently slower than the unbalance type (see Sec. 3-2).

B. Effects on Characteristics of Components

Characteristics of components in a measuring system may change due to environmental conditions or the deterioration of parts. In this section we show that the characteristics can be stabilized by means of feedback.

Consider the deterioration of G in the open-loop system shown in Fig. 3-39a. From Eq. 3-68 the effect of G on the output C is

$$\frac{dC}{C} = \frac{dG}{G}$$

This states that for a given R, the change in the output C is proportional to the change in G. If G is a control valve, a 10% change in its characteristic will cause a 10% change in the flow.

Now, examine the effect of G on the output C in a feedback system. Assume that R and H remain constant. Differentiating Eq. 3-70 with respect to G yields

$$\frac{dC}{R} = \frac{dG}{1 + GH} - \frac{GH\, dG}{(1 + GH)^2} = \frac{dG}{(1 + GH)^2}$$

Multiplying the left side by R/C and the right side by $(1 + GH)/G$ and assuming that $GH \gg 1$, we obtain

$$\frac{dC}{C} = \frac{1}{1 + GH} \frac{dG}{G} \simeq \frac{1}{GH} \frac{dG}{G} \tag{3-72}$$

For example, if $GH = 100$, a 10% change in G will result in a 0.1% change in the output C. Hence the performance of a system with feedback is insensitive to the changes in G.

The feedback, however, will not compensate for the changes in H. Assume that R and G remain constant. Differentiating Eq. 3-70 with respect to H yields

$$\frac{dC}{C} = -\frac{G\, dH}{1 + GH} \simeq -\frac{dH}{H} \tag{3-73}$$

Let us examine the implications of feedback from Eqs. 3-72 and 3-73. G is the process under control. It is at a high power level involving large components. H deals with the feedback signal at a low power level. H has small-size components usually with passive components. Hence H can be made accurate rather inexpensively. Moreover, H can be stabilized with its own feedback loop. The deductions from Eqs. 3-72 and 3-73 are correct, but the feedback enables the control of G more precisely and economically.

Example 3-16. The feedback system in Fig. 3-41a has $G = 10$, and $H = 0.1$. G is nonlinear as shown in Fig. 3-41b. Plot the overall characteristics of C versus R for $0 < R < 0.5$ to show that the feedback has a linearizing effect [18].
Solution: From Fig. 3-41a, we get

$$C = f(x) \tag{3-74}$$
$$x = G_1 E = G_1(R - HC) = G_1 R - G_1 HC$$

The last equation is rearranged to give a straight line

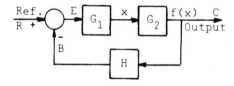

(a)

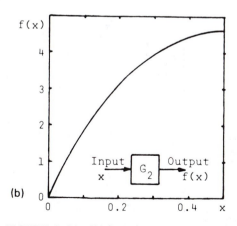

(b)

FIGURE 3-41 Reducing nonlinearity with feedback. (a) Feedback system. (b) Characteristic of G_2.

$$C = -\underbrace{\frac{1}{G_1 H}}_{} x + \underbrace{\frac{R}{H}}_{} \qquad (3\text{-}75)$$

$$\underset{\text{slope}}{} \quad \underset{\text{constant}}{}$$

The characteristic for C versus R is found graphically from the simultaneous solution of Eqs. 3-74 and 3-75 for $0 < R < 0.5$. Equation 3-74 is replotted in Fig. 3-42a. Assuming a value of R, Eq. 3-75 gives a straight line. For example, if $R = 0.2$, we get

$$C = -\frac{1}{10 \times 0.1} x + \frac{0.2}{0.1} = -x + 2$$

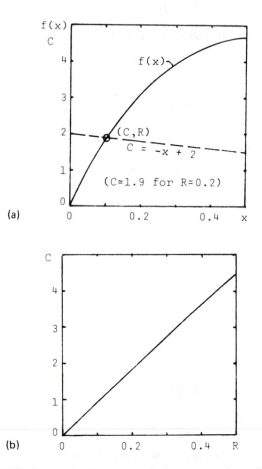

(a)

(b)

FIGURE 3-42 Reducing nonlinearity with feedback. (a) Graphical construction. (b) C versus R with feedback.

This is shown as a dashed line in Fig. 3-42a. The lines intersect at $(C, R) = (1.9, 0.2)$. The process is repeated for the range of values of R. The resulting characteristic for C versus R in Fig. 3-42b shows that the nonlinearity of G is minimized by means of feedback.

C. Effects on System Performance

In this section we generalize the effects of feedback presented in the preceding section. An ac amplifier (see Sec. 3-3D) with feedback is used to illustrate the discussions [19].

The model of an ac amplifier with voltage feedback is shown in Fig. 3-43. The feedforward element is a unilateral controlled-source device, with an input impedance R_i, an output impedance R_o, and a controlled voltage source Gv_1, where $G = v_o/v_1$ is the transfer function for the open-circuit voltage gain without feedback. The feedback element H is a potentiometer with a feedback voltage Hv_0 for $0 < H < 1$.

For the amplifier with negative feedback, the input loop gives

$$v_1 = v_i - Hv_o \tag{3-76}$$

This corresponds to $E = R - B$ in Fig. 3-39. From $G = v_0/v_1$ and Eq. 3-76, we obtain

$$v_o = Gv_1 = G(v_i - Hv_o) \quad \text{or} \quad \frac{v_o}{v_i} = \frac{G}{1 + GH} = G_f \tag{3-77}$$

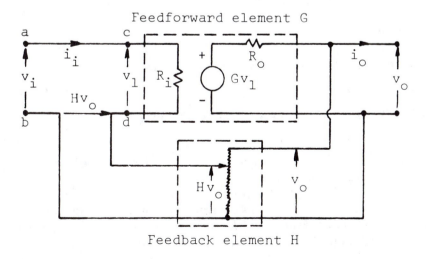

FIGURE 3-43   Amplifier with voltage feedback.

where G_f is the overall voltage gain with feedback and the subscript f denotes the feedback. Note that Eq. 3-77 is identical to Eq. 3-70 for the basic feedback system.

Case 1. Improving Dynamic Response

If the gain G in Eq. 3-77 is very large, such that $GH \gg 1$, and H, a constant, the overall gain G_f in Eq. 3-77 reduces to

$$G_f = \frac{1}{H} \quad \text{for} \quad GH \gg 1 \tag{3-78}$$

This states that the gain G_f for an amplifier with feedback is constant independent of the operating frequency, from zero to infinity. In reality, the gain G for the amplifier without feedback is not constant, and $GH \gg 1$ cannot be maintained for all frequencies. The implication, however, is that degenerative feedback will improve the overall dynamic response.

Case 2. Increasing Input and Decreasing Output Impedance

The input impedance for the amplifier without feedback, as measured across the terminals c and d in Fig. 3-43, is R_i.

$$v_1 = R_i i_i$$

The input impedance with feedback, as measured across the terminals a and b, is R_{fi}.

$$v_i = R_{fi} i_i$$

Hence the ratio of the impedances is

$$\frac{R_{fi}}{R_i} = \frac{v_i}{v_1}$$

Eliminating v_o between Eqs. 3-76 and 3-77 gives

$$v_i = (1 + GH)v_1 \tag{3-79}$$

Substituting v_i/v_1 from the equations above, we get

$$R_{fi} = (1 + GH)R_i \tag{3-80}$$

If $GH \gg 1$, the input impedance of the amplifier with feedback is increased by a factor of $(1 + GH)$.

From Fig. 3-43 the output impedance of the amplifier without feedback is R_o. Let us use the Thévenin's equivalent in Fig. 3-11 to illustrate the measurement of R_o. From $V = RI$, where $V = V_{eq} = Gv_1 = V_{oc}$ (open circuit), $R = R_o$, and $I = I_{sc}$ (short circuit), we obtain

$$V_{oc} = R_o I_{sc} \quad \text{or} \quad R_o = \frac{V_{oc}}{I_{sc}}$$

Similarly, the output impedance R_{fo} of the amplifier with feedback in Fig. 3-43 is defines as

$$R_{fo} = v_o/(i_o)_{sc} \tag{3-81}$$

where v_o is the open-circuit voltage with feedback from Eq. 3-77.

$$\frac{v_o}{v_i} = \frac{G}{1 + GH}$$

When the output of the amplifier in Fig. 3-43 is short circuited, the loop equation is

$$GV_1 = R_o(i_o)_{sc} \tag{3-82}$$

Furthermore, when the output is shorted, there is no feedback signal and $v_1 = v_i$. Substituting v_o and $(i_o)_{sc}$ into Eq. 3-81 and noting that $v_1 = v_i$, we obtain

$$R_{fo} = \frac{1}{(1 + GH)} R_o \tag{3-83}$$

Hence the output impedance R_{fo} with feedback is less than R_o without feedback by a factor of $(1 + GH)$.

Case 3. Reducing Distortion

Assume the amplifier in Fig. 3-44 has a distortion voltage v_d. The voltage output without feedback is

$$v_o = Gv_1 + v_d \tag{3-84}$$

The feedback signal is

$$Hv_o = H(Gv_1 + v_d) \tag{3-85}$$

Substituting Hv_o into Eq. 3-76 and simplifying, we get

$$(1 + GH)v_1 = v_i - Hv_d$$

Eliminating v_1 between Eq. 3-84 and the equation above and rearranging, we obtain

$$v_o = \frac{G}{1 + GH} v_i + \frac{1}{1 + GH} v_d \tag{3-86}$$

Hence the distortion voltage v_d in the amplifier without feedback is reduced by a factor of $(1 + GH)$ by means of feedback.

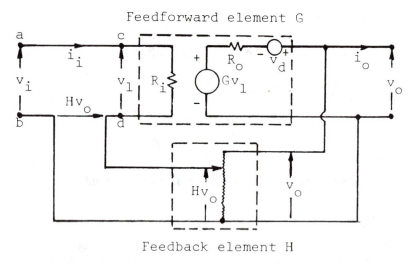

Feedforward element G

Feedback element H

FIGURE 3-44   Use feedback to reduce distortions.

In summary, for all the examples presented, it is advantageous to use a high gain for G in order to have a large value for GH. Thus the performances can be improved by a factor of $(1 + GH)$. Since components in a system will deteriorate with time, the technique presented is often used for stabilizing instruments.

3-7.   METHODS OF NOISE REDUCTION

Noise in a measurement may be from the undesired external inputs (see Eq. 2-4) or generated internally. Signal and noise are like quantities and they may or may not possess distinct characteristics. Methods of noise reduction for analog signals are examined in this section.

Had it not been for noise, instrumentation would be simple and straightforward. Once an experimental physicist was frustrated in an investigation of the slippage between crystals in metals. The amplification of an extremely weak signal was simple. Two operational amplifiers in cascade would amplify a million times or more. The noise, however, was amplified with the signal. It was the noise that "killed" the experiment.

A.   Noise Reduction at the Interface

External noise inputs can be reduced at the input interface or the output interface of an instrument, as shown in Fig. 3-45. The double arrows denote the multiple quantities and the vertical wavy lines the interface

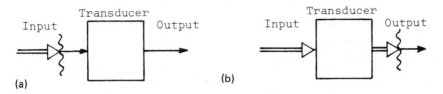

FIGURE 3-45 Noise reduction at input/output interface. (a) Reduction at input. (b) Reduction at output.

for the noise reduction. Schemes commonly used for noise reduction are filtering and shielding.

Whenever possible, corrective measures should be applied at the input interface rather than at the output. First, the input is at a low power level and is more susceptible to noise pickups. A noise of constant magnitude will degrade the low-level signal at the input more than the higher-level signal at the output. Second, when the signal/noise ratio is low, the noise can saturate the instrument and produce extraneous signals at the output (see Sec. 2-2D). Finally, even when the signal/noise ratio is high, the noise could be at a resonant frequency of the instrument and produce undesired effects, such as in the tool dynamometer described in Sec. 1-3A.

Isolation implies the separation of the instrument from a hostile environment and is often called a noise reduction technique. Isolation, however, is only loosely defined. The isolation mount in Fig. 3-46a separates an instrument from the high-frequency vibration at its base by means of soft springs. Thus the isolation mount is a frequency-selective filter. Another example is a pressure gage with a needle valve, shown in Fig. 3-46b. The gage will indicate slow pressure changes in a tank but will not respond to the high-frequency pulsating pressure of the air compressor. The thermocouple in Fig. 3-46c isolates the reference junction from the ambient temperature by means of a thermal shield. An isolation transformer is often used to eliminate the effect of ground loops in instruments. The effective use of isolation transformers requires a good knowledge of electrical shielding [20]. Since isolation is only loosely defined, we shall not belabor the subject.

1. Filters

When the signal and noise possess distinct characteristics, a filter can be used to separate the signal from the noise by means of selective filtering. For example, the air filter in a home furnace is size selective. A light filter for photography is frequency selective according to the wavelengths of light. Isotopes of chemical elements have identical chemical properties,

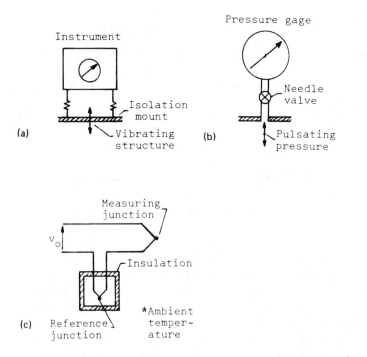

FIGURE 3-46 Examples of noise reduction. (a) Vibration isolation. (b) Throttling. (c) Thermal isolation.

but they can be separated mechanically by means of a centrifuge. The suspension system of an automobile is an example of frequency-selective filtering. It filters out the high-frequency road roughness to give passengers a smooth ride.

Electrical filters for instruments are frequency selective. They are classified as high-pass, low-pass, bandpass, and band-reject. Simple filters can be constructed from resistors and capacitors. This is described in Chapter 4. Design formulas for passive filters are found in the literature [21].

A *high-pass filter* passes the high-frequency components of an input signal and rejects the low-frequency components below the cutoff frequency f_1, as shown in Fig. 3-47a. The cutoff is not abrupt. The sharpness of the filter is described by the slope at cutoff. Active electronic filters with a slope of 16:1 or higher are common.

A *low-pass filter* passes the low-frequency components of an input signal and rejects the high-frequency components above the cutoff fre-

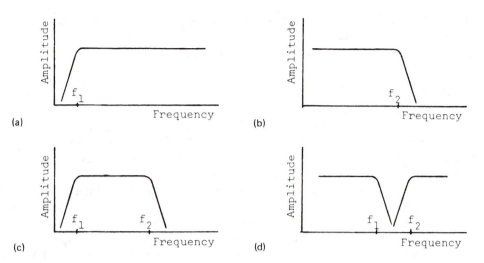

FIGURE 3-47 Characteristics of filters. (a) High-pass. (b) Low-pass. (c) Band-pass. (d) Band reject.

quency f_2, as shown in Fig. 4-37b. It is interesting to note that mechanical systems are inherently low-pass filters. Due to the inherent mass, a mechanical system does not respond to extreme high frequencies. This observation may serve to explain certain test results or can be capitalized in applications.

A series combination of a high-pass and a low-pass filter gives a *bandpass filter*. A broadband filter passes the harmonic components of its input signal from frequency f_1 to f_2, as shown in Fig. 4-37c. For example, the audio-frequency range is from 20 Hz to 20 kHz. A narrowband filter, with a bandwidth of a few hertz in the range of kilohertz, is often used for the detailed examination of the harmonic contents in a signal, such as in a vibration test. In this case, f_1 is very close to f_2 and the filter characteristic appers as a spike.

A parallel combination of the high-pass and low-pass filters gives a band-reject filter, as shown in Fig. 3-47d. The filter rejects the harmonics in the input signal within the frequency band between f_1 and f_2. This filter can be used in a recorder to prevent the 60-Hz noise voltage from saturating its amplifier.

Statistical filtering or signal averaging is used when the signal and noise occupy the same part of the frequency spectrum, but the signal is repetitive and the noise random. The method synchronizes the data sampling with the repetitive signal. If a large number of data samples are averaged, the random noise will average to zero, while the repetitive signal will converge to its mean value. The technique is also useful for low signal/noise ratio signals. This filtering technique can be done by analog or digital means.

2. Shields

Shielding blocks the noise inputs from a measuring system, while filtering separates the noise from the signal. Hence a shield is directed toward the noise source, while a filter is directed toward both the signal and noise. Shielding for noise reduction should be built into a system. In this section we describe the shielding of electrical fields and magnetic fields.

A noise voltage due to an *electric field* can loosely be called the *antenna effect*. For example, an oscilloscope will show a large 60-Hz noise pickup when an unshielded cable is connected to its input. The noise level increases when the lead is touched with a bare hand, particularly when the hand is moist. The person touching the signal lead has extended the antenna.

A *shielded cable* is shown in Fig. 3-48. The signal lead is shielded when the shield in the form of a braided copper mesh is connected to ground. The shield minimizes the voltage pickup between the signal lead and the surrounding electric field. This is called an *electrostatic shield*. First, the shield intercepts the surrounding electric field and a voltage is induced in the shield due to the antenna effect. The voltage in the shield will also induce a voltage in the signal lead by virtue of the distributed capacitance between the shield and the lead. Second, the ground connection drains the charge in the shield to ground to render the shield at the ground potential. Hence the signal lead of a shielded cable is free of the surrounding electric field. An instrument is shielded in like manner by means of a metallic chassis.

A *ground* is the common connection. An *earth ground* is analogous to the sea level for the measurement of elevation. A *chassis ground* or a floating ground is simply a common connection in an instrument, and it is not

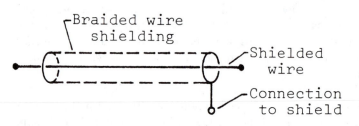

FIGURE 3-48 Shielded cable for electric field.

connected to earth. A ground is analogous to a sump in a hydraulic network. Not all sumps are at the same level.

Similarly, not all grounds in a measuring system are at the same potential. Even the grounds of electrical outlets in the same laboratory may have a potential differential. When a system has two grounds at different potentials, a current will flow from one ground to the other, thereby forcing a part of the system to complete the circuit. This is called a *ground loop*. By analogy, a flow will occur in a hydraulic network when two sumps of different potential are connected. A current flow in the signal lead due to the ground loop is equivalent to a voltage signal. For example, a 10-μA current in a 2-Ω line will appear as a 20-μV signal. When noise occurs in a low-level signal lead, it will be amplified together with the signal. Hence all signal grounds, shield grounds, and power grounds in a measuring system should be tied to only one common ground. By the same token, a shield should be ground at only one point.

The requirements for an electrostatic shield are a high degree of coverage of the low-level signal portion of an instrument, and a shield of high conductive material, such as copper. General rules for shielding and grounding are stated by Morrison [22].

A noise voltage from a *magnetic field* is due to the generator effect. For example, a voltage is induced in a conductor moving in a magnetic field, such as a vibrating signal lead or a strain gage in a shock test. Alternatively, a voltage is induced in a stationary signal lead in a changing magnetic field from a motor or a transformer. A magnetic noise pickup can be from a rotating machine part, the magnetic flux of which is from the magnetization in past inspection.

Magnetic shielding is achieved by minimizing the coupling between the magnetic flux and the instrument.

1. A first choice for reducing the coupling is by menas of a physical separation, since the field strength is inversely proportional to the distance squared.
2. The magnetic coupling can be reduced by avoiding loops in the signal leads or by a reorientation of the leads. It may be possible to divert the flux away from the instrument by providing an alternative path for the flux.
3. A shield can be used to attenuate the magnetic flux entering the intended shielded area. The shielding material must have a high magnetic permeability, such as a ferromagnetic material. A large amount of shielding material is generally required compared with electrostatic shielding.

In summary, electrostatic shields are more common than magnetic shields in instrumentation. Electric and magnetic fields are separate mechanisms and different criteria are required for their shielding. An electrostatic shield will not serve as a magnetic shield, and a magnetic shield in general is not a good electric shield. It is best to avoid a magnetic field than to attempt a shielding.

B. Noise Reduction by Insensitivity

Noise reduction by decreasing the sensitivity of instruments for the undesired inputs is discussed in this section (see Sec. 2-1A). The techniques presented are inherent insensitivity, cancellation, and information conversion.

1. Inherent Insensitivity

Physical laws for sensing were described in Sec. 2-1. The signal and noise component for the first-order effects are shown in Eq. 3-87 (see Eq. 2-5).

$$\Delta Y = \frac{\partial f}{\partial D}\bigg|_0 \Delta D + \frac{\partial f}{\partial U}\bigg|_0 \Delta U \tag{3-87}$$

output   signal   noise

where D is the desired input and U all the undesired inputs. The system is inherently insensitive to the noise inputs when $\partial f/\partial U = 0$. This is achieved by (1) the proper selection of material and/or operating conditions, or (2) a structural change in the system, such as feedback. The discussions presented previously will not be repeated here.

The selection of material and operating conditions should be considered together for the given application. Consider the operation conditions for $\partial f/\partial U = 0$ in a strain measurement, where U is the undesired temperature input. A strain gage made of constantan is temperature insensitive for the operating temperature at 75°F for the range 50 to 100°F, as shown in Fig. 3-49. It is also insensitive to temperature if when operating at 275°F for the range 250 to 300°F. As stated in Sec. 2-1, the conditions $\partial f/\partial U = 0$ is necessary only for the given range of operation.

Now, consider the selection of material for an application. The strain gage of the isoelastic material is more temperature sensitive, but it also has a higher gage factor and therefore greater sensitivity [24]. For dynamic measurements and over a relatively short period, the isoelastic gage may be more suitable than the constantan gage, particularly when the test setup has a long time constant. In other words, an undesired effect can be ignored if it does not interfere with the given application.

An example of inherent insensitivity by means of a structural change in a strain measurement is shown in Fig. 3-50, in which a non-noise-generating component is substituted for a noise-generating component. Signal from the strain gage on a rotating shaft is brought out by means of a pair of slip rings. A slip ring is a metallic ring on a shaft to transfer the signal from a rotating member to the stationary instrument. The ring is insulated from the shaft. A ballast circuit is used for the measurement, as shown in Fig. 3-50b. The circuit is unsatisfactory because of the unpredictable contact resistance at the brushes. The signal can be completely masked by the brush noise (see Prob. 1-4).

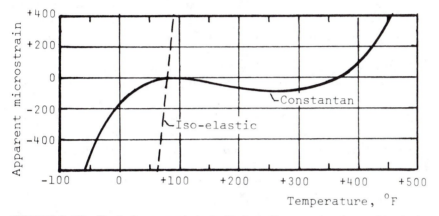

FIGURE 3-49 Typical apparent strain for two alloys commonly used in strain gages. (From Ref. 23.)

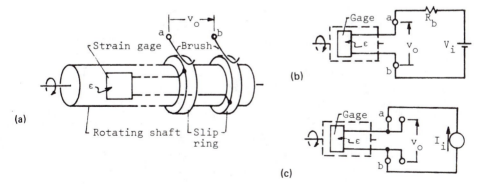

FIGURE 3-50 Noise reduction by insensitive structure. (a) Signal from rotating shaft. (b) Ideal voltage source V_i. (c) Ideal current source I_i. (From Ref. 25.)

The problem is solved by replacing the voltage source with a current source I and using two pairs of slip rings instead of one pair, as shown in Fig. 3-50c. Since I is constant, the contact resistance at the slip rings has no effect on the current flow. Since v_o is measured with a high-impedance voltmeter, there is no current flow in the voltage measurement and the brush noise has no effect on the signal transmission.

2. Cancellation

The differential method is used for noise reduction by cancellation. Assume two identical sensors A and B. When A and B are exposed to an undesired environment, they can be arrange for noise cancellation by means of opposing effects (1). When A is exposed to the environment and B only to the undesired inputs, A and B can be arranged for noise cancellation by means of equal effects due to the undesired inputs (2).

Noise cancellation by opposing effects is commonly used in instrumentation for transducers exposed to the same environment. For example, signal leads are often in twisted pairs to cancel the noise pickup in a measurement. When a magnetic field is undesired in a strain measurement, two strain gages can be made noninductive by means of a side-by-side mounting, as shown in Fig. 3-51a. The magnetic field produces equal effects in the gages, but the gages are mounted in series opposition to cancel the effect. In practice, identical transducers are not necessary for cancellation by opposing effects. For example, a transistor with a negative temperature coefficient can be used to compensate for the temperature effect on an electronic circuit with a positive temperature coefficient.

Noise cancellation by *equal effects* in a pressure measurement is shown in Fig. 3-51b. Transducers A and B are exposed to the same undesired effects, such as shock and temperature, but A is subjected to the pressure input and B is vented to the atmosphere. The differential output $A - B$ gives the desired output relative to atmospheric pressure.

Similarly, the adjacent strain gages A and B in Fig. 3-52a are exposed to the same environment, but only A is subjected to the strain input. Gage B is the "dummy" gage for temperature compensation. The differential output gives the strain signal. The resistances in the lead wires must also be compensated. The leads are of the same length and exposed to the same environment. The effects of the leads will cancel if they are on adjacent arms of the bridge circuit. The three-wire system has the advantage that there is no current in the center lead when the bridge is balanced. The four-wire system provides greater flexibility when using two sensors.

3. Information Conversion

Information conversion changes the characteristics of the signal without altering its information content (see Sec. 2-3). The input signal mod-

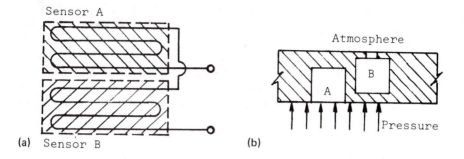

FIGURE 3-51 Noise reduction by cancellation. (a) Noninductive resistance wire sensors. (b) Pressure transducers.

ulates the carrier to produce a modulated output, which is at a higher power level as well as less susceptible to noise. Modulation is treated in Chapter 5. Mechanical modulation in an optical pyrometer is described to illustrate the process [26].

The optical pyrometer in Fig. 3-53a, used for measuring the temperature T from a source, can be made inherently insensitive to the ambient temperature T_a. Assume that the waveforms of T and T_a are as shown in Fig. 3-53b and c, respectively. If the output of the pyrometer is proportional to $T + T_a$, the signal T can be masked by the noise T_a.

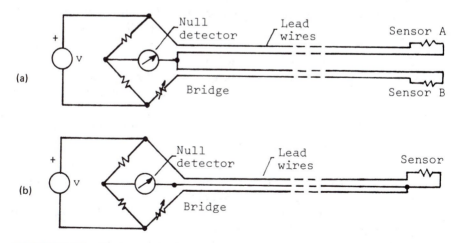

FIGURE 3-52 Compensation for resistance in leads. (a) Four-wire system. (b) Three-wire system.

The chopper in the pyrometer converts T into a high-frequency ac signal T_c. The chopper is a slotted disk and acts as a mechanical shutter. Thus the output of the sensor v_1 is the proportional to $T_c + T_a$, as shown in Fig. 3-53d. Since T_a is slowly varying, its frequency is below the cutoff frequency of the ac amplifier, and T_a is rejected by the amplifier. Hence the amplifier output v_2 is shown in Fig. 3-53e. The original signal T is shown as a dashed line in the figure. It can be recovered from v_2 by means of a demodulation process.

3-8. NOISE DOCUMENTATION

The method for noise documentation is similar to that for calibration described in Sec. 2-4. The principle of calibration is that a known input produces a known output, and a zero input produces a zero output. An output without a corresponding input is noise in the system. When the noise level is identified and the measuring system calibrated for in-service conditions, the experimenter can vouch for the validity and accuracy of the data. A method for noise documentation by Stein is described in this section [27].

The steps for the documentation are: (1) systematically identify the information paths in a measurement, and (2) sequentially reduce the noise from each path to the residual level. The noise levels for each signal path are thus documented.

The general model for transducers (see Fig. 2-17) is used for this discussion. The information paths 1 to 4 are identified in the model in Fig. 3-54a. The strain gage bridge in Fig. 3-54b is used to illustrate the process. The switches 1, 2, and 3 are added for the purpose of noise documentation and circuit calibration. The switches are shown in the normal operating positions. Since the switches can be operated before, during, and after a run, the noise level can also be documented before, during, and after a run.

It should be reiterated that the noise for each step should be reduced to a residual level and documented before the next step is executed.

Step 1: Component check. Close switch 3 to position d. This preliminary step checks the noise in the components downstream from the transducer. Since the input to the components is shorted, the output v_o should be zero. If not, the noise in the components should be reduced to a residual level before the execution of the next step.

Step 2: Information path 1. Switch 1 to position a. Switches 2 and 3 at normal positions. Measurand OFF; that is, no strain is applied to the gages. Since the measurand and the carrier voltage V are off, there is no non-self-generating signal. From Fig. 3-54a the output v_o must be from

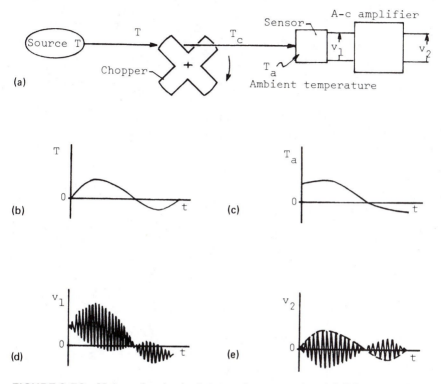

FIGURE 3-53 Noise reduction by information conversion. (a) Schematic of optical pyrometer. (b) Source temperature. (c) Ambient temperature. (d) Sensor output. (e) Amplifier output.

the self-generating part of the system. This is the noise from the undesired inputs in path 1.

Step 3: Information paths 1 and 2. Measurand OFF. All switches in their normal positions. Since the measurand is off, the output v_o must be from paths 1 and 2, as shown in Fig. 3-54a. The noise is from the self- and non-self-generating parts of the undesired input.

Step 4: Information paths 1 and 3. Switch 1 at a. Switches 2 and 3 at normal positions. Measurand ON. Since the carrier V is OFF, the output v_o can only be from the self-generating part of the transducer. Thus the residual noise in paths 1 and 3 are documented.

Step 5: All paths. Measurand ON. All switches in normal positions. The output v_o is due to the measurand and all the residual noise in the pre-

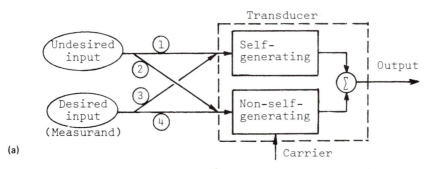

FIGURE 3-54 Transducer model and method for noise documentation. (a) Model of transducer and information channels. (b) Checking information channels.

vious steps. Thus the signal/noise ratio can be found and the amount of noise in the output is known.

PROBLEMS

3-1. With the aid of a sketch when necessary, briefly *discuss* each of the following:
  (a) Differential measurement.
  (b) Comparative measurements.
  (c) Comparison of methods of measurement.
  (d) One-, two-, and three-port devices.

(e) Basic circuits for non-self-generating transducers.
(f) Using feedback for correcting nonlinearities.
(g) Filtering and shielding.
(h) Methods for noise reduction.

3-2. A force transducer of stiffness k_m is inserted as shown in Fig. P3-1 for measuring the static force transmitted through k_1. Find the ratio force$_{measured}$/force$_{true}$ in the measurement.

3-3. A displacement gage of stiffness k_m is placed as shown in Fig. P3-2 for measuring the static deflection of the mass m. Find the ratio $x_{measured}/x_{true}$ in the measurement.

3-4. Find the ratio force$_{measured}$/force$_{true}$ for the static force in k_2 by means of a force transducer of stiffness k_m, as shown in Fig. P3-3.

3-5. The characteristic of a transducer is $Y = 2X^2$ for $X > 0$, and $Y = 0$ for $X \leqslant 0$. Two transducers are used in a differential measurement. (a) If the neutral position is at $X_0 = 0.3$, derive an expression for the output. (b) Show a plot of the output for $X_0 = 0.3$ and $\Delta X = \pm 0.3$.

3-6. Repeat Prob. 3-3 if $Y = 2e^x$ for $X > 0$, and $Y = 0$ for $X < 0$. (a) Assume that $X_0 = 1.0$ and $\Delta X = \pm 0.5$. Find the deviation from linearity. (b) Repeat for $X_0 = 2$ and $\Delta X = \pm 0.5$.

3-7. The strain gage bridge in Fig. P3-4 is used as a differential transducer. Initially, $R_1 = R_2 = R_3 = R_4 = R$. (a) Find the output voltage V_o if the bridge uses four active gages. (b) Repeat part (a) (1) if R_1 and R_2 are active, (2) if R_2 and R_3 are active, and (3), if only R_3 is active.

3-8. The output voltage of the strain gage bridge in Fig. P3-5 is measured by means of a galvanonometer of resistance $R_m = 100$ Ω. Initially, $R_1 = R_2 = R_3 = R_4 = 120$ Ω. Assume that $G_f = 2.0$, $V_i = 5$ V, and $\Delta R/R = 0.001$ for each of the four gages due to the applied strain. (a) Find the voltage across R_m. (b) Find the deflection of the galvanometer if its sensitivity is 0.1 μA/mm.

3-9. The strain gage bridge in Fig. P3-6 is calibrated by means of R_c. Assume that $R_1 = R_2 = R_3 = R_4 = 120$ Ω, $G_f = 2.0$, and V_o reads 150 units for the calibration. (a) Find the equivalent strain applied to R_2 in the calibration. (b) If the bridge uses two active gages for a measurement and $V_o = 500$ units, what is the applied strain? (c) If all four gages of the bridge are active and $V_o = 500$ units, what is the applied strain? (d) Propose a scheme for using four active gages in a tensile test. If V_o reads 500 units, what is the axial tensile strain?

3-10. An air gage is used for measuring the displacement x of the flapper valve in Fig. P3-7. Assume that the flow rate Q is

$$Q = f(P_0, P_1, d_1, d_2, x)$$

where P's are gage pressures and $P_0 = $ constant. (a) Derive an expression for P_1/P_0 in terms of d_1, d_2, and x. (b) Find the maximum sensitivity of the gage. (c) If $d_1 = d_2 = 0.02$ in., $P_0 = 10$ in. of water, and P_1 can be read to 0.1 in. of water, find x_{min} that can be detected. (d) Plot P_1/P_0 versus d_2x/d_1^2 for $0 < d_2x/d_1^2 < 1$. (e) Plot sensitivity versus d_2x/d_1^2 for $0 < d_2x/d_1^2 < 1$.

3-11. Draw the sketch of a differential transducer, using two of the air gages shown in Fig. P3-7. P_1 of gage 1 measures $x_1 = x_0 \pm \Delta x$ and P_2 of gage 2 measures $x_2 = x_0 \mp \Delta x$. (a) Using the data in Prob. 3-10 and assuming that the neutral position of the flapper valve is $x_0 = 0.01$ in. of water, plot $(P_1 - P_2)/P_0$ versus Δx for $-0.01 < \Delta x < 0.01$. (b) Find the sensitivity of the differential gage at $x_0 = 0.01$ in. of water.

3-12. An ac amplifier circuit and the static characteristics of its transistor are shown in Fig. P3-8. (a) Determine the operating point Q of the amplifier. (b) Given $h_{ie} = 3.6$ kΩ, $h_{re} = 3 \times 10^{-3}$, $h_{fe} = 150$, and $h_{oe} = 1.4 \times 10^{-4}$ S. Calculate the voltage and current gain and the input and output impedance of the amplifier.

3-13. (a) Draw the ac equivalent circuit for the transistor amplifier shown in Fig. P3-9. (b) Derive an expression for the voltage gain v_o/v_i. (c) Given $R_1 = 930$ kΩ, $R_2 = 250$ kΩ, $R_E = 3$ kΩ, $h_{re} = 3.6$ kΩ, $h_{fe} = 3 \times 10^{-3}$, $h_{fe} = 150$, and $h_{oe} = 1.4 \times 10^{-4}$ S. Find the voltage gain and the input impedance of the amplifier.

3-14. A two-stage amplifier is shown in Fig. P3-10. The first stage is represented by its Thévenin's equivalent. For the transistor of the second stage, assume that $h_{ie} = 2$ kΩ, $h_{re} = 0.001$, $h_{fe} = 70$, and $h_{oe} = 50$ μS. Derive the ac equivalent circuit and an expression of the overall voltage gain.

3-15. A feedback system with a nonlinear component G_2 is shown in Fig. P3-11. $G_1 = 10$, $H = 0.1$, and the characteristics of G_2 is

x	0	0.1	0.2	0.3	0.4	0.5	0.6	0.7	0.8	0.9	1.0
$f(x)$		1.85	3.3	4.7	5.8	6.9	7.8	8.5	9.1	9.4	9.5

Plot the overall characteristics of C versus R for $0 \leqslant R \leqslant 1.0$.

3-16. Repeat Prob. 3-15 for the characteristics of G_2 shown in Fig. P3-12a.

3-17. Repeat Prob. 3-15 for the characteristics of G_2 shown in Fig. P3-12b.

3-18. Referring to Fig. P3-15, the characteristics of G_2 is

$$f(x) = 20x - \frac{x^3}{2}$$

Design the feedback system and specify the values of G_1 and H such that $dC/dR = 20$ for $-2.5 < R < 2.5$.

3-19. Repeat Prob. 3-18 for $dC/dR = 10$ for $-5 < R < 5$.

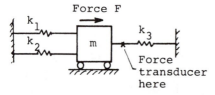

FIGURE P3-1

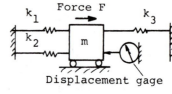

FIGURE P3-2

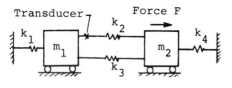

FIGURE P3-3

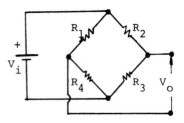

FIGURE P3-4

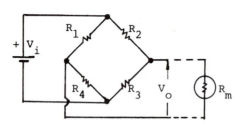

FIGURE P3-5

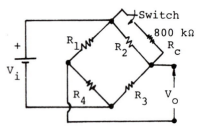

FIGURE P3-6

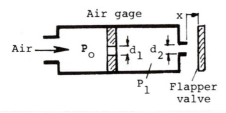

FIGURE 3-7

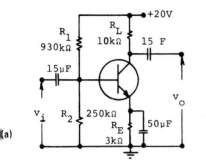

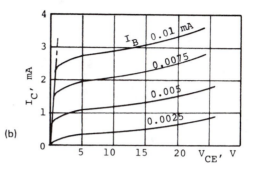

(a)

(b)

FIGURE P3-8

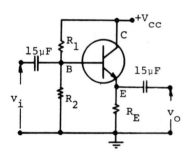

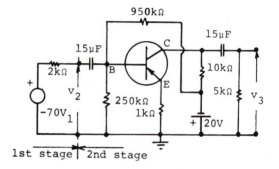

FIGURE P3-9

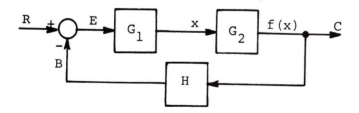

FIGURE P3-10

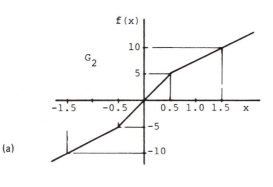

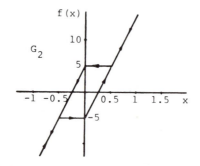

FIGURE P3-11

REFERENCES

1. Brophy, J., *Basic Electronics for Scientists,* 2nd ed., McGraw-Hill Book Company, New York, 1972, p. 228.
2. *ASTM Standards on Petroleum Products,* Method D8, American Society for Testing and Materials, Philadelphia.
3. Schlesinger, G., *Testing Machining Tools,* 8th ed., revised by F. Koenigsherger and H. Burdenkin, Pergamon Press, Inc., Elmsford, N.Y., 1978.
4. Stone, B. J., The Development of Dynamic Performance Tests for Lathes, *Proceedings of the Twelveth International Machine Tool Design and Research Conference,* Manchester, England, Sept. 1971, F. Koenigsberger and T. Tobias (eds.), Macmillan Publishing Company, New York, 1972, pp. 299–314.
5. Ramboz, D. J., Self-Contained Electronic Vibration Meters and Special Purpose Transducers, in *Shock and Vibration Handbook,* 2nd ed., C. M. Harris and C. E. Crede (eds.), McGraw-Hill Book Company, New York, 1976, Chap. 14.
6. Dranetz, A. I., and A. W. Orlacchio, Piezo-electric and Piezo-resistive Pick-ups, in *Shock and Vibration Handbook,* 2nd ed., C. M. Harris and C. E. Crede (eds.), McGraw-Hill Book Company, New York, 1976, Sec. 16.

7. Stein, P. K. Measuring System Performance Capabilities as Governed by Input Conditions Design., *Proceedings of Western Regional Strain Gage Committee Meeting,* Sept. 21, 1970, Society of Experimental Analysis, Westport, Conn.

8. Tse, F., I. Morse, and R. Hinkle, *Mechanical Vibrations,* 2nd ed., Allyn and Bacon, Inc., Boston, 1978, p. 48.

9. Flanagan, P. M., *Design of a Clinical Phono-angiography System,* Ph.D. dissertation, University of Cincinnati, 1980.

10. Tse, Morse, and Hinkle, Ref. 8, p. 176.

11. Pipes, L. A., *Matrix Methods for Engineers,* Prentice-Hall, Inc., Englewood Cliffs, N.J., 1963, Chap. 12.

12. Lynch, W. A., and J. G. Truxal, *Signal and Systems in Electrical Engineering,* McGraw-Hill Book Company, New York, 1962, Chap. 9.
    Brown, P. B., G. N. Franz, and H. Moraff, *Electronics for Modern Scientists,* Elsevier Science Publishing Co., Inc., New York, 1982, Chap. 2.

13. Cowles, L. G., *Transistor Circuits and Applications,* 2nd ed., Prentice-Hall, Inc., Englewood Cliffs, N.J., 1974, p. 11.

14. Horowitz, P., and W. Hill, *The Art of Electronics,* Cambridge University Press, New York, 1980, pp. 650-652.

15. Amey, W. G., et al., Bridges and Potentiometers, in *Process Instruments and Control Handbook,* 2nd ed., D. M. Considine (ed.), McGraw-Hill Book Company, New York, 1974, Sec. 14.

16. Stein, P. K., The Constant Current Concept for Dynamic Strain Measurement, *Strain Gage Read.,* Vol. 6, No. 3 (Aug.–Sept. 1963), pp. 53-72.

17. Perry C. C., and H. R. Lissner, *The Strain Gage Primer,* 2nd ed., McGraw-Hill Book Company, New York, 1962, p. 93.

18. West, J. C., *Analytical Techniques for Nonlinear Control Systems,* D. Van Nostrand Company, Princeton, N.J., 1960, Chap. 1.

19. Gray, P. E., and C. L. Searle, *Electronic Principles, Physics, Models, and Circuits,* John Wiley & Sons, Inc., New York, 1969, p. 619.

20. Morrison, R., *Grounding and Shielding Techniques in Instruments,* John Wiley & Sons, Inc., New York, 1972, pp. 81-85.

21. *The Radio Amateur's Handbook,* Amateur Radio Relay League, West Hartford, Conn., (published yearly).

22. Morrison, Ref. 20, Chap. 4.

23. Wilson, E. J., Strain Gage Instrumentation, in *Shock and Vibration Handbook,* 2nd ed., C. M. Harris and C. E. Crede (eds.), McGraw-Hill Book Company, New York, 1976, Chap. 13.

24. Perry and Lissner, Ref. 17, p. 19.

25. Stein, P. K., The Constant Current Concept for Dynamic Strain Measurement, *Instrum. Control Syst.,* Vol. 38, No. 5 (1965), pp. 154-155.

26. Magison, E. C., Radiation and Optical Pyrometry, in *Process Instruments and Control Handbook,* 2nd ed., D. M. Considine (ed.), McGraw-Hill Book Company, New York, 1974, Sec. 2, p. 2.

27. Stein, P. K., A Unified Approach to Handling of Noise in Measuring Systems, *Lecture Series No. 50,* Flight Test Instru. Adv. Group for Aerospace R & D, NATO, Edward Air Force Base, Calif., May 1973.

4
Dynamic Characteristics of Instruments

4-1. INTRODUCTION

The aspects for valid measurement are the transducer, the structure of the measuring system, and the system dynamics. The transducer and structure were discussed previously. Dynamics of linear systems are presented in this chapter.

The subject will be studied in the time and frequency domains. Since the same system is being examined, its behavior must be independent of the methods employed. The methods are therefore different views of the system and there must be a correlation between their results. Different methods are presented because each will bring out certain concepts. The method selected for testing depends on the convenience and equipment available, although the trend in recent years favors the transient method.

Time response describes the dynamics of a system by means of differential equations. Time response is intuitive, because all events occur in real time. Concepts of basic system parameters, such as the time constant, can best be introduced in the time domain, but the method is less suited for the study of more complex systems.

Frequency response describes the dynamics of a system by means of its sinusoidal steady-state behavior, similar to the ac circuit analysis. The method seems artificial, because few signals in actual applications are sinusoidal, and the interpretation of some parameters, such as the time constant, is less direct. The method is convenient for the analysis of more complex systems. The method can be extended readily for studies in transient response by means of Fourier methods.

4-2. MODELING

The model of a single-input, single-output linear instrument, shown in Fig. 4-1, is used for the study of system dynamics. The system equations are

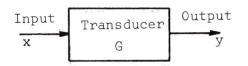

FIGURE 4-1 Model of transducer.

$$\text{output } y = (\text{transfer function } G)(\text{input } x) \tag{4-1}$$

$$\frac{1}{G}\, y = x \tag{4-2}$$

$$\frac{y}{x} = G \tag{4-3}$$

where x is the input or excitation, y the output, and G an operator or the transfer function, representing the system dynamics. Equation 4-2 is the system equation in the differential equation form and Eq. 4-3 in the transfer function form. The simplified model assumes that loading and noise are negligible and a three-port device can be reduced to a two-port (see Secs. 2-3C and 3-3D).

The model of an instrument for the time-response analysis is the general differential equation in Eq. 4-2, in which time t is the independent variable. Rewriting Eq. 4-2 explicitly gives

$$(a_n D^n + \cdots + a_1 D + a_0)y = (b_m D^m + \cdots + b_1 D + b_0)x(t) \tag{4-4}$$

where the a's and b's are constants, m and n integers, $D = d/dt$ is an operator, and n denotes the order of the differential equation. It will be shown that $m \leqslant n$ for the system to be realizable. Since $x(t)$ is a given time function, the right-hand side of Eq. 4-4 is also known. For convenience, Eq. 4-4 can be simplified to yield

$$(a_n D^n + a_{n-1}D^{n-1} + \cdots + a_1 D + a_0)y = b_0 x(t) \tag{4-5}$$

The transfer function in the operational form from Eq. 4-4 is

$$\frac{y}{x}(D) = G(D) = \frac{b_m D^m + b_{m-1}D^{m-1} + \cdots + b_1 D + b_0}{a_n D^n + a_{n-1}D^{n-1} + \cdots + a_1 D + a_0} \tag{4-6}$$

where $G(D)$ is an operator. It denotes that G is a function of the operator D but not the product of G and D. $G(D)$ is often called an output/input ratio.

It should be cautioned that $G(D)$ is not a ratio in the normal sense of the word, such as the ratio of two numbers.

1. A ratio of two numbers is nondimensional, but $G(D)$ is dimensional. If the input temperature of a thermocouple is $x(°C)$ and the output is $y(mV)$, the dimension of $G(D)$ is mV/°C.
2. $G(D)$ is not a ratio of the instantaneous values of y/x. If the value of $x(t)$ at a given time is known, the corresponding value of y can only be found from a solution of the system differential equation.
3. $G(D)$ is an operator. Similar to $D = d/dt$, it does not have a physical meaning in itself until it operates on some physical quantity.
4. $G(D)$ represents the dynamic behavior of the physical system only when loading and noise are neglected.
5. The numerator and denominator in Eq. 4-6 are polynomials in D. When the polynomials are factorized (see Sec. 4-6), some of the factors may cancel. The transfer function describes the system behavior only when there is no cancellation of the factors between the numerator and the denominator [1].

The model of an instrument for the frequency response study is a steady-state sinusoidal transfer function, in which both the input and the output are sinusoidal at the input frequency in $\omega(rad/s)$. Substituting $j\omega$ for $D (= d/dt)$ in Eq. 4-6 (see Sec. A-4 in App. A), we obtain the transfer function $G(j\omega)$.

$$\frac{y}{x}(j\omega) = G(j\omega) = \frac{b_m(j\omega)^m + b_{m-1}(j\omega)^{m-1} + \cdots + b_1(j\omega) + b_0}{a_n(j\omega)^n + a_{n-1}(j\omega)^{n-1} + \cdots + a_1(j\omega) + a_0}$$

$$(4\text{-}7)$$

where $G(j\omega)$ denotes that G is a function of $j\omega$ and not the product of G and $(j\omega)$.

The condition $m \leqslant n$ is necessary for the instrument to be realizable, that is, for the instrument to be constructed physically. For example, if $m > n$ and as the excitation frequency ω approaches infinity, the magnitude of y/x in Eq. 4-7 also becomes infinite. In other words, the value of y can be very large even when the value of x approaches zero. Note that this is not a resonance phenomenon. Furthermore, it can be shown that the power to maintain a reasonable amplitude of y at an infinite frequency is very large. The concept of realizability is found in the literature [2].

The behavior of linear systems is described by the amplitude ratio Y/X and the relative phase angle ϕ between the input/output quantities in Eq. 4-7. These quantities are frequency dependent as shown in Fig. 4-2. It can be shown that

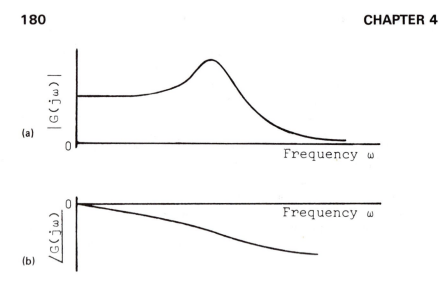

(a)

(b)

FIGURE 4-2 Transfer function of instrument. (a) Amplitude ratio versus frequency. (b) Phase angle versus frequency.

$$\frac{Y}{X} = \left|\frac{y}{x}(j\omega)\right| = |G(j\omega)| \tag{4-8}$$

$$\phi = \angle\frac{y}{x}(j\omega) = \angle G(j\omega) \tag{4-9}$$

The transfer function $G(j\omega)$ gives a convenient way for handling complex systems. Note that $G(j\omega)$ is a complex number at a given frequency ω, and it can be manipulated like an algebraic quantity. For example, the instrument in Fig. 4-3 has three components in cascade, and the overall transfer function is

$$\frac{y_3}{x_1} = \frac{y_3}{x_3}\frac{y_2}{x_2}\frac{y_1}{x_1}$$

$$G(j\omega) = G_1(j\omega)G_2(j\omega)G_3(j\omega) \tag{4-10}$$

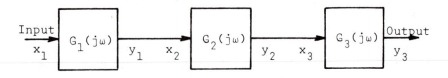

FIGURE 4-3 Components in cascade.

where $y_2 = x_3$ and $y_1 = x_2$. Another example is the manipulation of the transfer functions $G(j\omega)$ and $H(j\omega)$ in Fig. 3-39 in a feedback control system. This gives the overall transfer function

$$\frac{C}{R}(j\omega) = \frac{G(j\omega)}{1 + G(j\omega)H(j\omega)} \tag{4-11}$$

4-3. TIME RESPONSE OF INSTRUMENTS

Basic system parameters and their interpretations in the time domain are presented in this section. Time response gives the time history of the output of an instrument in response to an input. In other words, it is the solution of the system differential equation. To this end, the types of instruments can be classified by the order of the differential equation. Their responses to different inputs are used to interpret the system parameters.

From Eq. 4-5, the *types of instruments* are classified as follows:

Zero-order, $n = 0$: $\qquad a_0 y = b_0 x(t)$ $\qquad\qquad\qquad$ (4-12)

First-order, $n = 1$: $\qquad (a_1 D + a_0)y = b_0 x(t)$ $\qquad\qquad$ (4-13)

Second-order, $n = 2$: $\quad (a_2 D^2 + a_1 D + 1)y = b_0 x(t)$ $\qquad$ (4-14)

where the a's and b_0 are constants, $D = d/dt$, $x(t)$ is a known input, and y the output.

Arbitrary signals are commonly used for testing. This is justified for several reasons.

1. Any test signal is arbitrary because the input to which the instrument will be subjected is never known beforehand.
2. An arbitrary test signal can be used for comparative tests (see Sec. 3-1D). If an instrument is tested with a particular signal and it performs well in the field, the same test can be used for another instrument of similar design and for similar application.
3. For the same reason, the step function is commonly the basis for the time-domain specifications of instruments.
4. The step and impulse functions are simple and meet the requirements of certain test conditions. This will be justified in Sec. 4-8.
5. The step and impulse functions are related mathematically [3]. It will be shown that the response of a linear system can be deduced from its response to other signals.

The type of instruments enumerated in Eqs. 4-12 to 4-14 and their responses to the step, ramp, and impulse functions are summarized in Fig.

4-4. Our interest is in the interpretation of the results, and the steps for obtaining the solutions of the differential equations are omitted.

A. Test Signals

Common test signals are described in this section. The step function $f(t) = Au(t)$ shown in Fig. 4-4a is defined as

$$Au(t) = \begin{cases} 0 & \text{for } t < 0 \\ A & \text{for } t > 0 \end{cases} \tag{4-15}$$

where A is a scalar multiplier and $u(t)$ is a unit step function. For example, a unit step force is a constant force of unit magnitude applied at time zero.

The *ramp function* $f(t) = At$ in Fig. 4-4b is defined as

$$At = \begin{cases} 0 & \text{for } t < 0 \\ At & \text{for } t > 0 \end{cases} \tag{4-16}$$

where A is a scalar, or the slope of the curve. The function starts at time $t = 0$ and continues with a constant slope. For example, a ramp displacement is a constant velocity.

An *impulse function* (Fig. 4-4c) is defined as

$$A\delta(t) = 0 \qquad \text{for } t \neq 0$$

$$\int_{-\infty}^{\infty} A\delta(t)\, dt = A \tag{4-17}$$

where A is a scalar and $\delta(t)$ a unit impulse applied at time zero. Consider the rectangular pulse in Fig. 4-5a of duration (width) T and height $1/T$. The area under the pulse is unity. If the area remains constant while the pulse width T approaches zero, the pulse becomes a *unit impulse* $\delta(t)$ or the Dirac delta function. A unit impulse, shown symbolically in Fig. 4-5b, has zero duration and infinite height. In practice, an impulse is realized when the duration of the pulse is short compared with the time for an object to respond. For example, an impulse force can be applied to a machine with a mallet. The strength of a pulse is the area under the curve representing the impulse. The impulse defined in Eq. 4-17 is of strength A. If a unit impulse is delayed by the time $t = \tau$, as shown in Fig. 4-5b, it is denoted by $\delta(t - \tau)$ and is defined as

$$\delta(t - \tau) = 0 \qquad \text{for } t \neq \tau$$

$$\int_{-\infty}^{\infty} \delta(t - \tau)\, d\tau = 1 \tag{4-18}$$

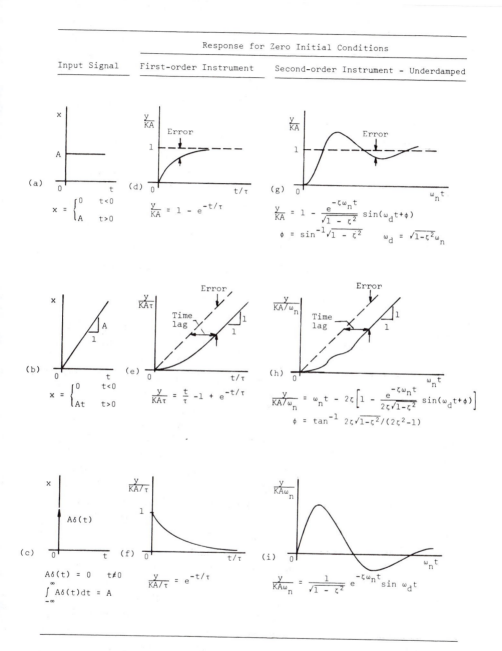

FIGURE 4-4 Normalized time response of first- and second-order instruments for various test signals. (a) Step input. (b) Ramp input. (c) Impulse input. (d) Step response. (e) Ramp response. (f) Impulse response. (g) Step response. (h) Ramp response. (i) Impulse response.

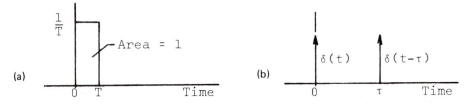

FIGURE 4-5   Derivation of unit impulse from rectangular pulse. (a) Rectangular pulse. (b) Unit impulse.

B.   Zero-Order Instruments

A *zero-order instrument* is defined by Eq. 4-12. Dividing the equation by a_0 gives

$$y = Kx(t) \tag{4-19}$$

where $K = b_0/a_0$ is the *sensitivity*. It is the basic parameter introduced by the zero-order instrument. A zero-order instrument is a "perfect instrument." Its response y is instantaneous and it differs from the input $x(t)$ by a scale factor. Its static characteristic is a straight line with the slope K. For dynamic applications, the output waveform is a perfect reproduction of the input. There is no distortion of any kind.

Note that sensitivity is a steady-state or "static" concept. It is a steady-state concept because the time derivatives in Eq. 4-5 are ignored in order to deduce the zero-order instrument in Eq. 4-12. An alternative view is that the solution from Eq. 4-5 has reached a steady state. Second, K is a static concept because Eq. 4-12 describes the static characteristics of an instrument. The instantaneous response is only relative because it takes time for an instrument to respond to an input. It implies that the input is slow varying compared with the speed of the response of the instrument. For example, a thermometer is sufficiently fast for indicating the room temperature, and it is zero-order for a room-temperature measurement. A signal of 100 kHz is slow compared with the speed of response of an oscilloscope, the frequency response of which generally exceeds 10 MHz.

An instrument with a *transport delay* is often called zero-order. For example, there is a time delay in drawing hot water from a water heater. The input (dashed lines) and output of an instrument with transport delay t_d are shown in Fig. 4-6. The delay may not be critical for some measurements, but its effect must be considered for systems with feedback or for multichannel measurements. Note that a transport delay and a time lag describe two different phenomena. Time lag is described in the sections that follow.

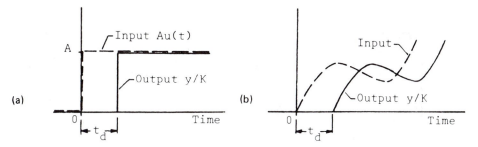

FIGURE 4-6 Transport delay: Zero-order instrument. (a) Response to a step input $Au(t)$. (b) Response to an arbitrary input.

C. First-Order Instruments

A *first-order instrument* is defined by the first-order differential equation in Eq. 4-13. Dividing the equation by a_0 gives

$$(\tau D + 1)y = Kx(t) \tag{4-20}$$

where $K = b_0/a_0$ is the sensitivity in Eq. 4-19 and $\tau = a_1/a_0$ is the time constant. The characteristic equation from Eq. 4-20 is

$$\tau D + 1 = 0 \tag{4-21}$$

Since $D = d/dt$ is an operator, the *time constant* τ is the basic parameter introduced by a first-order instrument. The response of first-order instruments to different inputs is described in this section. Let us first illustrate the concept of time constant with examples.

Example 4-1.   The sight glass shown in Fig. 4-7a for liquid-level measurement consists of a uniform glass tube of cross-sectional area C. It is connected to an open tank by a needle valve of resistance R. Find the time constant of the system.

Solution: Assume that the flow rate between the tank and the glass tube is proportional to the rate change of liquid level in the tube, and is proportional to the differential liquid level $(x - y)$.

$$\text{flow rate} = C(\text{rate of change in liquid level}) = C\frac{dy}{dt}$$

$$= \frac{1}{R}(x - y)$$

Combining the equations above and simplifying, we get

$$C\frac{dy}{dt} + \frac{1}{R}y = \frac{1}{R}x \qquad \text{or} \qquad \tau\frac{dy}{dt} + y = x$$

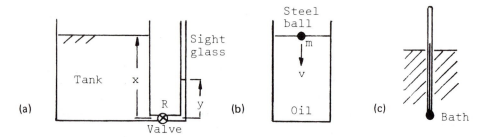

FIGURE 4-7 Examples of first-order instruments. (a) Liquid-level measurement. (b) Viscous friction. (c) Thermometer.

where $\tau = RC$ is the time constant of the sight glass.

Example 4-2.   A steel ball of mass m is released into a viscous oil shown in Fig. 4-7b. Derive the time constant for the system.

Solution: From Newton's second law, we have

$$m \frac{dv}{dt} = \sum (\text{forces in the vertical direction})$$

$$= -cv + mg$$

where v is the velocity of the ball, cv the viscous frictional force, and mg the gravitational force. Rearranging the equation yields

$$\frac{m}{c} \frac{dv}{dt} + v = \frac{m}{c} g \qquad \text{or} \qquad \tau \frac{dv}{dt} + v = KAu(t)$$

The examples show that a time constant is due to an energy storage and a resistance to the energy transfer. The elements performing these functions take many physical forms. In Example 4-1, potential energy is stored in the sight glass, and the resistance is due to the needle valve. In Example 4-2, kinetic energy is stored in the mass, and the resistance is due to the viscosity of the oil.

The thermometer in Fig. 4-7c is another example of the first-order instrument. When it is immersed in a constant-temperature bath, it takes time for the thermometer to assume the temperature of the bath. The thermometer has a thermal capacitance C and resistance R to the heat transfer. The RC time constant causes a time lag in the thermometer. An additional example is the servo-position controller in Fig. 3-40. In this case, the R and C elements in the system may not be obvious.

It is important to note the characteristics of the time constant.

1. A time constant of a physical system is not always time invariant. For example, a thermometer in a constant temperature bath will assume the temperature of the bath faster when the fluid is agitated than when not agitated. The time constant decreases when the resistance to the heat flow is reduced by means of the agitation. Similarly, the time constant of a thermocouple for measuring the exhaust temperature of a jet engine is not constant. The response time of the thermocouple is dependent on the flow condition.
2. The time constant τ and the sensitivity K in Eq. 4-20 are related, where $\tau = a_1/a_0$ and $K = b_0/a_0$. Since τ and K have the same denominator, it is not always possible to increase K without increasing τ. A large time constant means a slower response. In other words, the increase in sensitivity is often at the expense of the response time of an instrument.
3. The first-order instrument is only a model. The instrument is first-order when the time constant is relevant for the application. Under other conditions, the instrument could exhibit other behaviors, such as oscillations. Thus it may be more appropriate to choose another model for the representation.

1. Step Response

A first-order instrument with a step input is obtained by substituting the step function $Au(t)$ for $x(t)$ into Eq. 4-20:

$$(\tau D + 1)y = KAu(t) \tag{4-22}$$

The time response for zero initial conditions $y(0) = 0$ is

$$y = KA(1 - e^{-t/\tau}) \quad \text{or} \quad \frac{y}{KA} = 1 - e^{-t/\tau} \tag{4-23}$$

The error shown in Fig. 4-4d is due to the time lag in the response. An instrument with a smaller time constant has less time lag and therefore less dynamic error, as shown in Fig. 4-8. In the limit, when the time constant approaches zero, a first-order instrument becomes zero-order. This conclusion is also evident from Eq. 4-23.

Methods for finding the value of τ from the step response are illustrated in Fig. 4-8.

1. A time constant is defined as the time for $t/\tau = 1$.

$$\frac{y}{K} = A(1 - e^{-t/\tau})|_{t=\tau} = 0.632A \tag{4-24}$$

Thus τ_1 and τ_2 are determined as shown in the figure.

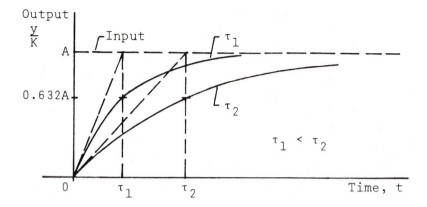

FIGURE 4-8 Effect of time constant: Step response of first-order instrument.

2. A second method uses the initial slope of the response curve.

$$\frac{d(y/K)}{dt}\bigg|_{t=0} = A\left(0 + \frac{1}{\tau}\,e^{-t/\tau}\right)\bigg|_{t=0} = \frac{A}{\tau} \qquad (4\text{-}25)$$

An extrapolation of the initial slope at $t = 0$ will intersect the constant input (dashed line). Since the slope is A/τ and the value of the ordinate at the intersection is A, a perpendicular from the intersection to the time axis gives the value of τ.

3. A preferred method (not shown) is to rearrange Eq. 4-23 and then take the natural logarithm of both sides of the equation.

$$1 - \frac{y}{KA} = e^{-t/\tau} \qquad \text{or} \qquad \ln\left(1 - \frac{y}{KA}\right) = -\frac{1}{\tau}\,t \qquad (4\text{-}26)$$

This gives the equation of a straight line with a slope of $-1/\tau$ in a semilog plot. The method is more accurate, because it is easier to "eyeball" a straight line than a slope or a curve.

2. Ramp Response

A first-order instrument with a ramp input is obtained by substituting the ramp function At for $x(t)$ into Eq. 4-20:

$$(\tau D + 1)y = KAt \qquad (4\text{-}27)$$

The time response for zero initial condition $y(0) = 0$ is

$$y = KA[t - \tau(1 - e^{-t/\tau})] \qquad \text{or} \qquad \frac{y}{KA\tau} = \frac{t}{\tau} - 1 + e^{-t/\tau} \qquad (4\text{-}28)$$

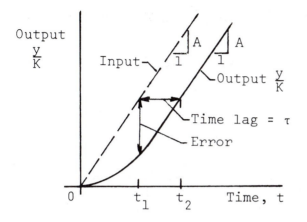

FIGURE 4-9 Effect of time constant: Ramp response of first-order instrument.

The time lag and error are indicated in Fig. 4-4e. It can be shown that the steady-state time lag is τ and the steady-state error is $A\tau$, as shown in Fig. 4-9.

3. Impulse Response

A first-order instrument with an impulse input is obtained by substituting the impulse function $A\delta(t)$ for $x(t)$ into Eq. 4-20:

$$(\tau D + 1)y = KA\delta(t) \qquad (4\text{-}29)$$

An impulse has zero width and infinite height. To avoid the infinite value, the equation is solved indirectly by converting Eq. 4-29 into a homogeneous equation with the appropriate initial condition. After the impulse is over at time $t = 0^+$, the system is unforced and is described by a homogeneous equation. Thus the energy input due to the impulse becomes the initial condition for the solution for $t > 0^+$.

The technique for finding the initial condition for $t = 0^+$ is described below. Integrating Eq. 4-29 for $0^- < t < 0^+$ gives

$$\tau[y(0^+) - y(0^-)] + \int_{0^-}^{0^+} y \, dt = KA \int_{0^-}^{0^+} \delta(t) \, dt = KA \qquad (4\text{-}30)$$

The integration of $\delta(t)$ is unity from Eq. 4-17. The integration of $y(t)$ over the infinitesimal duration is zero, because physically $y(t)$ cannot be infinite. Thus Eq. 4-30 yields

$$y(0^+) = \frac{KA}{\tau} + y(0^-) \qquad (4\text{-}31)$$

Assume, for convenience, that $y(0^-) = 0$. The system equation becomes

$$(\tau D + 1)y = 0 \tag{4-32}$$

with the initial conditions $y(0^+) = KA/\tau$. The solution is

$$y = \frac{KA}{\tau} e^{-t/\tau} \quad \text{or} \quad \frac{y}{KA/\tau} = e^{-t/\tau} \tag{4-33}$$

D. Second-Order Instruments

In this section we examine the time response of *second-order instruments* for various input functions and introduce natural frequency and damping ratio as additional parameters for dynamic studies [4]. The model of a second-order instrument is the second-order differential equation in Eq. 4-14. Dividing the equation by a_0 and defining

$$\frac{a_2}{a_0} = \frac{1}{\omega_n^2} \frac{a_1}{a_0} = \frac{2\zeta}{\omega_n} \quad \text{and} \quad \frac{b_0}{a_0} = K \tag{4-34}$$

the system equation becomes

$$\left(\frac{1}{\omega_n^2} D^2 + \frac{2\zeta}{\omega_n} D + 1 \right) y = Kx(t) \tag{4-35}$$

where $x(t)$ is the excitation, y the output, ω_n the *undamped natural frequency* in rad/s, ζ *the damping ratio,* and K the sensitivity. The instrument is assumed underdamped unless otherwise stated. The characteristic equation from Eq. 4-35 is

$$\frac{1}{\omega_n^2} D^2 + \frac{2\zeta}{\omega_n} D + 1 = 0 \tag{4-36}$$

Since $D = d/dt$ is an operator, a second-order instrument is characterized by ω_n and ζ.

Example 4-3. A seismic instrument for vibration measurements is shown schematically in Fig. 4-10a. The seismic mass m is supported by the base through a spring k and a damper c. The base is attached securely to a vibrating body, the motion of which is to be measured. Let $x_1(t)$ be the motion of the base, and x_2 the motion of m from its static equilibrium position. The relative motion $y = x_2 - x_1$ is conceptually as recorded by means of a pen and a rotating drum. Derive the differential equation of motion of the instrument.

Solution: From Newton's second law, we have

$$m\ddot{x} = \sum (\text{forces in the } x \text{ direction})$$

$$= -k(x_2 - x_1) - c(\dot{x}_2 - \dot{x}_1)$$

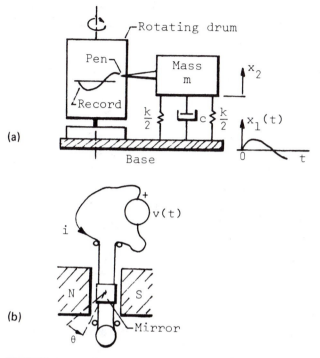

FIGURE 4-10 Examples of second-order instruments. (a) Seismic instrument. (b) Galvanometer.

where the ($\cdot$) denotes a time derivative, $k(x_2 - x_1)$ the spring force, and $c(\dot{x}_2 - \dot{x}_1)$ the viscous damping force. The relative motion $y = x_2 - x_1$ is recorded on the drum. Substituting y, $\dot{y}$, and $\ddot{y}$ into the equation and simplifying, we get

$$m\ddot{y} + c\dot{y} + ky = -m\ddot{x}_1(t)$$

Example 4-4. A galvanometer in a recording oscillosgraph is shown in Fig. 4-10b. It consists of a single-turn phosphor-bronze ribbon placed between the poles of a magnet. A small mirror is cemented to the conducting ribbon and tension is applied by means of a spring. When a current passes through the loop, the two sides of the ribbon move in opposite directions, and the mirror turns about a vertical axis. The applied voltage $v(t)$ is measured by the angle θ of the mirror. Neglecting the mechanical damping and the self-inductance and capacitance of the condutors, derive the differential equation of the instrument.

Solution: Assume (1) The torque T developed by the galvanometer is proportional to the current i in the conductor, and (2) the voltage v (back emf) induced by the loop is proportional to the angular velocity $\dot{\theta}$.

$$T = bi \quad \text{and} \quad v = b\dot{\theta}$$

where b is a constant. From Newton's second law, the torque equation is

$$J\ddot{\theta} + k\theta = bi$$

where J is the mass moment of inertia of the moving parts about the vertical axis and k the torsional spring constant of the galvanometer suspension. From Kirchhoff's voltage law, the loop equation is

$$Ri + b\dot{\theta} = v(t)$$

where R is the electrical resistance of the loop. Eliminating i between the equations above yields

$$J\ddot{\theta} + \frac{b^2}{R}\dot{\theta} + k\theta = \frac{b}{R}v(t)$$

1. *Step Response*

A second-order instrument with a step input is obtained by substituting the step function $Au(t)$ for $x(t)$ into Eq. 4-35:

$$\left(\frac{1}{\omega_n^2}D^2 + \frac{2\zeta}{\omega_n}D + 1\right)y = KAu(t) \tag{4-37}$$

For zero initial conditions and $\zeta < 1$, the solution is

$$y = KA\left[1 - \frac{1}{\sqrt{1-\zeta^2}}e^{-\zeta\omega_n t}\sin(\omega_d t + \psi)\right] \tag{4-38}$$

$$\psi = \sin^{-1}\sqrt{1-\zeta^2} \qquad \omega_d = \sqrt{1-\zeta^2}\,\omega_n \tag{4-39}$$

$$\frac{y}{KA} = 1 - \frac{1}{\sqrt{1-\zeta^2}}e^{-\zeta\omega_n t}\sin(\omega_d t + \psi) \tag{4-40}$$

where ω_d rad/s is the frequency of oscillation with damping. The exponential term $e^{-\zeta\omega_n t}$ can be compared with $e^{-t/\tau}$ in Eq. 4-23. Hence the quantity $1/\zeta\omega_n$ is often called the time constant of a second-order instrument. The step response is shown in Figs. 4-4g and 4-11.

The parameters ζ and ω_n can be estimated from the response curve in Fig. 4-11. The maximum overshoot occurs at the first peak of the response curve. It can be shown that

$$\text{maximum overshoot} = 100e^{-\zeta\pi/\sqrt{1-\zeta^2}} \qquad \text{percent} \tag{4-41}$$

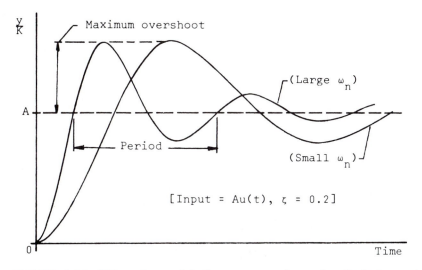

FIGURE 4-11   Effect of ω_n and ζ: Step response of second-order instrument.

The maximum overshoot is determined solely by ζ, as shown in the equation and in Fig. 4-12. Alternatively, ζ can be determined by means of the log-decrement method [5]. The natural frequency ω_n rad/s is estimated from the period of the damped oscillation in Fig. 4-11.

$$\omega_d = \frac{2\pi}{\text{period}} \quad \text{and} \quad \omega_n = \frac{\omega_d}{\sqrt{1 - \zeta^2}} \tag{4-42}$$

It is intuitive that ω_n determines the speed of response of an instrument. For the same damping ratio ζ, a system with a larger ω_n has a faster response, as shown in Fig. 4-11. For the same ω_n, a system with a smaller damping ratio ζ gives a larger overshoot, as shown in Fig. 4-13.

The parameters ζ and ω_n in an instrument are interrelated. It can be deduced from Eqs. 4-14 and 4-34 that when a_2, a_1, and b_0 are constants, an increase in a_0 in order to increase the natural frequency will also cause a decrease in damping and sensitivity. In other words, a high natural frequency transducer inherently has lower damping and lower sensitivity. The trade-off is similar to that between the time constant and sensitivity in a first-order instrument. Note that the discussion does not imply that a transducer with very high natural frequency and low damping, such as piezoelectric accelerometers, cannot be used [6]. This will be justified in the frequency-domain analysis.

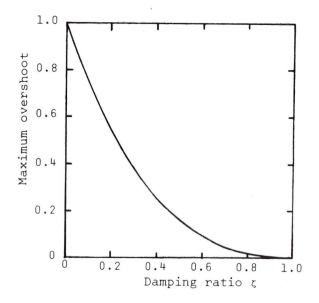

FIGURE 4-12   Maximum overshoot versus ζ: second-order instrument.

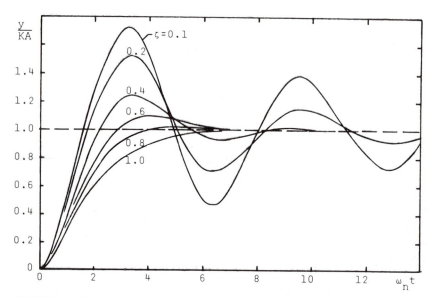

FIGURE 4-13   Effect of damping ratio ζ: step response of second-order instrument.

2. Ramp Response

A second-order instrument with a ramp input is obtained by substituting the ramp function At for $x(t)$ in Eq. 4-35.

$$\left(\frac{1}{\omega_n^2} D^2 + \frac{2\zeta}{\omega_n} D + 1\right) y = KAt \tag{4-43}$$

For zero initial conditions and $\zeta < 1$, the solution is

$$\frac{y}{KA/\omega_n} = \omega_n t - 2\zeta\left[1 - \frac{1}{\sqrt{1 - \zeta^2}} e^{-\zeta\omega_n t} \sin(\omega_d t + \psi)\right] \tag{4-44}$$

$$\psi = \tan^{-1} \frac{2\zeta\sqrt{1 - \zeta^2}}{2\zeta^2 - 1}$$

The response is shown in Fig. 4-8h, in which the time lag and error are indicated. It can be shown that the normalized steady-state response lags the input by the amount of time $2\zeta/\omega_n$.

3. Impulse Response

A second-order instrument with an impulse input is obtained by substituting the impulse function $A\delta(t)$ for $x(t)$ in Eq. 4-35:

$$\left(\frac{1}{\omega_n^2} D^2 + \frac{2\zeta}{\omega_n} D + 1\right) y = KA\delta(t) \tag{4-45}$$

Using the technique for obtaining the impulse response of first-order instruments, it can be shown that the equivalent system is

$$\left(\frac{1}{\omega_n^2} D^2 + \frac{2\zeta}{\omega_n} D + 1\right) y = 0 \quad \text{with} \quad \begin{cases} y(0^+) = y(0^-) \\ \dot{y}(0^+) = \dot{y}(0^-) + KA\omega_n^2 \end{cases} \tag{4-46}$$

Assuming that $y(0^-) = 0$ and $\dot{y}(0^-) = 0$, the impulse response, as shown in Fig. 4-4i, is expressed as

$$\frac{y}{KA\omega_n} = \frac{1}{\sqrt{1 - \zeta^2}} e^{-\zeta\omega_n t} \sin \omega_d t \tag{4-47}$$

The initial conditions in Eq. 4-46 can be deduced directly. Let an impulse $A\delta(t)$ be applied to a mass–spring system.

$$m\ddot{y} + c\dot{y} + ky = A\delta(t)$$

where m is the mass, c the viscous damping coefficient, k the spring constant, and y the displacement of m. Assume zero initial conditions $y(0^-) = 0$ and $\dot{y}(0^-) = 0$. When the impulse is applied at $t = 0$, the mass cannot move with zero time, that is, $y(0^+) = y(0^-) = 0$. The effect of the impulse is a momentum change. Thus

$$m[\dot{y}(0^+) - y(0^-)] = A \quad \text{or} \quad \dot{y}(0^+) = A/m$$

After the impulse is over, the equivalent system is

$$m\ddot{y} + c\dot{y} + ky = 0 \quad \text{with} \quad y(0^+) = 0 \quad \text{and} \quad \dot{y}(0^+) = A/m$$

4-4. ANALOG DATA: ERRORS AND CORRECTIONS

Analog data give the time history of a measurand, such as the temperature recording from an engine. If the response of an instrument is not instantaneous, a time lag and/or a waveform distortion are inevitable. In this section we discuss the requirements for "adequate" response, dynamic errors, and some considerations for correcting analog data.

A. Requirements for Adequate Response

The requirements for a perfect waveform reproduction without time lag is a zero-order instrument, that is, an instantaneous response with constant sensitivity K for all input amplitudes and frequencies. A time lag is equivalent to the introduction of a phase angle in the output relative to the input signal (see Sec. 2-2D). A constant sensitivity K is common for instruments, but the simultaneous requirement of zero phase angle over the frequency range is much more difficult.

The less stringent requirements of a constant sensitivity K with a time lag are used for most instruments. It will be shown in Sec. 4-8 that a time function or a transient can be expressed in the frequency domain by means of a Fourier transform, and it is possible to speak of the frequency content of a transient. Thus a constant K is equivalent to a constant-amplitude ratio for the harmonic components between the output and input. It is expedient to deduce the requirements of true waveform reproduction with time lag by means of a simple harmonic analysis.

An *amplitude distortion* is a distortion of the output waveform due to changes in the sensitivity K for different harmonic components. A *phase distortion* is the distortion of the output waveform due to phase angles introduced at the output. Assume that an input $x = x_1 + x_2$, shown in Fig. 4-14a, has two harmonic components x_1 and x_2. There is no phase distortion if the components in the output $y = y_1 + y_2$ lags the input by the same amount of time, as shown in Fig. 4-14b. A phase distortion occurs if the time lag between y_1 and x_1 is different from that between y_2 and x_2, as shown in Fig. 4-14c. Note that the same time lag does not imply the same

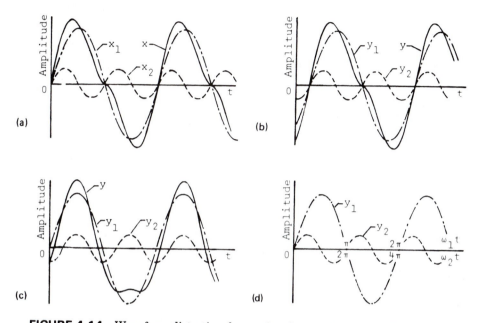

FIGURE 4-14 Waveform distortion due to time lag. (a) Input signal; $x = x_1 + x_2$. (b) Output; time lag $\phi/\omega = C$. (c) Output; time lag $\phi/\omega \neq C$. (d) Phase angle ϕ in radians.

phase angle for the components. If $\omega_2 = 2\omega_1$, a phase angle of π radians for y_1 corresponds to 2π radians for y_2, as shown in Fig. 4-14d.

For example, assume that a signal $x(t)$ has two harmonic components $x_1(t)$ and $x_2(t)$ with the frequencies ω_1 and ω_2.

$$x = x_1 + x_2$$
$$= X_1 \sin \omega_1 t + X_2 \sin \omega_2 t$$

where X_1 and X_2 are the amplitudes. The corresponding output is

$$y = y_1 + y_2$$
$$= Y_1 \sin (\omega_1 t - \phi_1) + Y_2 \sin (\omega_2 t - \phi_2)$$
$$= Y_1 \sin \omega_1 \left(t - \frac{\phi_1}{\omega_1} \right) + Y_2 \sin \omega_2 \left(t - \frac{\phi_2}{\omega_2} \right)$$
$$= Y_1 \sin \omega_1 (t - t_{d1}) + Y_2 \sin \omega_2 (t - t_{d2})$$

There is no amplitude distortion if $Y_1/X_1 = Y_2/X_2$. There is no phase distortion if each component lags by the same amount of time t_d, that is, $\phi_1/$

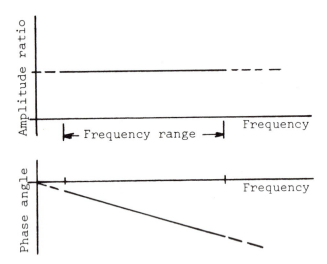

FIGURE 4-15   Criteria for waveform reproduction with time lag.

$\omega_1 = \phi_2/\omega_2 = C$, a constant. The waveforms are illustrated in Fig. 4-14a and b with $\omega_2 = 2\omega_1$.

The requirements for waveform reproduction with time lag are shown in Fig. 4-15 (see also Fig. 4-2).

1. The relative amplitude of the harmonic components in the input and output signals remain unchanged.
2. The phase angle of the harmonic components in the output is linear with frequency, that is, $\phi = C\omega$.

The criteria above are deduced from harmonic analysis. For a transient signal, the criteria must include the entire frequency range in which the Fourier transform of the transient has significant magnitude.

B. Dynamic Errors

Although criteria for true waveform reproduction with time lag can be stated, a quantitative dynamic error for measurements cannot be deduced. It cannot be stated that if the amplitude distortion, due to changes in sensitivity, is within $\pm x\%$ and the phase distortion, due to deviations from linearity of phase angle with frequency, is within $\pm y\%$, the dyanmic error will be within $\pm z\%$. It is suggested that the performance index [7] in control theory can be employed for the quantitative evaluation of dynamic errors.

A *dynamic error* is a transient, giving the deviation of the output waveform from that of the input as a time function, as shown in Fig. 4-4. The error can be observed by applying a known test signal (see Sec. 4-3A) to the instrument and noting the deviations at the output, or by comparing the output of a test instrument with that of a reference, as shown in Fig. 4-16. The observations may show the deviations but do not give a numerical value for the transient error.

The *performance index* PI is a scheme for assigning numerical values to dynamic errors. For example,

$$PI = \int_0^\infty (\text{error})\, dt \tag{4-48}$$

where error $= y/K - x$, x is the input, y the output, and K the sensitivity. The output y is normalized by K for dimensional homogeneity. The criterion above assumes that the error does not change sign, and the steady-state error is zero. Evidently, the lower the index, the better the performance of the instrument.

The index in Eq. 4-48 must be modified if the response is oscillatory and the steady-state error is not zero. When a positive and a negative error tend to cancel, it might be concluded that a highly oscillatory system will perform satisfactorily.

Many performance indices are devised to overcome the dilemma. Examples of common performance indices are

$$PI = \int_0^\infty |\text{error}|\, dt$$

$$PI = \int_0^\infty (\text{error})^2\, dt \tag{4-49}$$

$$PI = \int_0^\infty [(\text{error})^2 t]\, dt$$

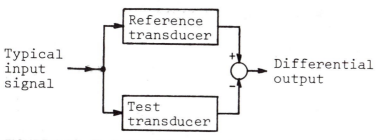

FIGURE 4-16 Evaluation of dynamic error.

The absolute error criterion penalizes a highly oscillatory system. The $(error)^2$ criterion tends to penalize the larger errors. The $[(error)^2 t]$ criterion penalizes the large errors as well as a persistent error. Evidently, the criterion employed for the PI for an application is subjective.

C. Corrections for Analog Data

It is expedient to select an instrument with adequate response as presented in the preceding section. Some considerations in correcting and interpreting analog data are discussed in this section.

Measurement is the reconstruction of the input x from the observed output y of an instrument (see Table 1-1). In theory, this can be achieved by using Eq. 4-3 in the reverse sense. The Fourier transform of the input is the quotient of the Fourier transform of the output and that of the system transfer function. Thus the input in the time domain is found by means of an inverse Fourier transform. The process can be implemented by means of a digital computer. In fact, it is not necessary for the instrument to meet the criterion of true waveform reproduction—that the phase angle vary linearly with frequency over a range of constant sensitivity K. However, the instrument has to respond fairly strongly to all frequencies present in the input. Otherwise, part of the input cannot be recovered from the output.

For multichannel systems, the unequal time lag in the channels may cause difficulty even when each channel satisfies the criterion for true waveform reproduction with time lag. Hence either the time lag in every channel is identical or suitable corrections must be applied in the data reduction.

A method for correcting the effect of the time constant in a first-order instrument is shown in Fig. 4-17. Assume that a thermocouple is calibrated in a constant-temperature furnace. It will indicate the correct temperature in the absence of temperature transients. Thus the flat portion of the response curve in the figure gives the correct temperatures.

During a startup or shutdown in an application, the thermocouple is described by the differential equation

$$\tau \frac{dT}{dt} + T = T_s(t) \tag{4-50}$$

correction

where T_s is the source temperature, such as in a gas turbine, and T the indicated temperature of the thermocouple. The correction for the transient is the product of the estimated time constant τ and dT/dt, and is added

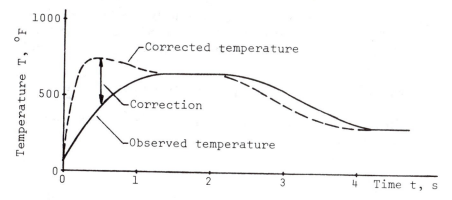

FIGURE 4-17 Correction of analog data: first-order instrument.

algebraically to the indicated temperature [8]. Note that the corrected value may greatly exceed the maximum temperature allowed for the process such as an aircraft engine.

Alternatively, compensation techniques can be applied when the primary sensor is inadequate (see Sec. 4-6C). Generally, the primary sensor is in cascade with an analog frequency-sensitive element, which continuously "reconstructs" the desired input.

An example of an instrument indicating an undershoot (negative value) for a positive pulse input is shown in Fig. 4-18a. Assume that an accelerometer with a voltage sensitivity $v_s/\ddot{x}$ is connected to an amplifier of input impedance R, as shown in Fig. 4-18b [6]. Let C be the capacitance of the accelerometer and the connecting cables. The undershoot is due to the RC time constant of the circuit.

Let us examine the undershoot from the time response of a simple RC

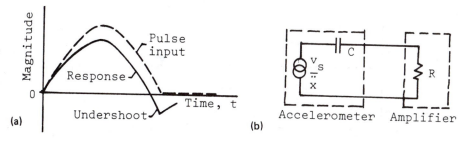

FIGURE 4-18 Undershoot due to time constant. (a) Negative output due to positive input. (b) Equivalent circuit.

circuit. A positive rectangular pulse in Fig. 4-19a is applied to the RC circuit in Fig. 4-19b. The response in Fig. 4-19c indicates the input has a negative part. The pulse is the superposition of two step functions, $Au(t)$ and $-Au(t - a)$.

1. For the input $Au(t)$ in Fig. 4-19d, the time response is $Ae^{-t/\tau}$ (solid line).
2. For the input $-Au(t - a)$ in Fig. 4-19e, the response is $-Ae^{-(t-a)/\tau}$ (solid line).

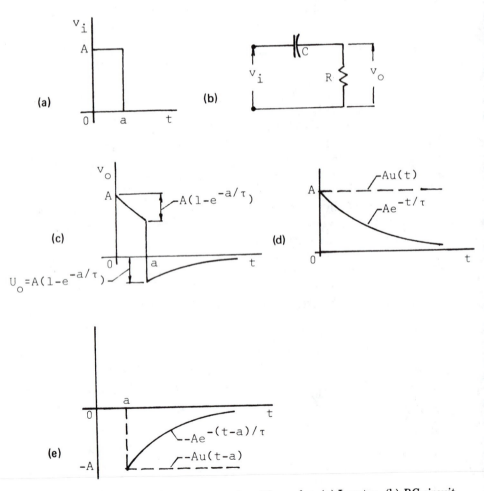

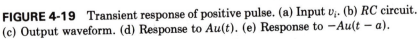

FIGURE 4-19 Transient response of positive pulse. (a) Input v_i. (b) RC circuit. (c) Output waveform. (d) Response to $Au(t)$. (e) Response to $-Au(t - a)$.

3. The response to the rectangular pulse, shown in Fig. 4-19c, is the superposition of the responses for the two step functions. It has an undershoot for a positive rectangular pulse.

A procedure for validating an undershoot for a first-order instrument is suggested by Stein [9].

1. Measure the time constant τ of the instrument and the maximum undershoot U_0 from the observed data, as shown in Fig. 4-19c.
2. Measure the pulse duration from 0 to a by means of the zero crossings. The zero crossings are determined in a like manner when the observed pulse is a smooth curve.
3. The undershoot U_0 anticipated from Fig. 4-19c is

$$U_0 = A(1 - e^{-a/\tau}) \simeq A\left(\frac{a}{\tau}\right) \qquad \text{for} \quad \frac{a}{\tau} < 0.1$$

4. If the observed undershoot $U_0 \gg A(a/\tau)$, the undershoot is a legitimate part of the data.

The speed of response of an instrument is often specified by its *rise time* to a step input. This may be defined as the time for the output to rise from 0.1 to 0.9 of the maximum value, as shown in Fig. 4-20. The observed rise time for components in cascade is [10]

$$t_{r0}^2 = t_{r1}^2 + t_{r2}^2 + \cdots$$

Let t_{r1} be the known rise time of the instrument, t_{r0} the observed value, and t_{r2} that of the process. The corrected rise time of the process is

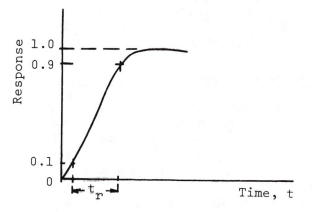

FIGURE 4-20 Rise time t_r.

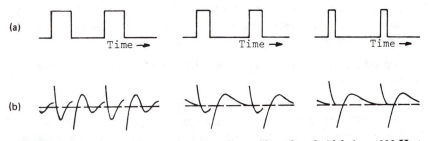

FIGURE 4-21   Output waveform of bandpass filter (bandwidth from 200 Hz to 200 kHz). (a) Pulse trains at 200 Hz. (b) Output waveforms.

$$t_{r2} = t_{r0} \left[1 - \left(\frac{t_{r1}}{t_{r0}} \right)^2 \right]^{1/2} \tag{4-51}$$

Finally, the phenomenon of "ringing" in a bandpass filter (see Fig. 3-47) is shown in Fig. 4-21. Assume that a periodic 200-Hz pulse train is fed to a filter of bandwidth from 200 Hz to 200 kHz. The Fourier series expansion of the pulse train consists of the fundamental at 200 Hz and its higher harmonics. The filter will pass the 200-Hz fundamental and harmonic components up to the 1000th harmonic. Thus the output waveform should be an approximation of the input pulse train. The input and output waveforms are as shown in the figure.

The distortions are similar to the undershoot described above. The bandpass filter has a positive slope at the 200-Hz cutoff. It will be shown that a positive slope corresponds to a differentiation (see Sec. 4-6). Since the frequency of the pulse train is at 200 Hz, the resulting waveform is due to the differentiations of the pulses. The problem is eliminated by moving the cutoff to a much lower frequency, say below 20 Hz, to avoid the differentiation. It can be demonstrated that the high-frequency cutoff of the bandpass filter has little effect on the output waveform.

4-5. LIMITATIONS OF TIME-DOMAIN ANALYSIS

Time-domain analysis suffers two disadvantages for the study of more complex systems. First, it is difficult to interpret the solution of a third- or higher-order differential equation. More important, it is difficult to relate the system behavior with its components or hardware. Note that the numerical solution of a high-order differential equation is routine by means of a digital computer.

For a first-order instrument, the time constant is defined by the parameters a_1 and a_0 in Eq. 4-13, and the a's can be related to the hardware in a system. It is possible to predict the effect of the parameters on system performance. Similarly, for a second-order instrument, the natural frequency is predictable from the equivalent mass and spring of the system. The relation for a higher-order system is not straightforward. Let us illustrate this with an example.

Example 4-5.   Repeat Example 4-4 for the galvanometer in Fig. 4-10b by including the self-inductance L of the coil in the analysis.
Solution: The torque equation remains unchanged.

$$J\ddot{\theta} + k\theta = bi$$

The loop equation from Kirchhoff's voltage law is

$$L\frac{di}{dt} + Ri + b\dot{\theta} = v(t)$$

Using the operator D (= d/dt), and eliminating the current i from the equations above, we get

$$[(JL)D^3 + (JR)D^2 + (k + b)D + (kR)]\theta = bv(t)$$

It will be difficult to interpret the effects of the parameters, such as J or R, on the overall system behavior.

4-6.   FREQUENCY RESPONSE OF INSTRUMENTS

Frequency response is a steady-state sinusoidal analysis. In this section we describe the method, discuss the elements of a transfer function, combine the elements to form a general transfer function, and briefly discuss system compensation.

Let us first describe the method. A frequency response test is performed by applying a sinusoidal input $x(t) = X \sin \omega t$ to a linear instrument G, shown in Fig. 4-22a. The output y is sinusoidal at the input frequency. From Eq. 4-7, the input/output relation is described by the equations

Transfer function: $\dfrac{y}{x}(j\omega) = G(j\omega)$ (4-52)

Amplitude ratio: $\dfrac{Y}{X}(j\omega) = |G(j\omega)|$ (4-53)

Phase angle: $\phi(j\omega) = /G(j\omega)$ (4-54)

The amplitude ratio and the phase angle are calculated for a particular input frequency ω. This gives one data point. The test is repeated over a frequency range to obtain the frequency response data. Generally, the test

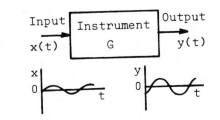

(a)

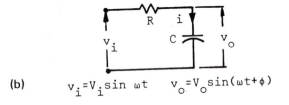

(b) $v_i = V_i \sin \omega t$ $v_o = V_o \sin(\omega t + \phi)$

(c)

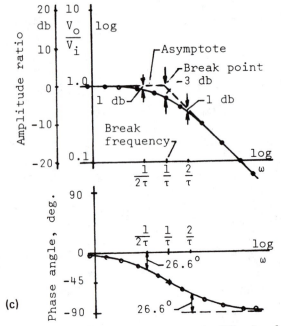

FIGURE 4-22   Frequency response of a RC network. (a) Test instrument. (b) Example of frequency response test. (c) Bode plots.

is mechanized in a *sweep-sine test,* in which the frequency is ranged and the amplitude ratio and phase angle are computed automatically.

For example, let the test be performed on the RC network shown in Fig. 4-22b. The data can be presented in two plots: (1) a log amplitude ratio versus log frequency plot, and (2) a phase angle versus log frequency plot, as shown in Fig. 4-22c. These are called *Bode plots.* Other methods for presenting the frequency response data are not discussed.

The amplitude ratio is commonly expressed in decibels (dB):

$$\text{decibel of a ratio, dB} = 20 \log_{10}(\text{ratio}) \qquad (4\text{-}55)$$

For example, the ratio of 1:1 is 0 dB, since $\log_{10}(1:1) = 0$. The ratio of 100:1 is 40 dB, since $20 \times \log_{10}(100:1) = 40$. The conversion from a ratio to dB shown in Fig. 4-23 is a semilog plot. The decibel notation originated from communication. If the power dissipation in a resistor R due to a current I_1 is $P_1 = I_1^2 R$, and that due to I_2 is $P_2 = I_2^2 R$, the power ratio is

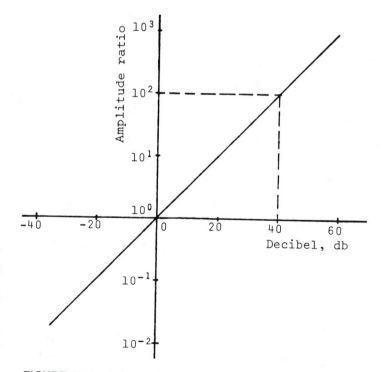

FIGURE 4-23 Ratio-to-decibel conversion.

$$\frac{P_2}{P_1} = \frac{I_2^2 R}{I_1^2 R} \quad \text{or} \quad \log_{10}\frac{P_2}{P_1} = 2 \times \log_{10}\frac{I_2}{I_1}$$

where $[2 \log_{10}(I_2/I_1)]$ is the bel, a tenth of which is a decibel.

Specifications are commonly expressed in decibels. For example, a voltage amplifier with an amplification of 1000:1 is said to have a gain of 60 dB. Note that ± 1 dB is not synonymous with $\pm 1\%$. In fact, $+1$ dB is a ratio of 1.12:1, and -1 dB a ratio of 0.89:1. If a specification states that the gain is 3 dB down at a certain frequency, the amplitude of the response attenuates by about 30% and not 3% at that frequency.

A. Elementary Transfer Functions

The Bode plots of elementary transfer functions and their interpretations are described in this section. The general transfer function in Eq. 4-7 is a quotient of two polynomials in $j\omega$. The numerator and the denominator can be factorized into elementary transfer functions $G(j\omega)$ with real coefficients.

Case 1. $G(j\omega) = K$, a constant
Case 2. $G(j\omega) = (j\omega)^{\pm 1}$
Case 3. $G(j\omega) = (1 + j\omega\tau)^{\pm 1}$
Case 4. $G(j\omega) = [(j\omega/\omega_n)^2 + 2\zeta(j\omega/\omega_n) + 1]^{\pm 1}$

The ($\pm$) signs indicate that the factors can be in the numerator or the denominator of a general transfer function.

1. Bode Plots

The plots of the elementary transfer functions above are summarized in Figs. 4-24 and 4-25. The Bode plots present the data in two plots: (1) the magnitude plot of $\log|G(j\omega)|$ versus $\log \omega$, and (2) the phase plot of $\underline{/G(j\omega)}$ versus $\log \omega$.

Case 1. $G(j\omega) = K$, a real constant. Since K is a constant, $G(j\omega) = K$ and $\underline{/G(j\omega)} = 0$, as shown in Fig. 4-24. K is the gain of an instrument, although its value can be greater or smaller than unity. The transfer function represents a zero-order instrument.

Case 2a. $G(j\omega) = 1/j\omega$. When $G(j\omega) = 1/j\omega$, we get $|G(j\omega)| = |1/j\omega| = 1/\omega$.

$$|G(j\omega)| = \omega^{-1} \quad \text{or} \quad \log|G(j\omega)| = -1 \log \omega$$

This is a straight line with a -1 slope in a log-log plot. Since $|G(j\omega)| = 1$ for $\omega = 1$, the line passes through the (1, (w = 1, |G| = 1) 1) point in the magnitude plot with a slope of -1, as shown in Fig. 4-25a. Evidently, the phase angle $\underline{/G(j\omega)}$ due to $1/j$ is $-90°$.

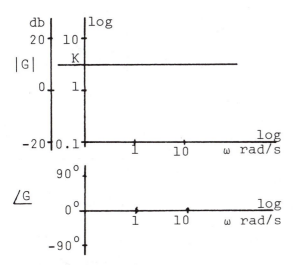

FIGURE 4-24   Bode plots of $G(j\omega) = K$.

Case 2b. $G(j\omega) = j\omega$. When $G(j\omega) = j\omega$, we get $|G(j\omega)| = |j\omega| = \omega^{+1}$.

$$|G(j\omega)| = \omega^{+1} \quad \text{or} \quad \log|G(j\omega)| = +1 \log \omega$$

This gives a straight line with a +1 slope in a log-log plot. Since $|G(j\omega)| = 1$ when $\omega = 1$, the line passes through the (1,1) point in the magnitude plot with a +1 slope, as shown in Fig. 4-25b. The phase angle due to j is $+90°$.

Note the similarity of the plots in Fig. 4-25a and b. The generalizations below are true for all the cases to follow.

1. The magnitude plots of the transfer functions in cases 2a and 2b are the mirror images about a horizontal line at $|G(j\omega)| = 1$.
2. The phase plots of the transfer functions in cases 2a and 2b are the mirror images about the zero-degree line.

Case 3a. $G(j\omega) = (1 + j\omega\tau)^{-1}$. The function $G(j\omega) = (1 + j\omega\tau)^{-1}$ is the transfer function of a first-order instrument (see Eq. 4-3). The Bode plots of $G(j\omega)$ in Fig. 4-25c are considered in two parts: when $\omega \ll 1$ and when $\omega \gg 1$.

1. For $\omega \ll 1$, we get $G(j\omega) = 1/1$. This is case 1 with $K = 1$. It is the low-frequency asymptote with $|G(j\omega)| = 1$ and $\underline{/G(j\omega)} = 0°$, as shown in the figure.
2. For $\omega \gg 1$, we get $G(j\omega) = 1/j\omega\tau$. Thus

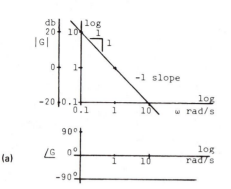

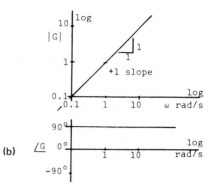

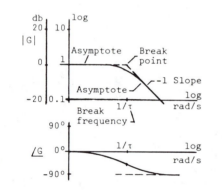

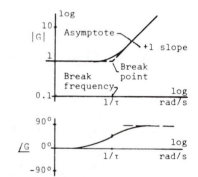

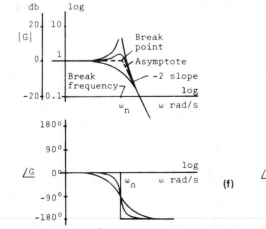

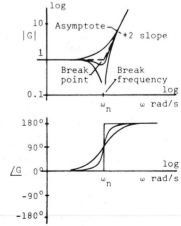

FIGURE 4-25 Bode plots of elementary transfer functions. (a) $G(j\omega) = 1/j\omega$. (b) $G(j\omega) = j\omega$. (c) $G(j\omega) = 1/(1 + j\omega\tau)$. (d) $G(j\omega) = (1 + j\omega\tau)$. (e) $G(j\omega) = 1/[(j\omega/\omega_n)^2 + 2\zeta(j\omega/\omega_n) + 1]$. (f) $G(j\omega) = (j\omega/\omega_n)^2 + 2\zeta(j\omega/\omega_n) + 1$.

$$|G(j\omega)| = \left|\frac{1}{j\omega\tau}\right| = \left(\frac{1}{\omega\tau}\right)^{-1} \quad \text{or} \quad \log|G(j\omega)| = -1\log\omega + C$$

where $C = \log(1/\tau)$. This is the equation of a straight line and the high-frequency asymptote in the magnitude plot. Since $|G(j\omega)| = 1$ when $\omega = 1/\tau$, the line passes through the $(1,1/\tau(|G| = 1, \omega = 1/\tau)$ point with a -1 slope. The asymptote of the phase angle due to $1/j$ is $-90°$.

Generally, only the high- and low-frequency asymptotes are used in the *magnitude plot*, shown in Fig. 4-25c. The asymptotes intersect at the break point at $(1,1/\tau)$. The *break frequency* is $\omega = 1/\tau$. The intermediate points can be obtained by substituting values of ω into the transfer function. It is expedient, however, to plot the asymptotes and then apply corrections, as illustrated in Fig. 4-22c. For example, the asymptotes can be sketched in the magnitude plot and corrections obtained for three intermediate points as described below.

1. When $\omega = 1/\tau$, $|G(j\omega)| = |1/(1 + j)| = 1/\sqrt{2}$, which is -3 dB from the low-frequency asymptote.
2. When $\omega = \frac{1}{2}\tau$, $|G(j\omega)| = |1/(1 + j0.5)| = 0.89$, which is -1 dB from the low-frequency asymptote.
3. When $\omega = 2/\tau$, $|G(j\omega)| = |1/(1 + j2)| = 1/\sqrt{5}$, or -7 dB. The high-frequency asymptote has a -1 slope; that is, attenuates by a ratio of 2:1 or -6 dB for a frequency ratio of 2:1. Since the high-frequency asymptote decreases by -6 dB for the freqeuncy from $1/\tau$ to $2/\tau$ and the actual change is -7 dB for the same frequency range, the correction is -1 dB from the high-frequency asymptote, as shown in Fig. 4-22c.

The *phase plot* can also be sketched from the $0°$ and $-90°$ asymptotes and the three intermediate points, shown in Fig. 4-22c.

1. When $\omega = 1/\tau$, $\underline{/G(j\omega)} = \underline{/1/(1 + j)} = -45°$.
2. When $\omega = \frac{1}{2}\tau$, $\underline{/G(j\omega)} = \underline{/1/(1 + j0.5)} = -26.6°$.
3. When $\omega = 2/\tau$, $\underline{/G(j\omega)} = \underline{/1/(1 + j2)} = -63.4°$, which is $26.6°$ from the $-90°$ line, as shown in Fig. 4-22c.

Case 3b. $G(j\omega) = 1 + j\omega\tau$. The Bode plots for $G(j\omega) = (1 + j\omega\tau)$ are the mirror images of those for case 3a and are shown in Fig. 4-25d.

Case 4a. $G(j\omega) = [(j\omega/\omega_n)^2 + 2\zeta(j\omega/\omega_n) + 1]^{-1}$. The function $G(j\omega) = [(j\omega/\omega_n)^2 + 2\zeta(j\omega/\omega_n) + 1]^{-1}$ is the transfer function of a second-order instrument (see Eq. 4-3). The Bode plots can be considered in two parts: (1) when $\omega \ll \omega_n$, and (2) when $\omega \gg \omega_n$.

When $\omega \ll \omega_n$, we have $(\omega/\omega_n)^2 \ll (\omega/\omega_n) \ll 1$ and $G(j\omega) \simeq 1$. Again this is case 1 with $K = 1$. The low-frequency asymptote has a value of 1 in the magnitude plot and zero phase angle in the phase plot, as shown in Fig. 4-25e.

When $\omega \gg \omega_n$, $(\omega/\omega_n)^2 \gg (\omega/\omega_n) \gg 1$, and $|G(j\omega)| \simeq 1/(\omega/\omega_n)^2$.

$$|G(j\omega)| = \omega^{-2}\omega_n^2 \quad \text{or} \quad \log|G(j\omega)| = -2\log\omega + C$$

where $C = \log (\omega_n)^2$. This gives a straight line and the high-frequency asymptote in the magnitude plot. Since $|G(j\omega)| = 1$ when $\omega = \omega_n$, the line has a -2 slope and passes the $[\omega = \omega_n, |G(j\omega)| = 1]$ point, as shown in Fig. 3-25e. The phase angle for $1/j^2$ is $-180°$.

The intersection of the high- and low-frequency asymptotes in the magnitude plot gives the breakpoint. The corresponding break frequency is at $\omega = \omega_n$. The asymptotes are generally used for the magnitude plot. Resonance occurs at $\omega = \omega_n$, and the phase angle at resonance is always $-90°$. An approximate phase plot can be sketched from the $0°$ and $-180°$ lines through the $-90°$ point.

The actual values can be calculated by substituting values of ω in the transfer function $G(j\omega)$ with ζ as a parameter. This can be programmed readily in a hand-held calculator. Alternatively, graphs for the magnitude and phase corrections are available. These graphs are not presented because the corrections are seldom used for Bode plots.

Case 4b. $G(j\omega) = (j\omega/\omega_n)^2 + 2\zeta(j\omega/\omega_n) + 1$. The Bode plots of this transfer function are the mirror images of case 4a about the appropriate axes and will not be discussed.

In summary, (1) the frequency response data of $G(j\omega)$ are shown in two plots, in which the magnitude and phase plots are symmetrical about the appropriate lines, (2) the asymptotic magnitude plot and a sketch of the phase plot are sufficient for the data presentation and interpretation, and (3) the actual values of the plots can be obtained by substituting values of ω in the transfer functions or applying corrections to the asymptotes.

2. Interpretations

In this section we interpret the elementary transfer functions and relate them to the types of instruments described in Eqs. 4-12 to 4-14. Let us illustrate this with examples.

Example 4-6.   (a) Derive the differential equation for the circuit in Fig. 4-26a. (b) Show that the circuit can be used as an integrator. (c) Derive its sinusoidal transfer function. (d) Show the frequency range in which the circuit can be used as an integrator.

Solution: (a) The equations from the RC circuit are

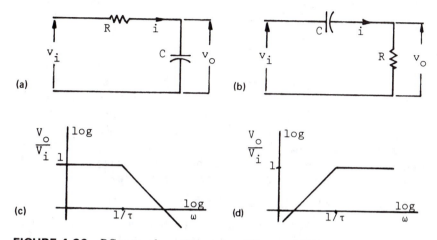

FIGURE 4-26 RC networks as integrator, filter, and differentiator.

$$Ri + \frac{1}{C} \int i \, dt = v_i(t) \tag{4-56}$$

$$v_o = \frac{1}{C} \int i \, dt \quad \text{and} \quad i = C \frac{dv_o}{dt}$$

$$\tau \frac{dv_o}{dt} + v_o = v_i(t) \quad \text{or} \quad (\tau D + 1)v_o = v_i(t) \tag{4-57}$$

where $\tau = RC$. The circuit describes a first-order instrument shown in Eq. 4-13.

(b) When $\tau \gg 1$, the equation becomes

$$\tau \frac{dv_o}{dt} \simeq v_i(t) \quad \text{or} \quad v_o \simeq \frac{1}{\tau} \int v_i(t) \, dt \tag{4-58}$$

The deduction is that when τ or RC is large, the circuit is an integrator. For example, if the input $v_i(t)$ is a step function $V_i[u(t)]$, the solution v_o from Eq. 4-57 is

$$v_o = V_i(1 - e^{-t/\tau})$$

$$\simeq V_i\left[1 - \left(1 - \frac{t}{\tau}\right)\right] = V_i\left(\frac{t}{\tau}\right) \quad \text{for} \quad \tau \gg 1$$

Since v_o increases linearly with time for a constant input V_i, v_o is proportional to the integration of the input. This property can be deduced more readily from the frequency domain analysis.

(c) The components R and C in Fig. 4-26a form a voltage divider for $v_i(t)$. Since $Z_C = 1/j\omega C$, the transfer function is

$$\frac{V_o}{V_i} = \frac{Z_C}{R + Z_C} \quad \text{or} \quad \frac{V_o}{V_i} = \frac{1}{1 + j\omega\tau} \quad \text{where} \quad \tau = RC \quad (4\text{-}59)$$

(d) The transfer function is identical to that in case 3a above, and the magnitude plot is shown in Fig. 4-26c. Note that $j\omega$ can be substituted for $D \,(= d/dt)$ and $1/j\omega$ for $1/D \,(= \int dt)$ for steady-state sinusoidal analysis. Hence the transfer function $G(j\omega) = 1/j\omega$ in case 2a denotes an integrator. The characteristics shown in Fig. 4-26b becomes an integrator when the input frequency is much greater than the break frequency at $\omega = 1/\tau$. In other words, when $\tau \gg 1$, the circuit becomes an integrator.

The deductions from Example 4-6 are as follows:

1. The circuit represents a first-order instrument, either in the differential equation form in Eq. 4-57 or in the transfer function form in Eq. 4-59. In fact, the transfer function can be derived directly from the differential equation by substituting $j\omega$ for $D \,(= d/dt)$. The system is first-order because it consists only one type of energy storage element and an energy dissipation element.

2. Physically, the circuit is an integrator because v_o is measured across C, which is a storage of electric charge.

3. The magnitude plot in Fig. 4-26c shows that the circuit is a low-pass filter with the break frequency at $\omega = 1/\tau$. When the input frequency is less than $1/\tau$, the response is flat, denoting a zero-order instrument. In other words, the impedance $1/j\omega C$ is large for the low-frequency range and C has no effect for low input frequencies. When the input frequency is greater than $1/\tau$, C becomes effective and the transfer function can be approximated by $G(j\omega) = 1/j\omega\tau$, which is an integrator.

Example 4-7. (a) Derive the differential equation for the circuit shown in Fig. 4-26b. (b) Show that the circuit can be used as a differentiator. (c) Derive its sinusoidal transfer function. (d) Indicate the freqeuncy range for which the circuit can be used as a differentiator.

Solution: (a) The equations from the RC circuit are

$$Ri + \frac{1}{C}\int i\,dt = v_i(t)$$

$$v_o = Ri \qquad i = \frac{1}{R}v_o \qquad \frac{1}{C}\int i\,dt = \frac{1}{\tau}\int v_o\,dt$$

$$v_o + \frac{1}{\tau}\int v_o\,dt = v_i(t) \qquad \text{or} \qquad \tau\frac{dv_o}{dt} + v_o = \tau\frac{dv_i}{dt} \qquad (4\text{-}60)$$

where $\tau = RC$. Note that Eq. 4-60 is in the form of Eq. 4-4.

(b) When $\tau \ll 1$, the circuit becomes a differentiator, as deduced from Eq. 4-60.

$$v_o \simeq \tau \frac{dv_i}{dt} \tag{4-61}$$

(c) The components in the circuit form a voltage divider for $v_i(t)$. Since $Z_C = 1/j\omega C$, the transfer function is

$$\frac{V_o}{V_i} = \frac{R}{R + Z_C} \quad \text{or} \quad \frac{V_o}{V_i} = \frac{j\omega\tau}{1 + j\omega\tau} \tag{4-62}$$

(d) The magnitude plot of the transfer function is shown in Fig. 4-26d. Since $j\omega$ in the frequency domain corresponds to D ($= d/dt$) in the time domain, the transfer function $G(j\omega) = j\omega$ represents a differentiation and its magnitude plot has a slope of $+1$, as shown in case 2b above. Thus, when ω is small, or the input frequency is less than the break frequency at $\omega = 1/\tau$, the circuit becomes a differentiator.

The deductions from Example 4-7 are as follows:

1. The circuit represents a first-order instrument, either in the differential equation form in Eq. 4-60 or the transfer function form in Eq. 4-62. In fact, the transfer function can be derived directly from the differential equation by substituting $j\omega$ for D ($= d/dt$). The system is first-order because it consists of only one type of energy storage element and an energy dissipation element.
2. Physically, the circuit is a differentiator because it is not possible to change the energy stored in the capacitor C with zero time; that is, the voltage across C does not change at time zero. To this end, C acts as a short circuit. For a step input, the entire input voltage $v_i(t)$ appears at v_o across R at time zero. Then v_o decreases exponentially. If $\tau = RC$ is small, v_o decreases rapidly and its waveform resembles a spike or a differentiation of $v_i(t)$.
3. The magnitude plot in Fig. 4-26d shows that the circuit is a high-pass filter. The flat portion of the plot shows that the circuit is a zero-order instrument; that is, the input frequency is high and C acts as a short circuit. At low input frequencies or when $\omega = 1/\tau$ is large, the input operates at the range below $1/\tau$ and on the $+1$ slope portion of the response curve. Thus the circuit acts as a differentiator for the low-frequency range.

Note that differentiation is noise amplifying. A differentiation will greatly magnify the irregularities in the signal to give a large noise component in the output. In contrast, an integration tends to smooth out the irregularities in a signal. Since integration is a summing process, the contribution of the irregularities to the sum is minimal.

In summary, the characteristics of an instrument depend on its operating frequency range.

1. A zero slope in the magnitude plot gives a zero phase angle, as shown in Fig. 4-25. This is the characteristics of a zero-order instrument. When an instrument has a flat response for a certain frequency range, the instrument is zero-order for that range of operation.

2. A −1 slope is associated with a −90° phase angle, as shown in Fig. 4-25a. This gives an integrator. A +1 slope is associated with a +90° phase angle, as shown in Fig. 4-25b. This gives a differentiator. Again, the behavior is contingent on the frequency range of operation.

3. Similarly, a −2 slope is associated with a −180° phase angle, as shown in Fig. 4-25e. This gives a double integration. For example, a displacement can be obtained from the double integration of an acceleration, and a −2 slope is associated with an inertia force. Evidently, a −3 slope is associated with a −270° phase angle, and so on.

4. The behavior of a system can be estimated from its asymptotic magnitude plot. For example, the combination of a flat response and a −1 slope at the higher frequencies gives the characteristics of a first-order instrument or a low-pass filter. The combination of a flat response and a −2 slope at the higher frequencies give the characteristics of a second-order instrument. The Bode plots of a general transfer function are described below.

B. Bode Plots of General Transfer Functions

In this section we show the procedure for obtaining the Bode plots of a general transfer function, such as shown in Eq. 4-7 in which the numerator and the denominator are polynomials in $(j\omega)$. The procedure is (1) to factorize the numerator and the denominator into factors or elementary transfer functions, (2) to plot each of the factors as shown in Figs. 4-24 and 4-25, and (3) to recombine the factors.

Transfer functions are complex numbers at a given frequency ω. The plotting of the elementary transfer functions and their recombination follows the rules of products of complex numbers. For the complex numbers A and B, the rules are

$$|AB| = |A| \cdot |B| \tag{4-63}$$

$$\underline{/AB} = \underline{/A} + \underline{/B} \tag{4-64}$$

The magnitude of the product of two complex numbers is the product of the individual magnitudes, and the phase angle of the product is the algebraic sum of the individual phase angles.

Example 4-8.   Obtain the asymptotic log magnitude versus log ω and the phase angle versus log ω plots of the transfer function

$$G(j\omega) = \frac{3.5(1 + j0.5\omega)}{j\omega(1 + j0.125\omega)}$$

Solution: Expressing $G(j\omega)$ in the product form gives

$$G(j\omega) = (3.5)\left(\frac{1}{j\omega}\right)(1 + j0.5\omega)\left(\frac{1}{1 + j0.125\omega}\right)$$

$$K \quad G_1(j\omega) \qquad G_2(j\omega) \qquad G_3(j\omega)$$

The elementary transfer functions $G_1(j\omega)$ to $G_3(j\omega)$ are shown in Fig. 4-27. The break frequencies are shown below.

$$G_1(j\omega) = \frac{1}{j\omega} \qquad \text{(see case 2a)}$$

$$G_2(j\omega) = (1 + j0.5\omega) \quad \text{with} \quad \omega = \frac{1}{\tau} = \frac{1}{0.5} = 2 \qquad \text{(see case 3b)}$$

$$G_3(j\omega) = \frac{1}{1 + j0.125\omega} \quad \text{with} \quad \omega = \frac{1}{\tau} = \frac{1}{0.125} = 8 \qquad \text{(see case 3a)}$$

Identify the frequency ranges $0 < \omega < 2$, $2 < \omega < 8$, and $\omega > 8$ from the elementary transfer functions. Define $G_c(j\omega) = G(j\omega)/K$ for convenience. For a given frequency ω, the terms in $G(j\omega)$ are complex numbers. Omitting $j\omega$ in the equations for ease of writing and using the rules of complex numbers above, we obtain

$$|G_c| = |G_1| \cdot |G_2| \cdot |G_3|$$
$$\log|G_c| = \log|G_1| + \log|G_2| + \log|G_3|$$

To obtain the asymptotic magnitude plot, we have:

1. For $0 < \omega < 2$, $\log|G_c| = \log|G_1| + 0 + 0$, since $\log(1.0) = 0$. This states that $|G_c| = |G_1| \cdot 1 \cdot 1 = |G_1|$. Hence the composite curve $|G_c|$ follows the $|G_1|$ plot.
2. For $2 < \omega < 8$, $\log|G_c| = \log|G_1| + \log|G_2| + 0$. This states that $|G_c| = |G_1| \cdot |G_2| \cdot 1 = |G_1| \cdot |G_2|$. The values of $|G_c|$ can be obtained graphically from $|G_1|$ and $|G_2|$ point by point. A simple procedure is to utilize the slopes of the individual plots. For the range $2 < \omega < 8$, the

slope of $|G_1|$ is -1, that of $|G_2|$ is $+1$, and that of $|G_3|$ is 0. Thus the slope of $|G_c|$ is $(-1 + 1 + 0) = 0$, and $|G_c|$ is a horizontal line. It starts from the endpoint of the previous range at $\omega = 2$ and terminates at $\omega = 8$.

3. For $\omega > 8$, the slopes of $|G_1|$, $|G_2|$, and $|G_3|$ are -1, $+1$, and -1, respectively. Since the combined slope is $(-1 + 1 - 1) = -1$, $|G_c|$ has a slope of -1, starting from the endpoint of the previous range at $\omega = 8$ and continuing to the higher frequencies.

4. Now, $|G_c|$ is multiplied by $K = 3.5$ to get $|G(j\omega)|$. Select a convenient point, such as $|G_c| = 1$, as shown in Fig. 4-27, and move the entire curve $|G_c|$ vertically by the amount $K = 3.5$. Since a distance along a logarithmic scale denotes a ratio, the procedure is equivalent to multiplying $|G_c|$ by a ratio of 3.5:1.

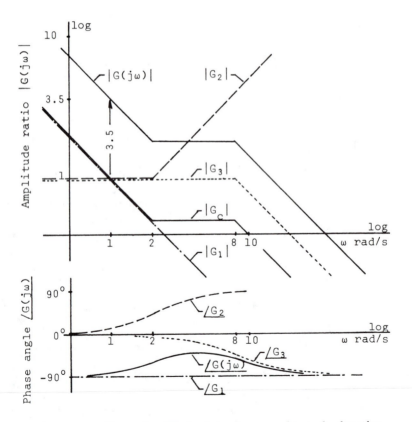

FIGURE 4-27 Example of Bode plots of a general transfer function.

The phase plots of the elementary transfer functions are shown in Fig. 4-27. The composite phase plot is obtained by means of the algebraic sum:

$$\underline{/G(j\omega)} = \underline{/G_1(j\omega)} + \underline{/G_2(j\omega)} + \underline{/G_3(j\omega)}$$

C. Dynamic Compensation

The use of hardware to improve the performance of instruments is a broad study. In this section we briefly describe applications of compensation for instruments.

Assume that an instrument and its compensator are in cascade for the *series compensation* [11]. The general scheme for compensation is shown in Fig. 4-28, where $G_1(D)$ is the transfer function of the instrument, $G_2(D)$ that of the compensator, and $D = d/dt$ that of an operator. Assuming zero loading, the input/output relation (see Eq. 4-10) is

$$y_2 = [G_1(D)G_2(D)]x_1 \tag{4-65}$$

If the compensator $G_2(D)$ has the transfer function

$$G_2(D) = \frac{K}{G_1(D)} \tag{4-66}$$

the compensated system is zero-order or a perfect instrument.

$$y_2 = G_1(D)\left[\frac{K}{G_1(D)}\right]x_1 \quad \text{or} \quad y_2 = Kx_1 \tag{4-67}$$

Unfortunately, the compensator in Eq. 4-66 is not realizable (see Sec. 4-2).

A practical compensator has the transfer function

$$G_2(D) = \frac{KG_c(D)}{G_1(D)} \tag{4-68}$$

Substituting this in Eq. 4-65 and canceling the $G_1(D)$ term gives

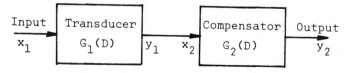

FIGURE 4-28 Series compensation.

$$y_2 = G_1(D) \frac{KG_c(D)}{G_1(D)} x_1 \quad \text{or} \quad y_2 = KG_c(D)x_1 \qquad (4\text{-}69)$$

where $[KG_c(D)]$ is the transfer function of the compenstated instrument with improved characteristics. Let us illustrate the compensation of first-order instruments with an example.

Example 4-9.   The compensating network of a thermocouple is shown in Fig. 4-29a. The differential equation and the transfer function of the thermocouple are

$$(\tau_1 D + 1)V_1 = K_1 Tu(t) \quad \text{and} \quad \frac{V_1}{T}(j\omega) = \frac{K_1}{1 + j\omega\tau_1} = G_1(j\omega)$$
$$(4\text{-}70)$$

where $Tu(t)$ is a step temperature input, $\tau_1 = 0.1$ s the time constant, and v_1 the output voltage. The uncompensated steady-state output voltage is 7 mV. (a) Derive the transfer function of the compensator shown in Fig. 4-29a. (b) Assume that $R = 100$ kΩ and $C = 1$ µF. Find R_2 for reducing the time constant of the compensated system by a factor of 5. (c) Sketch the time response of the thermocouple before and after the compensation. (d) Sketch the corresponding Bode plots.

Solution: (a) The impedance of R and C in parallel is Z.

$$Z = \frac{RZ_C}{R + Z_C} = \frac{R(1/j\omega C)}{R + 1/j\omega C} = \frac{R}{1 + j\omega RC} = \frac{R}{1 + j\omega\tau_2}$$

where $\tau_2 = RC$. R_2 and Z form a voltage divider for V_1. Hence the transfer function of the compensator is

$$\frac{V_2}{V_1} = \frac{R_2}{R_2 + Z} = \frac{R_2}{R_2 + R/(1 + j\omega\tau_2)} = \frac{R_2(1 + j\omega\tau_2)}{R_2 + R + j\omega R_2\tau_2} \qquad (4\text{-}71)$$

$$\frac{V_2}{V_1} = \alpha \frac{1 + j\omega\tau_2}{1 + j\omega\alpha\tau_2} = G_2(j\omega)$$

where $\alpha = R_2/(R + R_2) < 1$ and $\alpha\tau_2 < \tau_2$.

(b) The overall transfer function of the compensated system is obtained from direct application of Eq. 4-65.

$$\frac{V_2}{T}(j\omega) = G_1(j\omega)G_2(j\omega) = \frac{K_1}{1 + j\omega\tau_1}\left(\alpha\frac{1 + j\omega\tau_2}{1 + j\omega\alpha\tau_2}\right)$$
$$(4\text{-}72)$$

$$\frac{V_2}{T}(j\omega) = \frac{\alpha K_1}{1 + j\omega\alpha\tau_1} = KG_c(j\omega) \quad \text{or} \quad \tau_2 = \tau_1$$

The time constant of the compensated system is $\alpha\tau_1 = \tau_1/5$. Since $\alpha = 1/5$, the value of R_2 is

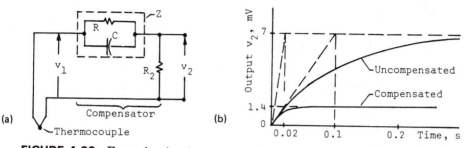

FIGURE 4-29 Example of series compensation. (a) Thermocouple with compensator. (b) Step response of thermocouple.

$$\alpha = \frac{R_2}{R + R_2} = \frac{R_2}{10^5 + R_2} = \frac{1}{5} \quad \text{or} \quad R_2 = 25 \times 10^3 \ \Omega$$

(c) The solution of the differential equation for the uncompensated system from Eq. 4-70 is

$$V_1 = K_1 T(1 - e^{-t/\tau_1})$$

where $K_1 T = 7$ mV. For the compensated system in Eq. 4-72, the differential equation and its solution are

$$(\alpha \tau_1 D + 1)V_2 = \alpha K_1 T u(t) \quad \text{and} \quad V_2 = \alpha K_1 T(1 - e^{-t/\alpha \tau_1})$$

where $\alpha K_1 T = (1/5)(7) = 1.4$ mV and $\alpha \tau_1 = 0.02$ s. The time response plots for the step input $T u(t)$ are shown in Fig. 4-29b. The initial slopes are used to show the τ's in the figure.

(d) The normalized asymptotic magnitude plots (dashed lines) of the uncompensated thermocouple $G_1(j\omega)$, the compensator $G_2(j\omega)$, and the compensated thermocouple $KG_c(j\omega)$ are shown in Fig. 4-30a. The phase plots are shown in Fig. 4-30b.

The following observations are made from the example:

1. The compensation improves the performance of the thermocouple by decreasing its time constant by a factor of 5 ($= 1/\alpha$). The penalty is a loss of sensitivity by a factor of 5. This may not be critical, because the signal can be amplified. An amplifier is usually placed between the transducer and the compensator to minimize loading. Due to the inherent noise in measurements, this technique cannot be carried on indefinitely, although an increase in the speed of response by a factor of 100 is possible.

2. The purpose of compensation is to reshape the transfer function in

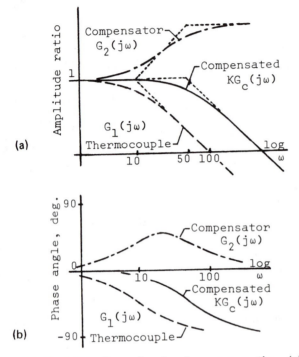

FIGURE 4-30 Example of series compensation. (a) Normalized amplitude ratio. (b) Phase angle.

order to obtain a "flat" response over the operating frequency range, as shown in Fig. 4-30a (see Sec. 4-6A). For example, the compensator extends the break frequency in the magnitude plot from 10 to 50 rad/s, and the phase shift is improved correspondingly. Thus, in addition to the shorter time constant (faster response), the compensated thermocouple has less distortions.

3. Since the time constant of the thermocouple is not time invariant, a mismatch between the compensator and the thermocouple will not yield the results predicted. The effect of the mismatch is left as an exercise.

The compensation for other instruments follows the same general principle. For example, the block diagram of a hydraulic servo valve is shown in Fig. 4-31a. A voltage is applied to the torque motor to actuate the valve. Assume that the valve has a drooping characteristic (dashed line) shown in Fig. 4-31b. The torque motor is designed to resonate (dashed-dotted

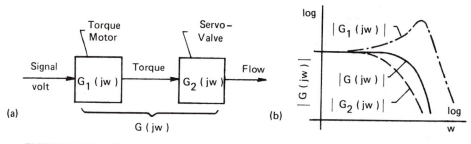

FIGURE 4-31   Compensation of servo-valve. (a) Hydraulic servo-valve. (b) Response characteristics.

line) at a frequency above the cutoff frequency of the valve. Thus the useful frequency range (solid line) of the assemble has extended beyond that of the valve itself.

Example 4-10.   The force transducer shown in Fig. 1-3 is lightly damped and the usable range is below its resonant frequency. The transfer function of the transducer is

$$\frac{V_1}{F}(j\omega) = \frac{K}{1 - (\omega/\omega_n)^2 + 2\zeta(j\omega/\omega_n)} = G_1(j\omega) \qquad (4\text{-}73)$$

Assume that the undamped natural frequency is $\omega_n = 10^4$, damping ratio $\zeta = 0.1$, and $K = 10^3$, and the units are consistent. Design the compensator shown in Fig. 4-32 for $\zeta = 0.6$.

Solution: Since R_1 and Z in the compensating network Fig. 4-32b form a voltage divider for v_1, we have

$$\frac{V_2}{V_1}(j\omega) = \frac{Z}{R_1 + Z} \qquad \text{and} \qquad Z = R + j\omega L + \frac{1}{j\omega C}$$

$$\frac{V_2}{V_1}(j\omega) = \frac{(R + j\omega L + 1/j\omega C)}{R_1 + (R + j\omega L + 1/j\omega C)} = \frac{1 - \omega^2 LC + j\omega RC}{1 - \omega^2 LC + j\omega(R_1 + R)C}$$

Defining $LC = 1/\omega_n^2$, $RC = 2\zeta_1/\omega_n$, $(R_1 + R)C = 2\zeta_2/\omega_n$, and $G_2(j\omega)$ the transfer function of the compensator, we obtain

$$\frac{V_2}{V_1}(j\omega) = \frac{1 - (\omega/\omega_n)^2 + 2\zeta_1(j\omega/\omega_n)}{1 - (\omega/\omega_n)^2 + 2\zeta_2(j\omega/\omega_n)} = G_2(j\omega)$$

$$\zeta_1 = \frac{1}{2} R\left(\frac{C}{L}\right)^{1/2} \qquad \text{and} \qquad \zeta_2 = \frac{1}{2}(R_1 + R)\left(\frac{C}{L}\right)^{1/2} \qquad (4\text{-}74)$$

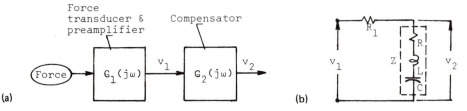

(a)

(b)

FIGURE 4-32 Compensation of second-order instrument. (a) Compensated force transducer. (b) Compensating network.

Assuming that $\zeta_1 = \zeta = 0.1$, and substituting Eqs. 4-73 and 4-74 into Eq. 4-65, the transfer function of the compensated system is

$$\frac{V_2}{F}(j\omega) = \frac{K}{1 - (\omega/\omega_n)^2 + 2\zeta_2(j\omega/\omega_n)} = KG_c(j\omega)$$

Let $L = 0.1$ H, $C = 0.1$ μF, and $\omega_n = (1/LC)^{1/2} = 10^4$ rad/s. For $\zeta_1 = 0.1$ and $\zeta_2 = 0.6$, we obtain

$$R = 2\zeta_1\left(\frac{L}{C}\right)^{1/2} = 2(0.1)\left(\frac{0.1}{0.1 \times 10^{-6}}\right)^{1/2} = 200 \; \Omega$$

$$R_1 + R = 2\zeta_2\left(\frac{L}{C}\right)^{1/2} = 2(0.6)\left(\frac{0.1}{0.1 \times 10^{-6}}\right)^{1/2} = 1200 \; \Omega$$

$$R_1 = 1000 \; \Omega$$

The magnitude plots of the transfer functions are shown in Fig. 4-33. If the criterion for amplitude distortion is 4%, the upper frequency limit of the original transducer is about 2×10^3 rad/s (320 Hz). The useful frequency is extended to about 8.3×10^3 rad/s (1320 Hz) by means of the compensator.

Series compensation may be inadequate for more complex problems. (1) It is difficult to match the characteristics of the compensator precisely with that of the instrument. (2) The characteristics of a transducer may change, such as a change in the time constant of a thermocouple. (3) The instrument may have many resonances to make series compensation impractical.

Feedback is often used for the automatic compensation of complex systems. For example, an electromechanical shaker many have many resonances, and the resonant frequencies also depend on the object tested and the fixture used for the tests.

Consider the feedback compensation of an electromechanical shaker

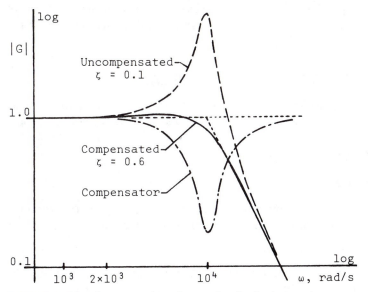

FIGURE 4-33   Compensation of second-order instrument.

shown in Fig. 4-34. The feedback loop maintains a constant output acceleration for a range of test frequencies. The reference $\ddot{x}_d$ is dialed into the system. The output $\ddot{x}$ is measured with an accelerometer, and the feedback signal $\ddot{x}_m$ is compared with $\ddot{x}_d$. Their difference $(\ddot{x}_d - \ddot{x}_m)$ is used to control the output voltage of the oscillator. The signal from the oscillator is then amplified to drive the shaker. Since the magnitude of the acceleration $\ddot{x}$ is set by the reference $\ddot{x}_d$, the output $\ddot{x}$ is controlled by means of the control loop. However, the frequency of the oscillator can be

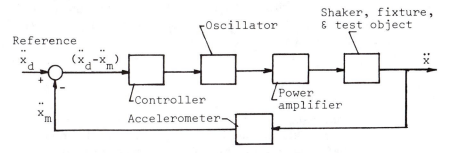

FIGURE 4-34   Feedback compensation of electromechanical shaker.

adjusted independently, and the vibration test can be automated to swept over a frequency range while maintaining a constant acceleration at the test object.

4-7. RESPONSE TO PERIODIC SIGNALS

Periodic signals are common in instrumentation. Signals from rotating machinery are periodic. The response of instruments to periodic inputs are examined in this section. Response to transient signals is treated in the next section.

A. Fourier Spectrum

The harmonics or the frequency content of a periodic signal are examined by means of the Fourier series. A function is periodic if

$$f(t) = f(t + \tau) \tag{4-75}$$

where τ is the period, the minimum time for the signal to repeat itself. The Fourier series expansion of $f(t)$ is

$$f(t) = \frac{a_0}{2} + \sum_{n=1}^{\infty} (a_n \cos n\omega t + b_n \sin n\omega t) \tag{4-76}$$

where $\omega = 2\pi/\tau$ rad/s is the fundamental frequency of $f(t)$. The coefficients of the series expansion are

$$a_n = \frac{2}{\tau} \int_{-\tau/2}^{\tau/2} [f(t) \cos n\omega t] \, dt \qquad n = 0,1,2,3, \ldots$$
$$b_n = \frac{2}{\tau} \int_{\tau/2}^{\tau/2} [f(t) \sin n\omega t] \, dt \qquad n = 1,2,3,4, \ldots \tag{4-77}$$

The integrations can be performed over one period from an arbitrary time t_0 to $(t_0 + \tau)$. The series converges to $f(t)$ if it satisfies the Dirichlet conditions [12]. For engineering problems, the Fourier series always converges. When $f(t)$ has a discontinuity, the series converges to the mean value at the discontinuity.

An alternative form of the Fourier series is

$$f(t) = d_0 + \sum_{n=1}^{\infty} d_n \sin (n\omega t + \gamma_n) \tag{4-78}$$

where $d_0 = a_0/2$, $d_n = (a_n^2 + b_n^2)^{1/2}$, and $\gamma_n = \tan^{-1} (a_n/b_n)$. The series expansion of a periodic function yields a *discrete spectrum*, shown in Fig. 4-35. It shows the frequency content or the harmonics of a periodic function, and has values only at isolated frequencies. The plot of magnitude d_n

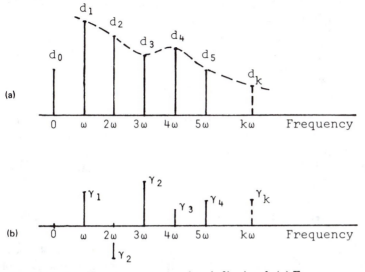

FIGURE 4-35   Fourier spectrum of periodic signal. (a) Frequency spectrum. (b) Phase spectrum.

versus frequency $n\omega$ in Fig. 4-35a is the frequency spectrum, and that of phase angle γ_n versus $n\omega$ is the *phase spectrum*. The two plots are usually called the *frequency spectrum* or the *Fourier spectrum*.

The Fourier series is an infinite series, but it is truncated for physical problems. First, the frequency range of a signal is limited. For example, the upper frequency limit for mechanical systems is on the order of kHz. Second, the harmonics of a signal attenuate at high frequencies. Due to the inherent noise in measurements, the signal/noise ratio is low at high frequencies, and their analysis is not meaningful.

The frequency spectrum of a signal can readily be observed by means of a spectrum analyzer. The Fourier coefficients in Eq. 4-76 can also be evaluated from a periodic waveform by means of numerical methods [13], but the hand calculation is tedious.

B.   Response to Periodic Signals

When a harmonic component of a periodic signal is applied to an instrument, the response to a periodic signal is a sinusoidal frequency response. The output is a composite response due to all the harmonics in the input. Since the harmonics at the input and output have the same frequencies, the output is also periodic with the same period as the input.

Let a periodic input $x(t)$ and the output $y(t)$ be expressed in the Fourier series in Eqs. 4-79 and 4-80.

$$x(t) = X_0 + \sum_{k=1}^{\infty} X_k \sin(k\omega t + \alpha_k) \tag{4-79}$$

$$y(t) = Y_0 + \sum_{k=1}^{\infty} Y_k \sin(k\omega t + \alpha_k + \phi_k) \tag{4-80}$$

where X_0 and Y_0 are constants, the sine terms are from the series expansion, ϕ_k are the phase angles of the harmonics in $y(t)$ relative to those in $x(t)$, and the phase angles α_k refer to the same reference.

Consider a harmonic component in $x(t)$ and $y(t)$. Following Eqs. 4-53 and 4-54, we get

$$\frac{Y_k}{X_k} = |G(jk\omega)| \qquad \text{and} \qquad \phi_k = \underline{/G(jk\omega)} \tag{4-81}$$

where $G(jk\omega)$ is the sinusoidal transfer function of the instrument. If the input frequency spectrum is as shown in Fig. 4-36a and the Bode plots of the instrument in Fig. 4-36b, the output frequency spectrum is as shown in Fig. 4-36c. For example, for a harmonic component at the frequency 5ω, the output magnitude Y_5 is the product of X_5 and $|G(j5\omega)|$ and the output phase angle is the algebraic sum of α_5 and $\underline{/G(j5\omega)}$.

The composite output $y(t)$ is the superposition of the responses due to all the harmonics, shown in Eq. 4-80. Thus the output $y(t)$ can be constructed from a knowledge of the input spectrum and the transfer function of the instrument.

Equation 4-81 can be used in three ways (see Table 1-1). (1) The output can be obtained from the input and a knowledge of the transfer function of the instrument. (2) The input $x(t)$ can be reconstructed by means of reversing the process. (3) The transfer function of an instrument can be measured from a knowledge of the input and output quantities.

4-8. RESPONSE TO TRANSIENT SIGNALS

The Fourier transform is employed for the analysis of transient signals in this section. The concept of the frequency content of a transient, the input/output relation in a measurement, and arbitrary signals for testing are presented.

A. Frequency Contents of Transients

A *transient* is a time function and nonperiodic, or a "one-shot" phenomenon, such as a mechanical shock. Once the concept of the frequency

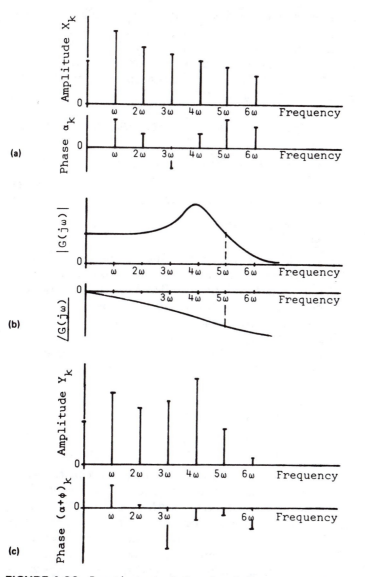

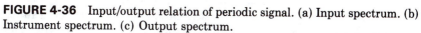

FIGURE 4-36 Input/output relation of periodic signal. (a) Input spectrum. (b) Instrument spectrum. (c) Output spectrum.

content of a transient is established, the analysis runs parallel to that for periodic signals.

The Fourier transform is used to represent a transient $f(t)$ in the frequency domain. Assume that $f(t)$ diminishes to zero within a finite time t_f, that is, $f(t) = 0$ for $t > t_f$. The Fourier transform $g(j\omega)$ of $f(t)$ is obtained from Eq. C-19 in Appendix C.

$$g(j\omega) = \int_{-\infty}^{\infty} f(t) \cos \omega t \; dt - j \int_{-\infty}^{\infty} f(t) \sin \omega t \; dt \qquad (4\text{-}82)$$

$$g(j\omega) = \text{Re}[g(j\omega)] + j \, \text{Im}[g(j\omega)] \qquad (4\text{-}83)$$

$$\text{Re}[g(j\omega)] = \int_{0}^{t_f} f(t) \cos \omega t \; dt \qquad (4\text{-}84)$$

$$-\text{Im}[g(j\omega)] = \int_{0}^{t_f} f(t) \sin \omega t \; dt \qquad (4\text{-}85)$$

where $\text{Re}[g(j\omega)]$ and $\text{Im}[g(j\omega)]$ are the real and the imaginary parts of $g(j\omega)$, respectively. The upper limit of the integrations is t_f instead of infinity because $f(t)$ becomes zero for $t \geqslant t_f$. The lower limit starts at a datum at $t = 0$. The computation is performed by means of a computer.

The Fourier transform of a transient from Eqs. 4-84 and 4-85 is a *continuous spectrum*. Since $\text{Re}[g(j\omega)]$ and $j \, \text{Im}[g(j\omega)]$ have the same physical meaning, only $\text{Re}[g(j\omega)]$ will be interpreted in the paragraphs to follow. Consider a pressure signal. The dimension of pressure is lb_f/in^2. From Eq. 4-84 we get

$$\text{Re}[g(j\omega)] = \int_{0}^{t_f} (\text{lb}_f/\text{in}^2) \underbrace{(\cos \omega t)}_{\text{nondimensional}} (\sec)$$

Since $(\text{lb}_f/\text{in}^2)(\sec)$ can be interpreted as $(\text{lb}_f/\text{in}^2)/(\text{rad}/\sec)$, $\text{Re}[g(j\omega)]$ denotes the amount of information per unit frequency increment. Comparing with a beam problem, the discrete spectrum, shown in Fig. 4-35, is analogous to the concentrated loads along a beam. The loads are discrete and are expressed as forces at isolated locations. Each Fourier coefficient gives the amount of information at an isolated frequency. On the other hand, the continuous spectrum of a Fourier transform is analogous to the distributed load along a beam. The distributed load is a continuous function in pound force per unit length. Hence the Fourier transform gives the information density, or the amount of information per bandwidth of frequency.

Let us compare the frequency contents of a slow and a fast transient to illustrate the concept further. It is intuitive that a fast transient has a higher-frequency content than that of a slow one. Applying Eq. 4-84 to the transients in Fig. 4-37, we have

$$\text{Re}[g(j\omega)] = \int_0^{t_f} f(t) \cos \omega t \, dt$$

Let $\omega = \omega_1$. The multiplication and integration for obtaining $\text{Re}[g(j\omega)]$ are shown in the figure. For the slow transient in Fig. 4-37a, $\text{Re}[g(j\omega_1)] = 0$, indicating that there is no information for $\omega \geqslant \omega_1$ in the slow transient. For the fast transient in Fig. 4-37b, $\text{Re}[g(j\omega_1)]$ has a steady-state value, indicating that there is information content for $\omega \geqslant \omega_1$ in the fast transient. The computation gives the value of $\text{Re}[g(j\omega)]$ at $\omega = \omega_1$. The process is repeated for a frequency range to obtain $\text{Re}[g(j\omega)]$ versus ω. $\text{Im}[g(j\omega)]$ is calculated similarly.

A typical spectrum $x(j\omega)$ for a transient $x(t)$ is shown in Fig. 4-38a. If the magnitude $|x(j\omega)|$ diminishes to zero beyond a certain frequency, the time function $x(t)$ has no information or frequency content beyond that frequency.

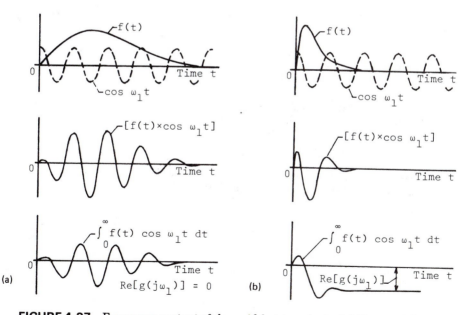

FIGURE 4-37 Frequency content of slow and fast transients. (a) Slow transient. (b) Fast transient.

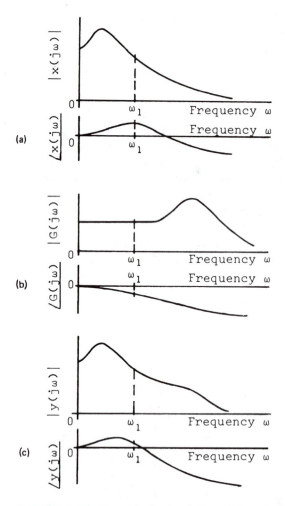

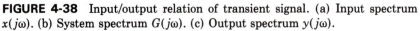

FIGURE 4-38 Input/output relation of transient signal. (a) Input spectrum $x(j\omega)$. (b) System spectrum $G(j\omega)$. (c) Output spectrum $y(j\omega)$.

B. Response to Transient Signals

The treatment of the response of an instrument to transient signals runs parallel to that for periodic signals. Let $x(j\omega)$ be the Fourier transform of the input $x(t)$, $y(j\omega)$ that of the output $y(t)$, and $G(j\omega)$ the transfer function of the instrument as shown in Fig. 4-38. For zero initial conditions, the input/output relation is

$$y(j\omega) = G(j\omega)x(j\omega) \tag{4-86}$$

$$\frac{y(j\omega)}{x(j\omega)} = G(j\omega) \tag{4-87}$$

The equations are analogous to Eq. 4-81 for periodic signals, except that the input and output quantities are Fourier transforms. Hence Eq. 4-86 can be used in three ways (see Table 1-1). (1) The output $y(t)$ is the inverse transform of $y(j\omega)$. It can be calculated from knowledge of the input $x(j\omega)$ and $G(j\omega)$ of the instrument. (2) Conversely, the input $x(j\omega)$ can be reconstructed from the output $y(j\omega)$ and $G(j\omega)$. (3) The transfer function $G(j\omega)$ can be determined from $x(j\omega)$ and $y(j\omega)$.

An inverse transform is obtained by the direct application of Eq. C-26 in App. C. For example, the inverse transform of $g(j\omega)$ is

$$f(t) = \frac{2}{\pi} \int_0^\infty \text{Re}[g(j\omega)] \cos \omega t \, d\omega \qquad \text{for} \quad t > 0$$

Let ω_f be the upper frequency limit for the integration; that is, $g(j\omega)$ has no information for $\omega > \omega_f$. For $t = t_1$, the equation becomes

$$f(t_1) = \frac{2}{\pi} \int_0^{\omega_f} \text{Re}[g(j\omega)] \cos \omega t_1 \, d\omega$$

This gives $f(t)$ at $t = t_1$. The process is repeated to obtain $f(t)$ versus t. Again, the inverse transformation can be automated.

The criteria for adequate response for transient signals are the same as those stated in Sec. 4-4A. For true waveform reproduction with time lag, the instrument has a constant gain and a phase lag linear with frequency. The frequency is interpreted as the frequency content in a transient.

C. Transient Test Signals

The impulse and the step function are examined in this section as test signals for frequency-domain analysis. These signals are used for time-domain analysis in Sec. 4-3, but time-domain data are difficult to interpret. The sinusoidal function is used to determine the transfer function in the frequency domain (see Sec. 4-6), but the process is time consuming. The trend in recent years favors transient testing.

1. *Impulse Function*

The Fourier transform of an impulse from Eqs. 4-17 and 4-82 is

$$g(j\omega) = \int_0^\infty A\delta(t) \cos \omega t \, dt - j \int_0^\infty A\delta(t) \sin \omega t \, dt$$

$$g(j\omega) = A - j \cdot 0 \quad \text{for all frequencies} \tag{4-88}$$

$$|g(j\omega)| = A \quad \text{and} \quad \underline{/g(j\omega)} = 0 \tag{4-89}$$

The frequency spectrum of $A\delta(t)$ is of constant magnitude A and zero phase for all frequencies, as shown in Fig. 4-39a [14]. This implies that an impulse test is equivalent to the simultaneous application of sinusoidal functions of equal magnitude and zero phase angle of all frequencies to a system. In other words, a frequency response test is accomplished by applying a single impulse to the system. This property of an impulse makes it a very attractive test signal.

In practice, a pulse of infinitesimal duration or pulse width with sufficient strength may not be attainable. If the duration of the pulse is short compared with the time for a machine to respond, the pulse is essentially an impulse for the application. The high-frequency components of a pulse of finite width decrease with increasing pulse width. It will be shown presently that this may be an advantage in the testing of heavy machines.

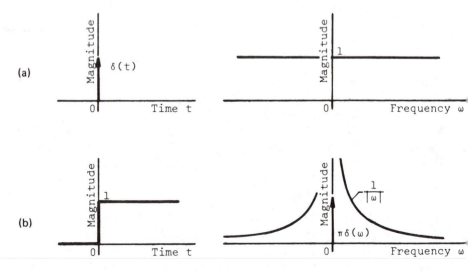

FIGURE 4-39 Frequency spectrum of $\delta(t)$ and $u(t)$. (a) Unit impulse function $\delta(t)$. (b) Unit step function $u(t)$.

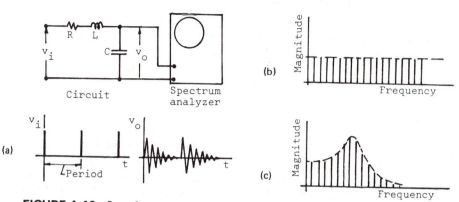

FIGURE 4-40 Impulse response of second-order system. (a) Simulation of second-order instrument. (b) Frequency spectrum of v_i. (c) Frequency spectrum of v_o.

When an impulse input $A(t)$ is applied to a system, its response from Eq. 4-86 is

$$y(j\omega) = G(j\omega)A \quad \text{when} \quad \text{input} = A\delta(t) \tag{4-90}$$

This shows that the output is the system transfer function and no additional computation is necessary. For example, an input votage $v_i(t)$, in the form of a pulse train, is applied to the RLC circuit shown in Fig. 4-40a. The output $v_o(t)$ is a train of impulse responses. The voltages can be measured with spectrum analyzer. The frequency spectra of $v_i(t)$ and $v_o(t)$ are shown in Fig. 4-40b and c, respectively. Discrete spectra are shown because the spectrum analyzer samples the voltages and treats them as periodic functions. The input spectrum shows the frequency content of $v_i(t)$, and the envelope of the output spectrum is the transfer function of the RLC circuit.

2. Step Function

A step function is attractive as a test signal for large structures because it has components of large magnitudes at the low end of the frequency spectrum, as shown in Fig. 4-39b. A test signal should provide sufficient energy to excite the resonances of a system, and energy is proportional to magnitude squared. Although an impulse is an "ideal" test signal, it spreads its energy equally over all frequencies. First, the high-frequency components of an impulse are not utilized because a large structure has low natural frequencies and the high-frequency excitations do not reveal this characteristics. Second, an impulse may not have sufficient energy for excitating the lower frequencies. Hence it is advantages to use a step to use a step function instead of an impulse for the testing of large structures.

In summary, any convenient test signal can be used for finding the transfer function $G(j\omega)$ of a system. From Eq. 4-86, $G(j\omega)$ can be determined from a knowledge of the input and output functions from the test (see Sec. 7-6).

PROBLEMS

4-1. The simplified model of a pressure transducer is shown in Fig. P4-1. Assume that the applied pressure p_1 is a step function, R is the resistance of the connecting tubing, and the cavity V is filled with fluid of bulk modulus B. Neglecting the inertia effect of the fluid, derive the differential equation of the system.

4-2. A mercury-in-glass thermometer is shown in Fig. P4-2. Derive a differential equation relating x and T_w and find the time constant of the thermometer. Given: A_c = cross-sectional area of the capillary, A_b = surface area of the bulb, C = differential coefficient of expansion between Hg and glass, T_b = temperature of the mercury, T_w = temperature of the hot bath, T_a = ambient air temperature, x = observed reading of thermometer due to $(T_w - T_b)$, c = specific heat of mercury, ρ = density of mercury. Assume that the heat loss from the stem to T_a is negligible.

4-3. A water supply system is shown in Fig. P4-3. Assume that the flow rate $\dot{q}$ is proportional to the differential head and inversely proportional to the resistance R of the pipe. Derive the differential equation relating $\dot{q}_i$ and the head h_2.

4-4. A hydraulic servo-position controller is shown in Fig. P4-4, where x is the displacement input and y the output. Assume that the flow rate $\dot{q}$ of the oil into the cylinder is proportional to the valve opening w and that the inertia effect of the masses is negligible. Derive the differential equation relating x and y.

4-5. The device in Fig. P4-5 is proposed for a pressure transducer, in which the pressure P is indicated by the displacement y. (a) Derive the differential equation relating y and P. (b) If the input P is sinusoidal, find the steady-state amplitude ratio Y/P and the phase angle of y relative to P.

4-6. Given the transfer function

$$G(D) = \frac{y}{x}(D) = \frac{6}{5 + 7D}$$

(a) For $y(t = 0) = 0$, find $y(t = 1)$ when $x = 2u(t)$. (b) Find the steady-state output y when $x = 2 \sin 4t$.

4-7. Write the transfer functions for the systems shown in Figs. P4-3 to P4-5. Assume that the input is a unit step function. (a) Is it possible to deduce the value of the output directly from the transfer function? (b) Find the steady-state output.

4-8. For the servo-position controller shown in Fig. P4-4, assume that the time constant is $\tau = 2$ sec, the sensitivity is $K = 5$ in./in., and the system is initially at rest. (a) Find $y(t)$ at $t = 0.5$ sec for the input $x = 2u(t)$. (b) Find the steady-state output $y(t)$ when $x = 2 \sin 3t$.

4-9. A first-order instrument is initially at rest. Its impulse response is $y = 4e^{-2t}$. Find the response $y(t)$ for the excitation shown in Fig. P4-6.

4-10. The half sine pulse in Fig. P4-7 is applied to a first-order instrument, $\tau \dot{y} + y = Kx(t)$. Assume zero initial conditions and $K = 1$. (a) Derive the solution $y(t)$. (b) Find the maximum amplitude of $y(t)$. (c) Show a sketch of $y(t)$ for $t > 2$.

4-11. A pressure transducer is described by the equation

$$\ddot{y} + 2\zeta\omega_n \dot{y} + \omega_n^2 y = KP(t)$$

Assume zero initial conditions and $\omega_n = 20$ rad/s, $\zeta = 0.6$, and $K = 1$. Derive the response $y(t)$ for the input shown in Fig. P4-8.

4-12. The pulse shown in Fig. P4-9 is applied to a pressure transducer, the equation of which is

$$m\ddot{y} + c\dot{y} + ky = KP(t)$$

Assuming zero initial conditions, outline the steps for finding the output $y(t)$.

4-13. A measuring system is described by the equation

$$\ddot{y} + 10\dot{y} + 100y = 5x(t)$$

Assume zero initial conditions and that the input $x(t)$ is as shown in Fig. P4-10. (a) Find the steady-state error and the time lag in the measurement. (b) Identify the error and the time lag in a sketch. (c) Find the value of $y(t)$ at $t = 1$ sec.

4-14. A mercury-in-glass thermometer is installed in a thermometer well for measuring the steam temperature as shown in Fig. P4-11 (see Prob. 4-2 for a description of the thermometer). Assume that T_b = temperature of Hg in the bob, T_w = oil temperature, and T_s = wall temperature of the well and steam temperature. Derive an equation relating T_b and T_s.

4-15. A clinical thermometer is described by the equation

$$\tau\dot{y} + y = Kx(t)$$

where $K = 1$ division/°F and $\tau = 20$ sec. Assume that the room is at 70°F. The thermometer is placed in the mouth of a patient and it reads 86°F at a time $t = 10$ sec. What is the temperature of the patient?

4-16. A balloon rises at a rate of 20 ft/sec, and it carries a thermometer. Assume that the time constant of the thermometer is $\tau = 15$ sec, the temperature at ground level is 70°F, and the atmospheric temperature decreases at a rate of 0.3°F/100 ft. What is the altitude of the balloon when the thermometer reads 40°F?

4-17. A thermocouple at room temperature of 70°F is plunged into a hot-water bath at 140°F. Find the time constant by the methods described in the text if the recorded temperatures are as follows:

Time (sec)	0	0.2	0.4	0.6	0.8	1.0	1.2
Reading (°F)	70	96	115	125	129	133	137

4-18. Given the following frequency response test data:

ω (rad/s)	0.02	0.08	0.2	0.8	1.0	2.0	4.0	8.0	20	80	200
$\lvert Y/X \rvert$	1	1	0.99	0.93	0.90	0.71	0.45	0.24	0.10	0.025	0.01
ϕ (deg)	−0.6	−2.3	−5.7	−22	−27	−45	−63	−76	−84	−89	−89

(a) Plot the $\log \lvert Y/X \rvert$ versus $\log \omega$ and ϕ versus $\log \omega$ curves. (b) Sketch the asymptotes in the $\log \lvert Y/X \rvert$ versus $\log \omega$ plot and find the time constant if the system is first-order.

4-19. Given the following frequency response of a second-order instrument:

ω (rad/s)	0.625	1.25	2.5	5.0	10	20	40	80	160
$\lvert Y/X \rvert$	1.0	1.01	1.06	1.29	2.5	0.32	0.07	0.016	0.004
ϕ (deg)	−1.4	−2.9	−6.1	−15	−90	−165	−174	−177	−178

(a) Plot the $\log \lvert Y/X \rvert$ versus $\log \omega$ and ϕ versus $\log \omega$ curves. (b) Draw the asymptotes in the magnitude plot and find the natural frequency.

4-20. Find the frequency range for which the pressure transducer shown in Fig. P4-5 will give an adequate response.

4-21. Sketch the waveform of

$$f(t) = 1 + 2 \sin (\omega t + 30°) + 1 \sin (3\omega t - 180°)$$

4-22. A transducer reproduces its input waveform with a time lag. The input $x(t)$ and output $y(t)$ are shown below. Plot the input and output waveforms.

$$x(t) = 2 \sin \omega t + 1 \sin 3\omega t$$

$$y(t) = 2 \sin (\omega t - 30°) + 1 \sin (3\omega t - \phi)$$

4-23. Sketch the asymptotic log amplitude ratio versus log ω and the phase angle ϕ versus log ω plots of the following transfer functions.

(a) $G(j\omega) = \dfrac{500(1 + j0.5\omega)}{j\omega[(j\omega)^2 + j8\omega + 100]}$

(b) $G(j\omega) = \dfrac{100(2 + j\omega)}{j\omega(5 + j\omega)[(j\omega)^2 + 10j\omega + 400]}$

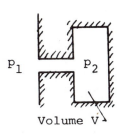
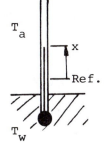

FIGURE P4-1   **FIGURE P4-2**

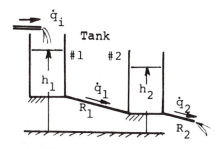

FIGURE P4-3

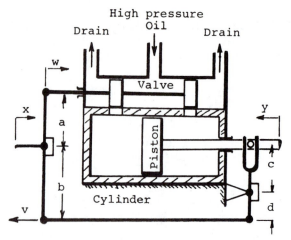

Drain　High pressure　Drain
Oil

w

Valve

x

a

Piston

y

c

b

Cylinder

d

v

FIGURE P4-4

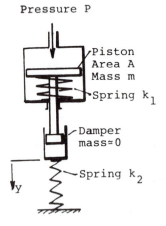

Pressure P

Piston
Area A
Mass m

Spring k_1

Damper
mass$\approx$0

Spring k_2

y

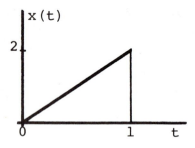

$x(t)$

2

0　　　1　　t

FIGURE P4-5

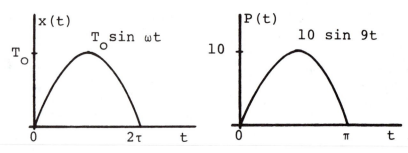

$x(t)$

$T_o \sin \omega t$

T_o

0　　　2τ　　t

$P(t)$

$10 \sin 9t$

10

0　　　π　　t

FIGURE P4-6

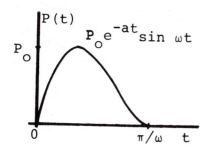

FIGURE P4-7

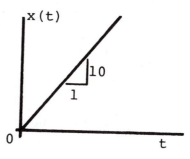

FIGURE P4-8

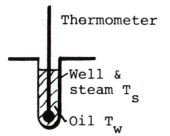

FIGURE P4-9

REFERENCES

1. Greensite, A. L., *Control Theory,* Vol. 1, *Elements of Modern Control Theory,* Spartan Books, New York, 1970, p. 146.
2. Truxal, J. G., *Automatic Control Feedback Systems Synthesis,* McGraw-Hill Book Company, New York, 1955, p. 424.
3. DeRusso P., R. J. Roy, and C. M. Close, *State Variables for Engineers,* John Wiley & Sons, Inc., New York, 1965, Chap. 1.
4. Tse, F., I. E. Morse, and R. Hinkle, *Mechanical Vibrations—Theory and Applications,* 2nd ed., Allyn and Bacon, Inc., Boston, 1978, Chap. 2.
5. Tse, Morse, and Hinkle, Ref. 4, pp. 77–80.
6. Dranetz, A. I., Piezoelectric and Piezoresistive Pick-Ups, in *Shock and Vibration Handbook,* 2nd ed., C. M. Harris and C. E. Crede (eds.), McGraw-Hill Book Company, New York, 1976, Chap. 6.
7. Greensite, Ref. 1, p. 757.
8. Bickle, L. W., and R. C. Dove, Numerical Correction of Transient Measurements, *ISA Trans.,* Vol. 13, No. 3 (1973), pp. 286–295.
9. Stein, P. K., Response of Systems to Step and Pulses, *I.F/MSE Pub.* 40, Stein Engineering Service, Inc., Phoenix, Ariz., 1972.

10. Gray, P. E., and C. I. Searle, *Electronic Principles—Physics, Models, and Circuits,* John Wiley & Sons, Inc., New York, 1967, p. 530.
11. D'Azzo, J., and C. H. Houpis, *Feedback Control System Analysis and Synthesis,* McGraw-Hill Book Company, New York, 1960, Chaps. 11–13.
12. Carslaw, H. S., *Fourier Series and Integrals,* 3rd ed., Cambridge University Press, New York, 1930, p. 207.
13. Scarborough, J. B., *Numerical Mathematical Analysis,* 5th ed., The Johns Hopkins Press, Baltimore, 1962, Chap. 19.
14. Hsu, H. P., *Fourier Analysis,* rev. ed., Simon and Schuster, New York, 1967, p. 107.

SUGGESTED READINGS

Doebelin, E. D., *Measurement Systems—Application and Design,* McGraw-Hill Book Company, New York, 1983.

Gardner, M. F., and J. L. Barnes, *Transients in Linear Systems,* Vol. 1, John Wiley & Sons, Inc., New York, 1942.

Papoulis, A., *Signal Analysis,* McGraw-Hill Book Company, New York, 1977.

Shearer, J. L., A. T. Murphy, and H. H. Richardson, *Introduction to System Dynamics,* Addison-Wesley Publishing Company, Inc., Reading, Mass., 1967.

5
Nonself-generating Transducers and Applications

5-1. INTRODUCTION

Transducers are broadly classified as self-generating or non-self-generating (see Sec. 2-3). The objective of this chapter is to present a general study of non-self-generating transducers and to integrate parts of the previous chapters.

A self-generating transducer detects an input and yields an output, as prescribed by the physical law. No auxiliary energy is required for its operation. Self-generating transducers and their applications are described in later chapters.

A non-self-generating transducer is impedance based and requires an auxiliary input (carrier) for its operation. Its output depends on the characteristics of the carrier, and it is possible to modify the signal for noise suppression or for signal transmission. The energy at the output of a non-self-generating transducer is mainly from the carrier. It is possible to have the output at a higher energy level than the self-generating type and there is less tendency for loading of the information source. A non-self-generating transducer is a three-port device, but it can be reduced to a two-port, as shown in Fig. 5-1, where the subscript (0) denotes values at an operating point (see Secs. 2-3 and 3-3).

The general study will give the reader a perspective on applications of non-self-generating transducers. For example, sensors, such as strain gages, are used in many types of transducers and for numerous measurements. By examining common components and techniques, it is possible to describe a class of problems and to avoid excessive repetitions on specific measurements in later chapters.

To integrate parts of the preceding chapters, it will be shown that the same phenomenon for sensing can be used in many ways for measurements (see Table 3-1). In other words, the topics presented in the preced-

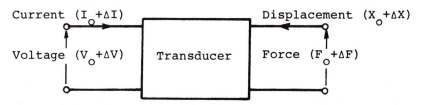

FIGURE 5-1 Block diagram of transducer with feedback.

ing chapters can be incorporated in transducers of different design but employing the same phenomenon for sensing. This echos an earlier statement that if an input produces a detectable effect on an object, the object is potentially a transducer for that variable. The manner in which the phenomenon is used depends on the application. Common resistive, inductive, and capacitive transducers for incremental motion measurements are presented for discussion.

In principle, a transducer is a simple device, as modeled in Sec. 2-1. In practice, it can be quite complicated, in order to protect against environmental conditions and other undesired effects. If it operates in an open loop, it can be described in terms of functional stages (see Fig. 1-2). It may operate in a closed loop, as shown in Fig. 5-2 (see Sec. 3-6). The double arrows indicate multichannel operation, such as simultaneous measurement of pressure and temperature. It is likely that the controller is "computer assisted" in some way. For simplicity, the power supply and functional blocks for calibration are omitted from the diagram.

The presentation in the chapter is mostly descriptive, although some analytical problems are given as exercises. A detailed study of the dynamics and feedback of electromechanical systems is beyond the intended purpose of the book.

5-2. DIFFERENTIAL TRANSFORMERS

The *linear variable differential transformer* (LVDT) is an inductive-electromechanical displacement sensor. Its voltage output is linear with a displacement input. LVDTs are used extensively for many measurements, because of the ease of application, versatility, and ruggedness. The construction and operation of LVDTs are described below.

The schematic of a LVDT, shown in Fig. 5-3a, consists of a movable core of magnetic material and three coils, the primary coil P and the secondaries S_1 and S_2. The coils of a LVDT for rectilinear motion form a hollow cylinder, shown in Fig. 5-3b. The core is slotted to minimize eddy currents.

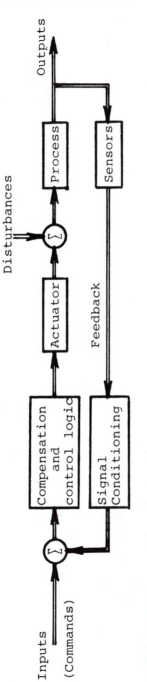

FIGURE 5-2 Incremental-motion transducer.

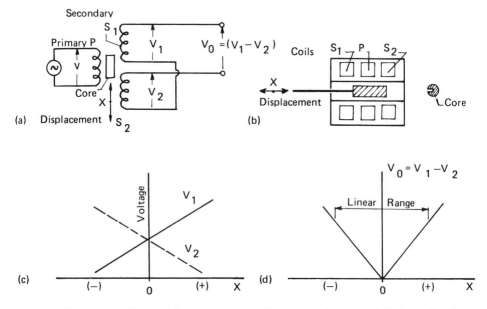

FIGURE 5-3 Differential transformer and output voltages. (a) Schematic of LVDT. (b) Cross-sectional view. (c) Induced voltages. (d) Differential voltages.

The LVDT is a non-self-generating transducer. The ac voltage V applied to the primary coil P is the carrier. The induced voltages in the secondary coils are V_1 and V_2. The displacement input x of the core determines the magnetic flux linkage between the primary and the secondary coils, and therefore the magnitudes of the induced voltages V_1 and V_2, shown in Fig. 5-3c. If the secondaries are in series opposition, as shown in Fig. 5-3a, the output is $V_o = V_1 - V_2$. The magnitude of V_o versus the core position x is as shown in Fig. 5-3d.

Now consider the operations in more detail. The waveforms of the carrier voltage v at the primary and the induced voltages v_1 and v_2 from the secondaries are shown in Fig. 5-4a, where the v's are instantaneous values. The phase shifts in the waveforms are due to the impedances in the coils. When the core is above the electrical neutral of the coils or the null position, v_1 is greater than v_2. The differential voltages $(v_1 - v_2)$ for the core above neutral, at neutral, and below neutral are shown in Fig. 5-4b. There is a 180° phase shift in $(v_1 - v_2)$ when the core goes from above to below neutral. When the core is nulled, the magnitudes of v_1 and v_2 are equal, that is, $|V_1| = |V_2|$. Since the secondaries S_1 and S_2 are not perfectly identical, there is a small phase angle between the voltages, and $|V_1 - V_2|$

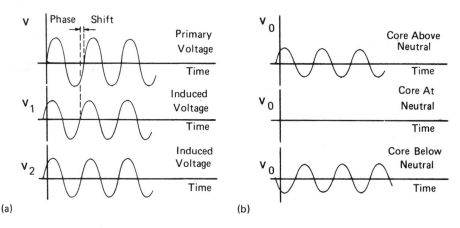

FIGURE 5-4 Voltage waveforms of a differential transformer. (a) Primary and induced voltages. (b) Differential voltages.

is not exactly zero at null. The null voltage is about 1% of the maximum output voltage. The phase shift and the null voltage can be corrected by means of external circuitry.

The LVDT in Fig. 5-3a is a *product modulator*. The low-frequency input signal x modulates the carrier voltage v, which is a high-frequency sine wave of constant amplitude. The modulated output $v_o = v_1 - v_2$ is proportional to the product of the input signal x and the carrier v. For example, when the waveforms of the inputs x's are as shown in Fig. 5-5a, those of the modulated outputs v_o's (solid lines) are as shown in Fig. 5-5b. Although the voltages v_o look alike, there is a 180° phase shift in v_o when the input x changes sign (dashed lines), that is, when the core goes from above to below null position. This phase shift cannot be detected directly by means of an ac meter or an oscilloscope. A phase-sensitive demodulator is necessary to discriminate the positions of the core.

A diode bridge is often used for the phase-sensitive demodulation. First, consider the *diode bridge-rectifier* in Fig. 5-6a. A diode is analogous to a check valve. If the pressure drop is in the direction shown in Fig. 5-6b, heavy flow (solid line) occurs and the valve offers little resistance to the flow. If the pressure drop is in the opposite direction (reverse bias), shown in Fig. 5-6c, only a small leakage flow (dashed line) occurs in the valve. A diode operates in a similar manner, with pressure analogous to a voltage and flow to an electric current. In other words, an ideal diode allows a current flow in only one direction. When an ac voltage is applied to terminals a and b of the diode bridge in Fig. 5-6a, it is deduced that the

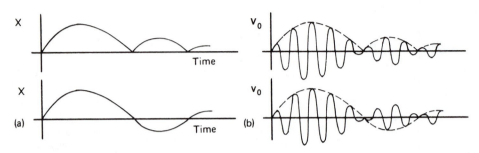

FIGURE 5-5 Waveforms of input x and output voltage v_o. (a) Displacement input waveforms. (b) Voltage output waveforms.

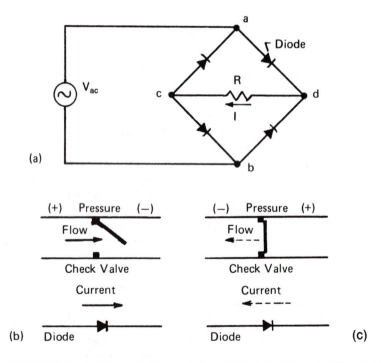

FIGURE 5-6 (a) Diode bridge rectifier. (b) Forward bias. (c) Reversed bias.

current I through the resistor R is always in the direction shown. Thus the voltage across R is the rectified voltage.

The *phase-sensitive demodulator* in Fig. 5-7a consists of two diode bridges. The currents through the resistors are always in the directions shown. The voltage from d to c is a voltage drop, because a voltage drop is in the direction of current flow. Hence the voltage V_{cd} from c to d is a voltage rise. Similarly, V_{gh} is a voltage rise. By connecting c and g, the differential voltage from h to d is the phase sensitive demodulated output, that is, $V_o = V_{hd} = (V_{cd} - V_{gh})$. For example, if the motion of the core x is as illustrated in Fig. 5-7b, the corresponding phase-sensitive demodulated output v_o (solid line) is as shown in the figure.

The demodulated signal v_o is at twice the carrier frequency and its envelope, shown in Fig. 5-7b as a dashed line, has the waveform of the input x. Since the carrier frequency is much higher than that of the signal, the

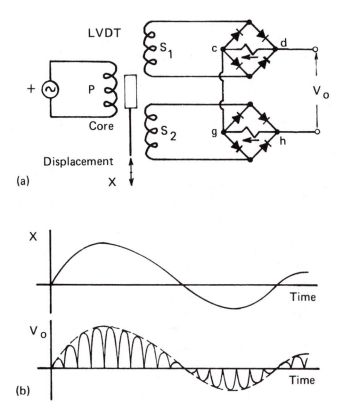

FIGURE 5-7 Waveforms in phase-sensitive demodulation. (a) Phase-sensitive demodulation. (b) Input/output waveforms.

carrier can be rejected by means of a low-pass filter and the input signal x recovered from v_o. The associated electronics can be incorporated in the transducer.

5-3. DIFFERENTIAL TRANSFORMERS: APPLICATIONS

The LVDT is a displacement sensor. In this section we describe some characteristics of LVDTs and then show that LVDTs can be used to measure any variable that generates a displacement, such as an acceleration, average temperature, or fluid flow [1].

The operating range of a LVDT may be from 0.005 to several inches. The sensitivity is of the order of 0.5 to 3 mV per 0.001 in./for 3 to 6 V applied to the primary coil. The higher sensitivity is for LVDTs of smaller operating ranges. The ac excitation ranges from 3 to 12 V and from 60 Hz to 25 kHz, depending on the design.

General-purpose LVDTs are inexpensive or LVDTs can be designed to meet stringent requirements. Some special features are: subminiaturization, high-pressure seals (3000 psi), spring loading or air actuated for gaging, contact or noncontact operation, large-bore to small-bore diameter ratios, long stroke-to-body lengths, cryogenic or high-temperature (-450 to $+1200°F$) operations, built-in or remote signal conditioning, imperviousness to hostile environments (e.g., nuclear radiation, dirt, grease, corrosion, electromagnetic or electrostatic interference), and shock and vibration resistance.

An electrical output is convenient for remote sensing and/or control. The output of strain gage sensors is also electrical, but the high-voltage output of LVDTs simplifies the instrumentation requirements. The chief disadvantage of LVDTs is the possible mass loading due to the mass of the core. There is also an axial force on the core when it is away from null, and a radial force on the core when it is not concentric with the bore of the transformer. The mass loading and the axial force can be neglected for most applications. The radial force is negligible if the core is guided centrally.

Examples of applications of LVDTs are described below. The schematic of a rotary variable differential transformer (RVDT) for angular measurements is shown in Fig. 5-8a. The core is an eccentric mass of magnetic material. Although a RVDT is capable of continuous rotation, its linear range is $±40°$ about null. For small angular displacements, resolutions to a fraction of a degree are common. Evidently, a RVDT can be adapted for torque and other measurements involving angular motions.

LVDTs are ideally suited for dimensional measurements. Gaging is

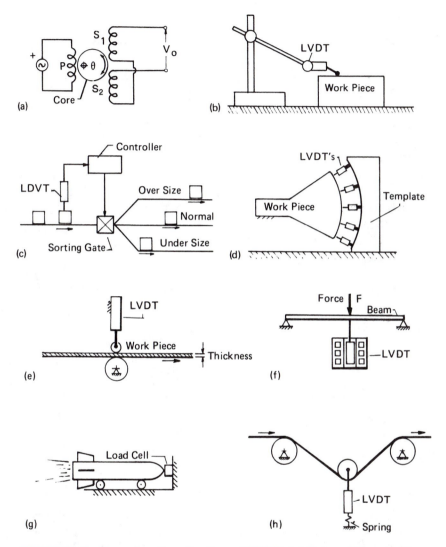

FIGURE 5-8 Examples of applications of LVDT. (a) Rotary differential transformer. (b) Dimentional gaging. (c) Gaging/sorting by size. (d) Multipoint gaging. (e) Thickness measurement. (f), (g) Load cells. (h) Tension measurement. (i) Force transducer. (j) Multipoint weighing. (k) LVDT accelerometer. (l) Velocity transducer. (m) Extensometer. (n) Liquid-level measurement. (o) Pressure transducer. (p) Noncontact thickness gage.

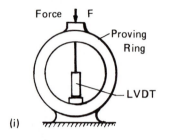

(i)

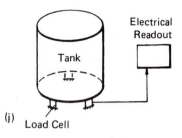

(j)

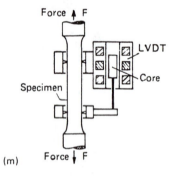

(k)

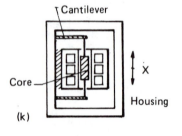

(l)

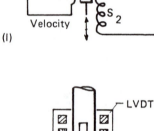

(m)

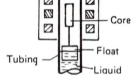

(n)

(o)

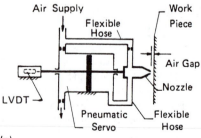

(p)

FIGURE 5-8 (*Continued*)

shown in Fig. 5-8b. The electrical output of the LVDT for gaging, shown in Fig. 5-8c, actuates a controller and a sorting gate. Multipoint gaging is illustrated in Fig. 5-8d. The LVDTs are mounted on a template, and are adjusted for zero output when the LVDTs are in contact with a master. Hence the readouts from the LVDTs during gaging give the dimensional variations from the master. The thickness measurement in Fig. 5-8e is a gaging process, although the workpiece is moving continuously.

The load cell in Fig. 5-8f uses the LVDT as a secondary sensor (see Fig. 1-3). The applied force is balanced by the reaction of a simply supported beam. The LVDT indicates the applied force by means of the elastic deflection of the beam. The applications shown in Fig. 5-8g to j for measurement of force and weight are self-explanatory. Multiple sensors are used in the weighing tank in Fig. 5-8j. Multiple sensors for mathematical operations and control will be discussed presently.

The schematic of an accelerometer is shown in Fig. 5-8k, in which the core is a part of the seismic mass and the coils are attached to the housing. The relative displacement between the core and the housing measures the acceleration of the housing. The LVDT in Fig. 5-8l is a velocity transducer. A dc excitation voltage is applied to the primary coil and the secondaries are in series. The output of the device indicates the velocity of the core relative to the coils.

LVDTs have numerous other applications. The LVDT in Fig. 5-7m is an extensometer in a tensile test. Due to the long-term stability of LVDTs, the scheme can be used for creep tests of materials. A liquid-level indicator is shown in Fig. 5-8n. The coils are outside the tubing and the core is inside. Conceivably, the device can be used for flow measurements (not shown). The output of the bourdon tube pressure gage in Fig. 5-8o can be used for remote pressure sensing and/or control. The noncontact thickness gage in Fig. 5-8p uses a pneumatic servo to maintain the air gap between the flow nozzle and the workpiece. The motion of the nozzle is transmitted to the core of the LVDT. The device is a proximity gage for nonmetallic materials. Moreover, the air gap is maintained by means of a servo and the nozzle tracks the profile of the workpiece. Hence the device can be used for a profile measurement. Finally, when temperature causes a deflection in a bimetallic device (not shown), a LVDT can be used as a secondary sensor in a temperature transducer.

Systems employing multiple LVDTs are illustrated in Fig. 5-9. The null balance system in Fig. 5-9a compares the voltage V_x of the input LVDT with V_y of the output LVDT. The differential voltage $(V_x - V_y)$ is applied to an amplifier and then to a servo motor. Let $V_x = Kx$ and $V_y = Ky$, where K is constant. If the system is initially nulled with $V_x - V_y = 0$, the servomotor is stationary. When the input x is altered,

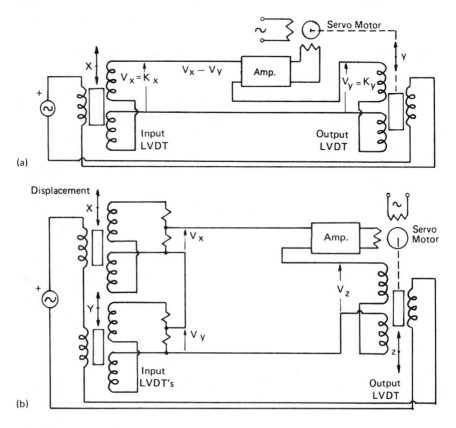

FIGURE 5-9   Multiple-LVDT circuits. (a) Null-balance system. (b) Summing circuit for $z = x + y$.

$V_x - V_y$ is not zero and the servomotor actuates the output LVDT until the system is again nulled. The system can be used for the remote indication of an x input, or for the control of a process by means of the x input.

Similarly, the circuit in Fig. 5-9b is a null balance system with two inputs. The z output is the average of the x and y inputs. When the system is nulled, the voltage to the amplifier is $V_z - (V_x + V_y) = 0$. Since $V_z = Kz$, $V_x = Kx/2$, and $V_y = Ky/2$, where K is a constant, it follows that $z = (x + y)/2$. Similar circuits can be devised to measure the sum or the difference of several input quantities.

5-4.   CARRIER SYSTEMS

The familiar AM and FM in radio are carrier systems at radio frequencies. Impedance-based transducers such as LVDTs are carrier systems (see Sec. 3-7B). Two mechanical systems are described here to introduce the concepts of amplitude modulation and demodulation. The analysis is given in the next section.

A pyrometer for infrared detection is shown in Fig. 5-10. The sensor is a thermocouple (TC). Instead of amplifying the low-level dc output of a TC, the temperature signal from the target is modulated to give a modulated signal v as shown in the figure. The mechanical modulator is a chopper in the form of a slotted rotating disk placed between the target and the TC. The scheme is often used in modern pyrometers [3]. The TC alternately sees the target and the back of the chopper. The datum of the signal v is the temperature of the chopper, which may be kept constant to improve the accuracy of the instrument. The ac amplifier eliminates the mean value of the chopped signal v to give an output v_1. The demoulator is a switch, which flips back and forth in synchronism with the chopper to rectify v_1 and to yield an output v_2. The mechanical connection for the synchronization is shown as a dashed line in the figure. A more practical mechanical demodulator is described below. The original signal is recovered from v_2 by means of a low-pass filter.

A more practical scheme for mechanical modulation and phase-sensitive demodulation is shown in Fig. 5-11. It operates on the same principle as the infrared detector. Signal from the strain gage bridge is modulated by means of a vibrator, which is an electromechanical "reed" switch. The first vibrator is a chopper-modulator and the second a switch-demodulator. Since both vibrators are driven from the same 400-Hz ac power source, the switching is synchronous and the demodulation is phase sensitive. The other components in the circuit are self-evident.

A.   Amplitude Modulation: Sine-Wave Carrier

The familiar *amplitude modulation* (AM) in radio is a process in which a sinusoidal carrier of constant amplitude is modified or modulated by an information signal, such that the amplitude of the carrier is proportional to the value of the signal. Product modulation and demodulation in AM are examined in this section.

The examples described in the preceding section are generalized in the AM process shown in Fig. 5-12. Assume a sinusoidal carrier $x_c(t) = A \cos \omega_c t$ of constant amplitude A and frequency ω_c rad/s. The modulated output (solid line) $x_1(t)$ is the product of the information signal $x(t)$ and $x_c(t)$.

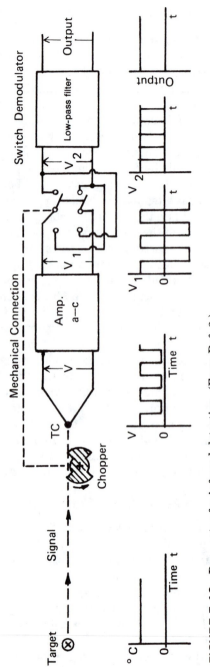

FIGURE 5-10 Pyrometer for infrared detection. (From Ref. 2.)

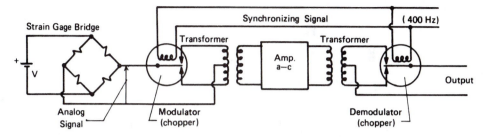

FIGURE 5-11   Amplitude modulation–demodulation system. (From Ref. 4.)

The signal $x_1(t)$ is amplified to become $x_2(t)$. The demodulated output (solid line) $x_3(t)$ is the product of $x_2(t)$ and the carrier $x_c(t)$. This is the rectified phase-sensitive demodulated signal. The original signal is recovered from $x_3(t)$ by means of a low-pass filter.

The modulated signal $x_1(t)$ from the product modulator is

$$x_1(t) = [x(t)](A \cos \omega_c t) \qquad (5\text{-}1)$$

For the time being, assume that the signal $x(t)$ is sinusoidal.

$$x(t) = B \cos \omega_s t$$

Substituting $x(t)$ in Eq. 5-1 and using the trigonometric identity for the product of cosines, we obtain

$$x_1(t) = (B \cos \omega_s t)(A \cos \omega_c t) \qquad (5\text{-}2)$$
$$x_1(t) = AB(\cos \omega_s t)(\cos \omega_c t)$$
$$x_1(t) = \frac{AB}{2}[\cos(\omega_c - \omega_s)t + \cos(\omega_c + \omega_s)t] \qquad (5\text{-}3)$$

The frequency spectra of the signal $x(t)$, the carrier $x_c(t)$, and the modulated signal $x_1(t)$ are shown in Fig. 5-13a. The spectrum of $x(t)$ has one spectral line at ω_s, and that of the carrier has a line at ω_c. Their product $x_1(t)$ from Eq. 5-3 contains neither ω_s nor ω_c. The spectrum of the modulated signal $x_1(t)$ consists of a pair of lines offset from ω_c by $\pm\omega_s$. In other words, the information $x(t)$ at the low-frequency ω_s has been translated to the higher frequencies at $\omega_c \pm \omega_s$.

Now, let the information signal $x(t)$ have a bandwidth W as shown in Fig. 5-13b. For example, the bandwidth of a human voice is from 20 Hz to 20 kHz. The corresponding spectrum of the modulated signal $x_1(t)$ will contain two sidebands, each of bandwidth W on either side of the carrier frequency ω_c. This is called a double sideband. It can be shown that the information $x(t)$ can be recovered from either of the sidebands.

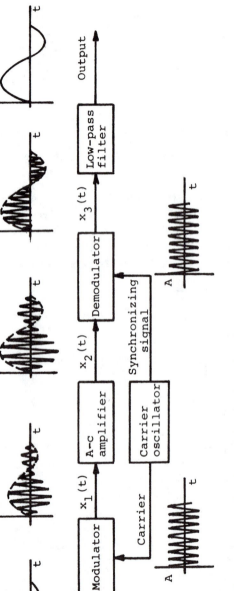

FIGURE 5-12 Steps in amplitude modulation–demodulation.

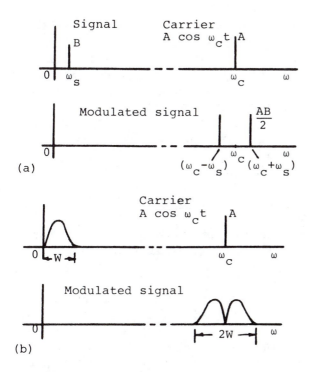

(a)

(b)

FIGURE 5-13 Frequency spectrum of amplitude-modulated signal. (a) Signal $x(t) = B \cos \omega_s t$. (b) Signal bandwidth $= W$.

The amplifier is not considered in this analysis, because $x_1(t)$ and $x_2(t)$ in Fig. 5-12 have identical characteristics except for amplitude. We assume, for convenience, that $x_2(t) = x_1(t)$.

The demodulated signal $x_3(t)$ is the product of $x_2(t)$ and the synchronizing signal $x_c(t)$, which is identical to the carrier as shown in Fig. 5-12. Hence output from the demodulator is

$$x_3(t) = [x_2(t)](A \cos \omega_c t) \tag{5-4}$$

Let $x_2(t) = x_1(t)$. Substituting $x_1(t)$ from Eq. 5-1 into Eq. 5-4 gives

$$x_3(t) = [x(t)](A \cos \omega_c t)^2 \tag{5-5}$$

$$x_3(t) = \frac{A^2}{2} [x(t)](1 + \cos 2\omega_c t)$$

$$x_3(t) = \frac{A^2}{2} [x(t) + x(t) \cos 2\omega_c t] \tag{5-6}$$

The first term in Eq. 5-6 is proportion to $x(t)$ and the second involves the frequency $2\omega_c$. Hence the high-frequency component in Eq. 5-6 can be rejected by means of a low-pass filter. That the demodulation is phase sensitive can also be deduced from Eq. 5-5. Since $x_3(t)$ is the product of the original signal $x(t)$ and $[x_c(t)]^2$ of the carrier and $[x_c(t)]^2$ is always positive, the product therefore retains the sign of $x(t)$.

The AM process described is called a *suppressed carrier system* because the carrier is absent in the modulated signal when the input $x(t) = 0$, as shown in Figs. 5-12 and 5-13 and Eq. 5-6. Suppressed carrier systems are often used in instruments but not in AM radio broadcasting.

B. Types of Carriers

A carrier is an auxiliary input that conveys information from an input to the output of a transducer. Carriers can be classified by their waveforms, which are (1) constant levels (generally called dc), (2) sine waves, and (3) pulse trains. These carriers are described briefly in this section.

Any form of energy can be used as a carrier. Electrical quantities are commonly used, but optical devices such as fiber optics are widely employed in communication. Digital computers of the future may well be optical [5]. A novel application of pulse modulation is the "measurement-while-drilling" technique used in offshore oil drilling. The drilling mud for cooling and lubricating the drill bit is pulsed, and the mud serves as a carrier of information to the surface [6].

Any property of a wave can be used for carrying information. A dc voltage is not usually called a carrier, because the only property of its "waveform" is a constant level. This property can be modulated by an information signal to yield an analog output proportional to the value of the signal. For example, the battery in Fig. 5-11 supplies energy to a strain gage bridge. Signal from the gages are "carried" downstream in the form of an analog output proportional to the applied strain.

The properties of a sine wave are amplitude, frequency, and phase angle. Each can be modulated for carrying information. Amplitude modulation has been described. Phase modulation is similar to frequency modulation FM. FM is described briefly below.

Frequency modulation (FM) can be demonstrated with a voltage-controlled oscillator, the output of which has a constant-amplitude sine wave with variable instantaneous frequency $f(t)$:

$$f(t) = f_c + mx(t) \tag{5-7}$$

where f_c is the center frequency of the carrier, $x(t)$ the information signal, and m the modulation index. The frequency of the FM signal, shown schematically in Fig. 5-14, is above or below the center frequency f_c in proportion to the magnitude of $x(t)$. The time-varying frequency in FM is more difficult to analyze than AM. The frequency translation for FM is similar to that for AM, shown in Fig. 5-13, but the sidebands for FM extend much beyond the bandwidth of the information signal.

A simplified scheme for information retrieval from a FM signal is shown in Fig. 5-15. The limiter eliminates the spurious amplitude variations due to noise and interferences in the incoming FM signal $x(t)$ to give $x_1(t)$. The demoulator or the FM-to-AM converter is essentially a high-Q resonant circuit (see App. A). The frequency response of the converter is as shown, where f_0 is the resonant frequency. By placing the center frequency f_c of the incoming FM signal at a linear portion (dashed line) of the frequency response curve, the voltage $x_2(t)$ becomes an AM signal, whose amplitude is proportional to the information in $x(t)$. Note that the information is carried by the envelope of $x_2(t)$. The envelope detector is a half-wave rectifier. The information in $x(t)$ is retrieved from the rectified voltage by means of filtering and eliminating the mean value so that the output is phase sensitive.

Pulse modulation is used in digital equipment. The pulse train in Fig. 5-16a consists of uniform pulses of equal height and equal width. The parameters of a pulse train are (1) pulse height, (2) pulse frequency, (3) pulse position, and (4) pulse width. Any one of these properties can be modulated for carrying information. This gives four methods of modulation. Additionally, a code can be devised according to the presence or absence of a pulse. This is called pulse code modulation or digital data.

An example of pulse height modulation is shown in Fig. 5-16b. A train of pulses of varying magnitude is obtained from the signal by means of

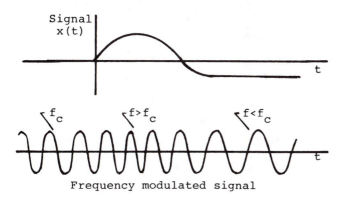

FIGURE 5-14 Waveforms of frequency modulation.

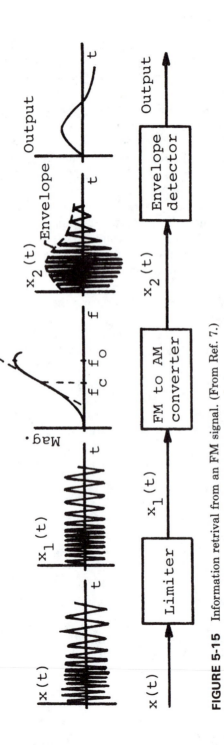

FIGURE 5-15 Information retrival from an FM signal. (From Ref. 7.)

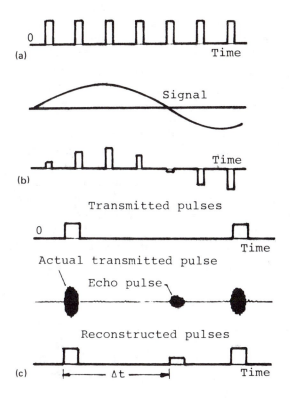

FIGURE 5-16 Examples of pulse signal transmission. (a) Pulse train. (b) Pulse height modulation. (c) Pulses in radar.

sampling (see Sec. 6-7). Since the information is carried by the magnitudes, pulse height modulation has the same disadvantage as AM in transmission. Accurate signal transmission is achieved if the receiving equipment only needs to detect the presence or absence of a pulse instead of the magnitude of the pulses. For example, pulse modulation is used in radar, as shown in Fig. 5-16c. The transmitted pulses modulate a sinusoidal carrier of extremely high frequency. The transmitted pulse and its echo can be reconstructed from the presence or absence of a signal. The time interval Δt between a transmitted pulse and its echo is the time for a round trip between the transmitter and the reflecting object.

Advantages of carrier systems are information conversion and multiplexing. The disadvantage is the added complexity, which is a trade-off for the advantages.

Information conversion allows a signal to be carried (1) at the desired waveform, such as AM, FM and pulse modulation; (2) at the desired frequency, such as the assigned frequency of a radio station; and (3) at a higher energy level, such as for radio telemetry in measurements. Instruments for stronger signals have less stringent requirements than those for weaker signals. In any case, a transducer is also an information processor [8].

Multiplexing allows the use of one transmitter for several signals. It is a saving on equipment and a necessity for some applications, such as in a spacecraft. Multiplexing of several signals is possible if each signal is given a distinct characteristic by means of its carrier. For example, if each signal is carried at a different frequency, it is possible to use the same channel for transmitting several signals. This is called *frequency-division multiplexing*. The composite signal can be separated by selective filtering at the receiving station. For example, the atmosphere carries numerous radio signals, yet it is possible to tune to a particular station. Time-division multiplexing is described in Chapter 6.

5-5.   INDUCTIVE TRANSDUCERS

Inductive transducers for mechanical measurements are electromechanical displacement sensors. The LVDT is a prime example. Transducers for different applications may employ variations of similar basic principles. In this section we discuss the general principles. Applications of inductive transducers are described in the next section.

The basis of an inductive transducer is *Faraday's law*. The induced current in a loop is in a direction as to produce a magnetic flux opposing a change in the flux density. That is, a changing magnetic field produces an electric field. When a flux $\phi(t)$ passes through an N-turn coil, shown in Fig. 5-17a, the induced voltage $v(t)$ across its terminals is

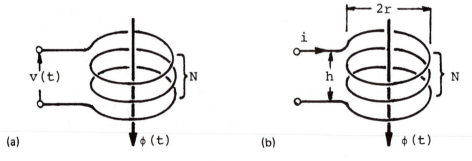

(a)   (b)

FIGURE 5-17   Self-inductance. (a) Induced voltage $v(t)$. (b) Flux in coil.

$$v(t) = -N \frac{d\phi}{dt} \tag{5-8}$$

The time rate of change of flux $d\phi/dt$ may be from a time-varying magnetic field or from the motion of the coil in a stationary field.

Let us relate Eq. 5-8 and the inductance L of a coil. The total flux ϕ due to current i in the coil of radius r and height h in Fig. 5-17b is

$$\phi = \mu Ni \frac{\pi r^2}{h} = \mu NiK \tag{5-9}$$

where μ is the permeability of the medium and K is a geometric form factor. Substituting Eq. 5-9 into Eq. 5-8 gives the familiar equation

$$v(t) = \mu N^2 K \frac{di}{dt} = L \frac{di}{dt} \tag{5-10}$$

Hence the inductance L of a coil is

$$L = \mu N^2 K \tag{5-11}$$

where L is in henrys and μ, the permeability, is in H/m. L is the self-inductance because it involves a single coil. Mutual inductance M exists between two adjacent coils, as in a transformer. A voltage is induced in coil 2 due to the changing flux from coil 1, and vice versa. For coils with N_1 and N_2 turns the mutual inductance is

$$M = \mu N_1 N_2 K \tag{5-12}$$

Note that the quantity $1/\mu K$ is called the *reluctance* of a magnetic circuit, and *permeance* is the reciprocal of reluctance. Inductive transducers are variously called variable-reluctance or variable-permeance devices. The terminology is not standardized. The magnetic path in an inductive transducer is neither uniform nor homogenous. It is difficult to calculate an inductance except for devices of very simple geometry.

Each of the parameters, μ, K, and N in Eq. 5-11 or 5-12 can be changed to yield a transducer for measurement. For example, a Taylor series expansion of Eq. 5-11 gives (see Eq. 2-4)

$$\Delta L = \frac{\partial L}{\partial \mu} \Delta \mu + \frac{\partial L}{\partial K} \Delta K + \frac{\partial L}{\partial N} \Delta N \tag{5-13}$$

Three basic types of inductive transducers are deduced from the terms on the right side of the equation. There are countless variations within each type, such as changes in K for different geometric configurations.

The types of transducers can be further subdivided by techniques of measurement, such as unbalance or the differential method (see Chap. 3).

Furthermore, the induced voltage in one transducer may be from a stationary coil in a time-varying magnetic field, and that in another may be from a moving coil in a stationary field. Both transducers work on the same basic principle, but they may appear different and serve different purposes. In view of the above, it is difficult to form a simple classification for inductive transducers [9].

5-6. INDUCTIVE TRANSDUCERS: APPLICATIONS [10]

To facilitate the discussion, let us divide inductive transducers into four groups: (1) inductive, (2) eddy current, (3) motor/generator, and (4) transformer transducers. This is neither a definitive classification nor a statement that knowing the principles alone is sufficient. In applications, the distinction between feasible and practical is often in the details.

The first group consists of *inductive devices* for the measurement of small displacements. The displacement gage in Fig. 5-18a consists of an iron core and an armature of ferromagnetic material. The flux is from a stationary coil. The input x changes the reluctance of the magnetic circuit. Since the reluctance is mainly in the air gap, a small cahnge in x can be detected with suitable circuitry. The device becomes a thickness gage if a layer of insulation (not shown), such as a coat of paint, separates the core and the armature. The gage can be modified for large displacements. For example, if a ferromagnetic material (not shown) is inserted into the air gap, the reluctance is determined by the amount of material inserted in the gap. This scheme is shown conceptually in Fig. 5-18b, in which the input is the amount of insertion of the tapered iron rod.

The *mutual-inductance transducer* for small displacements shown in Fig. 5-18c has two coils. The induced voltage in the second coil is a function of the change in reluctance of the magnetic circuit, as before. The thickness gage in Fig. 5-18d works on the same principle, but the reluctance is primarily in the test sheet. The coils are not shown, because either self- or mutual inductance can be used for the device.

A *differential* or a bridge circuit is often used when the change in inductance $\Delta L/L$ is small. The pressure transducer in Fig. 5-18e uses the deflection of a steel diaphragm to change the inductance of the coils. The coils and the components for balancing form a bridge circuit. A phase-sensitive demodulator is required if the diaphragm deflects on both sides of its neutral position. The differential method is used in the accelerometer in Fig. 5-18f. Note the similarity between this device and the LVDT.

The second group may be called *eddy-current devices*. The eddy currents i in Fig. 5-19a are local electric currents induced in a conducting material by a time-varying flux $\phi(t)$. The eddy currents, in turn, cause a magnetic field of a direction opposite to that produced by the coil. The result is a reduction of the inductance of the coil.

A differential is used for the eddy-current proximity transducer in Fig. 5-19b. The probe L and the balancing coil form two arms of a bridge. The target is of electrically conducting material. Flux from the probe generates eddy currents in the target, which in turn cause a measurable change in the inductance of the probe. If the excitation is of the order of 1 MHz, the eddy currents are mostly at the surface of the target. Thus this scheme also allows for nonconducting targets if aluminum foil is attached to the nonconducting surface. The system requires a phase-sensitive demodulator and a low-pass filter at the output.

The *drag-cup tachometer* in Fig. 5-19c consists of a rotating magnet, a conducting cup of nonmagnetic material, and a torsional spring for restoring the cup to its static equilibrium position. Flux from the rotating magnet induces eddy currents in the cup. The currents interact with the magnetic field, thereby producing a torque on the cup. The torque is proportional to the speed of rotation of the magnet. The pointer indicates the rotational speed.

The *eddy-current dynanometer* in Fig. 5-19d is the converse of the drag-cup tachometer, but it is much larger in size and can be designed to absorb thousands of horsepower. The rotor is of conducting material and is free to rotate within the stator. The stator is mounted on trunnion bearings but is restrained to small angular deflections by means of a spring. The magnetic field is from the stator. Eddy currents are induced in the rotor as it is driven by a prime mover. The stator produces an opposing torque equal to the product of force and arm length, shown in the figure. The power dissipation is calculated from the torque and the rotating speed.

The third group exemplifies the *motor–generator effects,* which are bilateral (see Sec. 3-3C and Fig. 5-1). In other words, either the motor effect or the generator effect can be used in a transducer. In principle, these effects are "two sides of the same coin."

The *proximity gage* in Fig. 5-20a utilizes the generator effect for measuring rotational speed by counting. It consists of a permanent magnet and a coil. When a gear tooth crosses the face of the gage, it interrupts the flux pattern from the permanent magnet, thereby inducing a voltage in the coil. The magnitude of the voltage generated is proportional to the rotational speed of the gear. Even at relatively low speeds, the induced voltage is sufficient for coutning. The output voltage for some of these pickups is sufficient to actuate a relay without amplification. The ac tachometer shown in Fig. 5-20b is an ac generator. It has a rotating magnet but works on the same principle as the proximity gage.

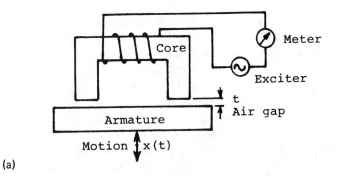

(a)

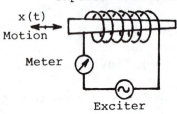

(b)

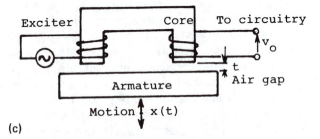

(c)

FIGURE 5-18 Examples of variable-inductance transducers. (a), (b) Self-inductance gages. (c) Mutual inductance gage. (d) Thickness gage. (e) Pressure transducer. (f) Accelerometer.

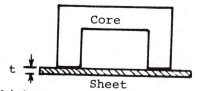

(d)

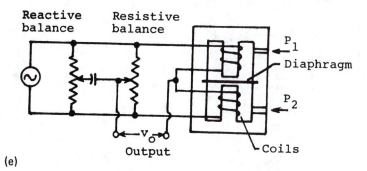

(e)

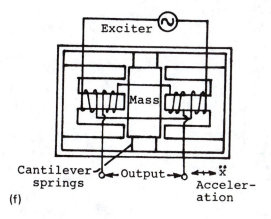

(f)

FIGURE 5-18 (*Continued*)

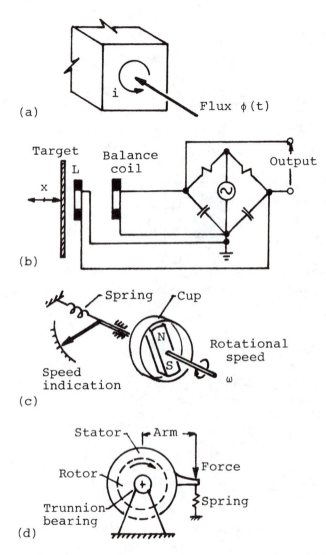

FIGURE 5-19 Examples of eddy-current transducers. (a) Flux and eddy current. (b) Proximity transducer. (c) Eddy current tachometer. (d) Eddy current dynamometer.

The *dc tachometer* in Fig. 5-20c is a dc generator. Voltages induced in the coils of the rotor are brought out by means of a split-ring commutator on the rotor shaft. An actual machine has many segments on the commutator. The split-ring is essentially a "switch-rectifier" since the current i reverses direction simultaneously with each field reversal. The output voltage is as shown in the figure, in which the "dips" occur at the instances of commutation. The ripple voltage due to the dips is about 2% of the average output voltage. The magnitude of the output voltage is on the order of 10 mV/rpm.

The *velocity pickup* in Fig. 5-20d has a moving coil in a stationary magnetic field. It is based on the generator effect, and the voltage output is proportional to the input velocity $\dot{x}$. It can also be used as a phonograph pickup. Conversely, if a voltage is applied to the coil, the device becomes a speaker by virtue of the motor effect, and the output is a velocity proportional to the applied voltage. Note that a speaker makes a good exciter for vibration tests, although it is necessary to cut out part of the diaphragm to reduce the acoustic output.

The *galvanometer* shown in Fig. 5-20e is based on the motor effect. It consists of a moving coil, a permanent magnet, a torsional-spring suspension, a light, and a mirror to indicate the deflection of the coil. The torque produced by a current through the coil is balanced by the restoring torque from the suspension. The deflection of the mirror is proportional to the current input. The deflection is magnified by means of an optical arrangement shown in the figure. The coil can be made very small in order to respond to high input frequencies and transient signals. Galvanometers are usually ganged in a high-speed oscillograph for multichannel recording.

The familiar *d'Arsonval movement* in Fig. 5-20f works on the same principle as the galvanometer. Due to the mass of the parts, it cannot respond to an ac or a transient signal. Hence the movement is basically a dc microammeter. With additional circuitry, however, it can be used in a multimeter to measure resistance or ac/dc voltage or current. (see Probs. 1-8 to 1-13). The deflection of the meter is proportional to product of the current through the coil and the strength of the field from its permanent magnet. If the magnetic field is from stationary solenoidal coil (not shown), the deflection is proportional to the product of the current through the coil and that through the solenoid. Thus the meter becomes an electrodynamometer and measures ac voltages or rms values. It can also be modified (not shown) as a wattmeter for electrical power measurements. The technique in using the product of two quantities has many applications in measurement.

The fourth group includes *transformers* or mutual-inductance devices for angular measurement, transmission, additions, and subtractions. The

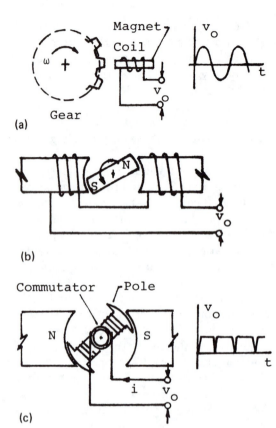

FIGURE 5-20 Transducers using motor–generator effects. (a) Counter. (b) Tachometer, ac output. (c) Tachometer, dc output. (d) Velocity pickup. (e) Galvanometer. (f) d'Arsonval meter movement.

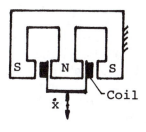

(d)

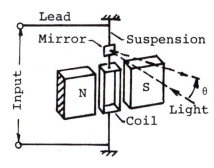

(e)

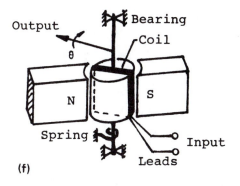

(f)

FIGURE 5-20 (*Continued*)

resolver shown in Fig. 5-21a has the input coil (primary) wound on the stator of a rotating machine and the output coil (secondary) on the rotor. Thus the coils form a transformer. If the axes of the coils are aligned, maximum voltage is induced in the secondary. Zero voltage is induced if the axes of the coils are perpendicular. The device is a resolver for which the output voltage is a sine function of the angular position of the rotor. Resolvers are used in navigation for generating accurate sine functions. If another coil in the rotor is perpendicular to the first, it generates a cosine function.

Conversely, let the input coil in Fig. 5-21a be on the rotor and the output coil on the stator. The voltage output is a linear function of the rotor for small rotations because $\sin \theta \simeq \theta$ for small angles. With proper design of the windings, the output voltage is linear with the rotation of the rotor up to $\pm 90°$ from null. The device is called an *induction potentiometer*.

Two *synchro transducers* for remote transmission of angular positions are shown in Fig. 5-21b. The devices are electrically identical, but the transformer has mechanical damping and other modifications. The

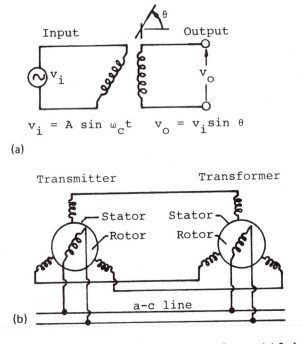

FIGURE 5-21 Transformer-type transducers. (a) Induction potentiometer. (b) Synchro transducers.

transmitter is the generator and the synchro transformer the motor, and both are exicted from the same ac line. It can be shown that the three-phase circuit for connecting the stators of the transducers is necessasry for reproducing the magnitude and direction of the induced voltages. If the rotor of the transmitter is turned, the voltage induced in its stator is transferred to that of the transformer. The torque developed in the transformer forces its rotor to turn the same amount for the remote transmission of angular positions. The transducers are rugged and can be used for continuous rotations up to 1200 rpm.

5-7. STRAIN GAGES [11]

The bonded electrical resistance strain gage is described in this section. Strain is defined as $\Delta L/L$, where L is the length of a resistance wire. Since a strain gage measures ΔL, it is a micro displacement sensor. Topics presented in this section are biaxial stresses, gage selection, and strain gage circuits (see also Sec. 1-3E). Applications will be described in Sec. 5-8.

Materials for the strain sensing element are either metallic or semiconductor. Metallic gages are of either the wire or the foil type (see Fig. 1-6). Foil-type gages have essentially replaced wire gages. Semiconductor gages employ the piezoresistive effect of silicon or germanium. Their chief advantage is high sensitivity, with a gage factor of the order of 150 compared with 2 or 3.5 for metallic gages. Their disadvantages are non-linearity, higher-temperature sensitivity, limited strain range, fragility, and higher cost. A special gage for transducers is the thin-film gage, which is deposited directly on the base material by means of a evaporation process. Other types of gages for stress and fatigue measurements are variations of the metallic gage. The discussions below focus mainly on metallic gages, although the general remarks are equally applicable for all gages.

Some knowledge of stress–strain relation is necessary in a strain measurement. Consider a few examples. Hooke's law is directly applicable only for the tensile specimen shown in Fig. 5-22a. If the gage is placed laterally, as shown in Fig. 5-22b, the strain is due to Poisson's effect, because no force is applied in the lateral direction. Stress concentration occurs at the edge of the hole in the test specimen in Fig. 5-22c. Since a gage measures average strain, the stress concentration is not revealed by a gage of reasonable size. Moreover, it will be shown that Hooke's law cannot be applied directly to biaxial principal stresses shown in Fig. 5-22d.

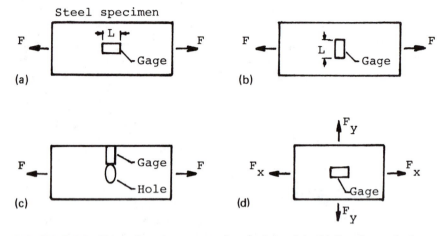

FIGURE 5-22 Examples of stress–strain relations. (a), (b) Specimens in tension. (c) Specimen with hole. (d) Principal stresses.

A. Biaxial Stresses

Biaxial stresses commonly occur in strain measurements, such as on the surface of a torsional shaft. In this section we derive the stress–strain relation for principal stresses and then state the relations for rosette gages.

Consider the strains due to the principal stresses σ_x and σ_y on an elemental surface in Fig. 5-22d. The stress σ_x produces strains in both principal directions. (1) The strain in the x direction is σ_x/E, where E is Young's modulus. (2) The strain in the y direction is $-\nu\sigma_x/E$, where ν is Poisson's ratio. Similarly, the strain produced by σ_y in the y direction is σ_y/E and that in the x direction is $-\nu\sigma_y/E$. The net strain ε_x and ε_y in the principal directions are the algebraic sums of the strains due to the stresses σ_x and σ_y.

$$\varepsilon_x = \frac{1}{E}\,(\sigma_x - \nu\sigma_y) \quad \text{and} \quad \varepsilon_y = \frac{1}{E}\,(\sigma_y - \nu\sigma_x) \tag{5-14}$$

The principal stress–strain relations are found from simultaneous solution of the equations

$$\sigma_x = \frac{E}{1 - \nu^2}\,(\varepsilon_x + \nu\varepsilon_y) \quad \text{and} \quad \sigma_y = \frac{E}{1 - \nu^2}\,(\varepsilon_y + \nu\varepsilon_x) \tag{5-15}$$

Rosette gages are used to determine principal stresses and the maximum shear stress from strain measurements. Since three strain measurements define a stress condition, rosettes generally have three gages packaged as a unit, although two- and four-gage rosettes are also used. A two-gage rosette is used only if the principal directions are known. The fourth gage in a four-gage rosette is for checking the results from the other three gages. Stress-train relations for rosette gages are given in Table 5-1, where the lines a, b, c, and d are the axial directions of the gages. The angle θ is the direction of the algebraically larger stress, measured from the a axais in the counterclockwise direction.

B. Gage Selection

The first step in a gage selection is to specify a realistic test profile. This defines the problem, the constraints, and the end results anticipated. The second step is the selection of gages and the instrumentation. In this section we outline the considerations for gage selection and then briefly examine the parameters, to bring the requirements for a realistic test profile into focus.

A gage is selected to match a number of parameters pertaining to the job. The parameters can be divided into two related groups: (1) the physical parameters of the gage itself, such as the strain sensing material; and (2) those pertaining to applications, such as the operating temperature. One gage is selected among numerous types of gages in a catalog, and the selection can be bewildering for a beginner. Moreover, the parameters can have conflicting attributes, and cannot be selected at random. Fortunately, the parameters are optimized, or compromized, by the manufacturer and gages are offered as types, or gage series, for common applications.

Steps for gage selection [12] and considerations are outlined as follows:

Step 1. Gage length considerations:

strain gradients	area of maximum strain
accuracy required	static strain stability
maximum elongation	cyclic endurance
heat dissipation	space for installation
ease of installation	

Step 2. Gage pattern considerations:

biaxiality of stress	strain gradients (in-plane and normal
heat dissipation	to surface)
space for installation	ease of installation
gage resistance	

Step 3. Gage series considerations:

operating temperature	type of strain measurement (static,
test duration	dynamic, etc.)
cyclic endurance	accuracy required
ease of installation	

TABLE 5-1 Stress–Strain Relation for Rosette Gages[a]

Type of rossette	Rectangular	Equiangular (delta)	T-delta
Principal strains, $\varepsilon_1, \varepsilon_2$	$\dfrac{1}{2}\left[\varepsilon_a + \varepsilon_c\right]$ $\pm \sqrt{2(\varepsilon_a - \varepsilon_b)^2 + 2(\varepsilon_b - \varepsilon_c)^2}$	$\dfrac{1}{3}\left[\varepsilon_a + \varepsilon_b + \varepsilon_c\right]$ $\pm \sqrt{2(\varepsilon_a - \varepsilon_b)^2 + 2(\varepsilon_b - \varepsilon_c)^2 + 2(\varepsilon_c - \varepsilon_a)^2}$	$\dfrac{1}{2}\left[\varepsilon_a + \varepsilon_d\right]$ $\pm \sqrt{(\varepsilon_a - \varepsilon_d)^2 + (\tfrac{4}{3})(\varepsilon_b - \varepsilon_c)^2}$
Principal stresses, σ_1, σ_2	$\dfrac{E}{1 - \nu^2}(\varepsilon_1 + \nu\varepsilon_2)$ $\dfrac{E}{1 - \nu^2}(\varepsilon_2 + \nu\varepsilon_1)$	$\dfrac{E}{1 - \nu^2}(\varepsilon_1 + \nu\varepsilon_2)$ $\dfrac{E}{1 - \nu^2}(\varepsilon_2 + \nu\varepsilon_1)$	$\dfrac{E}{1 - \nu^2}(\varepsilon_1 + \nu\varepsilon_2)$ $\dfrac{E}{1 - \nu^2}(\varepsilon_2 + \nu\varepsilon_1)$
Maximum shear, τ_{max}	$\dfrac{E}{2(1 + \nu)}$ $\times \sqrt{2(\varepsilon_a - \varepsilon_b)^2 + 2(\varepsilon_b - \varepsilon_c)^2}$	$\dfrac{E}{3(1 + \nu)}$ $\times \sqrt{2(\varepsilon_a - \varepsilon_b)^2 + 2(\varepsilon_b - \varepsilon_c)^2 + 2(\varepsilon_c - \varepsilon_a)^2}$	$\dfrac{E}{2(1 + \nu)}$ $\times \sqrt{(\varepsilon_a - \varepsilon_d)^2 + (\tfrac{4}{3})(\varepsilon_b - \varepsilon_c)^2}$
$\tan 2\theta$	$\dfrac{2\varepsilon_b - \varepsilon_a - \varepsilon_c}{\varepsilon_a - \varepsilon_c}$	$\dfrac{\sqrt{3}(\varepsilon_c - \varepsilon_b)}{2\varepsilon_a - \varepsilon_b - \varepsilon_c}$	$\dfrac{2(\varepsilon_c - \varepsilon_b)}{\sqrt{3}(\varepsilon_a - \varepsilon_d)}$
$0 < \theta < +90°$	$\varepsilon_b > \dfrac{\varepsilon_a + \varepsilon_c}{2}$	$\varepsilon_c > \varepsilon_b$	$\varepsilon_c > \varepsilon_b$

[a] θ is the angle of reference, measured positive in the counterclockwise direction from the a axis of the rosette to the axis of the algebraically largest stress.

Step 4. Options considerations:

type of measurement (static, dynamic, etc.)	installation environment (laboratory or field)
stability requirements	soldering sensitivity of subtrate (plastic, etc.)
space available for installation	installation time constraints

Step 5. Gage resistance considerations:

heat dissipation	lead-wire desensitization
signal/noise ratio	

Step 6. Self-temperature-compensation number considerations:

test specimen material	temperature range
accuracy required	

Parameters pertaining to the metallic gage are (1) strain sensing alloy, (2) backing material and bonding cement, (3) gage dimension and pattern, and (4) lead wires.

1. Constantan (advance) and isoelastic are typical alloys for the strain sensing element in gages. Constant is a generic copper–nickel alloy. Isoelastic is a nickel–chromium–molybdenum–iron alloy. Properties of the alloys depend on the composition, processing, and cold work.

A constantan gage nominally has a gage factor of 2 and a resistance of 120 Ω. It is relatively insensitive to ambient temperature changes for normal usage, as shown in Fig. 5-23 [13], and is often used for static strain measurements over long periods. It has good fatigue life and elongation capability, and can be processed into postyield or self-temperature com-

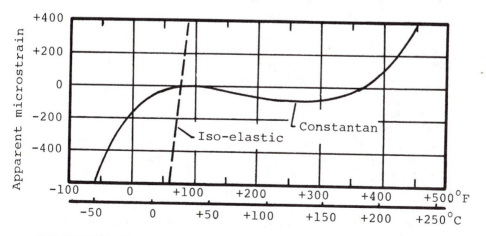

FIGURE 5-23 Temperature effect on strain-sensing alloys. (From Ref. 13.)

pensating gages. A postyield gage allows strains beyond the yield of the test specimen. A self-temperature compensating gage is selected to match the temperature coefficient of the specimen. The disadvantages of constantan are lower sensitivity (nominal gage factor = 2) and the tendency to drift continuously at temperatures above +65°C.

An isoelastic gage has a higher nominal gage factor of 3.5 and a higher resistance of 350 Ω. It is temperature sensitive, as shown in Fig. 5-23, but is good for dynamic strain measurements and for short durations, for which the effect of ambient temperature change is insignificant. It has a superior fatigue life compared with constantan gages. The disadvantages are nonlinearity at strains beyond 5000 μstrain and non-self-temperature compensation.

2. The strain-gage grid is supported on a backing material for handling, electrical insulation, and bonding to the test specimen. The backing may be very flexible to fit small radii, or extremely rigid for long-term stability or for high-temperature applications. The bonding cement adheres the backing to the surface of a test specimen so that the gage becomes an integral part of the specimen. If an installation is exposed to a higher-than-cure temperature and for long times, however, the bonding cement will receive an additional curing [14] and its characteristics are changed accordingly. The backing material and the bonding cement are dictated by the test profile and recommendations of the manufacturer rather chosen at random.

3. Gage pattern and dimensions are governed by the requirement of the application and space limitations. For example, a rosette is used for biaxial strains and a special pattern for the diaphragm of a pressure transducer [15]. Gage length and gage width of a rectangular strain gage are the length and width of the strain-sensitive grid pattern.

The dimensions of a gage determine its ability for heat dissipation. Note that heat dissipation and the "optimum" excitation level [16] of a gage depend on many factors, including the backing material. The consequence of excessive heating due to the excitation is a degradation of performance, such as a zero drift. A larger gage has more surface area for heat dissipation than a smaller one. A larger gage tends to have a greater averaging effect and should not be used for surfaces of severe strain gradients. The averaging effect of a large gage, however, is mandated for the strain measurement of nonhomogeneous materials, such as concrete. A small gage is difficult to handle. Gages shorter than 3 mm (0.125 in.) have lower maximum elongation, fatigue life, and stability.

4. Lead wires are part of a gage installation. Heavier wires are soldered to the tabs of a gage. Some alloys for the strain sensing element, such as karma, may be more difficult to solder [13]. Improper soldering and non-

uniformity could impair the performance of a gage, particularly in fatigue life. Small gage sizes and field installations in restricted areas could compound the difficulties. Options in lead constructions are offered by manufacturers to facilitate the soldering, including solder dots on tabs of a gage, or gages with copper ribbon leads.

External parameters pertaining to the operation usually include ranges of temperature, strain, and fatigue life. The parameters are related. For example, the fatigue life in terms of number of cycles is dependent on the strain level. Usually, a selection chart listing the ranges of parameters for various gage series is available from the manufacturer. Following the steps for gage selection above, a user may choose the gage for the application from a particular series.

C. Strain Gage Circuits

Circuits for strain gages and the associated components for signal transmission are described in this section.

A strain-gage circuit may consist of a single gage or a bridge. A *single-gage circuit* is generally used for dynamic measurements. There are two possible circuits when a single gage is driven by an ideal power source (see Fig. 3-35). With an ideal voltage source, only the current through the gage is available for the strain measurement, as shown in Fig. 5-24a. With an ideal current source, only the voltage across the gage is available for the strain measurement, as shown in Fig. 5-24b. A nonideal power source can be represented by means of its Thévenin's equivalent. Hence the ballast circuit in Fig. 5-24c is representative of a single gage driven by a nonideal power source (see Fig. 3-37 and Probs. 1-5 and 1-6).

A *bridge circuit* (see Sec. 3-4) may consist of one, two, or four active gages. Generally, all the gages in a bridge have the same resistance. From Eq. 3-60, the voltage output V_o of a basic bridge circuit in Fig. 5-25 with gages R_1 to R_4 is

$$\frac{V_o}{V_i} = \frac{V_1 - V_2}{V_i} = \frac{R_3 + \Delta R_3}{R_2 - \Delta R_2 + R_3 + \Delta R_3} - \frac{R_4 - \Delta R_4}{R_1 + \Delta R_1 + R_4 - \Delta R_4}$$

$$(5\text{-}16)$$

The resistors R_{b1} and R_{b2} are for the resistive balancing in a dc excited bridge. This is called *parallel balancing* and is most often used. The effect of this balancing is to desensitize gages R_1 and R_4 relative to gages R_2 and R_3. It can no longer be assumed that the R's in Eq. 5-16 are equal [17].

An ac excited bridge is adjusted for both the resistive and reactive un-

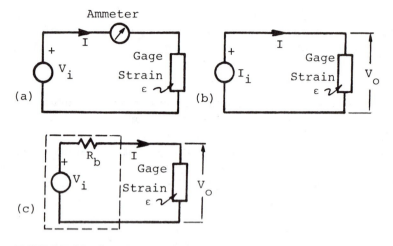

FIGURE 5-24 Circuits using a single gage. (a) Constant-voltage drive. (b) Constant-current drive. (c) Nonideal power source.

balance. The resistive unbalance is adjusted as before. The reactive unbalance is adjusted by means of R_{b3} and C, shown in Fig. 5-25. The balancing procedure depends on the equipment employed. R_s sets the excitation level of the bridge and is therefore a sensitivity adjustment.

A bridge is generally calibrated indirectly (see Fig. 3-33a) because, when bonded to a machine, a gage cannot be transferred for a direct calibration. When no strain is applied to the bridge and the calibrating resistor R_c is placed in parallel with one of the gages, shown in Fig. 5-25, the subsequent bridge output gives the equivalent strain for one active gage

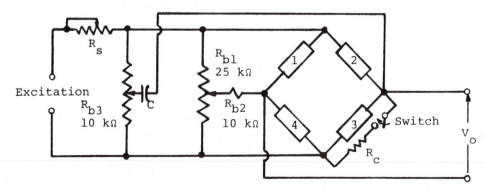

FIGURE 5-25 Bridge with null adjustment and calibration.

(see Prob. 3-9). If the bridge has more than one active gage, a bridge constant C_b is defined to calculate the measured strain:

$$C_b = \frac{\text{total bridge output}}{\text{output from a corresponding one-active-gage bridge}} \qquad (5\text{-}17)$$

Circuit components associated with signal transmission are lead wires, slip rings, and switches. If a component is in series with a gage, the resistance change in a component is an equivalent strain. Common techniques for reducing the undesirable equivalent strains are generalized in Sec. 3-7B. For example, by means of noise cancellation, the effect of resistance in the lead wires of a bridge can be canceled (see Fig. 3-53).

Strain signal from gages on a rotating shaft is conveyed to stationary instruments by means of (1) direct connections, (2) telemetering, or (3) slip rings. Direct connection can be used if the shaft rotates slowly and sufficient length is allowed for the leads to wrap on the shaft. Telemetering is practical if the additional cost can be justified. A common method is to use slip rings. The performance of commercial slip rings is quite satisfactory. To avoid the erratic resistance change in the slip rings, the rings must be external to the bridge, as shown in Fig. 5-26a, instead of in series as shown in Fig. 5-26b. Since the slip-ring assembly is generally mounted at a free end of the shaft, it may not be convenient to thread the leads from the gages to the free end of a shaft. Note that noise due to slip rings can be minimized by using a constant-current drive instead of a constant-voltage drive (see Fig. 3-51).

Switching is used for multipoint measurements when simultaneous strain readings are not required. Similar to slip rings, change in the contact resistance of a switch is an error in a strain measurement. Switches in series with the gages should be avoided. Hence the circuit in Fig. 5-27a is preferable to that in Fig. 5-27b, where gage 3 is for temperature compensation.

5-8. STRAIN GAGES: APPLICATIONS

Applications of bonded strain gages and sources of error are described in this section. Strain gages are routinely used in experimental stress analysis. This encompasses the analyses of new designs, troubleshooting, service load analyses, theoretical investigations, and special problems such as residual stresses [18].

Inasmuch as a bonded strain gage is a secondary sensor for a flexure to which the gage is attached, the applications are examples of the elastic deformation of flexures. Consequently, the errors are due to the coupled modes in the deformation of flexures, the instrumentation for strain

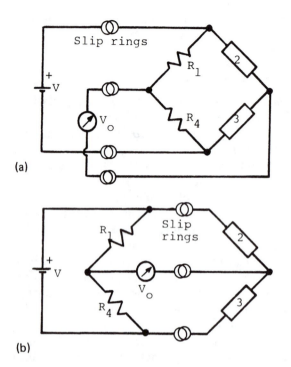

(a)

(b)

FIGURE 5-26 Equivalent resistance due to slip rings in series with strain gages. (a) Slip rings external to bridge. (b) Slip rings in series with gages.

measurement, and environmental effects. Some flexures in common use are very simple and others are extremely elaborate. Simple cases of flexures in tension, bending, and torsion, and some error considerations are described below.

Flexures in tension are shown in Fig. 5-28. The tensile specimen in Fig. 5-28b has one active gage. Since the tension and the bending modes are coupled when the forces F are not symmetrical, the bending mode gives an apparent strain in a tension test. The setup is also sensitive to changes in ambient temperature unless a dummy gage or a self-temperature-compensating (STC) gage is used. The STC gage selected must match the temperature coefficient of the material of the specimen.

Two or four of the gages in the bridge in Fig. 5-28a can be used on a tensile specimen. Gages 2 and 4 are placed on the opposite sides of the specimen shown in Fig. 5-28c. This alleviates the effect of the bending mode, due to the nonsymmetrical forces F, but the arrangement does not compensate for temperature changes. The tensile specimen in Fig. 5-28d

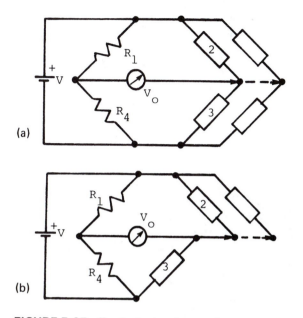

FIGURE 5-27 Equivalent resistance due to switches in series with strain gages. (a) Strain gage switching circuit. (b) Switch circuit with common compensating gage.

is compensated for both temperature and bending, where the gages 2A and 2B are in series at location R_2 of a bridge, and 3A and 3B are in series at location R_3. The bridge constant from Eq. 5-17 is about 1.3. Note that when two gages are in series, the strain indicated is not twice that of a single gage.

Flexures in bending are shown in Fig. 5-29. The cantilever in Fig. 5-29a is a force transducer. Gages 2 and 3 of the bridge are on the opposite sides of the beam. The configuration is insensitive to ambient temperature, axial load, and torsion. The cantilever in Fig. 5-29b is used in a differential pressure transducer. The same scheme is used in the flow transducer in Fig. 5-29c. The cantilevers in Fig. 5-29d form a torque table (e.g., Lebow Torque Sensor, Eaton Corp., Troy, Michigan). It can be used for measuring the output torque of an engine. Other applications (not shown) of flexures in the bending mode include weighing scales, seismic instruments, torque wrenches, and transducers for torque measurements of rotating shafts.

Other applications of flexures are illustrated in Fig. 5-30. Torque can be measured directly by means of the shaft shown in Fig. 5-30a. The

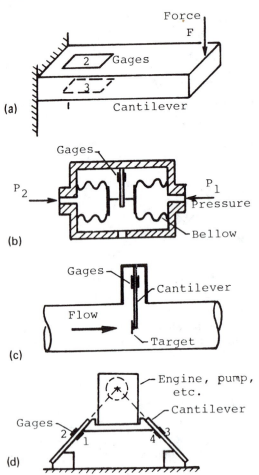

FIGURE 5-28 Flexures in tension. (a) Strain gage bridge. (b) Tensile specimen. (c) Force transducer. (d) Tensile specimen.

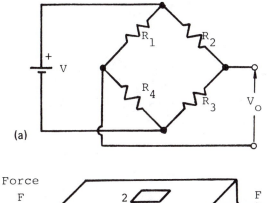

(a)

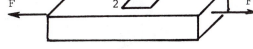

(b)

(c)

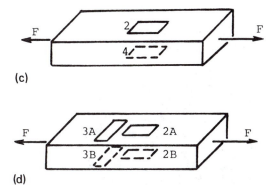

(d)

FIGURE 5-29 Flexures in bending. (a) Force transducer. (b) Differential pressure transducer. (c) Flow measurement. (d) Engine dynamometer.

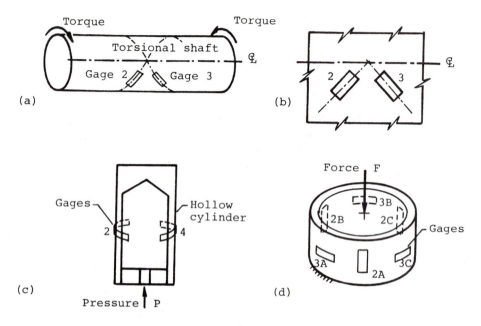

FIGURE 5-30 Applications of strain gages. (a) Torsional shaft. (b) Torque measurement. (c) Pressure transducer. (d) Load cell.

locations of the gages are shown in Fig. 5-30b. The arrangement is insensitive to temperature. The pressure transducer shown in Fig. 5-30c is a hollow cylinder with gages around the circumference. It is for higher-pressure measurements compared with the bellows in Fig. 5-29b. Its frequency response can be improved by inserting a plug in the cylinder to reduce the volume of the cavity. The load cell in Fig. 5-30d uses three series-connected axial gages and three series-connected Poisson-ratio gages to minimize the effects due to eccentric loading.

Common sources of error in strain gage applications are the coupled modes in the elastic deformation of a flexure, misalignment of gages, electrical noise pickups, and field environments. Errors due to the coupled modes are minimized (1) by placing the gages on a flexure such that they tend to compensate for each other, as shown in Fig. 5-30d, or (2) by designing the flexure to respond only to the desired input. The error due to the small misalignment of a gage in a uniaxial stress field is minimal. The Poisson effect in a unixial stress field is already accounted for in the gage factor and no correction is necessary. A correction for transverse sensitivity may be necessary in a biaxial stress field [19]. The principal stress directions are frequently predetermined by means of a stress coat analysis prior to applying the gages for the correct alignment.

Electrical noise in strain measurements due to electric and magnetic fields has been described in Sec. 3-7. Noise from the thermoelectric effect due to temperature differences in the materials in a circuit can be suppressed by using a carrier system. Low-noise cables can be used to minimize the triboelectric effect [20]. This is an erratic electrical noise in the vibrating leads, caused by the momentary separations between the conductor and the dielectric in the leads.

An apparent strain due to temperature is common in a field environment. The "dummy" gage for temperature compensation must be placed in an arm of the bridge adjacent to the active gage. This method for compensation is effective only if the dummy and the active gage are exposed to identical environments except for strain, including the thermal gradient of the active gage.

The model in Fig. 5-31a shows the variable series resistors as sources of

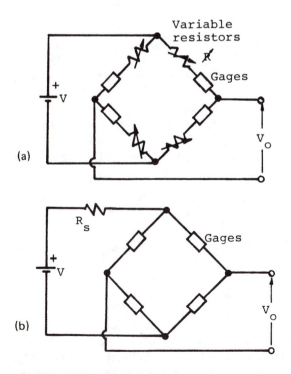

FIGURE 5-31 Errors due to temperature and compensation. (a) Model of apparent strains. (b) Bridge compensation.

apparent strain, including resistance changes in slip rings and other circuit components. The entire system, including the leads, the gage factor [21], and the modulus of elasticity of the material to which the gages are bonded, is susceptible to changes in temperature. If the modulus decreases with increasing temperature, a larger strain is indicated for the same applied force. A scheme for this compensation is to change the sensitivity of the bridge by means of a series resistor R_s, shown in Fig. 5-31b (see also Fig. 5-25).

Extreme field environments and high accuracies are special problems. For some applications, a gage and its leads must be protected. Moisture-proofing, lead-wire considerations, and other techniques can best be learned from hands-on experience or application notes from the manufacturer. The proper installation of a gage could mean a successful or unsuccessful measurement. Considerations such as creep at high temperatures or the long-term stability of gages and material in a transducer are special problems not discussed here.

5-9. CAPACITIVE TRANSDUCERS

A capacitive transducer for mechanical measurements is an impedance-based electromechanical incremental-motion device. In this section we present the basic equations of the capacitor. Applications are described in the next section.

A capacitor consists of two plates of conducting material of common area A separated by a distance d by a dielectric ε (insulation), as shown in Fig. 5-32. Leads from the plates are connected to a high-frequency electronic circuit. Neglecting the nonuniform fringe effect at the edges of the plates, we have

$$C = 0.0885 \, \varepsilon \, \frac{A}{d} \quad \text{(in pF for } A \text{ and } d \text{ in cm)} \tag{5-18}$$

$$C = 0.225 \, \varepsilon \, \frac{A}{d} \quad \text{(in pF for } A \text{ and } d \text{ in in.)} \tag{5-19}$$

or

$$C = k \, \frac{A}{d} \tag{5-20}$$

where ε is the dielectric constant. For example, $\varepsilon = 1$ for air, $\varepsilon = 9$ for mica, and $\varepsilon = 100$ for titanium dioxide. The dielectric constant may change with operating conditions, such as humidity. For air, ε changes about 5% from 1 to 100 atm.

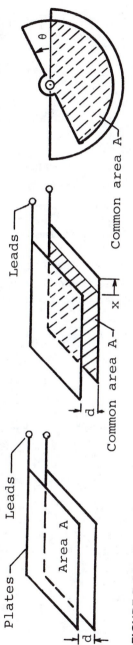

FIGURE 5-32 Capacitor $C = kA/d$.

The potential v across the plates of a capacitor C is due to the accumulation of charges q on the plates:

$$v = \frac{q}{C} = \frac{1}{C} \int i \, dt \qquad (5\text{-}21)$$

where v is in volts, q in coulombs, C in farads, and i in amperes. Note that an electric current does not flow conductively across the plates of a capacitor. A current flow is due to the flow of electrons into a plate, thereby inducing corresponding positive charges in the opposite plate.

The equivalent capacitance of two capacitors C_1 and C_2 in parallel and in series, shown Fig. 5-33a and b, are

$$C = C_1 + C_2 \qquad (C_1 \text{ and } C_2 \text{ in parallel}) \qquad (5\text{-}22)$$

$$\frac{1}{C} = \frac{1}{C_1} + \frac{1}{C_2} \qquad (C_1 \text{ and } C_2 \text{ in series}) \qquad (5\text{-}23)$$

The differential output shown in Fig. 5-33c is often used in transducers. The plates P_1 and P_2 are fixed, plate P is movable, and x is the displacement of the plate P. Since the plates in series are equally charged, we obtain from Eqs. 5-20 and 5-21

$$C_1 = k \frac{A}{d+x} \quad \text{and} \quad C_2 = k \frac{A}{d-x} \qquad (5\text{-}24)$$

$$\frac{V_1}{V_2} = \frac{C_2}{C_1} \qquad (5\text{-}25)$$

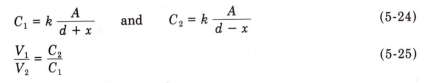

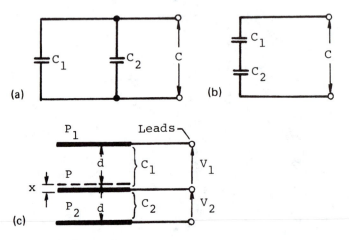

FIGURE 5-33 Capacitors in series and in parallel. (a) C_1 and C_2 in parallel. (b) C_1 and C_2 in series. (c) Differential output.

It can be shown that for $V = V_1 + V_2$, the differential output $\Delta V = V_1 - V_2$ is linear with the input x.

$$\Delta V = V_1 - V_2 = V\frac{x}{d} \tag{5-26}$$

5-10. CAPACITIVE TRANSDUCERS: APPLICATIONS

Capacitance transducers are widely used [22]. In this section we show the applications of basic types of capacitive transducers, describe a number of examples, and discuss the advantages and disadvantages.

Three types of transducers are deduced from Eq. 5-20, where the dielectric constant ε, the common area A, and the separating distance d are independent variables (see Example 2-2).

$$\Delta C = \underbrace{\frac{\partial C}{\partial k}\Big|_0 \Delta k}_{\text{type 1}} + \underbrace{\frac{\partial C}{\partial A}\Big|_0 \Delta A}_{\text{type 2}} + \underbrace{\frac{\partial C}{\partial d}\Big|_0 \Delta d}_{\text{type 3}} \tag{5-27}$$

$$\Delta C = \frac{A}{d}\Big|_0 \Delta k + \frac{k}{d}\Big|_0 \Delta A - \frac{kA}{d^2}\Big|_0 \Delta d \tag{5-28}$$

$$\frac{\Delta C}{C_0} = \frac{\Delta k}{k_0} + \frac{\Delta A}{A_0} - \frac{\Delta d}{d_0} \tag{5-29}$$

where the subscript (0) denotes the steady state about an operating point (see Sec. 2-1). Type 1 transducer is due to a change in the dielectric. Type 2 transducer is due to a change in the common area between the plates. Type 3 transducer is due to a change in the separation between the plates. This type is most frequently used. Bridge circuits (see Sec. 3-4) are often used with capacitive transducers, and a bridge may operate in the balanced or unbalanced mode.

An example of the type 1 capacitive transducer is the liquid-level gage shown in Fig. 5-34a. It consists of a metal cylinder and a concentric metal sheath. The ΔC is from a change in the level of a nonconducting liquid (dielectric) between the cylinder and the sheath. Due to its simplicity of construction and ease for cleaning, it is used for food processing when other transducers are less suitable. Two such transducers in parallel for measuring the average volume of fuel in the tank of an aircraft are shown in Fig. 5-34b (see also Fig. 5-33a).

An example of the type 2 capacitive transducer is the torque sensor shown in Fig. 5-34c. It consists of a serrated shaft and a sleeve, separated by a constant distance d. The capacitance is due to the common area between the teeth of the two components. A torque applied to the shaft

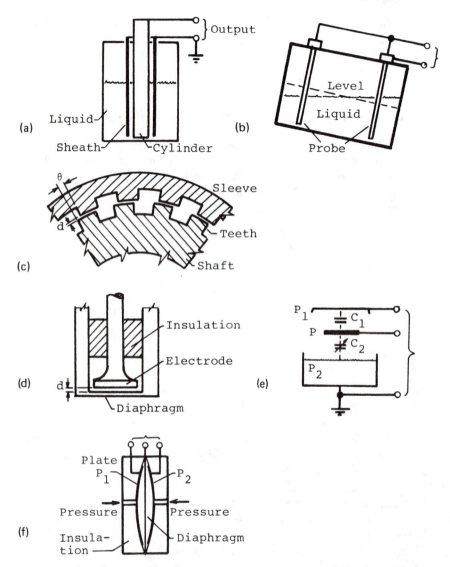

FIGURE 5-34 Examples of capacitive transducers. (a) Liquid-level transducer. (b) Average volume transducer. (c) Torque transducer. (d) Pressure pickup. (e) Liquid-level transducer. (f) Differential pressure pickup.

causes a relative rotation between the shaft and the sleeve, thereby changing the common area and the capacitance of the system. The tuning capacitor of a radio shown in Fig. 5-32 is another example of a capacitance change due to a change in the common area between the plates.

An example of the type 3 capacitive transducer is the pressure pickup for internal combustion engines, shown in Fig. 5-34d. The pickup consists of a diaphragm and a stationary insulated electrode. The motion of the diaphragm changes the distance between the diaphragm and the electrode. The resulting ΔC is used for the pressure measurement. Other examples of the type 3 capacitive transducer are described below.

The differential for a liquid-level measurement, shown in Fig. 5-34e (see also Fig. 5-33c), consists of the plates P_1, P, and P_2, where P_2 is the liquid at the ground potential. The ΔC is due to a change in the liquid level. The differential pressure transducer in Fig. 5-34f consists of a sensing diaphragm in a cavity between two stationary plates P_1 and P_2. It is capable of sensing small differential pressures in the presence of large pressures. An overpressure causes the diaphragm to bottom-out on P_1 or P_2. The diaphragm can be isolated from the working fluid by using two additional outer diaphragms, with the cavities between the transducer and the outer diaphragms filled with silicon fluid.

The common condenser microphone, shown in Fig. 5-35a, consists of a dc polarizing voltage V, a drop resistor R for measuring the output v_0 and the plates of the capacitor. Let d be the nominal separation between the plates and x the change in the separation due to an external input. The total instantaneous separation between the plates is $d \pm x$. It can be shown that the transfer function (see Sec. 4-6) of the linearized system is

$$\frac{v}{x}(j\omega) = \frac{K(j\omega\tau)}{1 + j\omega\tau} \tag{5-30}$$

where τ is the RC constant ($\tau = 0.0885 \times 10^{-12} AR/d$; see Eq. 5-18), and K ($= V/d$) is the sensitivity of the microphone. The frequency response plot of Eq. 5-30 is left as an exercise.

A more detailed diagram of the condenser microphone is shown in Fig. 5-35b. The stationary plate and the movable diaphragm form a variable capacitor. The relative displacement x is due to the motion of the diaphragm in response to the sound pressure. The diagram is deduced by neglecting the relation between the diaphragm and the sound pressure waves, the equivalent mass of the diaphragm, the spring constant, and the damping in the system.

The output of a capacitive transducer may be expressed in voltage, current, or frequency. For example, the circuit shown in Fig. 5-36a con-

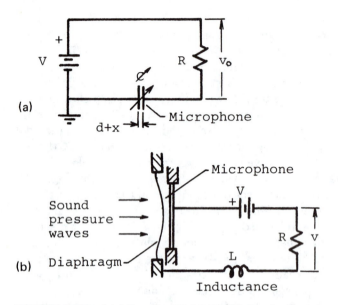

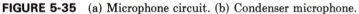

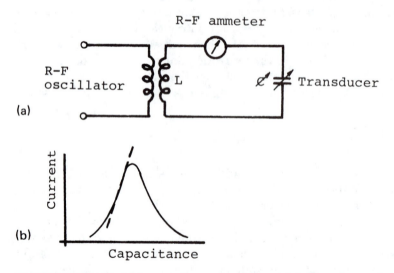

FIGURE 5-35   (a) Microphone circuit. (b) Condenser microphone.

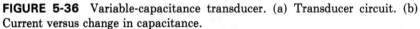

FIGURE 5-36   Variable-capacitance transducer. (a) Transducer circuit. (b) Current versus change in capacitance.

sists of an inductor L and a variable capacitor C. It is driven by means of a radio-frequency oscillator. Peak response occurs at the resonant frequency of the LC circuit, shown in Fig. 5-36b. A linear portion of the response curve (shown as a dashed line) can be used to indicate the output. Since the response is in terms of current versus ΔC, the output is measured with an ammeter. Alternatively, the frequency can be compared with that of a reference standard, such as a crystal oscillator, and the beat frequency used to indicate the output.

A capacitive transducer has many advantages.

1. It is conceptually simple. Its operation does not depend on the plate material, and one of the plates can be at the ground potential. In other words, the measurement is not between parts of a transducer, but between an element of the transducer and some other object. For example, in measuring the runout of a rotating shaft, the variable capacitance is between the insulated probe and the shaft.

2. It is capable of high sensitivity. For the type 3 transducer, the sensitivity depends on the distance separating the plates, which can be made extremely small (see Eqs. 5-20 and 5-27). The minimum separation is the breakdown voltage of the dielectric. High sensitivity also implies small displacements, which can be an advantage for some applications.

3. It is a versatile tool, as shown in the examples above. It is employed when other transducers are less suitable, such as in food processing. As a proximity pickup, it has the smallest effect on the object being measured, second only to an optical transducer. The force due to the electric field between the plates is negligible. The separation between the plates can be very large for some applications. For example, an altimeter for aircraft can be constructed for measuring heights between 3 and 30 m. [23].

4. Its limitations depend more on the mechanical construction than on the electric characteristics. A capacitive transducer can have excellent dynamic response, stability, repeatability, and resolution. With proper design it can withstand severe environmental conditions. For example, its high-temperature range is largely governed by the dielectric.

The main disadvantage of a capacitive transducer is its high output impedance and the associated problems listed below.

1. A capacitive transducer generally costs more than the inductive or resistive type, and the associated electronics for its operation are inherently more complex. For example, the capacitive transducer is

sensitive to stray capacitance, the shunt capacitance of the leads, and insulation leakage. These could be minimized, but the additional complexity would add to the cost of the system.

2. The inherent nonlinearity of a capacitive transducer (see Eq. 5-20) can be improved by means of electronic circuitry and feedback, by restricting the operation to small displacements, or by using a differential circuit.

3. Other limitations can be overcome by means of proper design. For example, the pressure pickup in Fig. 5-34d is inherently nonlinear. The space between the plates can be partially filled with mica. This improves the robustness of the device as well as its linearity. Limitations related to mateiral properties, such as changes in the dielectric constant with temperature or moisture, are inherent in all design problems.

PROBLEMS

5-1. The demodulated signal v_o from a LVDT shown in Fig. 5-7b is recovered by means of the RC filter shown in Fig. P5-1a. The excitation frequency of the LVDT is 3 kHz, the frequency band of the input $x(t)$ is from 0 to 400 Hz, and v_o is observed by means of an oscilloscope with an input impedance R_i of 1 MΩ. (a) Design a filter to give a ripple of less than 5% of the unfiltered signal. (b) If the highest-frequency content of $x(t)$ is 400 Hz, find the amplitude and phase distortion of the signal at this frequency. (c) Using the same RC filter, find the frequency of the signal for which the amplitude distortion is 5%. Find the corresponding phase distortion.

5-2. Repeat Prob. 5-1 for the two-stage RC filter shown in Fig. P5-1b.

5-3. A LVDT summing circuit for remote indicating is shown in Fig. 5-9. Sketch a similar diagram for summing three inputs, x, y, and z.

5-4. Referring to Fig. 5-9, sketch a LVDT subtracting circuit for $z = x - y$.

5-5. Repeat Prob. 5-4, but use a slide-wire balance (potentiometer) instead of a LVDT at the output.

5-6. A synchro pair (see Fig. 5-21) is used as an error detector in an instrument with feedback. The input to the transmitter is measured by means of the transformer, the output of which is amplified to drive a servomotor. The output of the motor is fed back to the transformer to null the system. Draw a schematic sketch of the system.

5-7. The steel cantilever shown in Fig. P5-2 is used in a force transducer. The spring constant of the cantilever is $k = 3EI/L^3$. Assume that $E = (30 \pm 1) \times 10^6$ psi, $w = 0.5$ in. ± 0.002 in., $h = 0.25$ in. ± 0.002 in., and $L = 7.0$ in. ± 0.04 in. Find the root-sum-square error in k.

5-8. The torque table shown in Fig. 5-29d is supported by means of four beams, which are at 45° from the vertical axis. Estimate the capacity of the torque table if the dimensions of the beams are as shown in Fig. P5-3. Assume that the effective length of the beam is 2.5 in. and that the strain gages are 8.0 in. from the center of rotation.

5-9. A circular shaft fixed at one end is used to sense the forces applied at the free end in the (x, y, z) directions by means of strain gages. Sketch the locations of the gages on the shaft.

5-10. Two adjacent strain gages in a bridge circuit are used in a torque transducer as illustrated in Fig. 5-30a and b. Owing to the axial locations of the gages, the transducer may be sensitive to the strain gradient along the shaft due to a bending load. Show the possible combinations of gage selection and locations. Select the one that is least sensitive to the strain gradient due to bending.

5-11. Repeat Prob. 5-10 when a full bridge is used for the torque transducer.

5-12. Derive the equations for (a) the rectangular, and (b) the delta strain-gage rosette shown in Table 5-1.

5-13. A rectangular rosette is attached to a steel plate having $E = 30 \times 10^6$ psi and $\nu = 0.3$. The strain measurements are: (a) $\varepsilon_a = 500$ μin./in., $\varepsilon_b = 300$ μin./in., and $\varepsilon_c = -100$ μin./in. (b) $\varepsilon_a = 305$ μin./in., $\varepsilon_b = 75$ μin./in., and $\varepsilon_c = 150$ μin./in. Calculate the principal strains and stresses and the maximum shear stress. Locate the axis of the principal stress.

5-14. A delta rosette is attached to a steel plate having $E = 30 \times 10^6$ psi and $\nu = 0.3$. The strain measurements are $\varepsilon_a = 500$ μin./in., $\varepsilon_b = 255$ μin./in., and $\varepsilon_c = -175$ μin./in. Calculate the principal strains and stresses and the maximum shear stress. Locate the axis of the principal stress.

5-15. The strain-gage bridge as shown in Fig. P5-4 is used for a torque wrench. A simple bridge is inadequate because of the uncertainty of the loation of the applied force F. Verify that the method illustrated is independent of the location of F.

5-16. Derive the transfer function (Eq. 5-30) for the simplified condenser microphone shown in Fig. 5-35a.

5-17. The torque transducer shown in Fig. 5-34c has 30 pairs of teeth between the shaft and the sleeve. The nominal common area between each pair of teeth is 0.5 in. × 0.5 in. and their separation is $d = 0.008$ in. Calculate the sensitivity of the transducer in pF/deg. of relative rotation between the shaft and the sleeve.

5-18. The capacitive transducer for motion measurement shown in Fig. P5-5 consists of vertical plates mounted on stable insulating material. The excitation frequency is 2 kHz, the common area between each pair of plates is 4 in. × 12 in., and the separation is $d = 0.01$ in. The plates are used in a bridge circuit. Show the bridge circuit and calculate the sensitivity of the bridge.

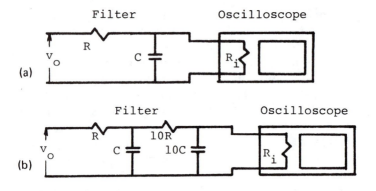

FIGURES P5.1a and b

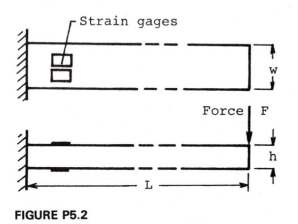

FIGURE P5.2

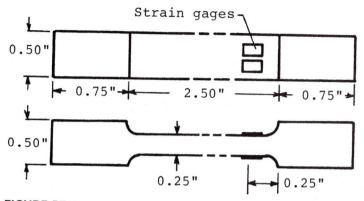

FIGURE P5.3

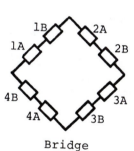

Bridge

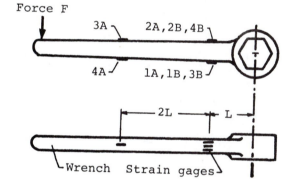

FIGURE P5.4

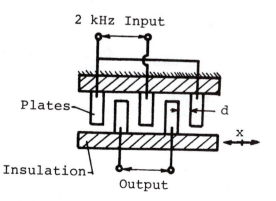

FIGURE P5.5

REFERENCES

1. Herceg, E. E., *Handbook of Measurement and Control*, Schaevitz Engineering, Pennsauken, N.J., 1972.
   Displacement Transducers, TB-1002C; *Velocity Transducers*, TB-1101; *Gage Heads*, TB-1506B; *Non-contact Gaging Heads*, TB-1505; *Inclinometers*, TB-4504C; *Force Transducers*, TB-5002A; *Pressure Sensor*, TB-3001A, *Accelerometers*, TB-4501D; Schaevitz Engineering, Pennsauken, N.J.
2. Wilson, E. B., Jr., *An Introduction to Scientific Research*, McGraw-Hill Book Company, New York, 1952, p. 103.
3. Tenney, A. S., III, Industrial Radiation Thermometry, *Mech. Eng.*, Vol. 108, No. 10 (Oct. 1986), p. 37; Barnes Engineering Co., Stamford, CT.
4. Gibson, J. E., and F. B. Tuteur, *Control System Components*, McGraw-Hill Book Company, New York, 1958, p. 265.
5. Abraham, E., C. T. Seaton, and S. D. Smith, The Optical Computer, *Sci. Am.*, Vol. 248, No. 2 (Feb. 1983), pp. 85–93.
6. Teleco Oilfield Services, Measurement-While-Drilling Instrument System, *Sonat*, Vol. 3, No. 2 (Spring 1984), pp. 3–7.
7. Carlson, A. B., and D. G. Gisser, *Electrical Engineering—Concepts and Applications*, Addison-Wesley Publishing Company, Inc., Reading, Mass., 1981, p. 519.
8. Stein, P. K., Sensors as Information Processors, *Res. Dev.*, Vol. 21 (June 1970), pp. 34–40.
9. Lion, K. S., *Instrumentation in Scientific Research*, McGraw-Hill Book Company, New York, 1959, p. 45.
10. Lion, Ref. 9, pp. 44–65.
    Roberts, H. C., *Mechanical Measurement by Electrical Methods*, The Instrument Publishing Co., Pittsburgh, Pa., 1951, Chap. 4.
11. Perry, C. C., and H. R. Lissner, *The Strain Gage Primer*, 2nd ed., McGraw-Hill Book Company, New York, 1962.
12. *Student Manual for Strain Gage Technology*, Bulletin 309, Measurements Group, Inc., Raleigh, N.C., 1983, p. 10.
13. *Strain Gage Selection, Criteria, Procedures, Recommendations*, TN-505, Measurements Group, Inc., Raleigh, N.C.
14. Stein, P. K., Thirty Years of Strain Gages: 1938–1968, *Instrum. Control Syst.*, Vol. 41 No. 2 (Feb. 1968), pp. 75–78.
15. *Design Considerations for Diaphragm Pressure Transducers*, TN-510, Measurements Group, Inc., Rayleigh, N.C.
16. *Optimizing Strain Gage Excitation Levels*, TN-502, Measurements Group, Inc., Raleigh, N.C.
17. Stein, P. K., Parallel Balancing of Strain Gage Bridges, *Strain Gage Read.*, Vol. 1, No. 2, (June–July 1958), p. 35.
18. *Measurement of Residual Stresses by the Hole Drilling Strain Gage Method*, TN-503-1, Measurements Group, Inc., Rayleigh, N.C.
19. *Errors Due to Transverse Sensitivity*, TN-509, Measurements Group, Inc., Rayleigh, N.C.

20. Eller, E. E., and R. W. Conrad, in *Shock and Vibration Handbook,* 2nd ed., C. M. Harris and C. E. Crede (eds.), McGraw-Hill Book Company, New York, 1976, pp. 12–26, Chap. 12.
21. *Temperature Induced Apparent Strain and Gage Factor Variation in Strain Gages,* TN-504, Measurements Group, Inc., Rayleigh, N.C.
22. Lion, Ref. 9, pp. 66–72, 98–100.
    Roberts, Ref. 10, Chaps. 3, 9, and 10.
23. Roberts, Ref. 10, p. 23.

SUGGESTED READINGS

Dally, J. W., and W. F. Riley, *Experimental Stress Analysis,* 2nd ed., McGraw-Hill Book Company, New York, 1978.

Dally, J. W., W. F. Riley, and K. C. McConnell, *Instrumentation for Engineering Measurements,* John Wiley & Sons, Inc., New York, 1984.

Doebelin, E. O., *Measurement Systems—Application and Design,* 3rd ed., McGraw-Hill Book Company, New York, 1983.

Gardner, M. F., and J. L. Barnes, *Transients in Linear Systems,* Vol. I, John Wiley & Sons, Inc., New York, 1956.

Seely, S., *Dynamic Systems Analysis,* Reinhold Publishing Corporation, New York, 1964.

Shearer, J. L., A. T. Murphy, and H. H. Richardson, *Introduction to System Dynamics,* Addison-Wesley Publishing Company, Inc., Reading, Mass., 1967.

6
Signal Conditioning and Output Devices

6-1. INTRODUCTION

The functions performed by a measuring system are signal detection, signal conditioning and transmission, and data presentation, as shown in Fig. 1-2. Signal detection has been described. Signal conditioning and data presentation in the analog and digital modes are examined in this chapter.

Signal conditioning deals with a host of topics. The instrumentation is mostly electrical, for ease of handling and amplification. A mechanical device can magnify a signal such as by means of leverage, but cannot conveniently increase the power level of a signal. Due to the lack of power amplification and the inherent mass loading, mechanical components, such as integrators, are for special applications. Mechanical devices for signal conditioning are omitted in this presentation.

Signal conditioning in the *analog mode* involves amplification, manipulation, filtering, modulation/demodulation (see Sec. 5-4), and dynamic compensation (see Sec. 4-6). In the analog mode, the value of the information is determined by the precise magnitude of a signal. Operational amplifiers and their applications for signal conditioning are examined in this chapter. Note that operational amplifiers are often used as components in digital equipment.

Signal conditioning in the *digital mode* by means of digital computers or microprocessors is almost without limit. In the digital mode, signals are basically binary pulses, in either the ON or the OFF state. The precise value of a pulse is not important (see Sec. 5-4B). The information is determined solely by the ON/OFF states and no information is lost as long as the states can be identified. Consequently, in addition to the employ of digital computers for data processing, digital transmission has the advantage of high tolerances for spurious noise signals. Fiber optic transmission is relatively new and but extremely important [1]. It will be shown that the basis for the digital operations rests on only a limited number of basic elements.

Modern measuring systems often operate in a *hybrid mode,* that is, a combination of analog and digital. Most transducers for signal detection are analog because physical events, such as a displacement, are continuous functions. Analog signals are converted to digital for storage and/ or manipulation in a computer. Owing to the widespread use of digital equipment, digital transducers are becoming more popular. Digital data are generally converted to analog for applications, such as in process control. Digital elements, applications, and analog–digital conversions are described in the chapter.

Equipment selected for signal conditioning must be compatible. For analog equipment, the main problem in compatibility is impedance matching to avoid loading or to transfer maximum power (see Sec. 3-3C). For digital equipment, comaptibility (interfacing) of equipment and software is more complex and not a task for the casual user.

Since all physical events occur in the time domain, there are time lags in both the analog and digital modes. For analog systems, time lag is manifested in the form of a phase angle in the dynamic response (see Chapter 4). For digital systems, "real-time" data processing is almost instantaneous. Let us not forget that it always takes time to manipulate data. There are time constraints such as in the digital real-time control of machine tools [2].

6-2. OPERATIONAL AMPLIFIERS: CHARACTERISTICS

Electrical signals from analog transducers are often at low voltages and/or low power levels. Signals are amplified prior to processing or for transmission. *Operational amplifiers* (op-amps) are widely used as building blocks in instruments. Ideal and actual characteristics of op-amps are described in this section [3]. Applications are examined in the sections to follow.

The open-loop op-amp circuit of a high-gain dc differential amplifier is shown in Fig. 6-1a. An op-amp is seldom used in an open loop and is generally used in a closed loop with simple external circuitry (see Sec. 3-6 for closed-loop operation and feedback). The voltages V_a, V_b, and V_o are measured with respect to ground, and the differential voltage is V_{ab}. The inverting input a is denoted by a $(-)$ sign, that is, V_a and V_o are of opposite signs. The noninverting input b is denoted by a $(+)$ sign, that is, V_b and V_o are of the same sign. Both V_a and V_b can be positive or negative voltages. For simplicity, the symbol of an op-amp is as shown in Fig. 6-1b, in which the ground and other connections are omitted.

An op-amp is generally in the form of an *integrated circuit* (IC). The

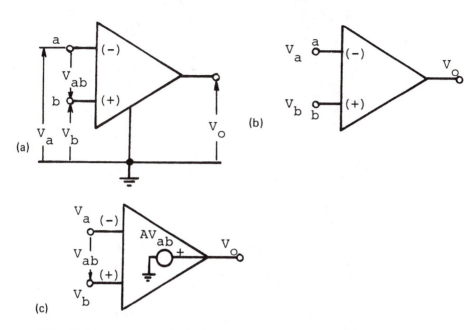

FIGURE 6-1 Operational amplifier in an open-loop circuit. (a) Open-loop circuit. (b) Symbol of op-amp. (c) Idealized model.

popular and inexpensive 741-type op-amp consists of about 20 transistors. The dual-in-line package (DIP) (¼ in. by ⅜ in.) shown in Fig. 6-2a has a notch or similar marking to identify the positions of the connecting pins. Pin 1 is adjacent to the notch, and the numbering is counterclockwise viewed from the top. The functions of the pins are shown in Fig. 6-2b. The pins for the inputs and output are readily identified. The output voltage is typically ±10 V. The dc supply voltage is approximately ±15 V applied to pins 7 and 4 (see Sec. 3-3D for amplifiers as three-port devices). Pin 8 is not connected (NC). Pins 1 and 5 are for the null adjustment of the amplifier by means of external circuitry (not shown).

The open-loop characteristics of an idealized op-amp shown in Fig. 6-1c are:

1. The inputs are perfectly symmetrical; that is, there is no offset voltages.
2. The input currents are zero; that is, input impedances at the terminals a and b are infinite.
3. The output impedance is zero.
4. The voltage gain for the differential input voltage V_{ab} from a to b is in-

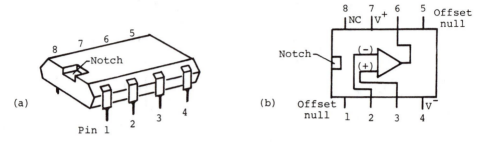

FIGURE 6-2 IC op-amp and schematic. (a) IC op-amp DIP package. (b) External connections.

finite for all frequencies; that is, the op-amp is a dc amplifier with an infinite gain A.

5. Since $V_o = AV_{ab}$, where V_o is finite and A is infinite, the differential voltage $V_{ab} = (V_b - V_a) = 0$.

The assumption $V_a = V_b$ is used in the derivations to follow. Note that the idealized model is a controller source differential amplifier. The idealized characteristics are tabulated below.

Open-loop differential gain, A	∞
Offset voltage and current, V_{os}, I_{os}	0
Bias current, I_a, I_b	0
Output impedance, Z_o	0
Input impedance, Z_i, Z_{cm}	∞
(differential or common mode)	

Inasmuch as many types of op-amps are available, the actual characteristics cover a wide range. Specifications are available from manufacturers.

Neglecting the internal dc offsets, the linear model in Fig. 6-3a shows the differential input impedance Z_i, the common-mode impedances Z_{cm1} and Z_{cm2}, and the Thévenin equivalent at the output. The output impedance is Z_o and the controlled voltage source is Av_d (see Sec. 3-3D), where A is the open-loop gain and v_d the differential input voltage.

The open-loop *gain–frequency characteristics* of an op-amp are illustrated in Fig. 6-3b (solid line). The dc gain A is of the order of 10^6. The gain is down by 3 dB at about 10 Hz. An op-amp can be compensated internally to approximate a 6-dB/octave roll-off [20 dB/decade or a -1 slope in a log-log plot (see Sec. 4-6)]; that is, the open-loop gain A decreases with increasing operating frequency. The open-loop *gain–bandwidth product*

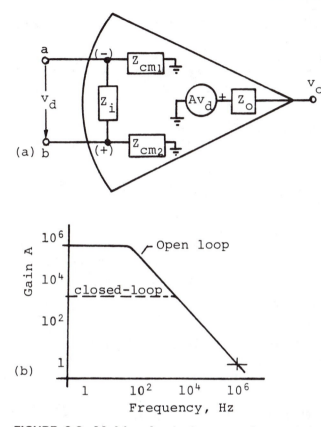

FIGURE 6-3 Model and gain–frequency characteristics of an op-amp. (a) Linear model (from Ref. 4). (b) Gain-frequency characteristics.

is about 1 MHz, as shown in the figure. A constant gain–bandwidth product means that the gain is the reciprocal of frequency. Knowing this, the gain–bandwidth product for closed-loop operations can be predicted. For example, if the closed-loop gain is 10^3 (dashed line), the corresponding bandwidth is from dc to 1 kHz.

Owing to manufacturing tolerances, the input stages of the amplifier are not perfectly symmetrical. When the input is shorted for $V_{ab} = 0$, an offset voltage V_{os} ($= 2$ mV) exists at the input of the op-amp, acting like a battery in series with one of the inputs. V_0 can be adjusted to zero by means of a simple external circuit. Small input currents also exist at the input terminals a and b. These currents are approximately equal. The offset current I_{os} is the difference in the input currents at the terminals a and b. Typically, $I_{os} \simeq 10$ nA. Due to the finite input current, a zero input voltage may not produce a zero output, even when the offset voltage is trimmed to zero.

The realized characteristics of a typical 741 op-amp are shown in the tabulation below. Values of offset voltage, and so on, are not repeated here. Detailed specifications, such as thermal drifts, are furnished by the manufacturer.

Open-loop differential gain, A	2×10^5
Input impedance	2 MΩ (10^{12} Ω for FET-input)
Output impedance	70Ω (open-loop)
Supplied voltage	± 15 V$_{dc}$
Output short-circuit current	20 mA
Output short time	indefinite
CMRR	80 dB
Slew rate	0.5 V/μs

The *common-mode rejection ratio* (CMRR; see sec. 3-1C) specifies the output V_o due to an average voltage (common mode) V_{cm} at the inputs.

$$\text{CMRR} = \frac{v_{cm}(\text{differential gain})}{v_o \text{ due to } v_{cm}} \qquad (6\text{-}1)$$

Slew rate is the rate of change in the output for an input unbalance. For example, a sine-wave input of amplitude A volts and frequency f (Hz) requires a minimum slew rate of $2\pi A f$ (V/s).

6-3. OPERATIONAL AMPLIFIERS: BASIC CIRCUITS

Op-amp are generally used in closed-loop mode for amplifying and signal conditioning. The basic inverting, noninverting, and differential closed-loop operations are described in this section. Op-amps are also used in open-loop mode as comparators for comparing the differential input voltage $V_{ab} = V_b - V_a$. Since the open-loop gain A is large, a small difference in $V_b - V_a$ will cause V_0 to change sign. Applications of the op-amp are almost limitless, and the discussions to follow cannot be inclusive.

Closed-loop operations are obtained by connecting external resistors and capacitors to an op-amp. Since the open-loop gain A is very high, the closed-loop characteristics of an op-amp with negative feedback depend mainly on the feedback network (see Sec. 3-6C). On the basis of this observation, the idealized model shown in Fig. 6-1c can be used to derive equations for most applications. The assumptions are that (1) the input currents are zero, and (2) the differential input V_{ab} is zero or $V_a = V_b$.

The basic *inverting amplifier* is shown in Fig. 6-4a. The input voltage

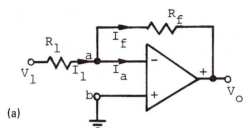

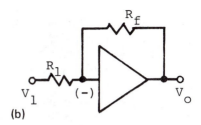

(a)

(b)

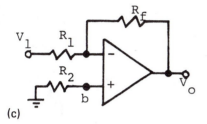

(c)

FIGURE 6-4 Op-amp as an inverting amplifier. (a) Inverted input amplifier. (b) Simplified diagram. (c) $R_2 = R_1 \parallel R_f$.

V_1 and the output V_o are of opposite signs. This is an advantage when a sign change is needed. The resistor R_f is the feedback element. Since terminal b is grounded, terminal a becomes a *virtual ground*, that is, $V_a \simeq V_b$ $= 0$. A virtual ground is essentially at the ground potential but is not directly connected to ground. Summing the currents at node a and using Ohm's law, we get

$$I_1 = I_f + I_a \qquad \text{where} \quad I_a \simeq 0$$

$$\frac{V_1 - V_a}{R_1} = \frac{V_a - V_o}{R_f} + 0 \qquad \text{where} \quad V_a \simeq 0 \qquad (6\text{-}2)$$

$$\frac{V_o}{V_1} = -\frac{R_f}{R_1} \qquad\qquad (6\text{-}3)$$

This is the basic equation for the inverting operation. For example, if the ratio R_f/R_1 is 1000:1, the closed-loop voltage gain is 1000:1 with a sign change, instead of the open-loop gain of 10^6. Note that (1) the response is flat from dc to about 1 kHz shown in Fig. 6-3b (dashed line), and (2) the input impedance of the closed-loop amplifier is no longer of the order of megohms as for the open-loop. For example, if R_1 is 10 kΩ, the input impedance is also 10 kΩ looking in from V_1 because terminal a is a virtual ground. The circuit diagram can be simplified as shown in Fig. 6-4b, and it is understood that the amplifier is inverting. It should be noted that if the suplied voltage is ±15 V_{dc}, the magnitude of V_o is less than ±15 V.

Due to the input bias current, the inputs in Fig. 6-4a are not balanced, because terminal a sees a driving impedance due to R_1 and R_f in parallel. This is corrected by connecting to terminal b a resistor R_2, equal in value to R_1 and R_f in parallel as shown in Fig. 6-4c.

The *generalized inverting amplifier* shown in Fig. 6-5 is obtained by substituting impedances Z's for the resistors R's.

$$\frac{V_o}{V_1} = -\frac{Z_f}{Z_1} \tag{6-4}$$

Since there are no restrictions on the Z's, Eq. 6-4 describes a general transfer function.

For example, the *integrator* shown in Fig. 6-6a consists of a feedback capacitor C_f and an input resistor R_1. The impedance of C_f is $Z_C = 1/j\omega C_f$. Substituting this in Eq. 6-4 gives the transfer function and the corresponding time response.

$$\frac{V_o}{V_1} = -\frac{1}{j\omega R_1 C_f} \tag{6-5}$$

$$v_o = -\frac{1}{R_1 C_f}\int v_1 dt \tag{6-6}$$

Due to the offset current in the op-amp, the integrator shown in Fig. 6-6a is bias unstable and will soon saturate. It can be stabilized by placing a large resistor R_f across C_f, as shown in Fig. 6-6b. R_f provides a leakage path so that the circuit does not keep integrating the small unbalance input. The circuit is an integrator only if the lowest frequency of the input is larger than the break frequency at $\omega = 1/R_f C_f$. The circuit also simulates a first-order instrument (see Sec. 4-6). The verification of these statements is left as an exercise.

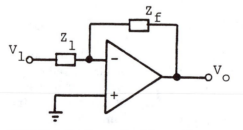

FIGURE 6-5   Generalized inverting op-amp circuit.

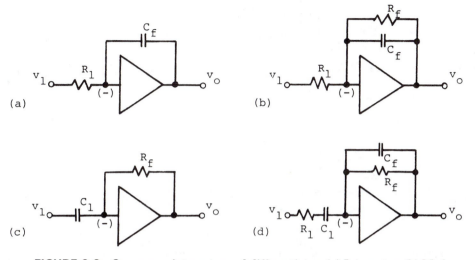

FIGURE 6-6 Op-amp as integrator and differentiator. (a) Integrator. (b) Modified integrator. (c) Differentiator. (d) Modified differentiator.

The *differentiator* shown in Fig. 6-6c has the R and C reversed. It can be shown from Eq. 6-4 that

$$v_o = -R_f C_1 \frac{dv_i}{dt} \tag{6-7}$$

Differentiators generally have instability problem with high-frequency noise, because the differentiation of small irregularities in the input is greatly magnified at the output. The problem can be corrected by using a small resistor R_1 and a small capacitor C_f, as shown in Fig. 6-6d. On the other hand, integration has the smoothening effect because integration is a summing process. Irregularities in the input have little effect on the sum at the output.

The inverting *summing amplifier* shown in Fig. 6-7a sums and amplifies the voltages v_1 and v_2. Following Eqs. 6-2 and 6-3, assuming that $i_a = 0$, and summing the currents at node a, we deduce that

$$v_o = -R_f \left(\frac{1}{R_1} v_1 + \frac{1}{R_2} v_2 \right) \tag{6-8}$$

Similarly, from Eq. 6-6, the equation for the *summing integrator* shown in Fig. 6-7b is

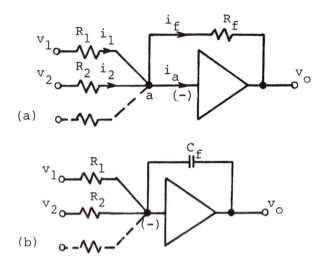

FIGURE 6-7   (a) Summing amplifier. (b) Summing integrator.

$$v_o = -\frac{1}{C_f} \int \left(\frac{1}{R_1} v_1 + \frac{1}{R_2} v_2 \right) dt \qquad (6\text{-}9)$$

The basic *noninverting amplifier* op-amp circuit is shown in Fig. 6-8a. The resistors R_1 and R_f form a voltage divider for V_o, that is,

$$\frac{v_a}{v_o} = \frac{R_1}{R_1 + R_f} \qquad (6\text{-}10)$$

Since $V_a \simeq V_b = V_i$, the closed-loop gain of the amplifier is

$$v_o = \left(1 + \frac{R_f}{R_1} \right) v_i \qquad (6\text{-}11)$$

If R_1 becomes infinite (an open circuit) and R_f remains finite ($R_f = 0$), we obtain a *voltage follower* or *preamplifier* (see Sec. 3-3A), with a gain of unity, as shown in Fig. 6-8b.

$$V_o = V_i \qquad (6\text{-}12)$$

A voltage follower has extremely high input impedance for v_i and very low output impedance. A follower with gain greater than unity is obtained from the circuit shown in Fig. 6-8c. The feedback voltage is nv_o, where $0 < n \leqslant 1$. Since $v_i = v_b = v_a = nv_o$, we have

$$v_o = \frac{1}{n} v_i \qquad \text{for} \quad 0 < n \leqslant 1 \qquad (6\text{-}13)$$

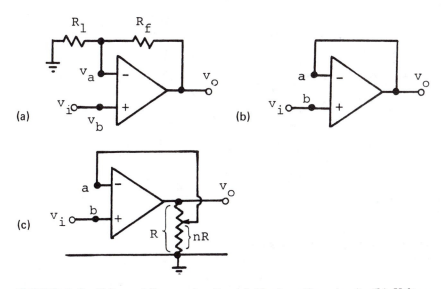

FIGURE 6-8 Voltage follower circuits. (a) Noninverting circuit. (b) Voltage follower. (c) Follower, gain $\geqslant 1$.

The basic *differential amplifier* op-amp circuit is shown in Fig. 6-9. The input voltages are $v_1 + v_{cm}$ and $v_2 + v_{cm}$, where v_{cm} is the common mode. Noting that R_2 and R_3 form a voltage divider for $(v_2 + v_{cm})$, and combining Eqs. 6-3 and 6-11 for the inverting and the noninverting inputs, we obtain

$$v_b = \frac{R_3}{R_2 + R_3}(v_2 + v_{cm}) \tag{6-14}$$

$$v_o = -\frac{R_f}{R_1}(v_1 + v_{cm}) + \left(1 + \frac{R_f}{R_1}\right)v_b \tag{6-15}$$

Combining the equations above and simplifying, we get

$$v_o = -\frac{R_f}{R_1}v_1 + \left(\frac{R_3}{R_2}\frac{R_f/R_1 + 1}{R_3/R_2 + 1}\right)v_2 + \left(\frac{R_3}{R_2}\frac{R_f/R_1 + 1}{R_3/R_2 + 1} - \frac{R_3}{R_2}\right)v_{cm} \tag{6-16}$$

The common-mode voltage v_{cm} is rejected if the resistors are trimmed such that $R_3/R_2 = R_f/R_1$. The differential output is

$$v_o = \frac{R_f}{R_1}(v_2 - v_1) \tag{6-17}$$

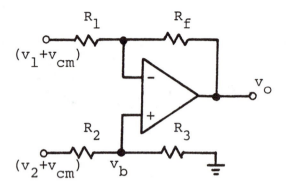

FIGURE 6-9   Differential amplifier.

In practice, the circuit described would require high-precision resistors. It will be shown that this requirement is alleviated by incorporating the circuit in the instrumentation amplifier.

6-4.   OP-AMP AMPLIFIERS AND APPLICATIONS

Op-amps are often used as components in instruments. Their applications in instrumentation are described briefly in this section.

A.   Amplifiers

Many types of amplifiers are used in instrumentation. They can be purchased as complete units or as IC packages. The charge amplifier and the commonly used instrumentation amplifier are described in this section. Understandably, there are many variations and refinements within each type of amplifier.

Charge amplifiers are widely used with *piezoelectric transducers*. The model of a piezoelectric crystal is a charge generator $q(t)$ in parallel with a capacitor C_d, as shown in Fig. 6-10a. When a mechanical strain is applied, an electric charge is generated and built up across the surface of the crystal. When the transducer is operated below resonance, $q(t)$ is proportional to the mechanical strain or the deformation of the crystal.

The output of the piezoelectric transducer is shunted by the capacitance C_s of the connecting cable, as shown in Fig. 6-10b. The Thévenin's equivalent of the transducer and its cable at the input to the charge amplifier is shown in Fig. 6-10c.

$$v_{in} = \frac{1}{C_d + C_s} q(t) \qquad\qquad (6\text{-}18)$$

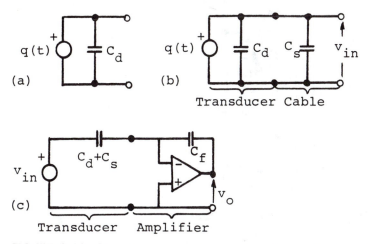

FIGURE 6-10 Piezoelectric transducer and charge amplifier. (a) Transducer. (b) Transducer and cable. (c) Transducer and amplifier.

The *charge amplifier* in Fig. 6-10c is basically an op-amp integrator. If its input current is zero, the net charge input is stored in the feedback capacitor C_f. From Eq. 6-4, we obtain

$$\frac{v_o}{v_{in}} = -\frac{Z_f}{Z_{in}} \tag{6-19}$$

where $Z_f = 1/j\omega C_f$ and $Z_{in} = 1/j\omega(C_d + C_s)$. Combining Eqs. 6-18 and 6-19 and canceling out the $C_d + C_s$ terms, we get the equation of a charge amplifier:

$$v_o = -\frac{1}{C_f} q(t) \tag{6-20}$$

Thus the output v_o of a charge amplifier is dependent only on C_f. It is neither affected by the capacitance of the transducer and the length of the connecting cable nor influenced by their change in values. In practice, the sensitivity of a piezoelectric accelerometer, in pF/g of acceleration, is somewhat temperature dependent.

The charge amplifier as shown is bias unstable because it is basically an op-amp integrator. It can be stabilized by means of a large feedback resistor R_f across C_f as shown in Fig. 6-11 (see also Fig. 6-6b).

Another example is the amplifier for a *variable-capacitance transducer* shown in Fig. 6-12. The variable capacitance C with a series resistor R is driven by a dc voltage V. If the input current of the amplifier is negligible, we have

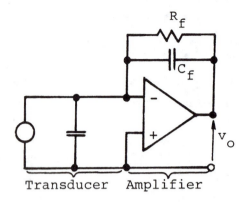

FIGURE 6-11   Stabilized charge amplifier.

$$q = CV \qquad\qquad \text{and} \qquad q = C_f V_o$$
$$dq = V\,dC + C\,dV \qquad \text{and} \qquad dq = C_f\,dV_o \tag{6-21}$$

Combining the equations above, the amplifier output $v_o = dV_o$ is

$$dv_o = \frac{V}{C_f}\,dC \qquad \text{for} \quad dV = 0 \tag{6-22}$$

Another example is the instrumentation amplifier. First, consider the differential amplifier (see Fig. 6-9) with two input buffers, or voltage followers, shown in Fig. 6-13a. Since the followers have a gain of unity, the desired signals v_1 and v_2 and the associated common-mode voltage V_{cm} are amplified equally. Thus the rejection of V_{cm} must be from the differential amplifier. This is inadequate for some applications, such as for strain

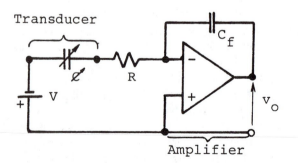

FIGURE 6-12   Charge amplifier with a variable-capacitance transducer.

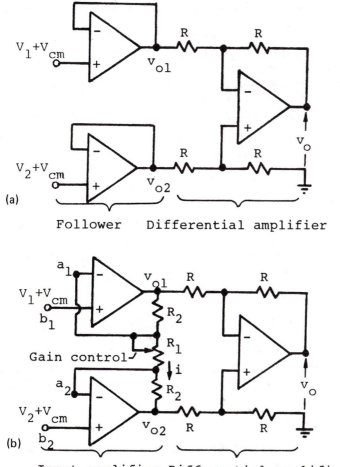

FIGURE 6-13 (a) Buffer and differential amplifier. (b) Instrumentation amplifier (from Ref. 5).

measurements, if the signal gain is unity and the common mode gain is 10^{-3}.

The *instrumentation amplifier* shown in Fig. 6-13b has a high CMRR. The signal/common-mode ratio is increased if the input op-amps provide a high gain for the differential signal and a low gain for the common mode. The remaining common mode can then be rejected by means of the differential amplifier. Assume that the voltage at a_1 is $v_{a1} = v_{b1} = v_1 + V_{cm}$ and that at a_2 is $v_{a2} = v_{b2} = v_2 + V_{cm}$. Since the input currents to the op-amps are negligible, the current i from v_{o1} to v_{o2} is

$$i = \frac{v_{o1} - (v_1 + V_{cm})}{R_2} = \frac{(v_1 + V_{cm}) - (v_2 + V_{cm})}{R_1} = \frac{(v_2 + V_{cm}) - v_{o2}}{R_2}$$

$$\text{(6-23)}$$

$$v_{o1} - v_{o2} = (v_1 - v_2)\left(1 + \frac{2R_2}{R_1}\right) \qquad \text{(6-24)}$$

Thus a high CMRR is obtained from the input op-amps. The input to the differential amplifier is $(v_{o1} - v_{o2})$. Assuming a gain of unity for convenience, we obtain from Eqs. 6-17 and 6-24,

$$v_o = (v_2 - v_1)\left(1 + 2\frac{R_2}{R_1}\right) \qquad \text{(6-25)}$$

An instrumentation amplifier has a gain from 1 to 1000, a CMRR of 90 dB, an input impedance of 10^9 Ω, and a moderate bandwidth. Note that the differential gain can be adjusted by means of the single resistor R_1 in Fig. 6-13b.

B. Applications

Examples of op-amp circuits for linear and nonlinear applications are described briefly in this section. Additional examples for digital operations will be shown in the sections to follow. There are many variations and refinements for each circuit.

1. Filters

Filters are passive or active. They can be classified as low-pass (LP), high-pass (HP), bandpass (BP), and band-reject (BR) filters. Their characteristics are shown in Fig. 3-47.

A *passive filter* consists of a combination of resistors R, inductors L, and capacitors C. For example, the simple RC circuits in Fig. 4-7 can be a LP or a HP filter with a cutoff of 6 dB/octave. The parallel resonant circuit in Fig. A-21 is a narrowband BP filter when R is small. Circuits for passive filters are provided in the literature [6].

Passive filters have several disadvantages compared with active filters. First, the inherent resistance in an inductor L makes it a nonideal component. Inductors are large and expensive for the audio-frequency range in which most mechanical measurements are made. Second, passive filters do not have power amplification. This tends to cause loading in measurements and to restrict the sharpness of the cutoff. Finally, passive filters are not designed for general-purpose instrumentation and lack flexibility in frequency-range selection.

An *active filter* consists of a combination of R, C, and amplifiers. Many are based on op-amp circuits. For example, the generalized inverting op-amp circuit shown in Fig. 6-5 can be used to simulate the transfer function of a filter. The circuit shown in Fig. 6-6b is a first-order LP filter. The differentiator shown in Fig. 6-6c becomes a first-order HP filter by inserting a resistor at the input in series with C_1. The circuit shown in Fig. 6-6d can be a simple BP filter, because it is a differentiator for low frequencies and an integrator for high frequencies.

The basic single feedback op-amp circuit shown in Fig. 6-14a is similar

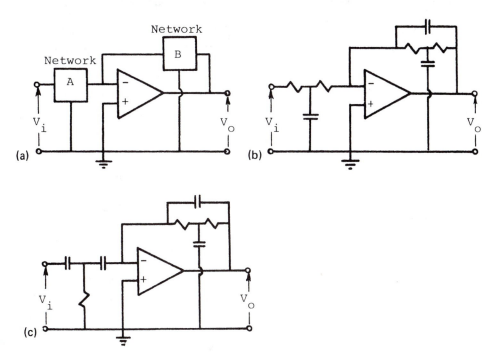

FIGURE 6-14 Examples of single-feedback active filters. (a) Basic single-feedback circuit. (b) Single-feedback low-pass filter (from Ref. 7). (c) Single-feedback high-pass filter (from Ref. 7).

to that in Fig. 6-5. The second-order LP and HP filters in Fig. 6-14b and c can be derived from this basic circuit. Higher-order filters may consist of multiple-feedback and multiple-stage circuits [8]. A detail discussion of active filters is beyond the intended scope of this book.

2. Bridge Circuits

A Wheatstone bridge for differential measurements is shown in Fig. 6-15a (see Fig. 3-33). Assume that $R_1 = R_2 = R_3 = R_4 = R$ and that the bridge is initially balanced. From Eq. 3-60 and for small changes from the balanced conditions, we get

$$\frac{V_o}{V_i} = (\text{constant})(\Delta R_1 + \Delta R_3 - \Delta R_2 - \Delta R_4) \tag{6-26}$$

The circuit in Fig. 6-15b can be used in lieu of the bridge [9]. The outputs of the inverting amplifiers are V_{o1} and V_{o2}. The output $V_o = (V_{o1} - V_{o2})$ can be measured with a differential amplifier (see Fig. 6-9) when a single-ended bridge output is desired. The main advantages of this circuit are that it is not restricted to small changes in resistances and is linear for all unbalanced conditions, but the values of the R's in the op-amp circuit are relatively higher than those for normal strain gage applications. It can be shown readily that the output $V_o = V_{o1} - V_{o2}$ is

$$\frac{V_o}{V_{\text{ref}}} = \frac{1}{R_{\text{ref}}} (R_1 + R_3 - R_2 - R_4) \tag{6-27}$$

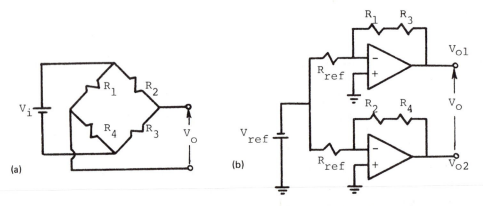

FIGURE 6-15   (a) Bridge circuit. (b) Substitution for bridge circuit.

3. Voltage and Current Measurements

Multimeters for voltage and current measurements were described in Chapter 1. The movement in a multimeter is a microammeter with input resistance R_m. The error in measurements with a multimeter can be unacceptably large (see Probs. 1-8 to 1-13). Op-amp circuits can be used to improve the accuracy.

The input resistance R_i of a *voltmeter* can be increased by several orders of magnitude by means of a voltage follower (see Fig. 6-8b). For a dc or a transient voltage measurement, the signal can be picked off from the output of the follower. For an ac voltage measurement with a multimeter, the ac is rectified so that the meter reads an average value, as shown in Fig. 6-16a (see Prob. 1-12). The disadvantages are that the diodes are not ideal and R_i of the meter is relatively low. The remedy is to use a bridge rectifier (see Fig. 5-6) and to place the meter in the feedback path, as shown in Fig. 6-16b. The high gain of the op-amp renders the diode ideal [10], and the R_i of the voltage follower is extremely high.

The input resistance R_i of an ideal *ammeter* is zero, and the circuit shown in Fig. 6-16c may not be satisfactory, particularly for low current measurements, such as from a photovoltaic diode [11]. The sensitive "ammeter" is the current-to-voltage converter shown in Fig. 6-16d. Load-

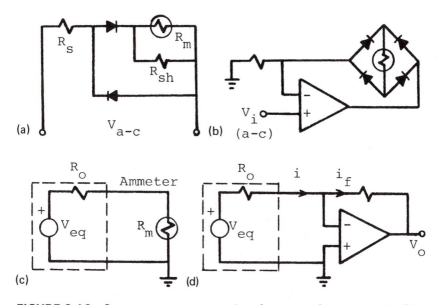

FIGURE 6-16 Op-amps as components in voltmeter and ammeter. (a), (b) Ac voltage measurement. (c) Current measurement. (d) Current-to-voltage converter.

ing is minimal because the energy to drive the output is from the op-amp instead of from the current source. The voltage drop across the "ammeter" is of the order of 1 mV, since terminal a is a virtual ground. The accuracy of the measurement is improved by an order of magnitude.

4. Nonlinear Operations

The absolute value of a signal may be desired in signal manipulation. The *absolute value circuit* shown in Fig. 6-17 is an active rectifier. Similar to the "active" voltmeter above, the amplifiers render the diodes ideal, because the rectification of signals smaller than a diode drop cannot be achieved with a simple diode–resistor circuit.

First, assume that the input v_i in Fig. 6-17b is negative. For the op-amp 1, the diode D_1 is ON and D_2 is OFF, the voltage v' is one diode drop above v_a or the ground potential, and its output is zero. Thus the output from op-amp 2 is $v_o = (R_f/R_1)|v_i|$. Second, assume that v_i is positive. For the op-amp 1, D_1 is OFF and D_2 is ON, D_2 makes it a unity-gain inverter, and its output is $v = -v_i$. Thus op-amp 2 has two inputs, and its output is $v_o = -(R_f/R_1)(v_i - 2v_i) = (R_f/R_1)v_i$.

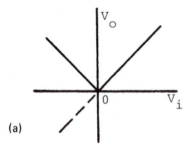

(a)

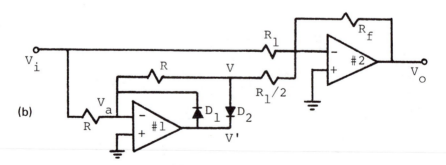

(b)

FIGURE 6-17 Example of a nonlinear application. (a) Absolute value. (b) Absolute value circuit. (From Ref. 12.)

6-5. BINARY NUMBERING SYSTEM

The *binary numbering system* counts by 2, while the decimal system counts by 10. Binary system is discussed briefly as a background for examining digital techniques. Various binary codes are used for digital computation. The discussion in this section is restricted to binary numbers and the binary-coded-decimal (BCD) code.

The position of a digit determines its value in a numbering system. For instance, the decimal number 138.75, expressed in the familiar decimal form, is

$$(1 \times 10^2) + (3 \times 10^1) + (8 \times 10^0) + (7 \times 10^{-1}) + (5 \times 10^{-2})$$
$$= 100 + 30 + 8 + 0.7 + 0.05$$
$$= 138.75$$

The decimal equivalent of the binary number 10011110.11 is

$$(1 \times 2^7) + (0 \times 2^6) + (0 \times 2^5) + (1 \times 2^4) + (1 \times 2^3) + (1 \times 2^2)$$
$$+ (1 \times 2^1) + (0 \times 2^0) + (1 \times 2^{-1}) + (1 \times 2^{-2})$$
$$= 128 + 0 + 0 + 16 + 8 + 4 + 2 + 0 + 0.5 + 0.25$$
$$= 138.75$$

The example shows that each position in a numbering system is given a weight. We are familiar with the weighting scale in the decimal numbering system. For binary numbers, the first position to the left of the decimal point has a weight of $2^0 = 1$, the second position a weight of $2^1 = 2$, the third position a weight of $2^2 = 4$, and so on. Similarly, the first position to the right of the decimal point has a weight of $2^{-1} = 0.5$, the second position $2^{-2} = 0.25$, and so on. The method for weighting binary numbers from position 1 to position 8 and the corresponding decimal numbers are shown in Table 6-1a. A comparison of binary numbers and their decimal equivalents from 0 to 15 is shown in Table 6-1b.

Numbers in a straight binary system are difficult to read. The *binary-coded-decimal* (BCD) *code* allows for easier reading. Each decimal number from 0 to 9 is expressed in the BCD code by a 4-bit sequence of binary code, as shown in Table 6-1b. The numbers below the dashed line in Table 6-1b are not allowed. For example, the decimal number 1685 in the BCD code is

0001	0110	1000	0101
1	6	8	5

TABLE 6-1 Decimal Equivalent of Binary Numbers

(a) Weight in binary system			(b) Decimal equivalent		
Position in binary system	Decimal equivalent	Binary number	Decimal equivalent	Binary number	Decimal equivalent
1	$2^0 =$ 1	0000	0	1000	8
2	$2^1 =$ 2	0001	1	1001	9
3	$2^2 =$ 4	0010	2	1010	10
4	$2^3 =$ 8	0011	3	1011	11
5	$2^4 =$ 16	0100	4	1100	12
6	$2^5 =$ 32	0101	5	1101	13
7	$2^6 =$ 64	0110	6	1110	14
8	$2^7 =$ 128	0111	7	1111	15

6-6. DIGITAL TECHNIQUES

The study of digital techniques can be divided itno three parts: gates, flip-flops, and functional logics. Gates and flip-flops are examined in this section. It will be shown that flip-flops can be constructed from gates, and functional logics from gates and flip-flops.

One tends to associate the study of digital techniques with digital computers and electrical engineering. Digital techniques are also used extensively in industry for switching, counting, and control. With the advent of integrated circuits, only a basic knowledge of circuitry is necessary for the understanding and use of digital techniques. The problem almost reduces to selection of the appropriate IC logic components and choice of the circuit to perform a given task.

A *digital signal* is conveyed by the ON/OFF *states* of pulses in a pulse train (see Sec. 5-4B). For the positive logic, a *binary* 1 is an ON state or high for any voltage between +2 and +5 V. A *binary* 0 is an OFF state or low for any voltage between 0 and +0.8 V.

The ON or the OFF state is a *bit,* and the time interval for the ON or OFF state is called a *bit interval.* An 8-bit word is a *byte*. Digital information is trasmitted and processed in bits. The time rate of the bits is closely controlled and synchronized by means of a clock (a crystal-controlled oscillator).

The advantage of a digital system is versatility in signal conditioning and computation by means of software. In the analog mode, information is carried by the magnitude of a signal as a time function, and therefore the precision is limited. In the digital mode, the number of bits in a digital word can be extended, and therefore a digital signal can be exceedingly precise. Double precision in computations involving iteration is especially desirable. One must not forget the distinction between precision and accuracy in a measurement (see Sec. 2-5). The ability of digital mode to use software, however, cannot be challenged.

A. Logic Gates

A *logic gate* selectively passes an input signal according to the purpose of the gate. There are three basic logic gates: AND, OR, and NOT. The complement of AND is NAND (not AND) and that of OR is NOR (nor OR). NAND and NOR gates are more often used than the AND, OR, and NOT gates because of economy. It will be shown that all functions of logic gates can be achieved using either NAND or NOR gates.

The basis of a logic circuit is the switching between the ON/OFF states. A switch is ON when closed, and OFF when open, corresponding to binary 1 and 0. The single-pole, single-throw (SPST) mechanical switch in Fig. 6-18a is self-explanatory. The switch can be solenoid operated, as shown in Fig. 6-18b. This switch is normally open. It closes when a voltage is applied to the solenoid. It is important to note that the ON/OFF state only implies conducting or nonconducting. An electronic "switch" does not physically open or close as a mechanical switch.

The transistor switch shown in Fig. 6-18c has two states. The transistor is driven to saturation (conducting) when v_i is high; it is driven to cutoff

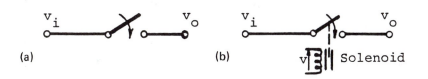

(a)   (b)   Solenoid

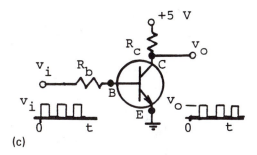

(c)

FIGURE 6-18 Mechanical and electronic switches. (a) SPST switch. (b) Solenoid switch. (c) Transistor switch.

(nonconducting) when v_i is low. Since v_o is almost at ground potential when the transistor is conducting, and v_o is 5 V when not conducting, the transistor is also a NOT gate, as shown by the waveforms of v_i and v_i in the figure.

1. AND Gates

A mechanical AND gate with two inputs is shown in Fig. 6-19a and its symbol in Fig. 6-19b. There can be more than two inputs to an AND. The operation is C "equals" A AND B, where the quantities A, B, and C can only assume logic levels of 1 or 0. The symbolic equation is

$$C = A \cdot B \qquad \text{(alternatively, } C = AB) \qquad (6\text{-}28)$$

The implementation of an AND using two diodes is shown in Fig. 6-19c (see Sec. 5-2 for explanation of diodes). Assume that ON (binary 1) is 5 V and OFF (binary 0) is 0 V.

1. When $A = 0$, diode D_1 conducts, thereby clamping C at 0. At the same time, B can be 0 or 1, and it has no effect on the circuit.
2. Similarly, C is clamped at 0 when $B = 0$.
3. The condition for $C = 1$ is when $A = 1$ and $B = 1$ at the same time.

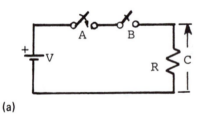

(a)

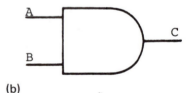

(b)

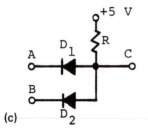

(c)

FIGURE 6-19 Examples of AND gates. (a) Switch AND gate. (b) AND gate. (c) Diode AND gate.

TABLE 6-2   Truth Table for AND, OR, XOR

| (a) AND, $C = A \cdot B$ | | | (b) OR, $C = A + B$ | | | (c) XOR, $C = A \oplus B$ | | |
A	B	C	A	B	C	A	B	C
0	0	0	0	0	0	0	0	0
0	1	0	0	1	1	0	1	1
1	0	0	1	0	1	1	0	1
1	1	1	1	1	1	1	1	0

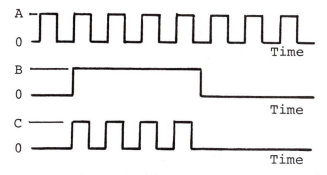

FIGURE 6-20 B enables or inhibits an AND gate for counting pulses in input A.

The statements above for the AND gate are summarized in the *truth table* in Table 6-2a. It gives the logic levels of the output C for all possible combinations of the inputs A and B. It is evident from the mechanical switches shown in Fig. 6-19 that C is high (logic level 1) only when both switches are closed (logic level 1). The truth table and Boolean algebra are useful tools for the analysis and design of digital logic circuits.

An application of the AND gate is illustrated in Fig. 6-20. Assume that the frequency of a pulse train A is being counted, and B controls the duration of the count. B enables or inhibits the passage of A through the AND gate, because both A and B must be ON for the AND gate to operate. The frequency of A is the number of pulses observed at C divided by the duration of B.

2. OR Gates

A mechanical OR gate with two inputs is shown in Fig. 6-21a and its symbol in Fig. 6-21b. There can be more than two inputs to an OR. The operation is C "equals" A OR B, where the quantities A, B, and C can only assume logic levels of 1 or 0. The symbolic equation is

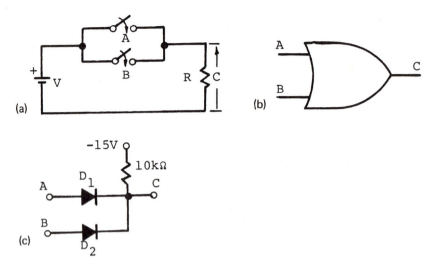

FIGURE 6-21 Examples of OR gates. (a) Switch OR gate. (b) OR gate. (c) Diode OR gate.

$$C = A + B \qquad\qquad (6\text{-}29)$$

The implementation of an OR gate using two diodes is shown in Fig. 6-21c.

1. When both A and B are low, that is, $A = B = 0$, both diodes D_1 and D_2 are conducting, because a 0 V is more positive than the negative supplied voltage of -15 V. Thus C is clamped to the logic level 0. The voltage across the 10-kΩ resistor is 15 V.
2. When $A = 0$ and $B = 1$ (+5 V), diode D_2 conducts and C is clamped at 5 V (logic level 1). At the same time, D_1 is reverse biased and not conducting. The voltage across the 10 kΩ is $15 + 5 = 20$ V.
3. Similarly, when both $A = 1$ and $B = 1$ (+5 V), C is clamped at 5 V.

The truth table for the OR gate in Table 6-2b gives the logic levels of C for all combinations of A and B. The mechanical switches in Fig. 6-21 show that C is at logic level 1 when switch A or B is closed (logic level 1), or when both A and B are closed. The characteristics of the OR gate are illustrated in Fig. 6-22 by means of the waveforms of the quantities A, B, and C.

The exclusive-OR XOR is a useful function, although less fundamental than the AND and OR. Its truth table is shown in Table 6-2c. An XOR never has more than two inputs. The symbolic equation is

$$C = A \oplus B \qquad\qquad (6\text{-}30)$$

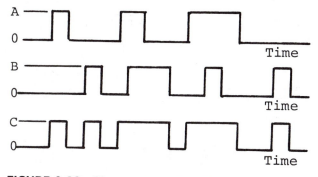

A

0

Time

B

0

Time

C

0

Time

FIGURE 6-22 Characteristic of an OR gate, $C = A + B$.

3. NOT Gates

A NOT gate has only one input. The output is the complement of its input; that is, a NOT is an inverter. If the input is at logic level 1 the output is at logic level 0, and vice versa. Two symbols for the NOT are shown in Fig. 6-23. The small circle in the figures denote an inversion. If the input is A, the output is $\bar{A}$ (read: not A).

4. NAND Gates

A NAND gate consists of an AND followed by a NOT, as shown in Fig. 6-24a. The symbol is that of an AND with a small circle at the output to indicate an inversion. If the inputs are A and B, the output of a NAND is $(\overline{A \cdot B})$ [read: not (A and B)]. The symbol in Fig. 6-24b is an equivalent form of a NAND. This is verified from columns 4 and 11 in the truth table in Table 6-3.

5. NOR Gates

A NOR gate consists of an OR followed by a NOT, as shown in Fig. 6-24c. The circle at the output denotes that a NOR is the complement of an OR. If the inputs of a NOR are A and B, the output is $(\overline{A + B})$ [read: not (A or B)]. The symbols for a NOR in Fig. 6-24c and d are equivalent. This is verified from columns 6 and 9 in the truth table in Table 6-3.

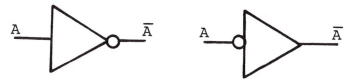

FIGURE 6-23 Symbols of a NOT gate.

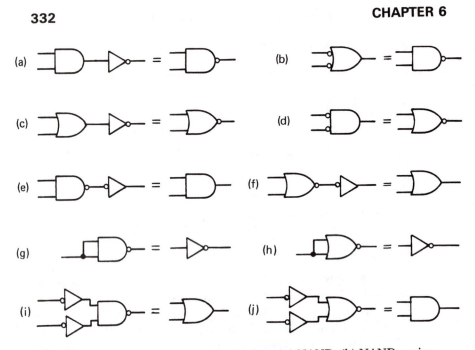

FIGURE 6-24 Logic gates and their equivalent. (a) NAND. (b) NAND equivalent. (c) NOR. (d) NOR equivalent. (e) AND. (f) OR. (g), (h) NOT. (i) OR. (j) AND.

TABLE 6-3 Truth Table of Equivalent Expressions

(1)(2)	(3) AND $A \cdot B$	(4) NAND $\overline{(A \cdot B)}$	(5) OR $A + B$	(6) NOR $\overline{(A + B)}$	(7)(8) $\bar{A}$ $\bar{B}$	(9) NOR $\bar{A} \cdot \bar{B}$	(10) OR $\overline{(\bar{A} \cdot B)}$	(11) NAND $\bar{A} + \bar{B}$	(12) AND $\overline{(\bar{A} + B)}$
0 0	0	1	0	1	1 1	1	0	1	0
0 1	0	1	1	0	1 0	0	1	1	0
1 0	0	1	1	0	0 1	0	1	1	0
1 1	1	0	1	0	0 0	0	1	0	1

Other combinations of gates are illustrated in Fig. 6-24. Some combinations are obvious. Others can be derived by means of a truth table.

B. Boolean Algebra

Boolean algebra deals with two value functions, the binary logic states of 1 and 0. The algebraic rules can be interpreted in terms of switching ON and OFF, where ON is 1 and OFF is 0. All quantities in a Boolean logic expression can only assume the states of 1 or 0.

A *Boolean logic expression,* such as $C = A \cdot B$, relates the output of a logic circuit to its inputs. Expressions for complex circuits are difficult to interpret. They can be manipulated and simplified by means of Boolean algebra for a better understanding of the circuit. Some rules for manipulation are shown in Table 6-4. Note the alternative forms of NAND and NOR in deMorgan's theorem. These are illustrated in Table 6-3. The equivalent forms for NAND are shown in columns 4 and 11, and those for NOR in columns 6 and 9.

Example 6-1.   (a) Show that the circuits in Fig. 6-24i are equivalent. (b) Repeat for the circuits in Fig. 6-24j.
Solution: (a) Let the inputs in Fig. 6-24i be A and B. Hence

$$C = (\overline{\overline{A} \cdot \overline{B}}) = \overline{\overline{A}} + \overline{\overline{B}} = A + B$$

TABLE 6-4   Boolean Algebra Rules

Equivalent expressions:

$\overline{\overline{A}} = A$ Double negative

$A \cdot B = B \cdot A$
$A + B = B + A$ } Commutative

$A \cdot (B \cdot C) = (A \cdot B) \cdot C$
$A + (B + C) = (A + B) + C$ } Associative

$A \cdot (B + C) = A \cdot B + A \cdot C$ Distributive

DeMorgan's theorem:

$$\overline{A \cdot B} = \overline{A} + \overline{B}$$
$$\overline{A + B} = \overline{A} \cdot \overline{B}$$

Let the inputs in Fig. 6-24j be A and B. Hence

$$C = (\overline{A + B}) = \overline{\overline{A}} \cdot \overline{\overline{B}} = A \cdot B$$

Example 6-2. Verify the following useful Boolean expressions.

(a) $A \cdot (A + B) = A$ (b) $A \cdot (\overline{A} + B) = A \cdot B$

(c) $A \cdot B + \overline{B} = A + \overline{B}$ (d) $A \cdot \overline{B} + B = A + B$

Solution: The expressions can be verified by means of the truth table. This is left as an exercise.

(a)
$$
\begin{aligned}
A \cdot (A + B) &= A \cdot A + A \cdot B \\
&= A + A \cdot B \\
&= A \cdot 1 + A \cdot B \\
&= A \cdot (1 + B) \\
&= A \cdot 1 \\
&= A
\end{aligned}
$$

(b)
$$
\begin{aligned}
A \cdot (\overline{A} + B) &= A \cdot \overline{A} + A \cdot B \\
&= 0 + A \cdot B \\
&= A \cdot B
\end{aligned}
$$

(c)
$$
\begin{aligned}
A \cdot B + \overline{B} &= A \cdot B + \overline{B} \cdot 1 \\
&= A \cdot B + \overline{B} \cdot (A + 1) \\
&= A \cdot B + \overline{B} \cdot A + \overline{B} \cdot 1 \\
&= A \cdot (B + \overline{B}) + \overline{B} \\
&= A \cdot 1 + \overline{B} \\
&= A + \overline{B}
\end{aligned}
$$

(d)
$$
\begin{aligned}
A \cdot \overline{B} + B &= A \cdot \overline{B} + B \cdot (A + 1) \\
&= A \cdot \overline{B} + A \cdot B + B \cdot 1 \\
&= A \cdot (\overline{B} + B) + B \\
&= A + B
\end{aligned}
$$

C. Flip-flops

There are three basic types of flip-flops, shown symbolically in Fig. 6-25. Operations and sample applications of flip-flops are illustrated in this section.

The *flip-flop* is literally the heart of a digital machine. Although the basic function of a flip-flop is to store one bit of binary information at logic level 1 or 0, it can perform a wide variety of tasks, such as counting and

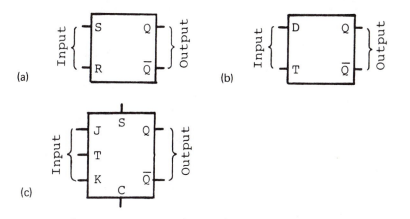

FIGURE 6-25 Symbols of basic flip-flops. (a) SR latch. (b) D flip-flop. (c) JK flip-flop.

multiplexing. For general applications, it is more economical to use integrated circuit IC packages instead of gates and discrete components.

1. Latch Flip-Flops

The SR *latch flip-flop* shown in Fig. 6-25a has two inputs, set S and reset R, and two outputs, Q and $\bar{Q}$, where $\bar{Q}$ is the complement of Q. The quantities S, R, Q, and $\bar{Q}$ can only assume logic levels of 1 or 0. A latch is set when $Q = 1$, and reset when $Q = 0$.

A SR latch using two cross-coupled NAND gates is shown in Fig. 6-26. In the truth table, Q_t is the state of Q before the inputs change states, and Q_{t+1} is the state of Q after the inputs change states.

Logic expressions for the SR latch are

$$Q = \overline{S \cdot Q} = S + Q \quad \text{and} \quad \bar{Q} = \overline{R \cdot Q} = R + \bar{Q} \quad (6\text{-}31)$$

Let us use the expressions above to examine the truth table in Fig. 6-26. It can be verified that if $S = R = 0$, Eq. 6-31 predicts no change from the previous states. If $S = 1$ and $R = 0$, then $Q = 1 + Q = 1$ and $\bar{Q} = 0 + \bar{Q} = 0$, since $Q = 1$. This defines the set state. Similarly, if $S = 0$ and $R = 1$, the latch is reset giving $Q = 0$ and $\bar{Q} = 1$. If $S = R = 1$, the equation predicts the noncomplementary outputs of $Q = 1$ and $\bar{Q} = 1$. This is not allowed, but it can be remedied by means of additional circuitry.

An application of the SR latch is in switch buffering, that is, the latch acts as a buffer between a mechanical switch and an electronic circuit. A mechanical reed-type switch would bound when making contact, giving repeated signals. If the mechanical switch sets the latch in order to switch

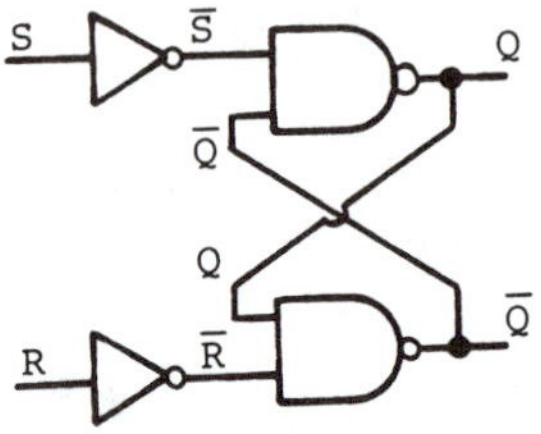

Truth table

S	R	$\bar{S}$	$\bar{R}$	Q_t	$\bar{Q}_t$	Q_{t+1}	$\bar{Q}_{t+1}$	
0	0	1	1	x	$\bar{x}$	x	$\bar{x}$	(No change)
0	1	1	0	x	$\bar{x}$	0	1	(Reset)
1	0	0	1	x	$\bar{x}$	1	0	(Set)
1	1	0	0	x	$\bar{x}$	–	–	(Not allowed)

(before) (after)

FIGURE 6-26 NAND latch.

an electronic circuit, the latch is set on the first contact of the switch. The bouncing attempts to set the latch repeatedly, but the latch remains set and the repeated signals from the switch bouncing are ignored. The latch can only be reset by means of a reset signal. Thus the latch gives clean ON/OFF switching, as shown in Fig. 6-27.

2. D-Type Flip-Flops

The D *flip-flop* shown in Fig. 6-25b has two outputs Q and $\bar{Q}$, as before. The T-input enables or inhibits the data transmission from D to Q; that is, T determines when data at the D-input is to be transmitted to the output. For example, the behavior of a D flip-flop (NAND type) is shown in Fig. 6-28a. The output Q tracks the input at D only when $T = 1$ or the flip-flop is enabled. The input at D is ignored when $T = 0$ or the flip-flop is inhibited, but the bit previously stored in the flip-flop is retained. In other words, T controls D, and Q is identical to D when $T = 1$ and Q remains unchanged when $T = 0$.

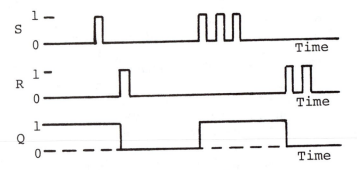

FIGURE 6-27 SR latch as a switch buffer.

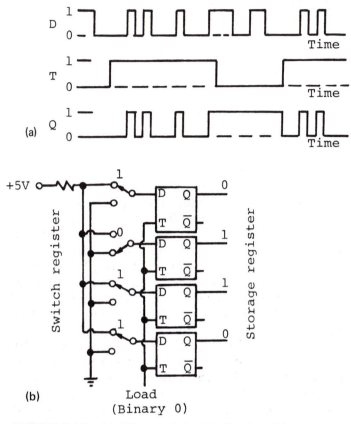

FIGURE 6-28   (a) Characteristic of D flip-flop. (b) Application of D flip-flop.

An application of the D flip-flop in a 4-bit storage register is shown in Fig. 6-28b. Assume that $T = 0$ and the binary data 1011 in the switch register are to be transferred into storage. The old data 0110 retained in storage have not yet changed, since $T = 0$. Data from the switch register are transferred when T goes from 0 to 1. Then the new data in storage are 1011, identical to that in the switch register.

3. JK Flip-Flops

The *JK flip-flop* shown in Fig. 6-25c consists of an input latch (master) and an output latch (slave), with feedback and logic components. The outputs are Q and $\bar{Q}$, as before. The quantities S, C, Q, and $\bar{Q}$ pertain to the output latch. The S and C inputs set the initial conditions for the Q

TABLE 6-5 Truth Table, JK Flip-Flop

Inputs		Outputs		
J	K	Q_t	Q_{t+1}	Operation
0	0	$\times$	$\times$	No change
0	1	$\times$	0	Reset
1	0	$\times$	1	Set
1	1	$\times$	$\bar{\times}$	Toggle

and $\bar{Q}$ outputs, and these inputs override all others in a JK flip-flop. Let us neglect the presetting by S and C inputs in this discussion.

The JK inputs of the input latch are similar to the SR inputs of a SR latch (see Fig. 6-25), with $J = S$ and $K = R$. The T-input controls the data transfer from the JK inputs to Q and $\bar{Q}$ of the output latch. The operations are (1) when $T = 0$, the JK inputs are inhibited, (2) when $T = 1$, the JK inputs are enabled, and (3) when T goes from 1 to 0, data stored in the input latch are transferred to the outputs Q and $\bar{Q}$. The T input is usually a train of clock pulses. Data from the input latch are transferred once for each clock cycle.

The truth table for a JK flip-flop in Table 6-5 is similar to that of a SR latch, where Q_t is the existing state of Q before T changes from 1 to 0, and Q_{t+1} is after the change. $\bar{Q}$ is not shown because it is the complement of Q. A distinctive feature of the JK flip-flop is the toggle mode, for which $J = K = 1$. The outputs Q and $\bar{Q}$ change states every time T goes from 1 to 0.

A 4-bit binary counter, shown in Fig. 6-29a, consists of four JK flip-flops in cascade, with the JK inputs in the toggle mode, that is, $J = K = 1$. The Q output of one flip-flop is connected to the T input of the next. The T input to the first JK flip-flop can be counted from decimal 0 to 15. The waveforms of the Q outputs are shown in Fig. 6-29b. Since data are transferred in a JK flip-flop when T goes from 1 to 0, a transition occurs only at the trailing (falling) edge of a pulse. Thus the frequency at the output of each JK is one-half of that of its input. Frequency division is the basis for binary counting. The binary number is read from the Q outputs, where Q_3 is the most significant bit (MSB) and Q_0 the least significant bit (LSB).

6-7. FUNCTIONAL LOGIC CIRCUITS

The possible combinations of digital components are innumerable. Certain combinations occur frequently and are called *functional logic cir-*

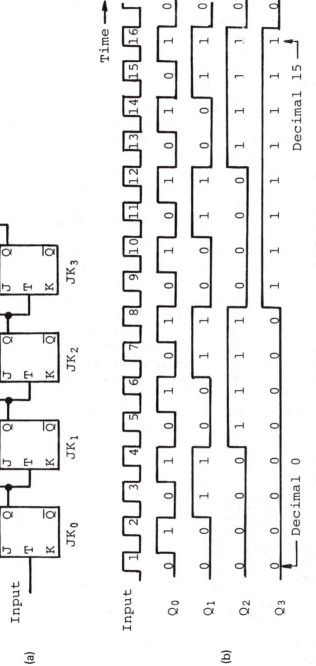

FIGURE 6-29 Waveforms of a 4-bit binary counter. (a) Binary counter. (b) Output waveforms.

cuits. The designer's job is often one of identifying the function and selecting the appropriate IC circuit from the many circuits capable of performing the task. Some functional logic circuits relating to instrumentation are described in this section.

A. Encoders–Decoders

An *encoder* expresses a given analog quantity into a digital code. The shaft encoder in Fig. 6-30a consists of a transparent disk attached to a rotating shaft. The disk has an opaque pattern of concentric bands. The light source and transducers are perpendicular to the disk. Light passing through a transparent portion is received by a transducer (photocell). This is an ON state, corresponding to a logic level 1. Light blocked by an opaque portion is not received. This is an OFF state and a logic level 0. Thus the position of the shaft is expressed in a digital number by means a bank of photocells. The shaft encoder is an *analog-to-digital converter*.

An encoding circuit for converting decimal 0 to 3 into binary 00 to 11 is shown in Fig. 6-30b. With all the switches opened, the output from the NAND gates is 00, that is, $A = 0$ and $B = 0$. The switches are closed one at a time. On closing switch S_1, the output is 01, that is, $A = 0$ and $B = 1$. The remaining operations can readily be traced.

A *decoder* is the converse of an encoder. It detects a code and expresses the coded information into another form. For example, a *digital-to-analog converter* is a decoder. A decoder is used for every digit in the readout of a pocket calculator. It converts a BCD word into a seven-segment LED display in decimal 0 to 9.

The basic decoding component is an AND gate. For example, the AND in Fig. 6-31a detects the coded word $\bar{A} \cdot B$ from a 2-bit storage register.

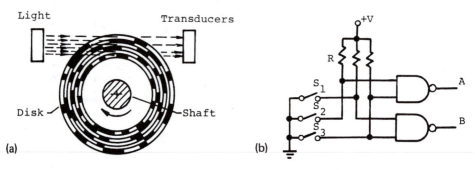

FIGURE 6-30 Examples of encoders. (a) shaft encoder. (b) Encoder for converting decimal 0 to 3 into binary 00 to 11.

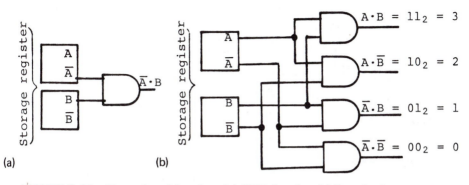

(a)   (b)

FIGURE 6-31 Examples of decoders. (a) AND decoder. (b) Decoder for converting binary 00 to 11 into decimal 0 to 3.

The circuit is expanded in Fig. 6-31b to decode binary 00 to 11 into decimal 0 to 3. Each and every combination from the storage register is detected by means of an AND gate. The operation of the decoder can readily be traced.

B. Multiplexers–Demultiplexers

A *multiplexer* is a selector. It selects data from parallel inputs and routes them to a serial output. A *demultiplexer* is a distributor, the converse of a multiplexer. The rotary switch in Fig. 6-32 may serve as a time-division multiplexer or a demultiplexer. In other words, a multiplexer is a parallel-to-serial converter and a demultiplexer a serial-to-parallel converter.

The simple multiplexing circuit in Fig. 6-33a uses a SR latch as a selector. When the latch is set ($Q = 1$), data from channel 1 are transmitted to the output OR gate, but data from channel 2 are ignored. Similarly, data from channel 2 are transmitted when the latch is reset ($\bar{Q} = 1$).

The multiplexing circuit in Fig. 6-33b selectively transmits the 2-bit binary word $A_1 B_1$ or $A_2 B_2$ to the output AB. (1) When $E = 1$, the ANDs are inhibited and no data transmission occurs, since $\bar{E}$ is applied to all the

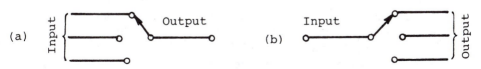

FIGURE 6-32 Rotary switch as a multiplexer or as a demultiplexer. (a) Multiplexer (selector). (b) Demultiplexer (distributor).

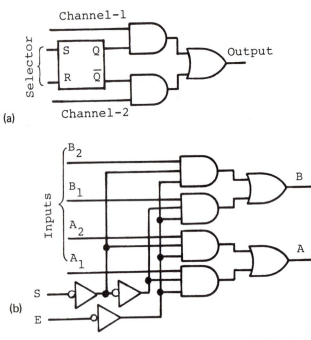

FIGURE 6-33 Simple multiplexer circuits. (a) Simple multiplexer. (b) Circuit for selecting A_1B_1 or A_2B_2.

AND gates. (2) When $E = 0$ and $S = 1$, the binary word A_1B_1 appears at the output AB. (3) When $E = 0$ and $S = 0$, the word A_2B_2 appears at the output.

The circuit in Fig. 6-34a uses an SR latch as a demultiplexer. Data go through a gate only when enabled. When the latch is set, the AND for output 1 is enabled, and input is routed to output 1. When the latch is reset, data are channeled to output 2.

A more complex demultiplexer is shown in Fig. 6-34b. The serial input is distributed to four flip-flops of the storage register. Hence a 4-bit serial word is converted to a parallel output. The distribution is controlled by a 2-bit counter (see Fig. 6-29a), synchronized with the serial input. Since the outputs from the counter are connected to the AND gates (connection not shown), they sequantially enable the AND gates. The SR latches are initially reset. (A latch is set when $S = 1$; see Fig. 6-26.) Since the gates are enabled sequentially in synchrony with the serial input, the latches are set sequentially by the serial input. Thus the parallel output from the storage register can be read simultaneously. It will be shown presently that the parallel output can be applied to a digital-to-analog converter.

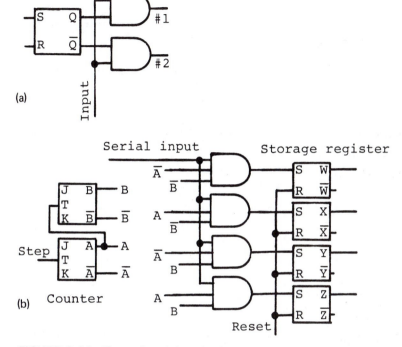

FIGURE 6-34 Examples of demultiplexer circuits. (a) Demultiplexer. (b) Demultiplexer with four parallel outputs.

C. Sample-and-Hold

A *sample-and-hold* (S/H) *circuit* samples and temporarily stores the value of an analog signal for subsequent processing, such as in a digital multimeter. Let us first describe the hardware, and postpone the applications until the next section.

The switch sampler in Fig. 6-35a is self-evident. The waveforms of the input voltage v_i and output v_o are shown in Fig. 6-35b. The interval of sampling is Δt and the sampling duration is D.

Similarly, the S/H circuit in Fig. 6-36a uses a FET switch (field-effect transistor). The first op-amp voltage follower is an input buffer. When the switch is closed (conducting), v_i charges the capacitor C, and the circuit tracks the input. When the switch is open (not conducting), C has no discharge path because the input impedance of the output follower is very high. Thus C holds the sampled voltage until the next sampling. The voltage for switching the FET and the S/H voltages are shown in Fig.

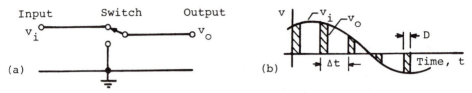

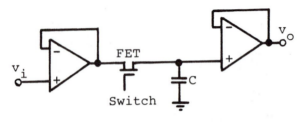

FIGURE 6-35 Mechanical switch as sampler. (a) Switch sampler. (b) Voltage waveforms.

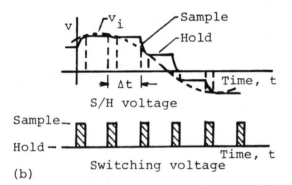

FIGURE 6-36 (a) Sample-and-hold circuit. (b) Voltage waveforms.

6-36b. The rate of sampling is governed by the RC time constant of the circuit. The R from the FET switch is of order of 50 Ω when conducting. The response due to the RC time constant is that of a first-order instrument (see Sec. 4-3). The effect of the time constant is noticeable in the voltage waveform for the sampling duration D.

D. Counters

Electronic counters are widely used in the laboratory. They are also components in instruments, such as in analog-to-digital converters. The decade counter is described in this section.

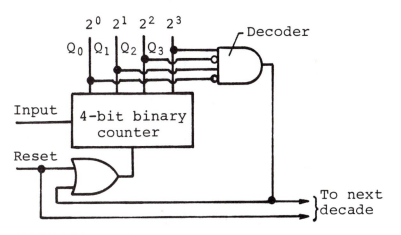

FIGURE 6-37 Decade counter.

The *decade counter* shown in Fig. 6-37 is a modification of a 4-bit binary counter (see Fig. 6-29). It resets for every 10 counts. When the binary code 1010 (decimal 10) is detected at the output $Q_3 Q_2 Q_1 Q_0$, the decoder gives a signal to reset the counter.

A number of decade counters can be used in cascade. For example, the decimal number 1682 is counted by cascading four decade counters. The counter described is a *ripple counter* because the signal passes sequentially from one flip-flop to the next like a wave. Since a time delay is introduced whenever a flip-flop changes state, the delays also propagate downstream. The total delay limits the speed of a ripple counter. This potential decoding error can be alleviated by means of a synchronous counter (not shown), in which the flip-flops operate in parallel.

E. Analog/Digital Converters

Digital-to-analog and analog-to-digital converters serve to interface between digital/analog devices. A wide variety of these devices with special features and a range of cost and performance are available as standard packages from many manufacturers. In all cases, the resolution depends on the number of bits employed. Units with 8 to 12 bits are most common.

The summing op-amp (see Fig. 6-7a) shown in Fig. 6-38a is a *digital-to-analog converter* (DAC). Assume that the values of the input resistors are weighted according to the 1–2–4–8 code. For a 4-bit DAC ($n = 4$), the value of a weighted resistor is $R_i = R/2^i$, for i from 0 to 3 ($= n - 1$). Voltages to operate the switches are from a storage register (see Fig.

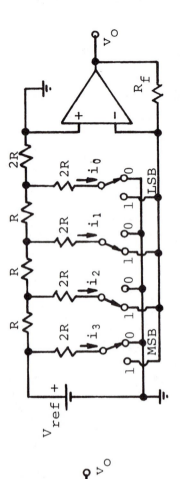

(b)

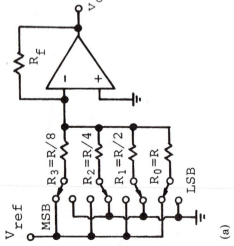

(a)

FIGURE 6-38 Digital-to-analog converters.

6-34b). Since this DAC requires a large range of precision resistors, it has problems of accuracy when n is large, say $n \geqslant 8$.

The DAC shown in Fig. 6-38b is more practical, because the R-$2R$ ladder network employed can be fabricated as an integrated circuit. The requirement is that the values of the resistors have a 2:1 ratio. The switches are OFF (binary 0) when connected to ground, and ON (binary 1) when connected to the virtual ground of the op-amp. The switches are operated according to a digital code from a storage register. It can be shown that the values of the weighted currents i_0 to i_3 have the ratio of 1:2:4:8, and the op-amp sums the currents. Variations of this DAC circuit are commonly used [13]. Evidently, the accuracy of the unit depends on the stability of the reference voltage.

Common techniques for *analog-to-digital converters* (ADC) are successive approximation and dual-slope integration [14]. The block diagram for *successive approximation* is shown in Fig. 6-39a. First, the analog voltage is quantized by means of a sample-and-hold (S/H) circuit (see Fig. 6-36). The comparator is an op-amp in an open loop (see Fig.

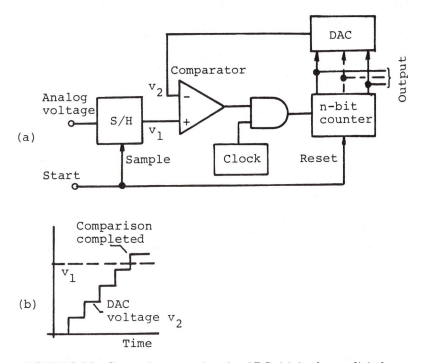

FIGURE 6-39 Successive-approximation ADC. (a) Analog-to-digital converter. (b) Comparison of voltages.

6-1); a sign change in its differential input ($v_1 - v_2$) is sufficient to cause the output to swing from high to low. The output of the comparator must be high or positive to enable the AND gate for counting the clock pulses. The S/H analog voltage v_1 is compared with a "staircase" voltage v_2 from the DAC, as shown in Fig. 6-39b. Since the DAC is driven by a counter, each clock pulse increments the counter and v_2. The measurement is completed when $v_2 \geqslant v_1$; that is, the output of the comparator swings from high-to-low to inhibit the AND gate and to stop the counting. The value of v_1 is proportional to the clock counts. Then the unit is reset for the next cycle. A successive-approximation ADC can operate at speeds up to 1 MHz.

The *dual-slope ADC* expresses the average value v_{av} of an analog input voltage in terms of clock pulses. The operating sequence is as follows:

1. The ADC is initially reset to zero. At time $t = t_1$, as shown in Fig. 6-40a, the analog input v_{in} is applied to an integrator. At the same time, a counter is started to count the clock pulses. The integrated voltage v_o is as shown in Fig. 6-40b.
2. After a number of preset counts for $t_2 - t_1$, the input to the integrator is switched from v_{in} to a reference voltage V_{ref} of opposite polarity, and the counter is reset to zero. Note that v_o at t_2 is proportional to v_{av} for the duration ($t_2 - t_1$), because integration is a summing process.
3. At $t = t_2$, the counter is restarted and V_{ref} is integrated until v_o crosses zero at $t = t_3$. Since v_o is linear with the slope of V_{ref}/RC, the counts for $t_3 - t_2$ are proportional to v_{av}, where RC is the time constant of the integrator. A conversion time of 4 ms with 17-bit resolution can be achieved.

Sources of error in analog/digital conversion are quantized error in the

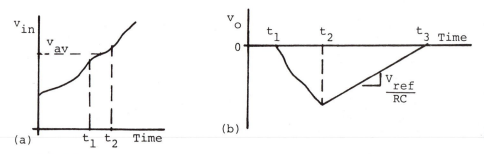

FIGURE 6-40 Operating sequence of a dual-slope ADC. (a) Analog voltage v_{in}. (b) Integrated voltage v_o.

digital mode, instabilities of the reference voltage and components and noise inputs. For example, if the coded word consists of n bits, there can be no more than 2^n possible output values. The quantized error is $\pm \frac{1}{2}$ LSB (least significant bit). This is 6.25% for a 4-bit word, about 0.39% for a 8-bit word, and 0.024% for a 12-bit word. It is interesting to note that the dual-slope ADC is almost immune to errors due to instabilities except for V_{ref}. The up-and-down integrations tend to cancel out inaccuracies of components, even the clock rate. High-frequency noise has little effect on integration. If the integrating duration $(t_2 - t_1)$ is an integer multiple of 1/60 s, the 60-Hz ac interference and its harmonics are suppressed. Of course, speed is sacrificed in favor of accuracy compared with the successive approximation ADC. Due to their accuracy and reliability, dual-slope ADCs are often used in digital voltmeters and other instruments.

6-8. OUTPUT DEVICES

Some common output devices are discussed in this section. Most of these are for voltage indicating or recording. Other instruments, such as galvanometers and meters, have been described and are not repeated here.

A. Cathode-Ray Oscilloscopes

The general-purpose analog oscilloscope is described in this section. A wide variety of oscilloscopes with special features and range of cost and performance are available as off-the-shelf items from many manufacturers. The digital oscilloscope has the potential for signal analysis and storage. This will be examined in the next section.

The *cathode-ray oscilloscope* (CRO) is a two-dimensional voltmeter for dynamic measurements. Its output is a bright spot on the screen of the cathode-ray tube. Voltages can be applied to the vertical and the horizontal inputs, and the deflection of the spot on the screen is proportional to an applied voltage. If the vertical input is a voltage from a transducer and the horizontal is linear with time, the trace of the bright spot on the screen of the cathode-ray tube gives the time history of the transducer output. This is the "normal" operating mode.

The *cathode-ray tube* shown in Fig. 6-41 is similar to those used in television sets. Electrons generated in the heated cathode are collected and accelerated by the anodes, and focused onto the face of the tube. The inner face of the tube has a grid with a series of hollow anodes. The electron beam impinging on the fluorescent screen gives a bright spot. The deflection of the spot is proportional to the voltages applied to the vertical and horizontal deflection plates.

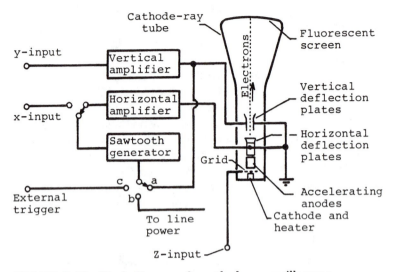

The block diagram shows that the y input gives the vertical deflection. The horizontal deflection is from an external x input or an internal saw-tooth generator. Since a sawtooth voltage increases linearly with time, it provides a linear time base. A combination of the y input and the linear time base gives the time history of the y input. The modulation by a z input is another method to provide a time base, but this is generally for special applications.

The sawtooth generator has three possible inputs, and therefore three modes for triggering or starting the time base for the x deflection. Upon connecting to position a as shown in Fig. 6-41, the time base is syn-chronized with the y input. When the y input is a repetitive fast transient, its trace appears stationary on the screen. This is due to the synchronized repetitive display of the trace, the retention of the phosphor on the screen, and the persistence of our vision. The time history of an event is a tremen-dous aid for dynamic studies. When the y input is not repetitive, the single-sweep mode can be used for a single display of the y input, the time history of which can be captured by menas photography.

If the sawtooth generator is connected to position b, the time base is synchronized with the power line. This is useful for measuring events related to the power-line frequency. If it is connected to position c, the time base is actuated by an means of an external trigger. For example, let the y input be the signal from a pressure transducer and the signal for the external trigger be from a detonation. The pressure–time history is thus synchronized with the detonation.

The *X–Y display* shows the functional relation between two events. For example, if the *y* input is a signal indicating pressure and the *x* input indicating volume, the trace on the screen gives a pressure–volume diagram. A dual-beam oscilloscope has two *y* inputs in one cathode-ray tube. The *y* inputs may be independent or time-shared by means of high-speed electronic switching, but there is usually only one common time base.

The accuracy of an analog oscilloscope is about 3%. The input impedance is from 1 to 20 MΩ, in parallel with 20 to 50 pF. The sensitivity is from 1 mV/cm to 20 V/cm for the vertical display. The horizontal sweep rate ranges from 20 ns/cm to 5 s/cm. The bandwidth is from dc to 20 MHz. Note that bandwidth and rise time (see Fig. 4-20) are related.

$$f_H t_r = 0.35 \tag{6-32}$$

where t_r is the rise time and f_H the high 3-dB point deduced from the RC circuit in Fig. 6-42. Since $2\pi f_H = 1/RC$, it can be shown that $t_r = (2.3 - 0.1)RC = 2.2/(2\pi f_H) = 0.35/f_H$.

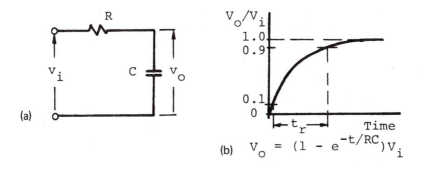

$$\text{(b)} \quad V_o = (1 - e^{-t/RC})V_i$$

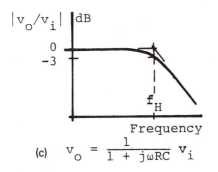

$$\text{(c)} \quad V_o = \frac{1}{1 + j\omega RC} V_i$$

FIGURE 6-42 Rise time and bandwidth. (a) rC circuit. (b) Indicial response. (c) Frequency response.

B. Digital Electronic Counters

Time, frequency, count, and phase angle are important measurements. Basic components in a digital electronic counter for these measurements are the counter (see Sec. 6-7D), the clock, circuits for waveform shaping, and gates.

The clock is a quartz-controlled oscillator that provides the frequency standard, usually at 10 MHz. The quartz may be placed in a constant-temperature oven to prevent thermal drifts. The 10-MHz oscillator gives a precise time base. For example, the period of 1 s is obtained by dividing 10 MHz by 10^7. This is accomplished by connecting the 10-MHz oscillator to seven decade counters in cascade. The output of the last counter is at 1 Hz.

The block diagram of a frequency counter is shown in Fig. 6-43a. The scheme uses a precise time base from the clock to control the AND gate (see Fig. 6-20) for counting the frequency of the input signal. The input is first shaped into pulses by means of an amplifier shaper. Many circuits can be used for the waveform shaping. The main criterion is that the input signal be above a threshold to avoid false counts due to noise superimposed on the signal. If the time base is 10 s and the count is 187,536, the frequency is 18,753.6 Hz. The same scheme is used for counting nonperiodic events. This mode counts events per unit time (EPUT) instead of frequency.

The scheme for time or period measurement, shown in Fig. 6-43b, uses the period of an input signal to control the counts from the 10-kHz oscillator. The time between two corresponding points of the input signal is its period, and the signal is not necessarily sinusoidal. The counts for one period of the input give the time of the period. If the frequency from the oscillator is 10 kHz and the count is 187,536, the period is 18.7536 s. It is preferable to measure period instead of frequency for measurements at low frequencies.

The scheme for time interval measurement shown in Fig. 6-43c is a modification of that for period measurement. The time interval between two events is used to control the gate for counting pulses from a 10-MHz oscillator. The inputs are shaped prior to triggering the gate control. Input 1 starts and input 2 stops the gate control for counting. If the count is 187,536, the time interval between the two inputs is 0.018,753,6 s.

Similarly, the phase angle between two sinusoids of the same frequency can be measured. In principle, the phase angle can be obtained from the time between the zero crossings of the events on the time axis. Since signals at zero crossings are susceptible to noise contamination, the two sinusoids are made equal in amplitude and their crossings at certain

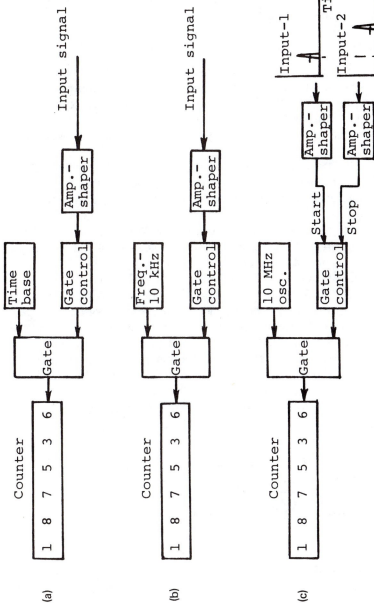

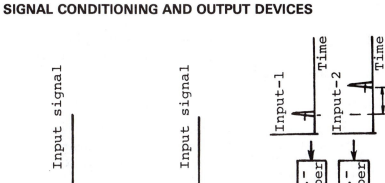

FIGURE 6-43 Digital electronic counters. (a) Frequency and EPUT measurement. (b) Period measurement. (c) Time interval measurement.

threshold are used for the phase angle measurement. Evidently, the same scheme can be used to measure the phase angle between any two periodic signals of the same frequency.

C. Magnetic Tape and Disk Recorders

The *magnetic tape recorder* is a recording and a mass storage device. Similar to a home tape recorder, it can record, erase, and reproduce. The storage and playback capability has many applications. For example, the effect of vibrations on airborne equipment can be simulated in the laboratory by playing a tape of aircraft vibrations into an electromechanical shaker.

The schematic diagram of a recorder shown in Fig. 6-44 consists of a tape transport, the record heads, and reproduce heads. The erase heads and electronics are not shown. The tape transport is servo controlled for constant speed. The tape speed ranges from 1⅞ to 120 in./sec. As the Mylar tape with magnetic coating passes under the record head, the head magnetizes the magnetic coating according to the mode of recording. The reproduce heads detect the magnetization.

The basic modes of recording are direct or AM, FM, and digital (see sec. 5-4). Each mode has a preferential field of application. AM and FM are for *analog recordings*. AM is more sensitive than FM to noise and imperfections of the tape. Its frequency range is from 100 Hz to 2 MHz, and the signal/noise ratio (S/N) is of the order of 25 dB (about 18:1). FM is more accurate than AM and is often used to record experimental data. Its freqeuncy range is from dc to 20 kHz at a tape speed of 120 in./sec. S/N is of the order 48 dB (250:1).

Multichannel recording is generally used in instrumentation. The head

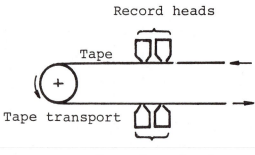

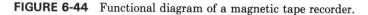

FIGURE 6-44 Functional diagram of a magnetic tape recorder.

assemblies are precisely positioned on a single base plate to ensure alignment with the tape. Even and odd channels are recorded separately to avoid crosstalk. Seven channels can be recorded on a ½-in.-wide tape and 14 channels on 1-in. tape.

Analog recordings can be used for time compression or time expansion; that is, information can be recorded at one speed and reproduced at a different speed. The time base of a recorder has a 64:1 ratio, since the tape speed ranges from 120 to 1⅞ in./sec. For example, an analog signal from the recorder can be fed to an ADC at an appropriate speed for digital processing. If a signal has high-frequency components, the tape speed can be reduced for playback into an oscillograph (frequency response to 2 kHz) for a "hard-copy" permanent record.

Digital recording stores data in the logic state of 1 or 0 by magnetizing the tape to saturation in one or the other direction. Data are stored as a series of pulses prescribed by a digital code. Digital recording requires high-quality tape and tape transport to ensure excellent head-to-tape contact and to avoid bit dropouts. The advantage of digital recording is the ability to communicate directly with a computer. For digital recording, up to 84 tracks can be accommodated on 2-in.-wide tape.

Both magnetic tapes and magnetic disks are used in digital processing. The common floppy disk in instruments is a 3½- or 5¼-in. flexible plastic disk with a magnetic coating. The advantage of magnetic tapes is the capability for mass storage, but the capacity of floppy disks is adequate for most applications in instrumentation. The capacity of a high-density 5¼-in. floppy disk exceeds 1 Mbyte. The disadvantage of the magnetic tape is its speed in random-access mode. In comparison, the random-access time of a tape cassette is on the order of 10 s versus 100 ms for a floppy disk.

D. Strip-Chart Recorders and X–Y Plotters

The strip-chart recorder and X–Y plotter are analog voltage recording instruments. The strip-chart recorder plots the time history of a voltage, while the X–Y recorder plots the functional relation between two voltages. The instruments perform the same tasks as the cathode-ray oscilloscope (CRO), but at slow speed. The slow speed is an advantage for numerous applications, such as in process control.

The schematic of a strip-chart recorder is shown in Fig. 6-45a. An applied voltage moves the pen in the y direction. The speed of the paper chart, from 8 in./min to 1 in./hr, gives the time base. The chart is advanced by means of sprockets to prevent slippage between the chart and the drive. The sprockets are driven by a stepping motor controlled by a

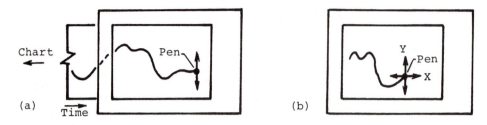

FIGURE 6-45 (a) Strip-chart recorder. (b) X–Y plotter.

variable-frequency oscillator or by a synchronous motor with mechanical pushbutton gear change. Alternatively, a timing marker is provided to indicate the chart speed.

An X–Y *plotter,* shown in Fig. 6-45b, is designed to operate at higher speeds than those of a strip-chart recorder. The pen moves in the (x,y) directions to record the functional relationship of two inputs on a stationary chart. A servo drives the pen in one direction. Another servo rides on a carriage of the first to move the pen in a second direction to give an X–Y plot.

The schematic of the servo drive for the recording pen of both instruments is shown in Fig. 6-46. The pen is driven by means of a position servo (see Sec. 3-6), and the pen and the wiper on the slide-wire potentiometer assume the same position through a mechanical linkage. The input voltage to the recorder is applied to the preamp. The voltage from the wiper at the potentiometer is fed back and compared with the input by means of a differential amplifier (see Fig. 6-9). Its output voltage is used to drive the servomotor in order to track the input voltage. The slew rate of the pen is of the order of 20 in./sec.

E. Stroboscopes

The *stroboscope* is widely used as a tachometer and in other applications. It emits a high-intensity light of very short duration, on the order of 10 µs. When the light flashes at a rate equal to the repetitive motion of an object, say a propeller, the object appears stationary. The upper frequency limit is about 2.5 kHz. For slow-moving objects below 10 Hz, the flicker due to flashing makes observations difficult.

The normal procedure for a speed measurement is to set the stroboscope above the estimated speed and then slowly decrease the flashing rate until a stationary image is observed. The rate of flashing is from high to low, because stationary images are also observed at multiples and submultiples of the actual speed, as shown in Fig. 6-47. When the speed of an

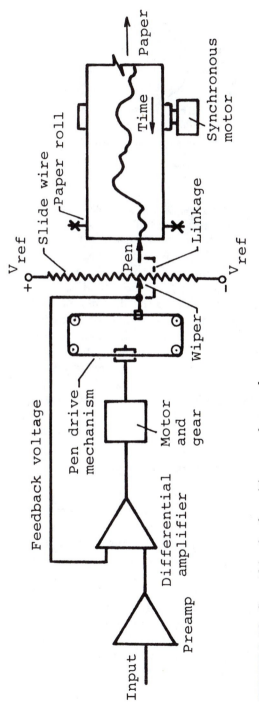

FIGURE 6-46 Servo-drive for the writing pen of a recorder.

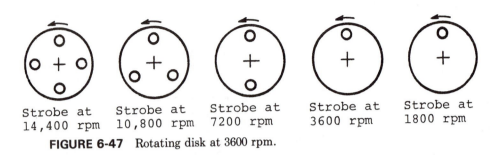

Strobe at   Strobe at   Strobe at   Strobe at   Strobe at
14,400 rpm 10,800 rpm 7200 rpm   3600 rpm   1800 rpm

FIGURE 6-47 Rotating disk at 3600 rpm.

object is above the range of a stroboscope, it can be shown that the actual speed can be calculated from consecutive submultiples.

Applications [15] of the stroboscope range from high-speed photography to slow-motion studies. The controlled emission of a high-intensity light is ideally suited for multiple exposures in photography. Thus the information can be quantitative for position and velocity measurements.

The stroboscope can be used for the slow-motion study of fast, repetitive events. For eample, the cavitation of hydraulic turbine blades due to air bubbles can be investigated by synchronizing the stroboscope with the speed of the turbine. Thus the path of the bubbles and the formation and collapse of bubbles can be observed and photographed. The malfunctioning of a machine can also be observed in slow motion. A flash delay, from 100 µs to 0.8 s, can be attached to the stroboscope. If the flashing is triggered at a reference point, the motion of one part of a machine relative to the other parts can be observed for the entire operating cycle.

6-9. OUTPUT/STORAGE DEVICES

Functions performed by the terminal stage of a measuring system (see Fig. 1-2) are signal monitoring, display, and data storage. Analog output instruments were described in the preceding section. A large group of instruments that converts analog input signals into digital data is examined in this section. This includes digital storage oscilloscopes, signal analyzers, data acquisition systems, and personal computer PC instrumentation.

The chief advantage of these instruments is the ability to store digital data. The data can be displayed, essentially without processing as in a digital oscilloscope, or processed for further applications as in a signal analyzer. Hence these instruments can be classified as data processing or nonprocessing. With the proliferation of accessories and equipment, it may be difficult to find the demarcation between the processing and the nonprocessing instruments.

A. Digital Storage Oscilloscopes

A *digital storage oscilloscope* stores the captured input signal in a digital mode and displays the signal without further processing. Most digital oscilloscopes can also be switched between the digital and analog modes. General-purpose analog oscilloscopes are useful and convenient. *Analog storage oscilloscopes* are available. Digital oscilloscopes have many desirable features. They will be more widely used as their cost becomes more competitive or as a part of a data acquisition system. It may be of interest to compare the two type of oscilloscopes.

An analog storage oscilloscope captures and displays a *transient signal* on the screen of a storage cathode-ray tube. The information cannot be saved and then retrieved for analysis. A digital storage oscilloscope converts the signal into digital data and stores the information in memory. The data saved can readily be recalled onto the screen. If the data are stored in a floppy disk, the signal saved in the past can be retrieved and compared with that being displayed today. The storage capacity is greatly expanded by means of the floppy disk.

With *repetitive signals,* the display of an analog oscilloscope gives the average of an input signal over many sweeps. When a signal is captured in a digital oscilloscope, the display can be scrolled to examine variations between the sweeps. Alternatively, multiple samples can be collected and averaged as a part of the acquisiton cycle.

A hard copy of the captured waveform can be obtained at the output of a digital oscilloscope by means of a digital plotter or an analog X–Y plotter. For the analog oscilloscope, this is achieved by means of photography. A digital oscilloscope can be interfaced with a computer. In fact, it is often a part of a personal computer PC instrumentation for data acquisition.

The accuracy of an analog oscilloscope is about 3%. The digital oscilloscope is capable of greater accuracy. Using a 12-bit analog-to-digital converter (ADC), a 0.1% accuracy is attainable. A cross-hair cursor can be positioned at a desired point on the waveform displayed to obtain its digital value. The waveform can also be "zoomed" by expanding the horizontal and/or the vertical scale in order to enlarge a small region for examination. Digital readouts, alphanumeric messages, operating menu, and interactive instructions can all be incorporated.

A digital-memory waveform recorder is also a nonprocessing periphery for digital data acquisition. It consists of a bank of ADCs at the input, the memory unit and controls, and the interface for digital output. Alternatively, digital-to-analog converters (DACs) can be added for an analog output.

B. Digital Signal Analyzers

A digital signal analyzer processes the acquired digital data by means of software. Applications of signal analysis are briefly introduced. A detailed study of Fourier transforms and applications for linear systems [16] is beyond the intended purposes of this book. Methods for obtaining digital data by fast Fourier transform (FFT) and sampling are examined briefly in the subsections to follow.

Digital Fourier analyzers are widely used for experimental study of dynamic systems [17]. Applications range from oil exploration to spacecraft. The scope of the study is extensive, from testing techniques [18] to computer algorithms [19].

For example, one of the output formats from a Fourier analyzer is the *Fourier spectrum* (see Fig. 4-36). The basic concept of spectral analysis is simple. A cantilever beam in the form of a vibrating reed, shown in Fig. 6-48a, will vibrate with large amplitudes at resonance. The speed indicator for a stationary steam turbine consists of a bank of reeds. If each reed is progressively tuned to a different resonant frequency, the operating speed of the turbine will excite one of the reeds into resonance, as shown in Fig. 6-48b, looking in from the front of the bank of reeds. The assembly of reeds is, in effect, a bank of filters. A spectrum analyzer works on the same "bank-of-filter" principle, whether mechanical, electromechanical, or electrical. The spectrum for all frequencies is displayed simultaneously in an electrical analyzer.

Applications of Fourier analyzers for mechanical systems are limited only by one's imagination. For example, the mathematical model of a complex structure can be determined experimentally. The model can then be used for predicting its field performance, examining the effects of

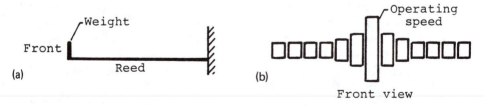

FIGURE 6-48   Vibrating reeds as a "spectrum analyzer." (a) Vibrating reed. (b) Speed indicator.

a design change prior to the change, or designing a machine from the model of its parts. Parametric estimation and signal analysis [20] are useful diagnostic tools in engineering. By means of a coherence function [21], the input/output causality, that is, the validity of test data, can be verified. Fourier analysis is based on linear theory, but nonlinearities can be detected when a system does not precisely follow linear theory.

A digital Fourier analyzer is versatile because the stored digital data can be processed by means of software. For example, the transfer function of a system can be obtained by means of Fourier transforms (see Sec. 4-8). It is more accurate to compute the transfer function from the ratio of the cross spectrum of the input/output signals and the power spectrum of the input. These operations can be performed by means of software. The same results can be achieved by means of analog devices, but the data are not stored. Thus every function performed requires a separate hard-wired component.

C. Fast Fourier Transform

The *fast Fourier transform* (FFT) is an algorithm for implementing the discrete Fourier transform (DFT). The Fourier series expresses a periodic function $x(t)$ in terms of its harmonic components. In effect, the series performs a mathematical filtering operation:

$$x(t) = \frac{a_0}{2} + \sum_{n=0}^{\infty} \left(a_n \cos \frac{2\pi nt}{T} + b_n \sin \frac{2\pi nt}{T} \right) \tag{6-33}$$

$$x(t) = \sum_{n=-\infty}^{\infty} c_n e^{j2\pi nt/T} \tag{6-34}$$

$$c_n = \frac{1}{T} \int_{-T/2}^{T/2} x(t) e^{-j2\pi nt/T} \, dt \tag{6-35}$$

where T is the period of $x(t)$, and Eqs. 6-33 and 6-34 are equivalent. The coefficients a_n and b_n can be calculated from sampled values of $x(t)$ [22]. Note that (1) a periodic function has a discrete spectrum, and (2) it takes a pair of coefficients a_n and b_n to obtain the amplitude and phase information in the frequency domain. The amplitude and phase angle of a harmonic component are

$$\text{amplitude} = (a_n^2 + b_n^2)^{1/2} \tag{6-36}$$

$$\text{phase angle} = \tan^{-1} \frac{b_n}{a_n} \tag{6-37}$$

When the period T of $x(t)$ becomes infinite, we obtain the Fourier transform pair (see Appendix C)

$$S_x(f) = \int_{-\infty}^{\infty} x(t)e^{-j2\pi ft}\, dt \qquad \text{(forward transform)} \qquad (6\text{-}38)$$

$$x(t) = \int_{-\infty}^{\infty} S_x(f)e^{j2\pi ft}\, df \qquad \text{(inverse transform)} \qquad (6\text{-}39)$$

A Fourier transform can also be viewed as a mathematical filtering operation, but the corresponding spectrum is continuous.

An analog signal $x(t)$ is converted to the digital mode by sampling at regular time intervals Δt, as shown in Fig. 6-49a. The time T for obtaining the samples must be finite to be practical. Letting dt in Eq. 6-38 become Δt, converting the integration to a summation and truncating the series, the equation becomes a *discrete Fourier transform* (DFT):

$$S_x'(m\,\Delta f) = \Delta t \sum_{n=0}^{N-1} x(n\,\Delta t)e^{-j2\pi m\,\Delta f n \Delta t} \qquad (6\text{-}40)$$

where N = number of samples, the time t in Eq. 6-38 becomes $n\,\Delta t$, and the frequency f becomes $m\,\Delta f$. The "window" for this sampling duration T, shown in Fig. 6-49b, is called a rectangular time window because the signal $x(t)$ is not modified for the duration of the observation.

The DFT has a discrete spectrum, because the function $x(t)$ observed through a time window is assumed to repeat itself for every T. This is assumed whether $x(t)$ as a whole is periodic or not. Corrections must be made when $x(t)$ from which the DFT is derived is not periodic within the window.

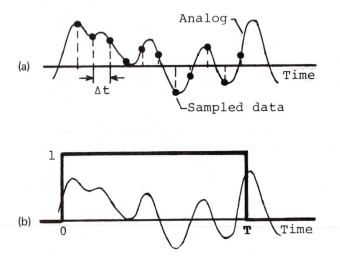

FIGURE 6-49 Time window in sampling. (a) Analog and sampled data. (b) Time window.

The fast Fourier transform (FFT) is an algorithm based on certain repetitive properties of the DFT in Eq. 6-40. Many FFT algorithms have been developed to simplify and to increase the speed of computation. The FFT algorithm reduces the number of calculations to about 2% of that required by the DFT [23]. The analysis is called "real time" because the operations are sufficiently fast for updating the information during a test.

D. Sampling

An analog signal is sampled (see Fig. 6-36) and converted to digital data by means of an analog-to-digital converter (ADC) (see Figs. 6-39 and 6-40). Each sampled value becomes a digital word, say of 12 bits. The sampling process must conform to constraints governed by sampling theory and the discrete Fourier transform. Sampling techniques are described in this section.

1. Sampling

Let us first relate the variables for sampling in the time and frequency domains. Sampling in the time domain, shown in Fig. 6-50a, gives

$$T = N \, \Delta t \tag{6-41}$$

where T is the duration of the time window, Δt the time increment in sampling, and N the number of samples for the duration T. N is some multiple or submultiple of 1024 (or 1K) of memory. It is also the block size allotted for a block of information in a computer. In the freqeuncy domain, shown in Fig. 6-50b, we have

$$F_{max} = \frac{N}{2} \, \Delta f \tag{6-42}$$

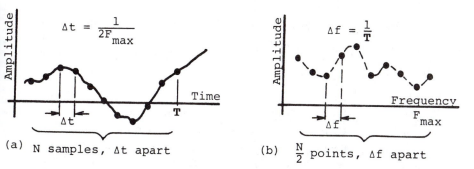

FIGURE 6-50 Sampling in the time and frequency domains. (a) Analog signal: $T = N \times \Delta t$. (b) Spectrum: $F_{max} = N/2 \times \Delta f$.

where Δf is the frequency increment and $N/2$ is the number of data points in the frequency domain, because it takes two time samples to get one point in the complex plane with amplitude and phase information. $N/2$ is also the number of spectral lines. Since the spectrum is discrete, frequencies between spectral lines are not shown. The Δf is the number of hertz between spectrum lines and is called the frequency resolution.

Time and frequency information is related by Shannon's sampling theorem [24], which in essence states that slightly more than two samples per period are required to define a sinusoid uniquely. Hence to resolve the maximum frequency of interest F_{max}, the sampling interval is $\Delta t \geqslant 1/(2F_{max})$. For simplicity, we use the equality sign for the expression:

$$\Delta t = \frac{1}{2F_{max}} \tag{6-43}$$

From Eqs. 6-41 to 6-43, we deduce that

$$\Delta f = \frac{1}{T} \tag{6-44}$$

The parameters for sampling are Δt, F_{max}, Δf, and T, as related by the four equations above. For example, if Δt is selected, F_{max} is automatically fixed by Eq. 6-43. Only one of the two remaining parameters, Δf or T, may be selected, since they are related in Eqs. 6-41 and 6-44. The selection of parameters is summarized in Table 6-6.

2. Aliasing

Aliasing occurs when the sampling control is set at a F_{max} and the incoming test signal contains frequencies higher than F_{max}. The higher frequencies will "fold back," appearing as a lower frequency. For example, if F_{max} is set at 2 kHz, shown in Fig. 6-51, a harmonic at 2.2 kHz will appear as a 1.8-kHz component in the display on the socilloscope. In other words, the input will appear to contain frequencies, which are "reflections" from higher frequencies about F_{max}.

Aliasing is caused by the sampling of signals above F_{max}. A technique for eliminating aliasing is to place an analog low-pass filter, called an *antialiasing filter*, before the sampling control to limit the input frequency to F_{max}.

3. Leakage

Leakage describes the result of the multiplication of the rectangular time window (see Fig. 6-49b) and the time function observed through that window [26]. Multiplication in the time domain corresponds to convolution in the frequency domain, but let us discuss windowing effects by intuitive reasoning.

TABLE 6-6 Selecting Parameters for Sampling

Selection	Fixed by selection	Select either one
Δt	$F_{max}\left(= \dfrac{1}{2\Delta t} \right)$	$T \, (= N \, \Delta t)$ $\Delta f\left(= \dfrac{1}{N \, \Delta t} \right)$
F_{max}	$\Delta t\left(= \dfrac{1}{2F_{max}} \right)$	$T \, (= N \, \Delta t)$ $\Delta f\left(= \dfrac{1}{N \, \Delta t} \right)$
Δf	$T\left(= \dfrac{1}{\Delta f} \right)$	$\Delta t\left(= \dfrac{T}{N} \right)$ $F_{max}\left(= \dfrac{N}{2} \Delta f \right)$
T	$\Delta f\left(= \dfrac{1}{T} \right)$	$\Delta t\left(= \dfrac{T}{N} \right)$ $F_{max}\left(= \dfrac{N}{2} \Delta f \right)$

Source: Ref. 25.

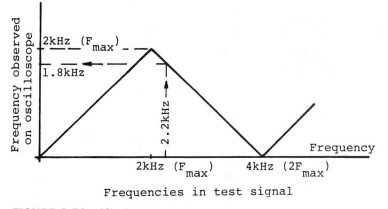

Frequencies in test signal

FIGURE 6-51 Aliasing error.

The discrete Fourier tranfsorm (DFT) assumes that the time function observed through a time window is periodic. A sine wave has only one harmonic component and a single spectral line, as shown in Fig. 6-52a. Now let the sine wave be observed with a rectangular time window. (1) If the sine wave has n periods within the window, where n is an integer, the DFT sees the function as a periodic sine wave, and it has a single spectral line, as shown in Fig. 6-52b. (2) If the sine wave does not have an integral number of periods within the time window, the DFT has several spectral lines, as shown in Fig. 6-52c. This is called a *leakage error*. In other words, the

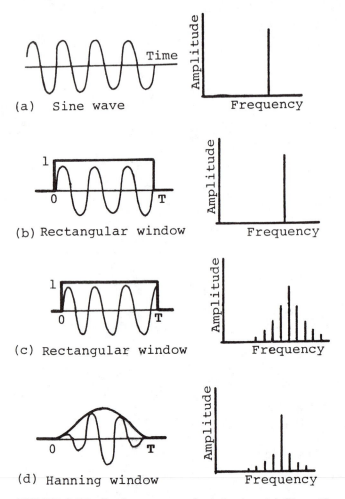

(a)   Sine wave

(b) Rectangular window

(c) Rectangular window

(d) Hanning window

FIGURE 6-52 Leakage error and windowing. (a) Sinusoid and spectrum. (b)–(d) Sinusoid observed and DFT.

spectral lines shown will indicate many frequencies, corresponding to the "periodic" function observed through the rectangular time window.

Leakage error is minimized by multiplying the analog input signal with a weighting function. As shown in Fig. 6-52d, the weighting function eliminates the discontinuities of the time function at the edges of the window, thereby forcing the function to be periodic for the observation. This is called *windowing*. For example, the weighting function $[1 - \cos(2\pi t/T)]/2$ shown in Fig. 6-52d is a Hanning window. The frequency spectrum of the corresponding DFT is more clearly defined, but the amplitude information is improved at the expense frequency resolution.

4. Windows

Many types of windows are commonly used for dynamic testing [27]. A special hammer with a force transducer is used for applying an impulse in the pulse testing of machines. Since the duration of the impulse input is very short and there is no signal after the impulse, a window with a sharp cutoff, as shown in Fig. 6-53a, can be used for a force window. This window effectively eliminates any noise input after the impulse.

The impulse response, shown in Fig. 6-53b, decays rapidly, and the signal/noise ratio is low at the tail end of the response. The noise at the tail end can be suppressed by means of an exponential window. Note that an increase in the rate of decay in the response is equivalent a larger damping in the system. The added damping introduced by windowing, however, can be removed by means of additional signal processing.

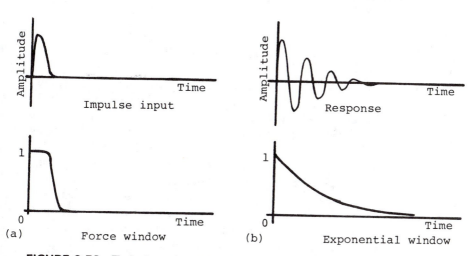

FIGURE 6-53 Techniques in windowing. (a) Force input and window. (b) Response and window.

E. Data Acquisition Systems

The share of industrial personal computers (PCs) to total PCs, which comprised less than 17% in 1986, is expected to increase to over 50% by 1991 [28]. In this section we briefly examine analog transducers and signal conditioning for digital data acquisition and personal computer (PC) instrumentation.

1. Transducers

The analog outputs of transducers are converted into electrical signals and conditioned to interface with an analog-to-digital converter (ADC), as shown in Fig. 6-54. Transducers are classified as self- and non-self-generating, and signal conditioning modules are available for every type of transducer.

Electrical outputs from transducers are shown symbolically in Fig. 6-55. Outputs from self-generating transducers are ac or dc voltage or current and frequency. A thermocouple is an example of a low-level voltage source. An IC (integrated-circuit) temperature sensor is an example of a current source. It is often used for measuring the temperature of the reference junction for thermocouple compensation. The output from a turbine flow meter is in the form of a pulse train or frequency. Non-self-generating transducers are impedance based and require auxiliary energy inputs for their operation. Examples of these transducers are resistive devices, such as strain gages and resistance thermal detectors RTDs, and differential transformers LVDTs.

2. Multiplexers

The logic circuit, shown in Fig. 6-54, directs the scanning or multiplexing (see Fig. 6-32) of a number of transducers. The output from the signal conditioner to ADC is usually from 0 to ± 10 V_{dc}.

Both solid-state (see Fig. 6-36) and electromechanical relay multiplexers are used. The former are capable of high-speed switching (up to 50 kHz) and high reliability. The disadvantages are that they are susceptible to static and transients and large offset voltages. Solid-state multiplexing are used for high-speed and preconditioned signals. The high-speed but less accurate successive-approximation ADC (see Fig. 6-39) can be used for the analog-to-digital conversion. The relay-type multiplexers are suitable for low-speed and low-level applications, such as signals from thermocouples. The switching rate is of the order of 15 channels per second. The low-noise dual-slope ADC (see Fig. 6-40) can be used for this conversion. The trade-off for the two types of multiplexers is between speed and noise rejection.

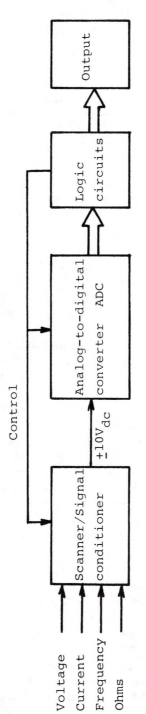

FIGURE 6-54 Simplified diagram of data acquisition system.

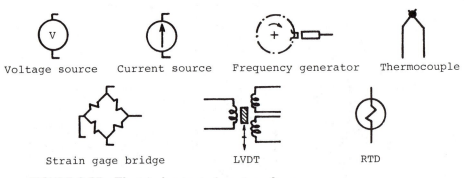

FIGURE 6-55   Electrical outputs from transducers.

3.   Signal Conditioning

Inputs to the signal conditioner are voltages, currents, frequency, or ohms, as shown in Fig. 6-54. A voltage is either amplified or attenuated in the signal conditioner. Low-level voltages, such as from a thermocouple, may require some consideration. For example, when the two "grounds" shown in Fig. 6-56a are not at identical potential, a current would flow in the LO lead and the subsequent IR voltage drop in the lead is added to the value to be measured. The guard shwon in Fig. 6-56b diverts the ground loop from the LO lead and is commonly used for low-level voltage measurements.

A current is measured by placing a "shunt" resistor R of known value in the current path as shown in Fig. 6-57a. The voltage measured across R gives the value of the current. Note that when a current loop is being scanned, the scanner switch should never be in series with the shunt resistor R, as shown in Fig. 6-57b. The shunt resistor R should always be on the unswitched side of the scanner, as shown in Fig. 6-57c. In other words, the current flow for controlling a process must not be interrupted.

A frequency or a pulse train can be measured in terms of total counts, frequency, or period, as shown in Fig. 6-43.

A resistance can be measured by sending a current of known value through the unknown resistor R, as shown in the "two-wire system" in Fig. 6-58a. Errors due to the resistance in the leads are negligible if R is large, such as in a thermistor. The effects of the lead wires are not negligible when R is small as in a RTD. The "four-wire system" shown in Fig. 6-58b (see also Fig. 3-51) is mandated for the accurate measurement of low resistance values. The resistance in the current leads have no effect because the current is constant; the resistance in the voltage leads is negligible compared with the input impedance of the voltmeter. Note that a bridge circuit is often used to measure small changes in resistances.

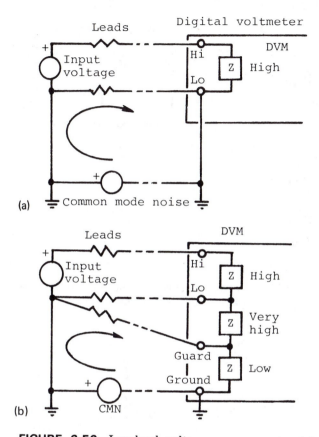

FIGURE 6-56 Low-level voltage measurements. (a) Common-mode noise source. (b) Guarded input voltage.

4. Data Acquisition

Data acquisition is the systematic recording or display of information. Some data reduction may be included in a data acquisition system. The terms *data logging* and *data acquisition* are often used interchangeably. A basic data logger consists of a scanner/signal conditioner, a digital voltmeter DVM, the associated logic circuits for control, and a printer, as shown in Fig. 6-54. The microprocessor and other components are shown more explicitly in Fig. 6-59. The components enumerated can be identified readily in the block diagram.

The microprocessor controls the operation of the system. It has a data bus for digital data transfer and an address bus (not shown explicitly) for memory addressing. The memory units are the ROM and RAM. In-

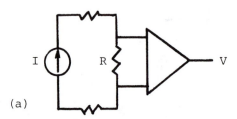

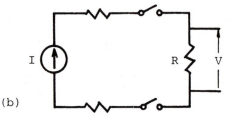

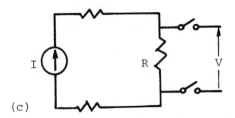

FIGURE 6-57 Use shunt resistor R for current measurement. (a) Current measurement. (b) Scanner switches in series. (c) Correct switching.

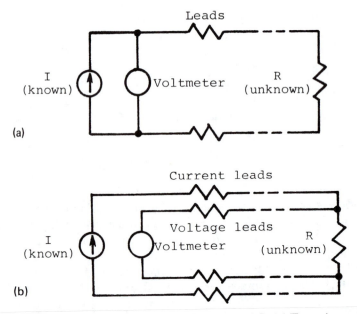

FIGURE 6-58 Methods to measure resistance R. (a) Two-wire system. (b) Four-wire system.

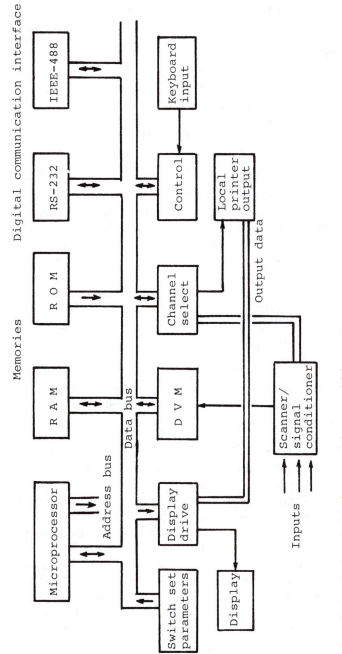

FIGURE 6-59   Simplified diagram of data acquisition system.

dividual channel parameters and other program routines can be entered through the keyboard. Some version of a high-level language, such as BASIC [29], is often used. The output is obtained from the local printer or recorded by means of other devices.

The IEEE-488 and the RS-232 are commonly used for the digital communication interface. The IEEE-488 bus can transmit digital data in either direction up to 14 instruments, with each instrument having its own address. Thus it is possible to talk to each separately. The use of one I/O card for 14 instruments is a saving. The RS-232 interface has no protocol standard and is defined only for electrical characteristics. It is defined up to 50 ft of cable separation and up to 20 kilobits/sec.

5. PC-Based Instrumentation

Personal-computer-based data acquisition and control systems are common in industry for specialized tests and general instrumentation. Historically [28], a number of PCs were provided with extra slots in the chassis for peripheral devices by the late 1970s. In 1981, the IBM-PC featured open slots that could accommodate analog interfaces. Hewlett-Packard introduced the IEEE-488 interface. There seems to be a never-ending stream of new data acquisition hardware and software products. Distributed control system manufacturers have made possible the integration of PCs into proprietary networks by developing necessary package interfaces. Process monitoring-control and data acquisiton software are available from more than 50 companies. The software can turn a PC into a process control station instead of a mere data logger. A PC can be used for interactive process monitoring and control, diagnostics, simulation, or operator training.

To interface "stand-alone" instruments via extra slots in a PC as mentioned above may not be the best method to take advantage of the PC. A PC has only a limited number of expansion slots, the slots are in a noisy environment within the computer, and the card configuration must be changed with each test setup. An alternative approach is to use a card cage with a large number of slots and a common power supply.

The pressure to use PCs as low-cost controllers of instruments is inevitable, because of the greatly improved cost-effectiveness of PCs and software and the upward trend in instrument cost. Interface cards are used to convert a PC into a digital oscilloscope, a spectrum analyzer, or other instruments (e.g., Rapid Systems, Inc., Seattle, Washington). The hard disk in a PC is ideal for fast data retrieval. The PC is well suited for integrating instruments with the storage, hard copy, and display equipment.

A PC can be utilized as a controller for a number of instrument modules [e.g., HP (Vectra) PC Instruments, Hewlett-Packard, Palo Alto, California], such as a digital multimeter, oscilloscope, and relay multiplexer. The stand-alone instruments are replaced by these modules, and the "hard" front panels of the instruments simulated by the interactive "soft" front panels displayed on the computer screen. In other words, a module functions as its traditional stand-alone counterpart. The appropriate soft panel, system status, and other information are shown simultaneously. Once interfaced, the modules are computer programmable. Furthermore, the instrument configuration for a test can be stored and recalled for subsequent applications, such as data logging, production tests, process monitoring and control, and component evaluation and inspection.

PROBLEMS

6-1. (a) Referring to Fig. P6-1a, $R_f = 100 \ k\Omega$ and $R_1 = 10 \ k\Omega$. (1) Estimate the input impedance at v_1. (2) Sketch the waveform of v_o if v_1 consists of a static signal of 0.6 V_{dc} and a dynamic sinusoidal signal of amplitude 0.1 V at 1 kHz. (3) A simple RC filter is used to extract the dynamic signal from v_o. Show a sketch of the entire circuit and specify the values of R and C if the attenuation of the dynamic signal is $\leqslant 2\%$.

(b) Referring to Fig. P6-1b, $R_f = 100 \ k\Omega$, $R_1 = 10 \ k\Omega$, and $R = 100 \ k\Omega$. Find v_o for $v_1 = 1.0$ V.

(c) Referring to Fig. P6-1c, $R_f = 100 \ k\Omega$, $R_1 = 10 \ k\Omega$, $R_2 = 5 \ k\Omega$, and the power supply to the op-amp is at $\pm 10 \ V_{dc}$. (1) Find v_o if $v_1 = 2.6$ V and $v_2 = -1$ V. (2) Repeat part (1) for $v_1 = 2.6$ V and $v_2 = -2$ V.

6-2. Derive the sinusoidal transfer function v_o/v_i for each of the op-amp circuits shown in Fig. P6-2.

6-3. The op-amp circuit shown in Fig. P6-2a is that of a modified integrator (see also Fig. 6-6b). Its transfer function from Prob. 6-2 is also that of a low-pass filter. A low-level signal of the order of millivolts at 0.5 Hz has a parasitic 60 Hz noise and the signal-to-noise ratio (S/N) is 1:5. Let $R = 1 \ M\Omega$, $C = 0.1 \ \mu F$, and $R_1 = 1 \ k\Omega$. Assume that a sign inverter is used at the output v_o. Calculate v_o and S/N.

6-4. The op-amp circuit in Fig. P6-2e is used to produce a phase angle lead or lag in v_o relative to v_i. Assume that a sign inverter is used at the output v_o. (a) Sketch the Bode plots of the transfer function (see Fig. P6-2e) for $R = 50 \ k\Omega$, $C = 0.1 \ \mu F$, $R_1 = 10 \ k\Omega$, and $C_1 = 0.1 \ \mu F$. Find the frequency in rad/s at which the maximum phase shift occurs and the value of the phase angle. (b) Repeat when the values of R and R_1 are interchanged.

6-5. The differentiator shown in Fig. 6-6c is modified by inserting R_1 in series with C_1 as shown in Fig. P6-2f. Let $R = 1\ \mathrm{M\Omega}$, $R_1 = 10\ \mathrm{k\Omega}$, and $C_1 = 0.1\ \mu\mathrm{F}$. What is the advantage of this modification? What is the frequency range that the circuit can be used as a differentiator?

6-6. The op-amp without feedback shown in Fig. 6-1 is a comparator. Sketch the output waveform for the input voltages if (a) $v_a = 3\ \sin \omega t$ and $v_b = 0$, (b) $v_a = (5 + 3\ \sin\ \omega t)$ and $v_b = 5$, and (c) $v_a = (4 + 3\ \sin\ \omega t)$ and $v_b = 5$.

6-7. A differential amplifier (see Fig. 6-9) is used as a bipolar potentiometer as shown in Fig. P6-3. The resistance of the potentiometer is $R_p = 100\ \mathrm{k\Omega}$. The range of the output voltage v_o is $(0 \pm V)$ volts for the supplied voltage V. The contact (wiper) of the potentiometer is at aR_p, where $0 \leqslant a \leqslant 1$. Show that the output $v_o = (2a - 1)V$ volts.

6-8. A signal v_s from a transducer is at $\omega = \omega_s$ rad/s. The RC low-pass filter shown in Fig. P6-4 is used to suppress the associated high-frequency noise v_p at the input at $\omega = \omega_p$. Let the attenuation of v_s be $x\%$ and the reduction of v_p be $y\%$. (a) For $y = 90\%$, plot the values of x versus $\log(\omega_p/\omega_s)$ for x from 0.5 to 10%. (b) Repeat part (a) for $y = 80, 70$, and 60%, respectively. (c) Let v_p be from a 60-Hz source. Find the maximum frequency allowed for the signal v_s when $x = 2\%$ and $y = 80\%$. (d) If the output impedance of the transducer is 10 kΩ, find the values of R and C in part (c).

6-9. A signal v_s from a transducer is at $\omega = \omega_s$ rad/s. The RC high-pass filter shown in Fig. P6-5 is used to suppress the associated low-frequency ambient noise v_a at $\omega = \omega_a$. Let the attenuation of v_s be $x\%$ and the reduction of v_a be $y\%$. (a) For $y = 90\%$, plot the values x versus $\log(\omega_s/\omega_a)$ for x from 0.5 to 10%. (b) Repeat part (a) for $y = 80, 70$, and 60%, respectively.

6-10. The high-pas filter described in Prob. 6-9 is also the ac couple at the input of an oscilloscope. Let $R = 1\ \mathrm{M\Omega}$ and $C = 0.1\ \mu\mathrm{F}$. (a) If the attenuation of the input to the oscilloscope is to be less than 2%, find the minimum frequency allowed for the input signal. (b) An ac coupled oscilloscope is used to measure the dynamic strain ($\geqslant 10\ \mathrm{Hz}$) from a ballast circuit (see Prob. 1-3). The strain amplitude is about 200 μstrain, the resistance of the strain gage is $R = 120\ \Omega$, the gage factor $G_f = 2.0$, $R_b = R$, and the dc source for driving the ballast circuit is 6 V with a slight ripple. Justify whether the oscilloscope is adequate for this measurement.

6-11. Derive the transfer function for the bandpass filter shown in Fig. P6-6. Plot $|v_o/v_i|$ versus log ω for $R_1 = R_2 = 1$ kΩ, and $C_1 = C_2 = 0.1$ μF.

6-12. The Wien bridge shown in Fig. P6-7 is an example of a notch filter, which is often used to suppress the 60-Hz noise in instruments. Derive its transfer function. Plot the characteristics in $|v_o/v_i|$ versus log(ω/ω_c), where ω_c is the critical frequency at (2π) (60 Hz) rad/sec.

6-13. Express the binary numbers in their decimal equivalents.
(a) 110101_2 (b) 110.011_2 (c) 10111.101_2
(d) 101101.11_2 (e) 100000.0111_2

6-14. Express the decimal numbers in their binary equivalents.
(a) 13_{10} (b) 144_{10} (c) 225_{10}

6-15. Express the decimal numbers in their binary equivalents.
(a) 13.625_{10} (b) 12.75_{10} (c) 53.421875_{10}

6-16. Find the sum or difference for the binary operations.
(a) $10111_2 + 10111_2$ 　　　　　　(b) $1101_2 + 1110_2$
(c) $10000_2 - 101_2$ 　　　　　　(d) $101110_2 - 10101_2$
(e) $1011_2 + 1001_2 + 1111_2 + 1011_2$

6-17. Convert the decimal numbers into the 8421 BCD code.
(a) 1049_{10} (b) 627_{10} (c) 583_{10} (d) 4957_{10}

6-18. Convert the 8421 BCD code to decimal numbers.
(a) 0010　0110　1001　　(b) 1000　1001　1010
(c) 1000　1001　0010　　(d) 0101　0000　1001

6-19. Verify the Boolean identities in Example 6-2 by means of truth tables,
(a) $A \cdot (A + B) = A$ 　　　(b) $A \cdot (\bar{A} + B) = A \cdot B$
(c) $A \cdot B + \bar{B} = A + \bar{B}$ 　　　(d) $A \cdot \bar{B} + B = A + B$

6-20. Verify the Boolen identities by means of algebra and truth tables.
(a) $A \cdot B + A \cdot \bar{B} = A$
(b) $A + A \cdot B = A$
(c) $A + B = A + \bar{A} \cdot B$
(d) $A \cdot B + A \cdot \bar{B} + \bar{A} \cdot B + \bar{A} \cdot \bar{B} = 1$
(e) $(A + B) \cdot (\overline{A + B}) = A \cdot \bar{B} + \bar{A} \cdot B$
(f) $A \cdot (B + C) = A \cdot B + A \cdot C$
(g) $A + B \cdot C = (A + B) \cdot (A + C)$

6-21. (a) Show the implied logic functions in a diagram if a car cannot start until the gears are in neutral or park, the seat belts are latched, and the front doors are closed.
(b) The home security system is set to sound an alarm between 10 P.M. and 4 A.M. when the garage or the front or back door is open or

any of the windows are open. Show the implied logic functions in a diagram.

6-22. Derive the logic functions for the circuits shown in Fig. P6-8a and b.

6-23. (a) Show that the circuits in Fig. P6-9a and b are equivalent. (b) Show that the circuits in fig. P6-9c and d are equivalent.

6-24. (a) The waveforms of pulse A and pulse B are as shown in Fig. P6-10a. Construct the waveforms of (A AND B) and (A OR B). (b) The waveforms of input D and input T of a D flip-flop are as shown in Fig. P6-10b. Construct the waveform of the output Q.

6-25. The D flip-flop shown in Fig. P6-11a is a modification of the SR latch (see Fig. 6-26). (a) Following Eq. 6-31, write the Boolean expression for the output Q and $\bar{Q}$. (b) Construct the truth table. (c) Verify that the circuits for the flip-flops shown in Fig. P6-11a and b are equivalent.

6-26. (a) The waveforms of the input J, K, and T of a JK flip-flop are as shown in Fig. P6-12a. Let $Q = 0$ initially. Construct the waveform of the output Q. (b) Repeat for the waveforms shown in Fig. P6-12b. (*Hint:* See Table 6-5.)

6-27. A rotating disk with a marker is observed with a stroboscope. The disk is at 3600 rpm and the patterns observed are as shown in Fig. 6-47. Give three speeds for each of the observed patterns. Justify the answers.

6-28. (a) The rotational speed N of a high-speed machine is above the range of a stroboscope. Let the observed speed X be the mth submultiple of N, where m is unknown. Let the observed speed Y be the $(m + n)$th submultiple, where n is known. Derive the equation

$$N = n \frac{XY}{X - Y}$$

(b) Deduce the value of N for $X = 59{,}700$ rpm, $Y = 29{,}500$ rpm, and $n = 3$. (c) Find the root-sum-square rss error for the measurement when the error for X is 2% and that for Y is 1.5%.

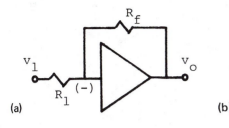

(a)

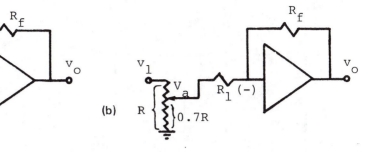

(b)

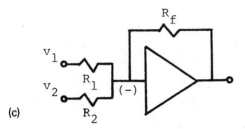

(c)

FIGURE P6-1

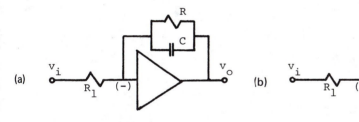

(a)

(b)

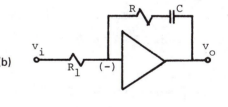

(c)

(d)

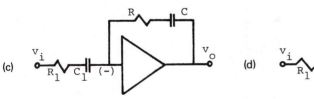

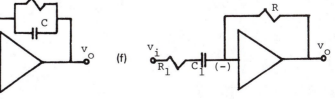

(e)

(f)

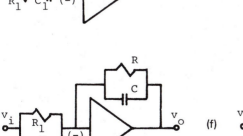

FIGURE P6-2

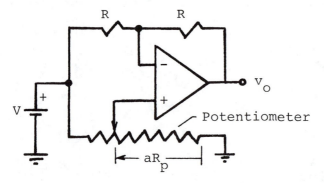

FIGURE P6-3

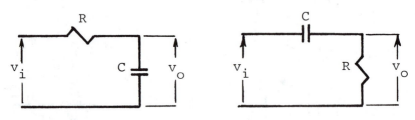

FIGURES P6-4 and P6-5

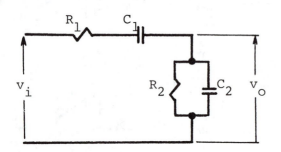

FIGURE P6-6

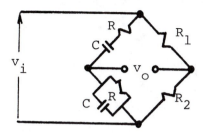

FIGURE P6-7

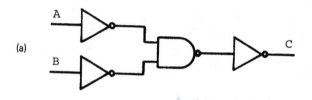

(a)

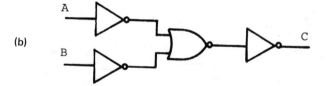

(b)

FIGURE P6-8

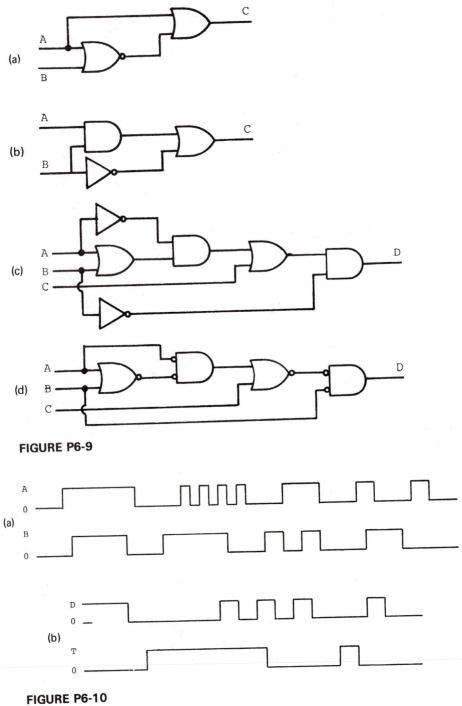

FIGURE P6-9

FIGURE P6-10

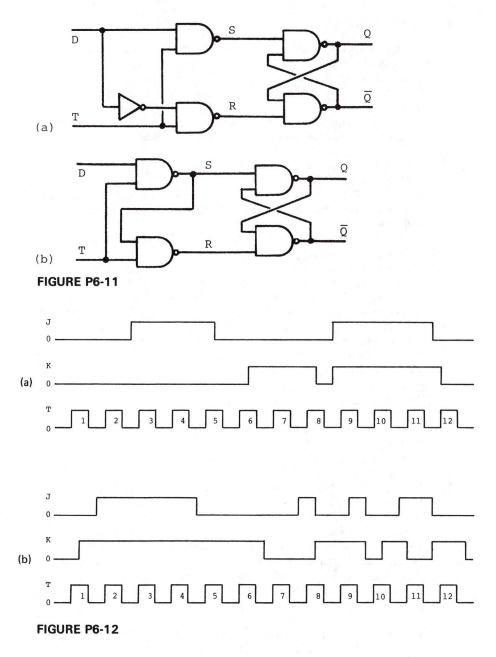

FIGURE P6-11

FIGURE P6-12

REFERENCES

1. Noda, K. (ed.), *Optical Fiber Transmission,* Studies in Telecommunication, Vol. 6, North-Holland Publishing Company, Amsterdam, 1986.
2. Frazier, J. R., Jr., *The Analysis and Design of a Minicomputer Closed Loop Controller for Machine Tools,* Ph.D. dissertation, University of Cincinnati, 1973.
3. Meiksin, Z. H., and P. C. Tackray, *Electronic Design with Off-the-Shelf Integrated Circuits,* Parker Publishing Company, Englewood Cliffs, N.J., 1980, Chap. 1.
    Helms, H. (ed.), *Operational Amplifiers Source Book,* Prentice-Hall, Inc., Englewood Cliffs, N.J., 1987.
4. Wait, J. V., L. P. Huelsman, and G. A. Corn, *Introduction to Operational Amplifier Theory and Applications,* McGraw-Hill Book Company, New York, 1975, p. 6.
5. Clayton, G. B., *Operational Amplifiers,* 2nd ed., Butterworth & Company (Publishers) Ltd., London, 1979, p. 144.
6. Wicker, D. (ed.), *The Radio Amateur's Handbook,* 45th ed., The American Radio Relay League, Newington, Conn., 1968, p. 50 (published yearly).
7. Huelsman, L. P., *Handbook of Operational Amplifiers, Active RC Networks,* 2nd ed., Burr-Brown Research Corp., Tucson, Ariz., 1966, Sec. II.
8. Johnson, D. E., J. R. Johnson, and H. P. Moore, *A Handbook of Active Filters,* Prentice-Hall, Inc., Englewood Cliffs, N.J., 1980.
    Wait, Huelsman, and Corn, Ref. 4, Chap. 4.
    Meiksin and Thackray, Ref. 3, Chap. 7.
9. Smith, J. I., *Modern Operational Circuit Design,* Wiley-Interscience, New York, 1971, p. 58.
10. Smith, Ref. 9, p. 73.
11. Horowitz, P., and W. Hill, *The Art of Electronics,* Cambridge University Press, Cambridge, 1981, p. 99.
12. Wait, Huelsman, and Corn, Ref. 4, p. 173.
13. Horowitz and Hill, Ref. 11, p. 411.
14. Sheingold, D. H. (ed.), *Analog-Digital Conversion Notes,* Analog Devices, Norwood, Mass., 1977.
15. Van Veen, F., *Handbook of Stroboscopy,* General Radio Co., West Concord, Mass., 1966, p. 38.
16. Hsu, H. P., *Fourier Analysis,* Simon and Schuster, New York, 1967.
17. Ewing, D. J., Whys and Wherefores of Modal Testing, *Report* No. 78005, Imperial College of Science and Technology, London, England.
    Sisson, T. R. Zimmerman, and J. Martz, Determination of Modal Properties of Automotive Bodies and Frame Using Transient Techniques, *SAE Paper,* No. 730502, May 18, 1973.
18. Ramsey, K. A., Effective Measurements of Structural Dynamics Testing, Part 1, *Sound Vib.,* Vol. 9, No. 11 (1975), pp. 24–35; Part 2, *Sound Vib.,* Vol. 10, No. 4 (1976), pp. 18–31.
19. Enochson, L. D., and R. K. Otnes, *Programming and Analysis for Digital*

Time Series Data, SVM-3, U.S. Department of Defense, Washington, D.C., 1968.

20. Papoulis, A., *Signal Analysis,* McGraw-Hill Book Company, New York, 1977.

21. Halvorsen, W. G., and J. S. Bendat, Noise Source Identification Using Output Power Spectra, *Sound Vib.,* Vol. 9, No. 8 (1975), pp. 15–24.

22. Scarborough, J. B., *Numerical Mathematical Analysis,* 3rd ed., The Johns Hopkins Press, Baltimore, 1958, Chap. 17.

23. *Fundamental Concepts of Vibration Testing,* Rockland Systems Corp., pp. 2–29.

24. Shannon, C. E., Communication in the Presence of Noise, *Proc. IRE,* Vol. 37, No. 10 (1949), pp. 10–21.

25. *Fourier Analyzer Training Manual,* Application Note 140-0, Hewlett-Packard, Palo Alto, Calif., p. 2–7.

26. Ramsey, K. A., Effective Measurements for Structural System Testing, *Sound Vib.,* Vol. 10, No. 4 (1976), pp. 18–31.

27. Brown, D., G. Carbon, and K. Ramsey, Survey of Excitation Techniques Applicable to the Testing of Automotive Structures, *SAE Paper,* No. 770029, March 4, 1977.

28. Baur, P. S., PCs in Control: An Undate of Industrial Software Packages, *In-Tech,* Vol. 34, No. 7 (1987), pp. 9–15.

29. *Complete BASIC Programmable Stand-Alone Measurement and Control System, μMAC-5000,* Analog Devices, Norwood, Mass.

7
Displacement, Motion, Force, Torque, and Pressure Measurements

7-1. INTRODUCTION

Length and time are basic measurements. In this chapter we describe common methods for measuring displacement, velocity, acceleration, force, torque, and pressure. These topics are displacement related, and flexures are often used as primary sensors for their measurements.

The descriptions cannot be inclusive, since numerous usual and unusual methods of measurement are described in the literature. For example, 40 methods for thickness measurements are presented in a NBS publication [1]. Basic principles are described, but the engineer should be well aware of details, inherent errors, capabilities, limitations, and sources of loading in applications.

Length is a distance between two points when one of the points is the datum. It may describe the dimension or the size of a machine part. Displacement implies a motion. A linear displacement describes the distance traversed between two points when some other point is often the datum. Velocity and acceleration refer to the motion of an object, assuming rigid body motion.

Velocity and acceleration are the first and second time derivatives of displacement, as expressed in Eq. 7-1 for rectilinear motions. A motion may be rectilinear, angular, periodic, transitory, one-dimensional, or a vectorial quantity in three-dimensional space.

Displacement: $\quad x = x_2 - x_1$

Velocity: $\qquad \dot{x} = \dfrac{dx}{dt}$ $\hfill$ (7-1)

Acceleration: $\ddot{x} = \dfrac{d^2x}{dt^2}$

7-2. DIMENSION AND DISPLACEMENT MEASUREMENTS

A dimension refers to a static measurement. It is as essential as a motion or a dynamic measurement in engineering. Without precision dimensional measurements, or metrology, parts would not be interchangeable and machines could not be mass produced.

Although somewhat intuitive, dimensional measurements should be exercised with care. First, the problem must be clearly defined. Before measuring the length of a bar, it should be asked: "Are the ends plane?" Before measuring the diameter of a cylinder, the "diameter" must be defined. It should be asked; "Is the cylinder round?" This will be illustrated presently. Furthermore, temperature and extraneous inputs must be considered in a more precise measurement. Some skill or feel is required in using simple tools, such as a go/no-go plug gage.

Dimension measuring devices can be classified conveniently by their resolution.

1. *Low resolution (0.01 in. or 0.25 mm):* calipers, dividers, etc.
2. *Medium resolution (10^{-4} in. or 2.5×10^{-3} mm):* micrometers in various forms, dial indicators, plug gages, measuring microscope, etc.
3. *High resolution (microinches or 2.5×10^{-5} mm):* comparators (mechanical, electrical, pneumatic), optical flats, interferometers, etc.

A. Gage Blocks: The Working Standard

Gage blocks (e.g., C. E. Johansson Gage Co., Dearborn, Michigan) are common dimensional standards in industry. They are hardened, highly polished, and dimensionally stable steel blocks, about ⅜ by 1⅜ in. with varying thicknesses. Any dimension between 0.1000 and 8.0000 in. in increments of 0.0001 in. can be obtained from a set of 81 blocks. Gage blocks are available in the classes shown in Table 7-1.

Blocks are stacked for obtaining specific dimensions. They are "wrung" together by means of a sliding motion with steady pressure to eliminate all but the thinnest oil film between the surfaces. The thickness of the oil film, about 0.2×10^{-6} in., is considered an integral part of the block itself.

Temperature may be a problem in accurate dimensional gaging. There is no error (1) if the workpiece and the gage blocks are at the reference

TABLE 7-1   Gage Blocks in English Units

Class	Type	Tolerance
B	Working blocks	±8 µin.
A	Reference blocks	±4 µin.
AA	Master blocks	±2 µin. for all blocks up to 1 in., and 2 µin./in. for larger blocks

temperature of 20°C (68°F), or (2) if the workpiece and the blocks have the same temperature and identical thermal coefficient of expansion. Corrections must be made for other conditions. For example, the thermal coefficient of steel is about 11.2 ppm/°C. A one degree temperature difference between the workpiece and the gage blocks may cause a dimensional change on the order of the gage tolerance. The difference in temperature may be from handling or from radiation due a nearby heat source, such as a light bulb.

B. Examples: Mechanical Methods

The vernier caliper, shown in Fig. 7-1a, measures a linear dimension between the jaws of the caliper. The fixed scale has 40 divisions/inch and the movable vernier has 25 units in 39/40 inch. Thus each division of the scale is subdivided into 25 units, or 0.001 in. can be measured. For purpose of illustration, the fixed scale shown in Fig. 7-1b has 10 divisions/inch and the vernier has 10 units in 0.90 in. The reading shown is 0.23 in., since the 3 on the vernier is aligned with one of the lines of the fixed scale.

A vernier increases resolution by means of a subdivision. Conversely, the dial gage shown in Fig. 7-1c increases resolution by means of magnification with gears and levers. It is often used in a comparator for dimensional inspection, as shown in Fig. 7-1d. The comparator is first set by means of a reference standard, such as gage blocks. Thus deviation of a part from the standard can be compared.

Returning to the diameter measurement. The out-of-roundness of an odd-lobed cylinder cannot be detected by means of a two-point measurement (see Fig. 2-23). The three-point measurement shown in Fig. 7-1e can be used to detect this out-of-roundness. The "diameter," however, must depends on the lobing characteristics of the cylinder as well as the angle of the V-block. The pitch diameter of a screw is commonly measured with the over-the-wire method, as shown in Fig. 7-1f. What if the wires are lobed and the angle of the screw thread differs from that of the V-block used for the calibration?

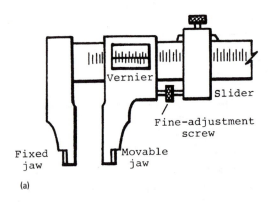

(a)

Fixed scale

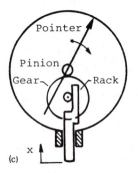

Vernier

(b)   (Vernier reads 0.23 units)

(c)

FIGURE 7-1 Examples of dimensional measurement. (a) Vernier caliper. (b) Detail of a vernier. (c) Dial gage (indicator). (d) Dial gage as a comparator. (e) Measurement of variations in diameter. (f) Over-wire measurement of pitch diameter of screw thread.

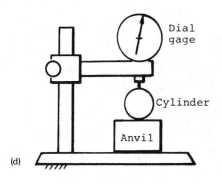

(d)

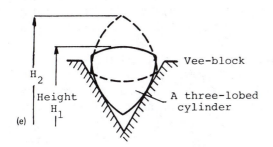

(e)

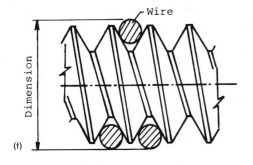

(f)

C. Example: Other Methods

The input to the primary sensor for most "mechanical" measurements is often a displacement or motion. Examples of displacement measurements and their applications are described in this section. Some examples from earlier chapters are reiterated to complement the discussion.

A potentiometer as a displacement sensor is shown in Fig. 7-2a. An op-amp voltage follower (see Fig. 6-8b) is used to interface the potentiometer R_p and the readout instrument R_m in order to minimize loading in the circuit (see Prob. 1-7). The follower has a very high input impedance and converts R_p into a low-resistance source (about 50 Ω) with sufficient power to drive a low-resistance R_m.

The pneumatic comparator shown in Fig. 7-2b is a variation of the orifice-type air gage (see Probs. 3-10 and 3-11). Gage blocks are used as reference standards for its calibration. The device is widely used in industry.

Dimension and displacement can be measured precisely by means of "interference" of light waves, or interferometry. When two monochromatic (same wavelength) light beams are superposed as shown in Fig. 7-3a, (1) the waves reinforce when they are in phase, producing a light more intense than either wave, and (2) the waves cancel when 180° out of phase, reducing the intensity to zero. Note that a wave repeats itself for every wavelength and is out of phase for every half wavelength.

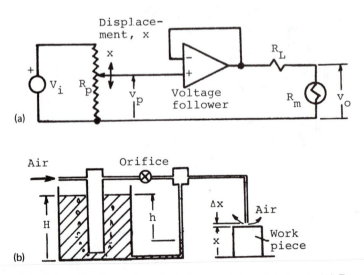

FIGURE 7-2 Examples of displacement measurement. (a) Potentiometer circuit with voltage follower interface. (b) Schematic of pneumatic gage in comparator.

Light waves in-phase.   Light waves out-of-phase.

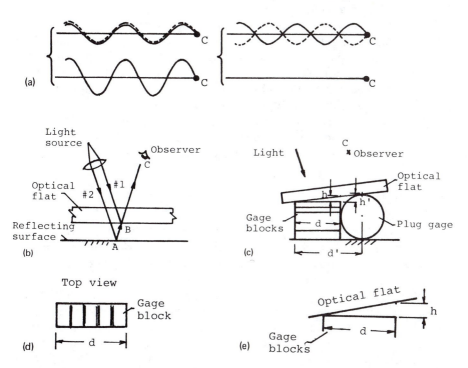

FIGURE 7-3   Interferometry by superposition of two light waves. (a) Superposition of two light waves from the same source. (b) Light paths 1 and 2. (c) Measuring diameter of plug gage. (d) Alternate fringe pattern on surface of gage block. (e) Geometry of the thin wedge.

Let L be the path of a monochromatic light beam of wavelength λ between a source and the observer. Two optical paths from a light source are shown in Fig. 7-3b. The beams are reflected at surfaces A and B and observed at C. The number n of wavelengths for a path L is

$$n_1 = \frac{L_1}{\lambda} \quad \text{and} \quad n_2 = \frac{L_2}{\lambda} \tag{7-2}$$

The difference in number of wavelengths between two beams is

$$n_1 - n_2 = \frac{1}{\lambda}(L_1 - L_2) \tag{7-3}$$

When $n_2 - n_1$ is an integer, or a multiple of the wavelength, the two waves are in phase. When $n_2 - n_1$ is an integer $\pm \frac{1}{2}$, the waves differ by half a wavelength and are out of phase.

The diameter of a plug gage is measured as shown in Fig. 7-3c. Beams of monochromatic light are reflected and observed at C. The small difference in dimension between the plug gage and the gage blocks causes a thin wedge to form between the optical flat and the gage blocks. Consequently, the beams from the light source to the observer have different optical paths, thereby causing the waves to cancel or to reinforce as described above. Fringes, or alternate dark and bright bands, are observed as shown in Fig. 7-3d. The interval between fringes is $\lambda/2$. If N is the number of fringes, the distance h illustrated in Fig. 7-3e is

$$h = \frac{\lambda}{2} N \tag{7-4}$$

The diameter of the plug gage is deduced from the similar triangles shown in Fig. 7-3c, where $h/d = h'/d'$. It is necessary to determine whether the diameter of the plug gage is larger or smaller than the height of the gage blocks. This is left as an exercise.

Interferometry is utilized to measure static, dynamic, small, or large displacements. Since a displacement is measured in terms of light fringes, the fringe counting can be automated. For example, instruments are available to measure (1) a distance of 200 ft with the precision of microinches [2], (2) the position of the cutter of a machine tool [3], (3) the resonance of a large building [4], and (4) the vibration of an electromechanical shaker up to several hundred hertz [5].

Displacement measurements have numerous applications in the sensing of other variables. The bellow shown in Fig. 7-4a measures pressure by means of an elastic deformation (see the Bourdon tube in Fig. 3-3). The bimetallic strip shown in Fig. 7-4b is used in thermostats for the temperature control of homes. The simple switch in Fig. 7-4c is an ON/OFF displacement device. In the form of a reed switch, it is found in digital instruments [6] for automated data acquisition systems.

Transducers utilizing displacement for their operations were described in previous chapters. Some examples are the tool dynamometer (Fig. 1-3), capacitance micrometer (Fig. 2-2), dial gage (Fig. 2-20), optical pyrometer with chopper (Fig. 3-54), galvanometer (Fig. 4-10), LVDT (Fig. 5-3), target flow meter (Fig. 5-29), pressure pickup (Fig. 5-34d), microphone (Fig. 5-35), optical encoder (Fig. 5-30), and pen recorder (Fig. 6-46).

Displacement transducers can also be classified according to the type of secondary sensor employed, such as a resistive or an inductive type. The

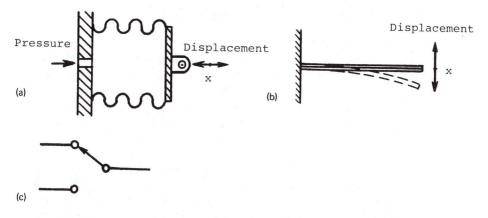

FIGURE 7-4   Applications of displacement measurement. (a) Pressure bellow. (b) Bimetallic strip. (c) Switch.

strain gage is a resistive device and has numerous applications (see Figs. 5-29 and 5-30). The LVDT is a prime example of the inductive type (see Figs. 5-3 and 5-8). Its application is virtually unlimited. Examples of inductive devices are the thickness gage (Fig. 5-18), proximity gages (Fig. 5-19), tachometers (Fig. 5-20), and synchros (Fig. 5-21). Similarly, capacitive transducers have numerous applications (see Figs. 5-32, 5-34, and 5-35).

7-3.   MOTION MEASUREMENTS

Motion involves displacement and time. The problem is (1) to examine the medium or device by means of which the displacement information is transferred from a moving object to a stationary instrument, and (2) to measure the time associated with the displacement. Time and frequency measurements are described in Sec. 6-7D and are not repeated here. As in other measurements, a motion measurement is valid when it has only the required information, which is not necessarily "all" the information of an event. For example, the peak amplitude may be the only required data in a shock test.

It is important to note that motion can be analyzed in the time as well as in the frequency domain. A sinusoidal motion can be treated as a steady-state phenomenon. The steady-state amplitude, phase angle, and frequency are sufficient to identify the motion. A transient motion can be analyzed by means of a fast Fourier transform (FFT) technique (see Sec. 6-9), the time and frequency analyses can be related in "real" time [7].

7-4. VELOCITY MEASUREMENTS

Information transferred from a moving object to stationary instruments by means of optics and electrical devices is examined briefly in this section. Examples of other transfer media are slip rings, telemetry, magnetic devices, electrostatic devices (capacitive), sound, and resistive transducers (strain gages and potentiometers). The information can be analog or digital. Signal conditioning with feedback can be incorporated in the sensor [8].

The vibrating wedge shown in Fig. 7-5a is used for monitoring sinusoidal motion of shakers [9] for relatively large amplitudes. Light is the transfer medium. The apparent intercept at c is 6.5; therefore, the peak-to-peak amplitude x_{p-p} is 0.065 in. The error is 10% for a larger wedge if the amplitude is greater than 0.1 in. and the frequency is over 30 Hz. The error also depends to a large extent on the experience of the observer. Similarly, optical methods are used to track the peak-to-peak amplitude of a moving object. Two moving edges can be used as an aperture between a light source and a photodetector. Alternatively, a moving aperture of constant area can be placed between a light source and a photodetector.

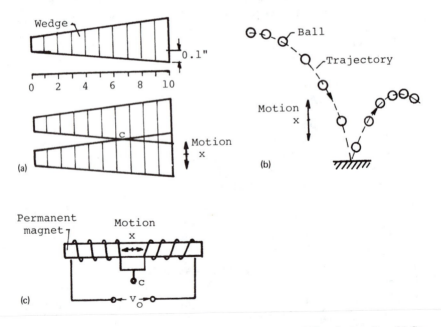

FIGURE 7-5 Examples of velocity measurement. (a) Vibrating wedge. (b) Stroboscopic positions. (c) Linear velocity transducer.

The motion of the aperture changes the amount of light on a photodetector. This is an optical "potentiometer."

Optical methods are often used in instruments for motion studies. Holography is applied for studies in vibration [10]. An optical encoder can be used as a tachometer (e.g., DRC Model TK-702, Dynamics Research Corp., Wilmington, Massachusetts). An electronic distance-measuring device for surveying can be adopted for measuring velocity by comparing the phase of a reference wave with that of the reflected wave from a moving object [11]. The speed limit of this device is from 40 to 160 km/s, depending on the wavelength employed and the resolution desired. Commercial fiber optic sensors have resolutions from 10×10^{-6} to 25×10^{-6} in. with dynamic response from 100 Hz to 2 MHz (e.g., Fotonic Sensors, Mechanical Technology, Inc., Latham, New York). Fiber optic sensors promise many applications and advantages for hostile environment, such as high temperature, fire hazard, and high radiation [12]. An appropriate reflective target can be attached to a moving object, and the motion of the object is tracked by means of feedback in an instrument (e.g., Optron Corp., Woodbridge, Connecticut). The optical image of the target in the instrument is converted to an electron image and the motion displayed by means of an oscilloscope. The freqeuncy response of the instrument exceeds 100 kHz.

Photographic methods are used for transient studies. In high-speed photography, the displacement of an object in one frame can be compared with that in another. If a stroboscopic light is used to obtain multiple exposures of an object on the same film, as shown in Fig. 7-5b, the positions of a falling ball and its velocity can be determined. The technique is used to study cavitation in hydraulic turbines and ship propellers [13] (see also Sec. 6-8E).

An electromagnetic device in the form of a linear velocity transducer (Fig. 7-5c) is suitable for a wide range of operation. The permanent-magnetic core moves in a shielded concentric coil to produce a sizable voltage by means of the generator effect (see Fig. 5-20). Hence the transducer is self-generating and no external excitation is required. It can be connected for a single-ended output v_o, or for a differential output by connecting the terminal c to ground. Its equivalent circuit is simply that of a voltage generator in series with an internal inductance L, resistance R, and an external resistance R_m of the recorder. If v_o is the output voltage, V the input velocity in inches per second (ips), and K_v the sensitivity in mV/ips, it can be shown that the transfer function is

$$\frac{v_o}{V} = \frac{R_m K_v}{(R_m + R) + j\omega L} \qquad (7-5)$$

7-5. ACCELERATION MEASUREMENTS

Uniaxial seismic instruments are generally used for acceleration measurements. A triaxial accelerometer simply consists of three uniaxial instruments, each of which is aligned in the x, y, or z direction. In this section we discuss seismic instruments, construction of accelerometers, equivalent circuits, environmental effects, and methods of calibration.

The model of a seismic instrument is a one-degree-of-freedom vibratory system shown in Fig. 7-6a, where x is the displacement of the base, y that of the seismic mass m, and $z = x - y$ the relative displacement between m and the base. Assume a linear spring k and a viscious damper c. The seismic mass m and spring k form the primary sensor shown in Fig. 7-6b. The secondary sensor can be piezoelectric or any of the secondary sensors illustrated in Chapter 5. Bonded or unbonded strain gages, potentiometers, differential transformers, capacitive, and other devices are all suitable for this purpose. In other words, the primary sensor detects the motion of an object, and the secondary sensor can be any suitable device for transferring this signal to stationary instruments.

A. Seismic Instruments

Applying Newton's second law to the seismic mass m shown in Fig. 7-6a, the equation of motion is

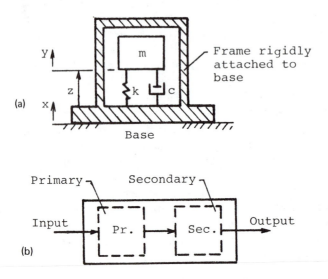

FIGURE 7-6 Model of seismic instruments. (a) Schematic of a seismic instrument. (b) Primary and secondary sensors.

$$m\ddot{y} = -c(\dot{y} - \dot{x}) - k(y - x) \tag{7-6}$$

Substituting $z = (y - x)$ and rearranging, we obtain

$$m\ddot{z} + c\dot{z} + kz = -m\ddot{x} \tag{7-7}$$

The corresponding sinusoidal transfer function is

$$\frac{z}{x}(j\omega) = \frac{-m(j\omega)^2}{m(j\omega)^2 + c(j\omega) + k} \tag{7-8}$$

Divide the numerator and demoninator by k, and define $m/k = 1/\omega_n^2$, $c/k = 2\zeta/\omega_n$, where ω_n is the undamped natural frequency in rad/s and $\zeta = c/(2\sqrt{km}$, the damping ratio comparing with a critically damped system. By substitution, we get

$$\frac{z}{x}(j\omega) = \frac{-(j\omega/\omega_n)^2}{(j\omega/\omega_n)^2 + 2\zeta(j\omega/\omega_n) + 1} \tag{7-9}$$

Hence the magnitude ratio and the phase angle $\phi(j\omega)$ are

$$\left|\frac{z}{x}(j\omega)\right| = \frac{(\omega/\omega_n)^2}{\{[1 - (\omega/\omega_n)^2]^2 + [2\zeta(\omega/\omega_n)]^2\}^{1/2}} \tag{7-10}$$

$$\phi(j\omega) = -\tan^{-1}\frac{2\zeta(\omega/\omega_n)}{1 - (\omega/\omega_n)^2} \tag{7-11}$$

The equations above are plotted as shown in Fig. 7-7. Since the output z is measured relative to the input x, a stationary reference is not required. Seismic instruments are especially useful for measurements in moving vehicles.

The seismic instrument shown in Fig. 7-6a can be used for displacement, velocity, and acceleration measurements. The range of a vibrometer for displacement measurements is as shown in Fig. 7-7a. When $\omega \gg \omega_n$, the output z approaches the input x and is independent of damping. Vibrometers are generally used for special situations. Since ω_n is low for a reasonable input frequency, the instrument tends to be heavy, thereby causing a mass loading of the vibrating object under investigation.

A practical method to measure velocity is to use a velocity transducer (see Fig. 7-5c) as the secondary sensor in a vibrometer. Since the voltage output $v_o = K_v\dot{z}$, we have

$$\frac{v_o}{\dot{x}} = \frac{K_v(j\omega/\omega_n)^2}{(j\omega/\omega_n)^2 + 2\zeta(j\omega/\omega_n) + 1} \tag{7-12}$$

This also requires that $\omega \gg \omega_n$, which may be a disadvantage. The most

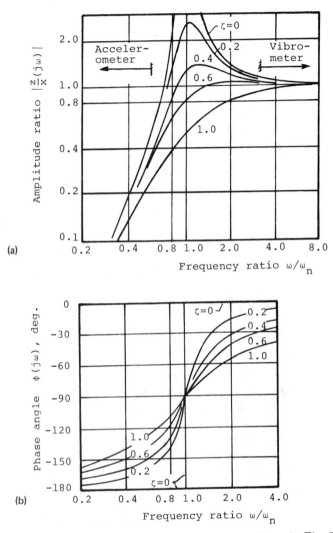

FIGURE 7-7 Frequency response of system shown in Fig. 7-6. (a) Amplitude ratio versus frequency ratio. (b) Phase angle versus frequency ratio.

common method is to first measure acceleration. Velocity and displacement can be deduced readily from acceleration by means of integrating circuits. For example, this technique can be used to measure the piston displacement of an internal combustion engine.

The range for acceleration measurements is shown in Fig. 7-7a, for which $\omega \ll \omega_n$. From Eq. 7-8 we get

$$\frac{z}{\ddot{x}}(j\omega) = \frac{-m}{m(j\omega)^2 + c(j\omega) + k} \tag{7-13}$$

or

$$\frac{z}{\ddot{x}}(j\omega) = \frac{-(1/\omega_n)^2}{(j\omega/\omega_n)^2 + 2\zeta(j\omega/\omega_n) + 1} \tag{7-14}$$

$$\left|\frac{z}{\ddot{x}}(j\omega)\right| = \frac{(1/\omega_n)^2}{\{[1 - (\omega/\omega_n)^2]^2 + [2\zeta(\omega/\omega_n)]^2\}^{1/2}} \tag{7-15}$$

$$\angle \frac{z}{\ddot{x}}(j\omega) = \tan^{-1}\frac{2\zeta(\omega/\omega_n)}{1 - (\omega/\omega_n)^2} \tag{7-16}$$

The equations relate the displacement output z to the acceleration input $\ddot{x}$. Since ω/ω_n is small or ω_n is large, a small mass with a very stiff spring can be used in an accelerometer. Thus the mass loading of a test object is minimal.

The sensitivity of an accelerometer, as shown in Eq. 7-15, is $1/\omega_n^2$ a constant. The input/output amplitude ratio $z/\ddot{x}$ and phase relation are as shown in Fig. 7-8. Since the response depends on the input frequency as well as the damping ratio ζ, accelerometers are designed with $\zeta = 0.7$ to minimize amplitude and phase distortions. When $\zeta = 0.7$, it is useful up to $\omega/\omega_n = 0.6$, as shown in Fig. 7-8a. Note that when an accelerometer is lightly dampled, it is useful only up to $\omega/\omega_n = 0.2$. Hence a lightly damped accelerometer with high ω_n can also be used successively.

B. Accelerometers

To follow up the theory on seismic instruments above, construction of accelerometers for rectilinear motions is examined in this section. Specifications and detail construction of accelerometers can be obtained from manufacturers. Directory of sales representatives are found in technical publications [14].

The seismic mass–spring–damper system shown in Fig. 7-9a is the primary sensor of an accelerometer, where m is the mass, k the spring, and c the damper. The base is rigidly attached to a test object from which $\ddot{x}$ is the acceleration input. The output z is the relative motion between

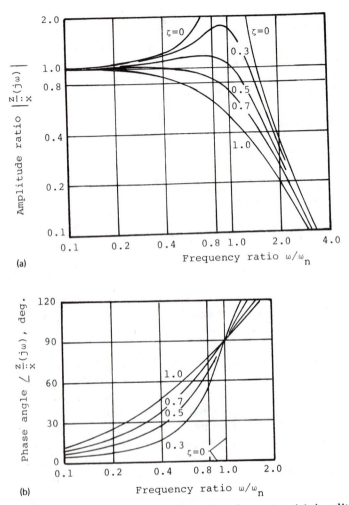

FIGURE 7-8 Frequency response of accelerometer. (a) Amplitude ratio versus frequency ratio. (b) Phase angle versus frequency ratio.

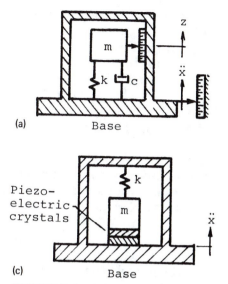

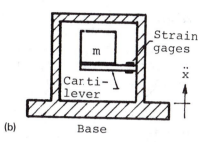

(a)   Base   (b)   Base

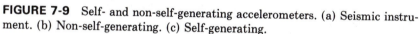

(c)   Base

FIGURE 7-9 Self- and non-self-generating accelerometers. (a) Seismic instrument. (b) Non-self-generating. (c) Self-generating.

the mass and the base. Damping can be achieved by immersing the moving mass in a silicon fluid. The cantilever shown in Fig. 7-9b represents the spring element. The relative motion z is picked off by means of strain gages. This type of accelerometer can be made very small by using an integral mass–beam assembly with the strain gages deposited directly onto the beam. Many types of flexure can be used for the spring element. A "cantilever" in the form of a spiral in a disk would be soft axially but stiff in its lateral direction. The secondary sensor can be any displacement transducer, such as a LVDT shown in Fig. 5-8k or an inductance pickup shown in Fig. 5-18f.

Accelerometers with strain gages as secondary sensor are called *piezoresistive* and are non-self-generating. The commonly used *piezoelectric* accelerometers are self-generating. They are available with a wide range of specifications and from a large number of manufacturers. The schematic of a basic compression-type piezoelectric accelerometer is shown in Fig. 7-9c. The spring k preloads the system so that the mass and the piezoelectric crystal are always in compression. The mass of the crystal and its stiffness contribute to the equivalent mass and equivalent spring of the system. Piezoelectric accelerometers are very stiff, with high resonant frequencies. They are essentially undamped. Their performance is satisfactory when used below one-fifth of their resonant frequencies (see Fig. 7-8a).

When a piezoelectric crystal is deformed, it generates within itself an electric charge. Thus the piezoelectric effect is the basis for piezoelectric sensors for acceleration, force, and pressure measurements. The effect is reversible and bilateral. If a charge is applied, the crystal deforms mechanically. Thus a piezoelectric device can also be used as a high-frequency shaker for a mechanical output by applying a voltage to the device.

A piezoelectric crystal with a deformation w due to a compressive force f is shown in Fig. 7-10a. The charge q generated is proportional to the deformation w with a sensitivity K_q. If f is the inertial force from a seismic mass, the charge q generated is proportional to acceleration $\ddot{x}$ with a sensitivity K'.

$$q(t) = K_q w \quad \text{or} \quad q(t) = K'\ddot{x} \tag{7-17}$$

Hence the model or the equivalent circuit of a piezoelectric element is a charge generator $q(t)$ in parallel with its inherent capacitance C_{pz} as shown in Fig. 7-10b.

Piezoelectric crystals can be cut to respond to different modes of mechanical deformation, such as compression, bending, and shear. The effect depends on the shape and orientation of the crystal relative to its

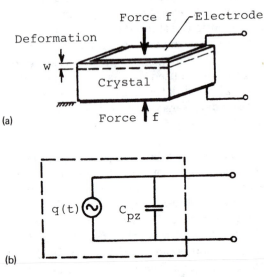

(a)

(b)

FIGURE 7-10 Piezoelectric crystal as a charge generator. (a) Piezoelectric crystal. (b) Equivalent circuit.

axes and the location of the electrodes. Natural crystals (quartz, rochelle salt), synthetic crystals (lithium sulfate, etc.), and polarized ferroelectric ceramics (barium titanate, etc.) are commonly used. The sensitivity values of piezoelectric accelerometers are shown in Table 7-2.

The sensing element in an accelerometer should not be affected by the strain applied to its base for mounting or inputs applied to the case that houses the sensing element. The accelerometer shown in Fig. 7-9c is not completely isolated from the case. The influence of the case is eliminated in the single-ended compression-type construction shown in Fig. 7-11a. The mass and the sensing element are under compression by means of a center post rather than by direct contact with the protective case. The effect of strain on the base is minimized in the isobase design by mounting the sensing element on a pedestal as shown in Fig. 7-11b. The isoshear type shown in Fig. 7-11c combines the advantages of the isobase and shear design. It consists of a balance beam with two separate masses and two separate crystal-stacks bolted together to form a single unit.

A servo-accelerometer utilizes feedback and the force-balance principle for its operation. First, the inertia force of the seismic mass due to acceleration is sensed by means of a displacement sensor such as a LVDT. Its output is amplified and fed back to drive a torque motor, which restores the mass to its null position. The electric current for the restoring force is proportional the acceleration. Since the seismic mass is restored to its null position for each measurement, the servo-accelerometer is a null-balance device (see Secs. 3-1 and 3-2). Consequently, servo-accelerometers have slower response but are capable of higher accuracy than and the more common piezoelectric type. For example, its maximum useful frequency is of the order of 500 Hz, while that of the piezoelectric type is about 500 kHz.

TABLE 7-2   Sensitivity of Piezoelectric Accelerometers

Material	Design mode	Capacitance (pF)	Sensitivity Charge (pC/g)	Sensitivity Voltage (mV/g)
Lead-zirconate	Shear	1000	1–10	1–10
titanate	Compression	1000–10,000	10–100	10–100
Quartz	Compression	100	1	10
Endevco P-10	Shear	100	1	1
	Compression	1000	1–10	1–10

Source: Ref. 15.

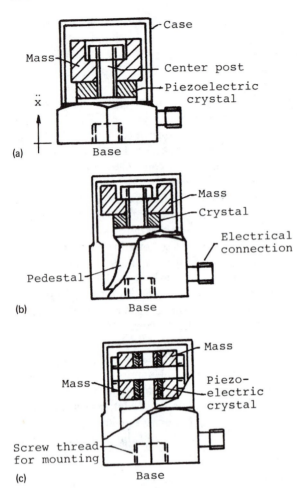

FIGURE 7-11 Typical construction of piezoelectric accelerometers. (a) Compression type. (b) Isobase type. (c) Isoshear type.

C. Equivalent Circuits

Piezoelectric transducers are charge generators. They are examined as a voltage and a current source in this section. The charge q generated is proportional to the deformation w of the crystal. Neglecting the inductance and the leakage resistance, the model of a piezoelectric crystal is that of a charge generator shunted by a capacitance C_{pz}, as shown in Fig. 7-10b. Since the transducer is a high-impedance device, the analysis must include the transducer, the cabling, and the amplifier as an integral unit.

A piezoelectric transducer as a voltage source is shown in Fig. 7-12a, where the op-amp is used as a voltage follower (see Fig. 6-8b). C_{cab} is the capacitance of the cable. C_a and R_a form the input impedance of the amplifier. The equivalent circuit in Fig. 7-12b has C and R in parallel, where $C = (C_{pz} + C_{cab} + C_a)$ and R is the combined effect of the input impedance of the amplifier and any load resistance that may be put in parallel with the amplifier. The current i is the flow rate of the charge q. From $q(t) = K_q w$ (Eq. 7-17) and Fig. 7-12b, and we get

$$i = \dot{q} = K_q \dot{w} \tag{7-18}$$

$$i = i_1 + i_2 \tag{7-19}$$

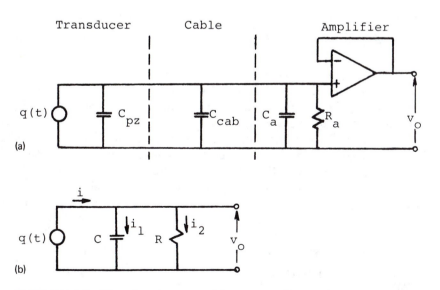

FIGURE 7-12 Piezoelectric sensor with op-amp voltage follower. (a) Piezoelectric sensor as a voltage source. (b) Equivalent circuit.

$$K_q \dot{w} = C \dot{v}_o + \frac{1}{R} v_o \tag{7-20}$$

Defining $\tau = RC$ and $K = K_q/C$, and substituting $j\omega$ for the time derivative, the transfer function from Eq. 7-20 is

$$\frac{v_o}{w}(j\omega) = \frac{j\omega\tau K}{1 + j\omega\tau} \tag{7-21}$$

The time and frequency response of the system can be analyzed as shown in Chapter 4. The system cannot be used for static measurements, because from Eq. 7-21 the output is zero for zero frequency. Neither the sensitivity K nor the time constant τ is constant, because $K = K_q/C$ and $\tau = RC$, where $C = C_{pz} + C_{cab} + C_a$ and C_{cab} is dependent on the length of the connecting cable.

A piezoelectric transducer as a current source is shown in Fig. 7-13a, where C is defined above and the op-amp is used as a charge amplifier (see Fig. 6-10). Summing the currents at node b, we get

$$i + i_f = i_1 + i_a \tag{7-22}$$

where $i = K_q \dot{w}$. Since the node b is a virtual ground, i_1 is negligibly small. Owing to the high impedance of the FET-input op-amp, i_a is also negligible. From $i_f = C_f \dot{v}_o$, we get

$$K_q \dot{w} = -C_f \dot{v}_o$$

$$v_o = -\frac{K_q}{C_f} w \tag{7-23}$$

This indicates that the system in Fig. 7-13a is zero-order. The output v_o is instantaneous and linear with the deformation w of the crystal. The circuit, however, is not practical. A small bias current will cause the amplifier-integrator to saturate.

The simplified circuit for a practical charge amplifier is shown in Fig. 7-13b (see Figs. 6-6b and 6-11). The R_f in parallel with C_f provides a leakage path to prevent the small bias current of the op-amp from developing a significant charge on C_f. Using ac circuit analysis and from Eq. 7-22, we get

$$I = -I_f \tag{7-24}$$

Applying Ohm's law, $V_o = Z_f I_f$, and simplifying, we obtain

$$j\omega K_q w = -\frac{1 + j\omega R_f C_f}{R_f} V_o$$

or

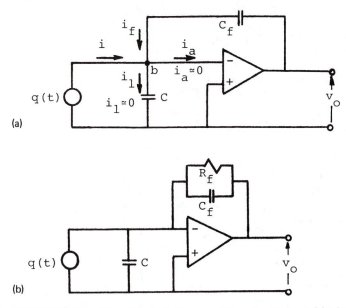

FIGURE 7-13 Simplified charge-amplifier circuits. (a) Idealized charge amplifier. (b) Practical charge amplifier.

$$\frac{v_o}{w}(j\omega) = -\frac{Kj\omega\tau}{1 + j\omega\tau} \tag{7-25}$$

where $\tau = R_f C_f$ and $K = K_q/C_f$.

Note that Eq. 7-21 for the voltage-follower circuit and Eq. 7-25 for the charge-amplifier circuit are of identical form. For Eq. 7-21, both K and τ change with the length of the cable between the sensor and the amplifier, because $C = (C_{pz} + C_{cab} + C_a)$ and C_{cab} depends on the length of cable. For Eq. 7-25, both K and τ are constants, because $C = C_f$ a constant. Hence the charge-amplifier circuit is preferred unless the effect of the connecting cable is eliminated by means of a built-in amplifier in the transducer.

Transducers with built-in amplifiers are less susceptible to environmental conditions. The capacitance C_{cab} of the cabling between the sensor and the amplifier is constant, because the miniaturized integrated circuit is within the transducer casing. The external components are the power supply and the readout device as shown in Fig. 7-14a. The external connections are at the low-impedance side of the output and have negligible effect on the system performance.

The equivalent circuit for the analysis with built-in electronics is shown

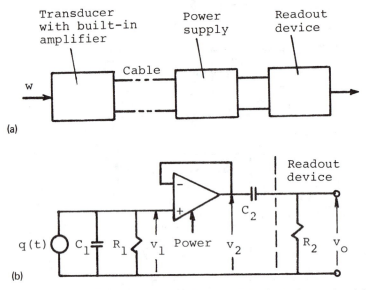

FIGURE 7-14 Piezoelectric transducer with built-in electronics. (a) Transducer with built-in electronics. (b) Equivalent circuit.

in Fig. 7-14b. For convenience, C, R, and v_o in Fig. 7-12b are renamed as C_1, R_1, and v_1 in Fig. 7-14b. The voltage follower forces $v_2 = v_1$. C_2 is a blocking capacitor and R_2 is the input resistance of the readout device.

From Eq. 7-21 we deduce that

$$\frac{v_2}{w}(j\omega) = \frac{j\omega\tau_1 K}{1 + j\omega\tau_1} \tag{7-26}$$

where $\tau_1 = R_1 C_1$ and $K = K_q/C_1$. R_2 and C_2 form a voltage divider for v_2. Thus

$$\frac{v_o}{v_2}(j\omega) = \frac{R_2}{R_2 + 1/j\omega C_2} = \frac{j\omega\tau_2}{1 + j\omega\tau_2} \tag{7-27}$$

where $\tau_2 = R_2 C_2$. Combining Eqs. 7-26 and 7-27, the overall transfer function for the system in Fig. 7-14 is

$$\frac{v_o}{w}(j\omega) = K\frac{j\omega\tau_1}{1 + j\omega\tau_1}\frac{j\omega\tau_2}{1 + j\omega\tau_2} \tag{7-28}$$

D. Environmental Effects

A transducer must interact with its environment in order to sense the measurand, or the desired input. At the same time, a transducer is susceptible to all the environmental effects (see Fig. 1-1), including the desired input and all the undesired inputs. The undesired inputs are sources of noise (see Eq. 2-4). A listing of requirements and conditions for motion measurements can be fairly extensive [16]. In this section we identify some of these inputs and briefly describe the effect of mounting on the performance of accelerometers.

For piezoelectric accelerometers, the common environmental inputs are temperature [17], pressure and humidity, acoustic noise [18], vibrating floors, cabling, electromagnetic fields, high-frequency noise and spikes, nuclear radiation, electrical ground loops [19], and mounting [20]. Each of these topics is a separate study.

Mounting is a common problem for accelerometers. It includes the effects of (1) the interface between an accelerometer and the vibrating structure, and (2) the fixture, if any, for attaching an accelerometer to the test structure. The mounting must transmit the motion faithfully to the accelerometer, and have negligible effect on the motion of the test structure.

Several methods for mounting accelerometers are shown in Fig. 7-15a. Effects of interfacing on the frequency response characteristics are illustrated in Fig. 7-15b. A transducer must operate on the flat portion of the response curve, far below the associated resonances. The threaded stud mount is the most ideal and gives the best frequency range. Beeswax and epoxy give comparable results, but beeswax is temperature sensitive. A magnetic mount or a hand-help probe for a quick look is a convenience, but the former should not be used above 2 kHz and the latter, 1 kHz. Duct seal is only for low-frequency testing up to about 500 Hz. Proper installation is important. For example, machined mating interface should be used for stud mounting to minimize the base distortion of an accelerometer (see the isobase design shown in Fig. 7-11b). Dirt and grit may cause a magnet mount to rock and to add modes to the test data. When an excessive amount of beeswax is used for mounting, the result is no better than that for duct seal.

Fixtures, as shown in Fig. 7-16, are used when a direct mount is not possible. Obviously, a fixture would add modes to the vibrating structure. A fixture must be rigid and light. The major resonant frequency of a fixture must be well above the test frequency. Mass loading due to a fixture may be large, although the mass of the accelerometer itself is small. In particular, loading must be considered when mounting an accelerometer on a delicate structure as shown in the figure. A mounting base can be used to distribute the load and to avoid stress concentration. In other words, loading is due to a mass loading and/or a structural change in the test structure (see Sec. 1-3A). The accelerometer, the fixture, and the test structure must be regarded as an integral unit for the test.

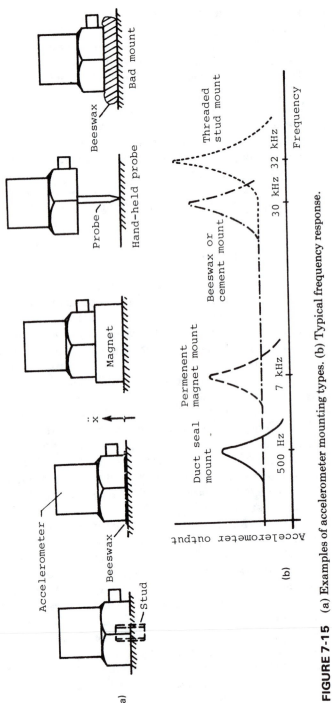

FIGURE 7-15 (a) Examples of accelerometer mounting types. (b) Typical frequency response.

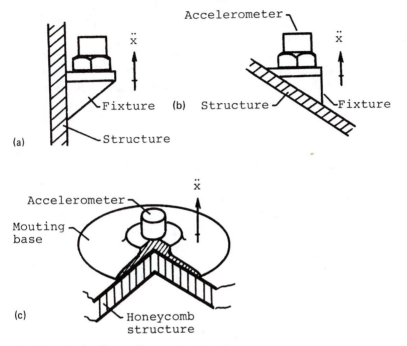

FIGURE 7-16   Examples of mounting fixtures.

E.   Calibration

Many techniques are used for the calibration of accelerometers [21]. Historically, calibration may be broadly classified as constant-acceleration, absolute, and comparative. A "calibration constant" certified by the manufacturer or a standards laboratory is inadequate, because (1) it does not account for the effects of the variables on the accelerometer under in-service conditions, and (2) the accelerometer is only one of the many components in a measuring system. Error is from the overall system, including the loading between components (see Sec. 3-3). Historical methods are described briefly here. More practical methods are examined in later sections on calibration of accelerometers and force transducers.

Constant-acceleration calibration includes the tilting method and the centrifuge. The tilting method uses the earth's local gravity (1*g*) as the constant input. The sensitive axis of an accelerometer is first aligned with the gravitational field. By reversing the axis 180°, the change in input is 2*g*. The method is suitable for accelerometers with very long time constants. Errors may arise from misalignment and vibrations. For the centrifuge method, the sensitive axis of an accelerometer is aligned radially with respect to the rotating axis of a horizontal disk. Up to 60,000g can be achieved by this method.

Absolute calibration, shown schematically in Fig. 7-17a, is used in standards laboratories to obtain "secondary standards." The precision input motion and the laser interferometer shown in the figure are seldom available except in standard laboratories. The secondary standards can be used as references to calibrate working accelerometers by a comparative method.

A *comparative calibration* simultaneously applies "identical" excitation to both the test accelerometer and a secondary standard, as illustrated in Fig. 7-17b. Differences in excitation may be observed by reversing the location of the accelerometers. When the test accelerometer is mounted on top of the standard, there may be a minor change in sensitivity due to mass loading. A sinusoidal excitation is commonly used. The magnitude and phase angle of the outputs provide a comparison of the relative frequency response of two accelerometers. Portable instruments are available for the field calibration of a working accelerometer.

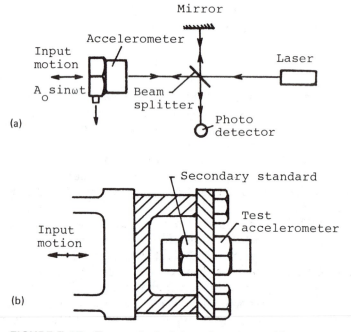

FIGURE 7-17 Two methods for accelerometer calibration. (a) Absolute method. (b) Comparison method.

Calibration methods for special applications are useful. For example, a calibrating voltage may be injected into an "in-place" piezoelectric accelerometer-amplifier system to simulate in-service conditions. These are special techniques and require detail considerations [22].

7-6. FORCE MEASUREMENTS

Force can be measured by means of the reaction of a component to an applied force. The reacting component is often a flexure. In this section we describe common methods for force measurement. Dynamic considerations and field calibration of force transducers are discussed later.

A physical variable can be measured by means of many physical laws (see Table 1-2). Although one equation may be more convenient than the other for a given application, any of the following force-related equations can be utilized for a force measurement:

$$\text{force} = (\text{mass})(\text{acceleration}) \tag{7-29}$$

$$\text{force} = (\text{pressure})(\text{area}) \tag{7-30}$$

$$\text{force} = (\text{constant}) \frac{(\text{mass}_1)(\text{mass}_2)}{(\text{distance})^2} \tag{7-31}$$

$$\text{force} = (\text{constant})(\text{spring deformation}) \tag{7-32}$$

When a mechanical spring is used to implement Eq. 7-32, the output is a displacement, but when a piezoelectric crystal is used, instead of a spring, the deformation of the crystal gives an electric charge. Hence different hardware and techniques can be used for each of the equations above. Moreover, an unbalanced or a null-balance measuring system can be used for the measurement (see Secs. 3-1 and 3-6). Only general principles are described here, but details are important in applications. Variations in applications are left for the reader as exercises.

Examples of force transducers or load cells using the force-balance principle are shown below. The cantilever in Fig. 7-18a is the spring in a force transducers described by Eq. 7-32. The associated deflection can be detected by means of many types of secondary sensors (see Chapter 5). Owing to good dynamic characteristics, load cells with "spring elements" are widely used.

The chemical balance shown in Fig. 7-18b uses gravitational force in a null-balance system for force measurement. A multilever platform scale uses the same force-balance principle. Similarly, the principle is applied in the hydraulic load cell shown in Fig. 7-18c and the pneumatic load cell in Fig. 7-18d. An electromagnetic or other type of force can be used in a force-balance system, as in a servo-accelerometer described above.

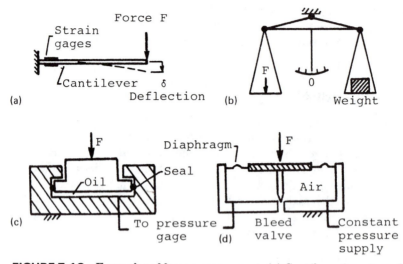

FIGURE 7-18   Examples of force measurement. (a) Cantilever force transducer. (b) Chemical balance. (c) Hydraulic load cell. (d) Pneumatic load cell.

A.   Dynamic Force Measurements

The *mass cancellation technique* [23] for dynamic force measurements is described in this section. The technique attempts to eliminate the effect of mass loading in a dynamic test.

The difference between a static and a dynamic measurement is in the inherent inertia effect of the masses in a test setup under dynamic conditions. The static characteristics of a linear transducer can be specified by a single number or calibration constant. The dynamic characteristics of the transducer and the overall system are frequency dependent, even when the transducer itself is zero-order. The dynamic response is described by a frequency-dependent transfer function.

Mass loading may cause errors in resonant frequencies, damping, amplitude, and phase angle in dynamic measurements. The effective mass of a system at a mode of vibration may be much lower than its rest mass, because resonances occur at frequencies of maximum dynamic weakness. In other words, a stiff system at rest is no assurance of its stiffness at resonances.

A dynamic force is applied to a test system by means of a shaker and a force transducer (load cell), as shown in Fig. 7-19a. This is an idealized condition. Any loading mass between the load cell and the test system, as shown in Fig. 7-19b, will decrease the force transmitted. Thus the actual

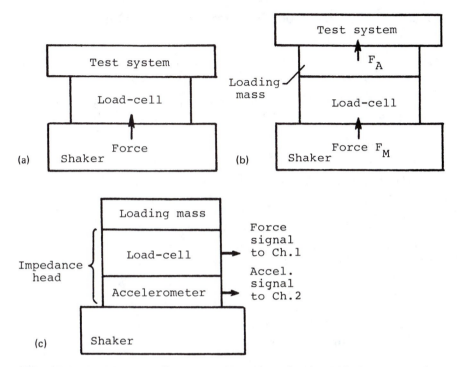

FIGURE 7-19 Mass cancellation using impedance head. (a) Ideal test setup. (b) Setup with mass loading. (c) Setup with impedance head.

force F_A applied to a test system is less than the force F_M measured by the force transducer. The difference is the inertia force of an equivalent mass M_M, where M_M is an equivalent rigid mass that gives the same resulting force under the same linear acceleration.

Components required for the mass cancellation technique are an impedance head, a two-channel digital signal analyzer, and a shaker. An impedance head consists of a load cell and an accelerometer, as shown in Fig. 7-19c. First, the effect of an equivalent mass M_M is determined. M_M includes the mass of the impedance head and the loading mass for attaching the impedance head to a test system. The transfer function H_I for the impedance head without the test system is

$$H_I = \frac{A}{F_I} = \frac{1}{M_M} \tag{7-33}$$

where A is the acceleration and F_I the indicated input force. Using the same mounting device (loading mass), the measured transfer function with the test system is

$$H_S = \frac{A}{F_M} \qquad (7\text{-}34)$$

where F_M is the measured force. Note that the actual applied force to the test system is $F_A = (F_M - F_I)$. Hence the actual transfer function is

$$H_A = \frac{A}{F_A} \qquad (7\text{-}35)$$

Substituting Eqs. 7-33 and 7-34 into Eq. 7-35 and simplifying, we get

$$H_A = \frac{H_I H_S}{H_I - H_S} \qquad (7\text{-}36)$$

The mass loading effect is small for massive structures but may be appreciable for light structures at resonances. The way to determine whether the correction by Eq. 7-36 is necessary is to make the correction and see if it is significant.

B. Transient Calibration of Force Transducers and Accelerometers

A drop test and an impact test for the transient calibration of force transducers and accelerometers are described in this section. Both methods calibrate the entire information channel for in-service conditions.

Let us first examine briefly the composition of an information channel, or a measuring system, to justify the calibration of an entire channel rather than the individual components in a channel separately. Components in an information channel may consist of a transducer and signal processing equipment, such as signal conditioning electronics, amplifiers, filters, switches, and multiplexers. Errors may be from individual components as well as from the interaction between components within the channel. The problem is further complicated when two channels are used simultaneously, such as in a force and an acceleration measurement for the transfer function of a test system.

The calibration certificate, furnished by the manufacturer or a standards laboratory, certifies only an individual component, not the interaction between components in a channel. Furthermore, the configuration of components in a channel may change with test setups. Environmental conditions also change with time. If a channel has five components in cascade, a 2% error in each component may result in a 10% overall error. A simple scale change from ×1 to ×10 may change from ×0.98 to ×10.2.

The error from this scale change is 4%. Due to the variables in an actual test, it is futile to estimate errors for in-service condition from the calibration of individual components.

The *drop test* shown in Fig. 7-20a can be used for the "in-place" calibration of an accelerometer. It is an updated version of the tilting method (see Sec. 7-5E) and uses the earth's local gravity (1*g*) as the constant input. The setup [24] consists of an elastically suspended impact platform, a monofilament line, a mass, and a test accelerometer. An impact applied squarely to the impact platform causes the monofilament line to go slack, but not enough to cause the transducer assembly to wobble after the impact. The duration of the free fall is from 30 to 50 ms. Thus a step change of 1*g* is applied to the test accelerometer. Similarly, the calibration of a load cell is shown in Fig. 7-20b. Initially, a constant force due to

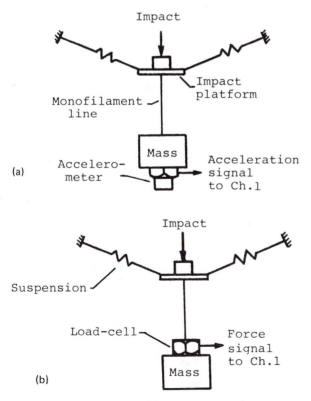

FIGURE 7-20 Drop calibration of accelerometer and load cell. (a) Accelerometer calibration. (b) Load-cell calibration.

the calibration mass is applied to the load cell. This force is relieved during the free fall. Thus a step force is applied to the load cell.

An ideal step input is shown in Fig. 7-21a. In reality, the impact is not instantaneous and has a duration of about 3 ms. Since the frequency spectrum of a step function is approximately proportional to the reciprocal of frequency (see Appendix C), the high-frequency response of the transducer is not strongly excited. If the test transducer does not have a long time constant, its time response has a drooping characteristic, as shown in Fig. 7-21b. This resembles the response of a RC first-order high-pass filter with the output measured across the resistor R (see Fig. 4-19). The response is typical of most piezoelectric transducers.

An *impact test* [25] for the simultaneous calibration of force and acceleration channels for in-service conditions is shown in Fig. 7-22a. A large known calibration mass M_C is suspended from long wires or supported by means of air bearings. From Newton's law—force F equals mass M_C times acceleration A—the expected frequency response is $1/M_C$:

$$\frac{A}{F} = \frac{1}{M_C} \tag{7-37}$$

This equation holds for frequencies above the rigid-body modes of the calibration mass. If F_I is the indicated force of the load cell and A_I the in-

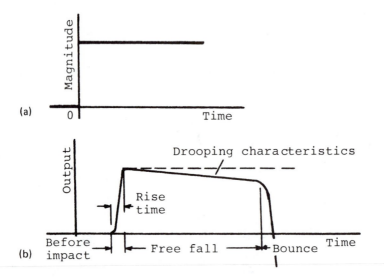

FIGURE 7-21 Accelerometer signal in drop test. (a) Step function. (b) Typical response.

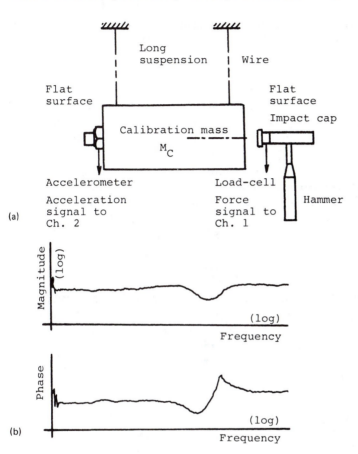

FIGURE 7-22   Obtaining calibration function for structural test. (a) Setup to find calibration function H_C. (b) Typical calibration function H_C.

dicated acceleration of the accelerometer, the indicated frequency response is A_I/F_I. The deviation of the indicated response from the expected response is due to mass loading. The ratio of the two frequency response functions is the calibration function H_C, shown in Fig. 7-22b. H_C includes the effects of the hammer, load cell, accelerometer, the impact cap on the hammer, and the dynamics of the hammer.

$$H_C = \frac{1/M_C}{A_I/F_I} \tag{7-38}$$

The same hammer and accelerometer are used for the test system. If the measured force is F_M and the measured acceleration A_M, the measured frequency response H_M is

$$H_M = \frac{A_M}{F_M} \qquad (7\text{-}39)$$

Assuming system linearity, the actual or corrected frequency response H_A is the product of the correction function and the measured response.

$$H_A = H_C H_M \qquad (7\text{-}40)$$

Finally, the accelerometer may cause a mass loading of the test system. This is not considered in the procedure above. The only solution is to substitute a different mass load at the same location of the test system, to repeat the frequency response test, and to compare the test results in order to verify whether the mass loading is insignificant.

7-7. TORQUE MEASUREMENTS

Torque measurements are often associated with the determination of mechanical power: either the output developed by a power source or the power transmitted to another machine. Power absorption and power transmission dynamometers are examined in this section.

The *absorption dynamometer* shown in Fig. 7-23a measures the power developed by an engine. Since the power absorbed by the dynamometer is

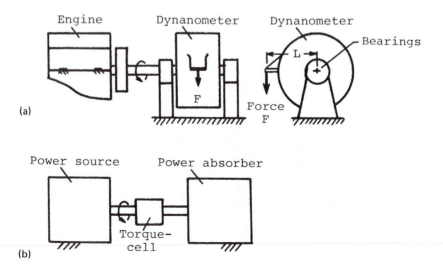

FIGURE 7-23 (a) Absorption dynamometer. (b) Torque "dynamometer."

dissipated, there must be provision for cooling in the system. Power or the rate of doing work is the product of torque and revolutions per unit time. The dynanometer shown is cradle mounted. The trunnion bearins allow measurement of the reactive torque, which is the product of the measured force F and the arm length L, as shown in the figure. If the dynanometer is an electric generator, the power absorbed is dissipated externally. If it is a dc generator, it can be reversed easily as a motor to drive the engine for finding its frictional horsepower. Other common types of absorption dynamometers (not shown) are the prony brake, water brake, and eddy-current dynanometer.

A *power-transmission dynanometer* is a torque cell for measuring the torque delivered. It is placed between a power source and a power absorber as shown in Fig. 7-23b. Generally, the primary sensor in a torque cell is a flexure and there is no power dissipation. The secondary sensor is a displacement-type sensor, such as strain gages. Slip rings, transformers, telemetry, or some hardware are necessary to transmit the torque signal from the rotating member to a stationary readout device.

Typical examples of flexures in a torque cell are shown in Fig. 7-24. The schemes are self-explanatory. The dynamic characteristics of a torque cell are essentially those of a lightly damped second-order system.

An interesting application of flexures is the multicomponent dynanometer shown in Fig. 7-25a. It consists of a mounting platform and four rings as flexures. It is capable of measuring three force components independent of location and three moment components. When the top and bottom surfaces of the ring shown in Fig. 7-25b are restrained from rotation, the force components F and P can be measured independently. The ring is modified, as shown in Fig. 7-25c, for greater stability and for meeting the requirement of zero rotation of the top and bottom surfaces. Strain gages are used as secondary sensors. Wiring diagrams of the gages for measuring F, P, and the moment M are as shown in Fig. 7-25d.

7-8. PRESSURE MEASUREMENTS

Pressure is defined as the force exerted by a fluid on a unit area of a medium. The force is measured by means of an reaction, as described in the preceding section. The area is the effective area of the sensor. Pressure transducers, applications, installations, and calibration are described in this section.

The datum for pressure measurement is either the atmospheric pressure or a "perfect" vacuum. Pressure above atmospheric is called a *gage pressure* and below atmospheric, a *vacuum*, as illustrated in Fig. 7-26. The datum for *absolute pressure* is a perfect vacuum. Units are pascal (1

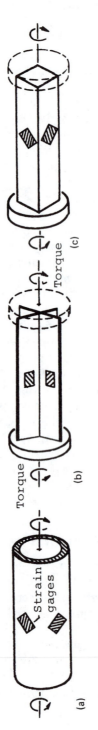

FIGURE 7-24 Examples of flexures in torque-cell. (a) Hollow shaft. (b) Cruciform. (c) Square shaft.

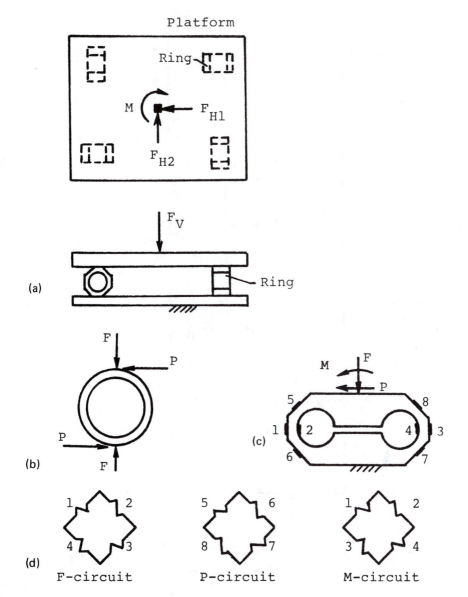

FIGURE 7-25 Multicomponent dynanometer. (a) Dynanometer (from Ref. 26). (b) Ring to measure forces F and P. (c) Octogonal ring for stability. (d) Strain gage wiring diagrams.

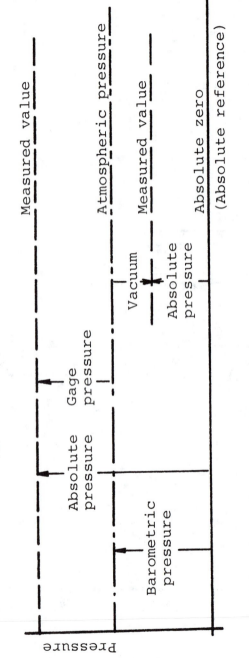

FIGURE 7-26 References for pressure measurements.

Pa = 1 newton/m^2) and pound-force per square inch, such as psig (gage), psia (absolute), and psid (differential). Other common units are:

Standard atmospheric pressure = 760 mm Hg = 29.921 in. Hg = 14.696
  psi = 101.325 × 10^3 Pa = 1 bar
1 μm (micron) = 10^{-6} m Hg = 19.34 × 10^{-6} psi = 0.133 Pa
1 torr = 1000 μm = 1 mm Hg

Pressure measurements can be divided in three ranges:

1. Moderate pressures (10^{-1} to 10^4 psi)
2. Very high pressures (greater than 10^4 psi)
3. Very low pressures (vacua)

Moderate pressures are more commonly encountered in engineering. Pressures in this range are usually measured by means of (1) the height of a liquid column raised by the pressure or (2) the elastic deformation of a "spring" element. Piezoresistive and piezoelectric elements are used for high-pressure measurements. Vacuum is generally a special study.

Pressure can also be measured by its effect on a medium. The problem is open ended. For example, the variation of thermal conductivity of a gas with pressure is used to measure a vacuum. The damping effect of a gas on a vibrating diaphragm is utilized to measure absolute pressure between 10^{-6} to 10^3 mm Hg [27]. This reinforces the statement that if an object has a property influenced by a physical variable, that object is potentially a transducer for the variable.

A. Sensing Elements

Manometers and "spring" elements for pressure sensing are described in this section. A *manometer* indicates pressure by the height of a column of fluid, such as mercury. The height H of the liquid column in the U-tube manometer in Fig. 7-27a gives the differential pressure $(P_1 - P_2)$. The height H is zero when $P_1 = P_2$, because water seeks its own level. An interesting application of this principle is in the layout of a very large machine. If a piece of long flexible tubing is substituted for the U-tube, the device can be used to seek locations of parts of the machine at the same level. A well-type manometer is shown in Fig. 7-27b. Similar construction is used for a barometer, in which mercury is the working fluid, P_1 is atmospheric pressure, and P_2 a vacuum. Greater resolution is obtained by using the inclined tube shown in Fig. 7-27c. The micromanometer in Fig. 7-27d works on the same principle. The micrometer gives greater resolution by means of subdivision, and the inclined tube gives greater resolution by means of magnification (see Sec. 7-2B).

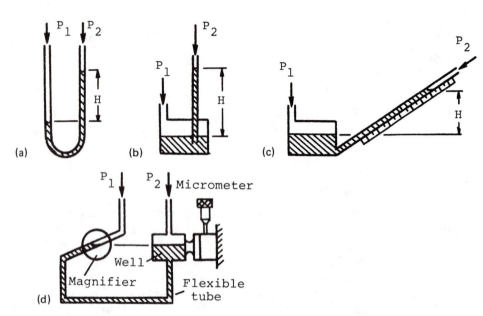

FIGURE 7-27 Examples of manometers. (a) U-tube. (b) Well type. (c) Inclined tube. (d) Micromanometer.

The "*spring*" *element* for pressure sensing is a flexure. It may be a part of an enclosure or the entire enclosure, such as a bourdon tube (see Fig. 3-2). The twisted bourdon tube shown in Fig. 7-28a is another example. Its output is an angular displacement. Two bellows for a differential pressure measurement are shown in Fig. 7-28b. The flattened tube shown in Fig. 7-28c and the hollow cylinder in Fig. 7-28d have a higher frequency response than does the bourdon tube. The internal volume of the cylinder can be reduced by using a solid filler to reduce the capacitance of the transducer in order to improve its frequency response.

It is implicit that the many types of secondary sensors can be utilized for detecting the elastic deformation of the flexure. For example, strain gages are secondary sensors, as shown in Fig. 7-28c and d. A differential transformer (see LVDT) as the secondary sensor for a hollow-tube pressure transducer is shown in Fig. 7-28e. A vibrating diaphragm pressure transducer is shown in Fig. 7-28f. A thin metallic diaphragm under radial tension is forced to vibrate at resonance in a gas chamber. The amplitude of vibration is determined by the damping effect of the gas under pressure. The output is from the varying capacitance between the vibrating diaphragm and the stationary plates [27].

Additional examples of flexures for pressure sensing are shown in Fig. 7-29. A pressure diaphragm with strain gages is shown in Fig. 7-29a (see also Figs. 5-18e, 5-34d and f, and 5-35). Special rosette gages with spiral pattern for pressure diaphragms are available. Chemical etching is used to produce *silicon pressure diaphragms* [29] for low-cost and high-volume industrial applications. "By 1985, Detroit will use more pressure transducers in one year than any domestic pressure transducer suppliers has manufactured in its whole history" [30]. The statement definitely shows the trend. Different configurations can be etched into a silicon chip. Semiconductor strain gages can be diffused at appropriate locations on the silicon. Furthermore, the electronics can be built in as an integral part of the transducer.

Pressure-to-force conversion is shown in Fig. 7-29b. The diaphragm merely isolates the sensor from the environment. A short column with strain gages is the force sensor. Similarly, a piezoelectric load cell for the pressure-to-force conversion is shown in Fig. 7-29c.

The pressure-bar transducer, shown in Fig. 7-29d, is suitable for measurement of pressure pulses with short rise time [31]. When a pressure pulse is applied to the end of a thin rod, strain waves propogate along a metal bar at the speed of sound through the material, and the strain pulse induced will propagate essentially unchanged in shape along the rod. A secondary sensor is used to pick up the signal. The effects of temperature and reflected waves in the bar must be considered in this type of transducer.

High pressures up to 50,000 psi are used in chemical and metallurgical plants, and pressures up to 100,000 psi can be measured with strain gage pressure cells or helical bourdon tubes [32]. A Bridgeman-type gage for very higher pressures is shown in Fig. 7-30a. The sensing element is a manganin or gold-chrome wire in a loose coil, enclosed in a kerosene-filled bellow [33]. The resistance of a wire changes with elastic deformation. Following Eq. 1-5 for strain gages, it can be shown that the pressure sensitivity of a wire is

$$\frac{\partial R/R}{P} = \frac{2}{E} + \frac{\partial \rho/\rho}{P} \tag{7-41}$$

where R is the electrical resistance, P the hydrostatic pressure, E is Young's modulus, and P is the resistivity. Another bulk-modulus pressure cell is shown in Fig. 7-30b. The sensor is a steel cylinder and its output a displacement. The upper range is 200,000 psi and the accuracy 1%. Both of these characteristics are inferior to the Bridgeman gage. However, it is more rugged, suitable for viscous fluids, and has a faster response.

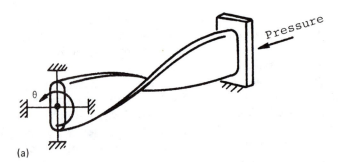

(a)

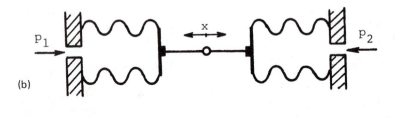

(b)

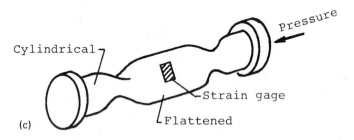

(c)

FIGURE 7-28 "Spring" elements in pressure transducers. (a) Twisted bourdon tube. (b) Bellows for $(p_1 - p_2)$. (c) Flattened tube. (*Continued on page 431.*)

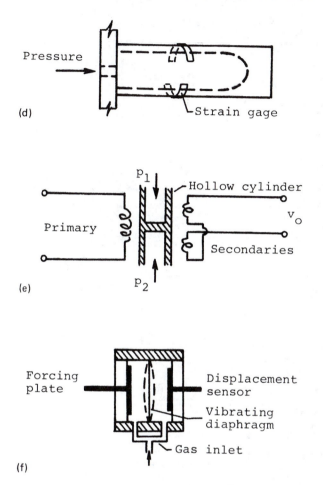

FIGURE 7-28 (*Continued*)   (d) Hollow cylinder. (e) Differential transformer (from Ref. 28). (f) Vibrating diaphragm transducer.

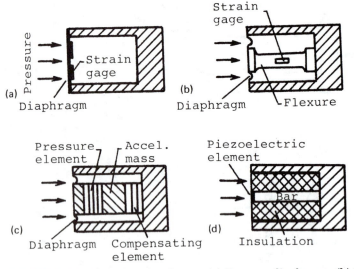

FIGURE 7-29 Pressure transducers. (a) Pressure diaphragm. (b) Force flexure. (c) Piezoelectric. (d) Pressure bar.

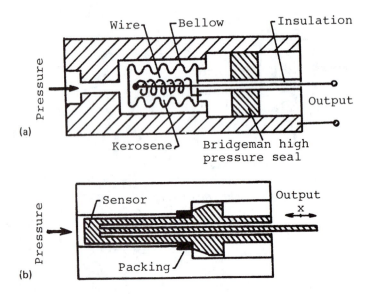

FIGURE 7-30 High-pressure transducers. (a) Bridgeman-type gage. (b) Bulk modulus gage (from Ref. 34).

B. Applications

As physical laws are used for measurements (see Sec. 2-1), any pressure-related equation can be used to measure pressure by means of other physical variables in the equation (see also Eqs. 7-29 to 7-32 for force measurement). Conversely, every pressure-related equation can be applied to measure other variables by means of pressure sensing. This section will explore some of these applications. It is understood that there are many variations and details in using the same law.

Several physical laws relating pressure and other variables are shown below. It is also possible to combine the laws in an application.

$$\text{Force} = (\text{pressure})(\text{area}) \qquad\qquad (7\text{-}42)$$

$$\text{Force} = (\text{mass})(\text{acceleration}) \qquad\qquad (7\text{-}43)$$

$$q + h_i + \tfrac{1}{2}V_i^2 + gZ_i = h_e + \tfrac{1}{2}V_e^2 + gZ_e + w \qquad\qquad (7\text{-}44)$$

$$v(P_e - P_i) + \tfrac{1}{2}(V_e^2 - V_i^2) + g(Z_e - Z_i) = 0 \qquad\qquad (7\text{-}45)$$

$$Pv = RT \qquad\qquad (7\text{-}46)$$

$$\text{Pressure} = g(\text{density})(\text{liquid column}) \qquad\qquad (7\text{-}47)$$

The first two equations are self-evident. The other equations are in SI units. The acceleration due to gravity is $g = 9.807$ m/s^2. The first law of thermodynamics in Eq. 7-44 is for a reversible steady-state steady-flow process, where q is the heat transfer per unit mass, h the enthalpy, V the velocity, Z the elevation, w the work, and the subscripts i and e refer to the inlet and exit. For an incompressible flow for which $q = 0$, $w = 0$, and the specific volume (volume per unit mass) v is constant, the first law reduces to Eq. 7-45, or the Bernoulli equation. The equation of state of an ideal gas is stated in Eq. 7-46, where P = absolute pressure, T = absolute temperature, and R = universal gas constant. Pressure due to a liquid column is shown in Eq. 7-47, where density $\rho = 1/v$ kg/m^3.

Applications of *hydrostatic head* [35] by Eq. 7-47 are described below. Measurement of a liquid level H, shown in Fig. 7-31a, assumes that density is constant. A switch can be incorporated for a high- and low-level alarm. The pneumatic purge system or "bubbler" shown in Fig. 7-31b is used when the liquid is corrosive or at a high temperature. The liquid density measurement shown in Fig. 7-31c assumes that H is constant. Alternatively, when the liquid level fluctuates, as shown in Fig. 7-3ad, density is measured by maintaining a constant H between two sensors. If the pressure transducer can be raised or lowered, as shown in Fig. 7-31e, the pressure transducer becomes a displacement gage. When the pressure in a closed tank can be both positive and negative, the level is measured by means of a differential pressure transducer, as shown in Fig. 7-31f. The

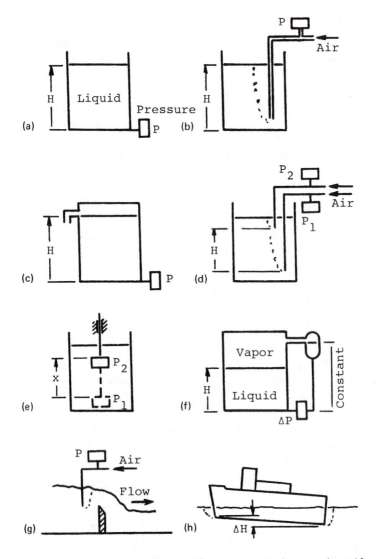

FIGURE 7-31 Applications of Equation 7-47 (pressure) = (density) (liquid column). (a), (b) Level. (c), (d) Density. (e) Displacement. (f) For (±) pressure. (g) Flow. (h) Draft and trim.

bubbler shown in Fig. 7-31g is used to measure flow in an open channel. The draft and trim of a boat can be measured as shown in Fig. 7-31h. The trim is the ΔH of the fore and aft of the boat.

A pressure-to-force conversion (see Fig. 7-18c) by Eq. 7-42 allows a pressure transducer to measure force-related quantities. Hence this type of load cell is susceptible to errors due to an undesired acceleration input. By the same token, the pressure transducer can be used to measure acceleration. Three accelerometers employing this principle are schematically shown in Fig. 7-32. The seismic mass for the high-g version is the fluid in a closed container. The output is 15 psi/1000g. The cross-sensitivity between the (x, y, z) axes is governed by the dimension of the container along the three axes. This may be an advantage or a disadvantage. Medium g is from 10 to 100g and low g is below 10g. These accelerometers are applicable from 0 to 100 Hz. They are used for automotive braking systems to prevent jackknifing of trailers.

Differential pressure transducers are often used for flow measurements. For adiabatic flow at constant elevation, Bernoulli equation (Eq. 7-45) in SI units reduces to

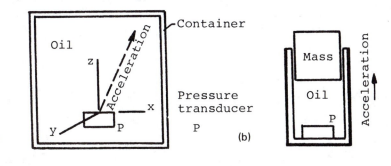

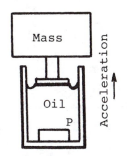

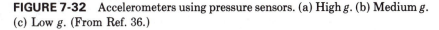

FIGURE 7-32 Accelerometers using pressure sensors. (a) High g. (b) Medium g. (c) Low g. (From Ref. 36.)

$$P_1 + \frac{\rho}{2} V_1^2 = P_2 + \frac{\rho}{2} V_2^2 \qquad\qquad (7\text{-}48)$$

where ρ is the fluid density. The pitot tube is an example of this application (see Fig. 3-6). An obstruction, such as the orifice shown in Fig. 7-33a, can be placed in a closed pipe to induce a large velocity change for a flow measurement by means of differential pressure. The venturi and flow nozzle work on the same principle. Similarly, the centrifugal flow meter shown in Fig. 7-33b utilizes a differential pressure due to the flow in a curved pipe.

Some other applications are described below. The equation of state (Eq. 7-46) is used in the constant-volume gas thermometer shown in Fig. 7-33c. The bulb is nitrogen filled and the temperature is indicated by means of a pressure gage. Alternatively, if the bulb is partially filled with liquid, its vapor pressure can be used to indicate temperature. Differential pressure is used to monitor the condition of a filter, as shown in Fig. 7-33d. A method for the on-line viscosity measurement is shown in Fig. 7-33e. The constant-flow pump draws an oil sample from the main stream and pumps it through a temperature-controlled capillary tube a–b. Differential pressure and viscosity are related by Poiseuille's equation.

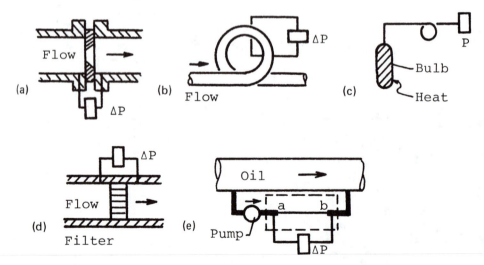

FIGURE 7-33 Applications of pressure measurement. (a) Flow orifice. (b) Centrifugal flow meter. (c) Constant-volume thermometer. (d) Filter monitoring. (e) Viscosity measurement.

C. Installations

The dynamic characteristics of a flush-mounted pressure transducer are simply those of the transducer itself. A remote-mounted transducer is used when a flush mount is not possible. Hence the transducer and the transmission line must be considered as an integral unit, or as a measuring system.

Errors for a flush-mounted pressure transducer are due mainly to vibrations and temperature. A pressure transducer becomes an accelerometer when the inertia force of the masses in the transducer is large compared to the force due to the pressure. An acceleration-compensating element is often found in piezoelectric pressure transducers (see Fig. 7-29c). Water cooling is used to minimize the effect of temperature, or a thermocouple can be incorporated in the pressure sensor for the purpose of temperature compensation. In a hostile environment, such as during the firing of a rocket engine, severe vibration and heat transfer rate are common [37]. Thermal isolation of the pressure sensor can be achieved by means of a continuous flow of helium gas around the sensor [38]. These are special situations.

The dynamic characteristics of a flush-mounted pressure transducer is examined by means of its frequency response, as shown in Fig. 7-34a. For a transducer with reasonable damping, the usable range is about 60% of its resonant frequency. For a remote-mounted system, shown in Fig. 7-34b, each part of the system has its resonance. A sharp transient can induce a ringing in the parts as well as the support for the system. Hence the signal transmission path and the transducer act as a complex filter. The dynamic characteristics may depend on the installation as much as on transducer itself. For example, the resonant frequency can be increased by reducing the internal volume, as shown in Fig. 7-34c, but this also reduces the damping. Thus the overall gain in usable frequency may not be appreciable. Coarse fibrous cord can be used in the tubing to increase the acoustic damping and to improve the frequency response characteristic [39].

Simplified assumptions for remote mounting are usable only for quasistatic and low-frequency applications. Pneumatic signal transmission lines may vary from inches to several hundred feet in process industries [40]. Using lumped-parameter approximations, a system may be considered as first- or second-order [41]. When inertia effects are negligible, pressure is slow varying, and damping is heavy, the system response is first-order with a time delay, shown in Fig. 7-35a. The delay time is due to the velocity of sound in the transmission line. For air, the velocity is approximately 1120 ft/sec. The system is second-order when inertia effects in the system cannot be neglected.

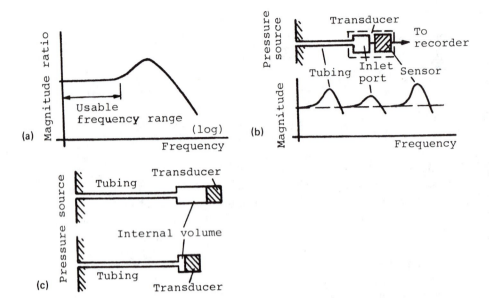

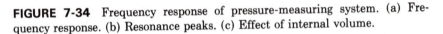

FIGURE 7-34 Frequency response of pressure-measuring system. (a) Frequency response. (b) Resonance peaks. (c) Effect of internal volume.

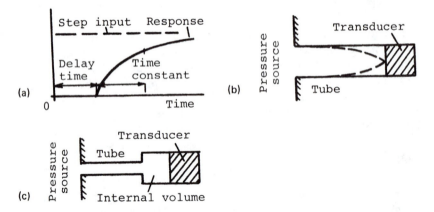

FIGURE 7-35 Characteristics of fluids in tubing. (a) First-order response. (b) Organ-pipe resonance. (c) Helmholtz resonator.

For high-frequency applications, a remote mount will degrade the performance of a transducer. The overall dynamic performance is governed by many variables and determined by the installation rather than the pressure transducer itself. In other words, the specifications of the transducer are inadequate to predict the performance of the installation. Acoustic resonance may occur as in a pipe organ, shown in Fig. 7-35b. The system may become a Helmholtz resonator, shown in Fig. 7-35c, when the internal volume acts as a spring and the fluid in the tube as a mass. It is difficult to ascertain whether (1) the flow is laminar or turbulent, and (2) the system is linear or nonlinear. For in-service conditions, the fluid in the transmission path may be a gas, a gas–vapor mixture, a liquid with entrapped gas, or a liquid. The velocity of sound in a medium is temperature dependent. It would be difficult to predict the resonance of a gas-filled tube when the composition of the gas is unknown.

There are no hard-and-fast rules for remote mounting of pressure transducers. Flush mounting is always preferred, but a transmission line may be used for frequencies up to 100 Hz. Never fill a line from a gas-pressure source with liquid "to increase frequency response" [42]. A liquid-filled line will introduce acoustic ringing from the characteristic velocity of sound in the liquid. The water-hammer effect of a liquid-filled line may damage the transducer.

D. Calibrations

Dynamic calibrations [43] of pressure transducers by means of a comparative test (see Fig. 4-16) and shock tube are described in this section. The accuracy from a static calibration by means of a dead-weight tester (see Fig. 2-3) can be stated as a number. The results from a dynamic calibration are interpreted in terms of time and frequency response (see Sec. 4-4B).

A comparative test compares the performance of a test transducer with that of a secondary standard on an all-things-equal basis. The parameters considered are (1) frequency, (2) waveform, (3) pressure, and (4) type of fluid. Mechanical calibrators, such as piston-cylinder and rotating valves, are useful for low-frequency ranges. Sirens, electromechanical exciters, acoustic shock generators, burst diaphragms, closed bombs, and the like are for medium and higher frequencies. For example, a piezoelectric and a burst diaphragm calibrator are shown in Fig. 7-36.

The waveform for calibration may be periodic or aperiodic. The waveform of an electromechanical pressure generator can be programmed by a controlling signal to the shaker. Sinusoidal signals cannot be generated in a gas with appreciable amplitudes [44]. If the amplitude is

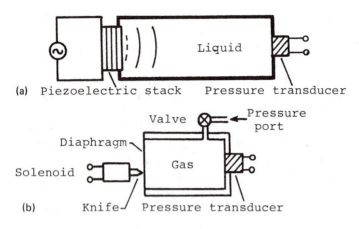

FIGURE 7-36 Pressure-transducer calibrators. (a) Piezoelectric exciter. (b) Burst diaphragm.

kept constant and the frequency increased, the waveform degenerates from a sinusoid to a sawtooth.

A shock tube, shown schematically in Fig. 7-37a, gives the time response of a pressure transducer to a step input. First, a pressure differential, shown in Fig. 7-37b, is built up across the diaphragm separating the high-pressure P_h and low-pressure P_l chambers. Bursting the diaphragm causes a pressure discontinuity, or a shock wave of pressure P_s, shown in Fig. 7-37c. Pressure distribution after the reflection of the shock wave at the closed end is shown in Fig. 7-37d. The shock velocity is measured by transducers placed longitudinally along the tube.

Shock tubes are used widely for testing at the high-frequency range. Pressures from a few psi to 600 psi with a step of 10^{-9} rise time are obtainable. The chambers must be properly proportioned in order to obtain a step of reasonable duration and known amplitude over a pressure range. Test data from shock tubes of different diameter may differ because the effect of boundary layer is more pronounced in a smaller-diameter tube. If the diaphragm of a pressure transducer is recessed from the pressure source, data from the recessed transducer are repeatable, but the results can be anywhere from valid to invalid [45].

A short rise time for the shock tube could be detrimental for some tests. To reproduce a transient signal faithfully, the frequency content of the signal must be below the transducer's lowest resonant frequency (see Fig. 7-34). A pressure-measuring system is a complex filter with many lightly damped resonances. The rise time should be short enough to excite the

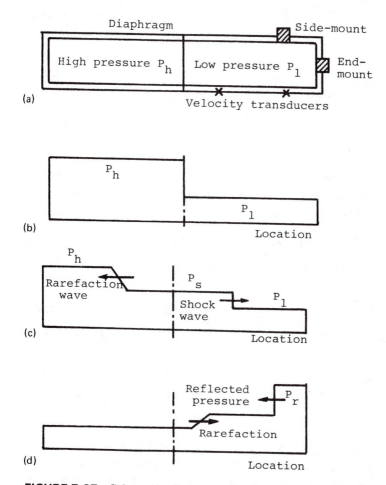

FIGURE 7-37 Schematic of pressure distribution in shock tube. (a) Schematic of shock tube. (b) Initial pressures. (c) After diaphragm burst and before reflection of shock wave and rarefaction wave. (d) Shock wave reflected by the closed end.

resonances of interest, but not to excite the higher resonances beyond the application anticipated. An approximate relation between rise time and frequency is

$$\text{rise time} = \frac{1}{4}\left(\frac{1}{\text{frequency}}\right) \qquad (7\text{-}49)$$

PROBLEMS

7-1. The vernier for the caliper in Fig. 7-26 has 10 units in 0.90 in. and the fixed scale has 10 divisions in 1.00 in. With the aid of a sketch, device a vernier with 10 units in 1.10 in. (Suggest a 1:0.2 scale for the sketch.)

Use the following typical data for Probs. 7-2 to 7-6.

Material	Temperature coefficient of expansion, a		Light source	Wavelength	
	ppm/°C	ppm/°F		µm	µin.
High-carbon steel	11.6	6.5	Mercury (^{198}Hg)	0.546	21.5
Stainless steel	16.7	9.3	Helium	0.589	23.2
Aluminum	23	13	Cadmium (red)	0.644	25.38
Gage blocks	11.2	6.4			

7-2. An aluminum machine part of nominal diameter $D_p = 8$ cm at 25°C is measured by means of gage blocks. Find the diameter at the reference temperature of 20°C. Compare the error with the tolerance of the "working-standard" gage blocks.

7-3. The diameter of a high-carbon-steel plug gage at 76°F is measured by means of gage blocks at the same temperature. The measured diameter is $D_p = 2.7855$ in. Determine whether a temperature correction is necessary.

7-4. The diameter D of a cylinder is compared with the dimension L of gage blocks as shown in Fig. 7-1. D is deduced from the N number of fringes observed over the top surface of the gage blocks (see Fig. 7-3d). (a) Describe a method for finding whether $D > L$ or $D < L$ and deduce D from L. (b) The point B on the diameter does not coincide with the point of contact C between the cylinder and the optical flat. Deduce the order of magnitude of this error.

7-5. The diameter of two aluminum cylinders are measured as shown in Fig. P7-1. The setups are at 68°F and $L = 1.7500$ in. Find the diameters if six fringes are observed using a helium light source.

7-6. Find N the number of fringes observed for the measurement described in Prob. 7-5 if the setups are at 80°F.

7-7. A rubber ball with a horizontal velocity $V_x = 4$ ft/sec rolls off a table onto a hard floor. Assume that the height of the table is $H = 3$ ft, V_x remains unchanged, and the coefficient of restitution is 0.70. Plot the trajectory of the ball for $0 < t < 1.0$ using $\Delta t = 0.1$ sec.

7-8. A 300-kg table for repairing instruments is isolated from the floor by springs with $k = 20$ kN/m and dampers with $c = 4$ kN $\cdot$ s/m. If the floor vibrates vertically ±2.5 mm at a frequency of 8 Hz, find the motion of the table.

7-9. The characteristics of vibrometers for displacement measurements are shown in Fig. 7-7. A vibrometer is used to measure the vibrations of a variable-speed engine, the operating speed of which ranges from 500 to 1500 rpm. If the vibrations consist of a fundamental and a second harmonic and the amplitude distortion is to be less than 4%, find the natural frequency of the vibrometer if (a) its damping is negligible, and (b) its damping ratio is 0.6.

7-10. An accelerometer is used to measure the vibrations described in Prob. 7-8. Characteristics of accelerometers as shown in Figs. 7-7 and 7-8. If the amplitude distortion of each of the harmonic components is to be less than 4%, find the natural frequency of the accelerometer if its damping ratio is (a) negligible and (b) 0.70. Determine the corresponding phase angles at the output relative to the input.

7-11. Referring to Fig. 7-14 and Eq. 7-28 for piezoelectric transducers with built-in electronics, the internal time constant τ_1 ranges from 0.5 to 2000 sec and the external τ_2 is R_2C_2. C_2 is built in, but R_2 depends on the input resistance of the readout instrument. (a) Assuming that $\tau_2 = \tau_1$, sketch the Bode plots for $\log |V_0/w|$ versus $\log \omega\tau_1$ and $\underline{/V_0/w}$ versus $\log \omega\tau_1$. (b) Find the minimum value of $\omega\tau_1$ for 2 and 5% error in $|V_0/w|$, respectively. (c) Repeat part (a) for $\tau_2 = 10 \tau_1$.

7-12. A chemical balance is schematically shown in Fig. 7-2, where W_{cg} is the weight of the beam assembly through the center of gravity and J_0 is it mass moment of inertia about the pivot 0. (a) Find the sensitivity S of the balance in terms of $\theta/\Delta W$. (b) With the aid of a sketch, show a scheme for which S is independent of W. (c) Derive the equation of motion of the system, assuming that the system is lightly damped. (d) Show that a balance with higher S has a the longer period of oscillation. Since it take time for the oscillation to die out, the implication is that it takes more time for a more precise measurement.

7-13. A platform scale with an applied weight W is shown schematically in Fig. P7-3. It consists of platforms with knife edges, a tie rod for transmitting the force to a balancing beam, and the pan weights W_p. The scale is initially balanced with counterpoise (not shown). If $W:W_p = 100:1$, find the relative proportion of the knife edges such that the measurement is independent of the position of W on the weighing platform.

7-14. Use the root-sum-square rss error criterion to estimate the error for each of the force sensing transducers shown in Fig. P7-4.

(a) The spring constant k of the cantilever shown in Fig. P7-4a.

$$k = \frac{3EI}{L^3}$$

E = Young's modulus = $(20 \pm 0.5) \times 10^{10}$ Pa
$L = 9 \pm 0.2$ cm
w = width of cantilever = 2.5 ± 0.005 cm
h = thickness = 0.9 ± 0.005 cm

(b) The spring constant of the coil spring shown in Fig. P7-4b.

$$k = \frac{E_s D_w^4}{8 D_m^3 N}$$

E_s = torsional elastic modulus
= $(80 \pm 3.5)10^6$ kPa
D_w = wire diameter = 3 ± 0.02 mm
D_m = coil mean diameter = 4 ± 0.1 cm
N = number of coils = 30 ± 0.5

(c) The equivalent spring constant k_{eq} of two springs. Spring k_1 is as described in part (b). Spring k_2 is similar to k_1 but has a smaller mean diameter, with $D_m = (2.5 \pm 0.1)$ cm and is inserted in k_1.

(d) The sensitivity (v_o/force) of the proving-ring-LVDT load cell shown in Fig. P7-4c.

Proving ring: $\qquad$ *LVDT:*

$$\text{Deflection } \delta = \text{force}\left[\frac{1}{16}\left(\frac{\pi}{2} - \frac{4}{\pi}\right)\frac{D^3}{EI}\right] \qquad v_o = (S_t v_i)\delta$$

I = moment of inertia about centroidal axis of bending section
$w = 1.0 \pm 0.003$ in. $\qquad t = 0.2 \pm 0.003$ in.
$D = 3.0 \pm 0.01$ in. $\qquad E = (30 \pm 0.5) \times 10^6$ psi
V_o = output voltage, V $\qquad V_i = 4 \pm 0.01$ V
S_t = sensitivity of LVDT = 6.2 ± 0.01 (mV/V_i)/0.001 in.

7-15. With the aid of a sketch, describe the operation of a water-brake dynamometer.

7-16. A cylindrical-shaft torque sensor (see Fig. 7-24) is calibrated as shown in Fig. P7-5. Two strain gages of a full bridge are on the top surface of the shaft and the other two are on the bottom surface. The gages are aligned on opposite spirals, as shown in the figure. Assume that the sensor is subjected to torque, bending, and axial load. (a) Find all the configurations in which a full bridge can be implemented. (b) Show analytically the configuration(s) that are not susceptible to bending and axial load.

7-17. Repeat Prob. 7-16 if a half bridge is used for the torque sensor.

7-18. Calculate the Young's modulus and Poisson's ratio of the material for each of the setups shown in Fig. P7-6. The strain measured are as shown below.

(a) A cylindrical tensile specimen:

$$F = 35 \times 10^3 \text{ N} \qquad \varepsilon_a = 980 \text{ μstrain} \qquad \varepsilon_b = -290 \text{ μstrain}$$

(b) A simple supported beam:

$$F = 3 \text{ kN} \qquad \varepsilon_a = 600 \text{ μstrain} \qquad \varepsilon_b = -157 \text{ μstrain}$$

7-19. A torque cell with a full bridge is calibrated as shown in Fig. P7-5. The applied torque is $T = 2000$ lb$_f$-in., the shaft diameter is $D = 1.50$ in., Young's modulus $E = 30 \times 10^6$ psi, and Poisson's ratio $\nu = 0.3$. Predict the strain output measured with a strain indicator.

7-20. A torque T is applied to a hollow steel shaft and the angular deflection is $2.00 \pm 0.05°$. Calculate the uncertainty in the torque transmission, using the root-sum-square criterion. Data:

$$r_o = 0.625 \pm 0.001 \text{ in.} \qquad r_i = 0.500 \pm 0.002 \text{ in.}$$

$$L = 6.00 \pm 0.01 \text{ in.} \qquad G = 12 \times 10^6 \text{ psi (constant)}$$

7-21. The power transmission for an engine-pump assembly is measured as shown in Fig. P7-7. The shaft diameter of the torque dynamometer is $D = 4$ cm with $E = 20 \times 10^{10}$ Pa and Poisson's ratio $\nu = 0.3$. The resistance of each of the strain gages in a full bridge is $R = 120 \ \Omega$. The gage factor is $G_f = 2.1$ and calibrating resistor $R_c = 100 \text{ k}\Omega$ (see Prob. 3-9). When R_c is shunted across one of the gages, the oscilloscope trace shows a deflection of 3.5 cm. The engine is at 1500 rpm and the trace due to the torque is 4.3 cm. Find the horsepower delivered to the pump.

7-22. (a) The differential head in a U-tube water manometer, shown in Fig. 7-27a, is 1.276 m. Find the corresponding pressure if the local gravitational constant is 9.75 m/s^2 and the water is at 27°C. Convert the data to English units, find the pressure in psi, and convert the psi into SI units for comparison.

(b) The differential head of a U-tube mercury manometer for a hydraulic pump test is 3 in. Assume that the density of mercury is 846 lb_m/ft^3 with a specific gravity of 13.55. Find the pressure indicated in psi.

(c) A sudden change in the applied pressure will cause the fluid in a U-tube manometer to oscillate. Estimate the frequency of oscillation if the total length of the fluid in a U-tube is 2 ft.

(d) Find the differential pressure due to the head H in a well-type manometer shown in Fig. 7-27b. Assume that the cross-sectional area of the well to that of the glass tube is 10:1.

7-23.   Two fluids with densities (ρ_1, ρ_2) are used in a manometer shown in Fig. P7-8. (a) For $\rho_2 > \rho_1$, find the sensitivity S_2 in terms of ρ_1, ρ_2, and d/D. (b) The sensitivity of a simple U-tube manometer in Fig. 7-27a is $S_1 = H/\Delta P$. Assume that its working fluid has density ρ_2. Define a relative sensitivity ratio $R = S_2/S_1$, and find the condition(s) for which R is maximum. (c) Let the working fluids be kerosene and water (specific gravities 0.80 and 1.00) and $d/D = 1/10$. Find R.

7-24.   Find the nominal measured pressure P and estimate its root-sum-square error for the pressure transducer shown in Fig. 7-28d. Hollow steel cylinder:

$D = 0.875 \pm 0.001$ in. $d = 0.750 \pm 0.005$ in.

$E = (30.0 \pm 0.5)$ psi   Poisson's ratio = 0.30 ± 0.02

Strain gages: Output of the half-bridge is 500 ± 5 µstrain.

$R = 120.0 \pm 0.5\ \Omega$ Gage factor $G_f = 2.1 \pm 0.1$

7-25.   The pressure transducer shown in Fig. 7-28d is constructed from steel tubing with 20 mm outside diameter and 2 mm wall thickness. The allowable design stress is 2.5×10^8 Pa. Assume that $E = 20 \times 10^{10}$ Pa and Poisson's ratio $\nu = 0.3$. Strain gages with $R = 120\ \Omega$ and gage factor $G_f = 2.1$ are used in a half-bridge for the measurement. Find (a) the pressure limit of the transducer, (b) the output in terms of V_o/V_i, and (c) the advantage in using a full bridge with the gages arranged in the longitudinal and circumferential directions.

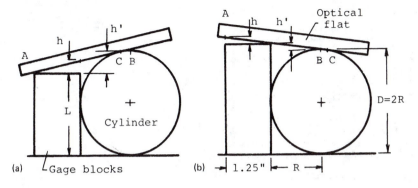

FIGURE P7-1

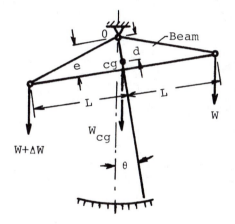

FIGURE P7-2

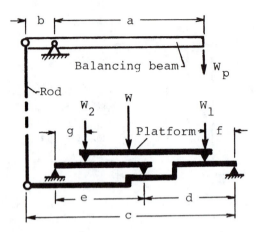

FIGURE P7-3

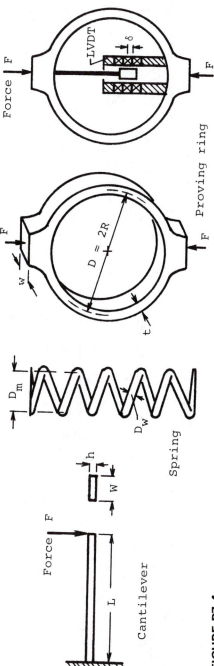

FIGURE P7-4

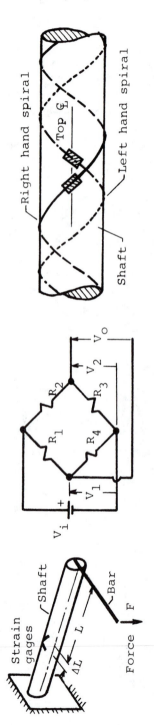

FIGURE P7-5

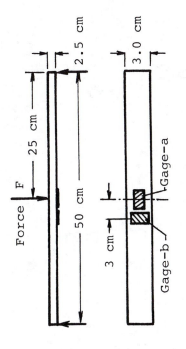

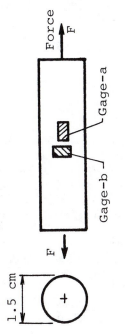

FIGURE P7-6

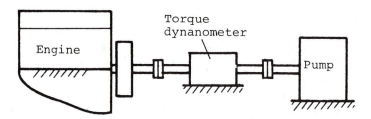

FIGURE P7-7

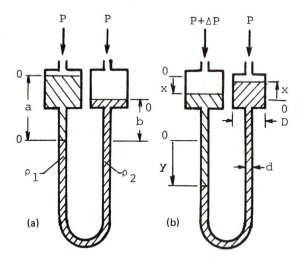

FIGURE P7-8

REFERENCES

1. Keinath, G., *Measurement of Thickness,* NBS Circular 585, Department of Commerce, Washington, D.C., Jan. 1958.
2. Dukes, J. N., and G. B. Gordon, A Two-Hundred-Foot Yardstick with Graduations Every Microinch, *Hewlett-Packard J.* (Aug. 1970).
3. Andre, F. R., and J. W. Michael, Linear Transducer System for High-Accuracy Machine Positioning, *Hewlett-Packard J.* (Feb. 1976).
4. Baldwin, R. R., G. B. Gordon, and A. F. Rude, Remote Laser Interferometry, *Hewlett-Packard J.* (Dec. 1971).
5. Ramboz, J. D., Calibration of Pickups, in *Shock and Vibration Handbook,* 2nd ed., C. M. Harris and C. E. Crede (eds.), McGraw-Hill Book Company, New York, 1976, p. 18–11.
6. *Measurement-Computation-Systems,* Hewlett-Packard Catalog, Hewlett-Packard, Palo Alto, Calif., 1986, p. 297.

7. Hoog, B. C., FET Implementation, *Hewlett-Packard J.* (Dec. 1984).

8. Perry, C. D. (ed.), *An Intelligent, Self-Adaptive Robot by IBM, Epsilonics III-3*, Measurements Group, Inc., Rayleigh, N.C., Dec. 1983.

    DeSilva, C. W., Motion Sensors in Industrial Robots, *Mech. Eng.*, Vol. 6 (1985), pp. 40–51.

9. Judd, J. E., The Optical Wedge, *Instrum. Control Syst.*, Vol. 34 (Feb. 1961), p. 237.

10. Fagebaum, J., The Laser—From Scientific Novelty to Practical Industrial Tool, *Mech. Eng.*, Vol. 105 (Sept. 1983), pp. 30–37.

11. Smith, D. E., Electronic Distance Measurement for Industrial and Scientific Applications, *Hewlett-Packard J.* (June 1980).

12. ME Staff Report, The Exciting Promise of Fiber-Optic Sensors, *Mech. Eng.*, Vol. 106 (May 1984), pp. 60–65.

    Krigman, A., Fiber Optics in Instrumentation: Is There Light at the End of the Tunnel? *InTech*, Vol. 32, No. 8 (Aug. 1985), pp. 35–38.

    Caro, R. H., Fiber Optics: Is It Cost Effective for Distributed Control?, *InTech*, Vol. 32, No. 8 (Aug. 1985), pp. 41–43.

    Kingsley, S. A., Fiber-Optic Sensors: Opportunities for Distributed Measurement, *InTech*, Vol. 32, No. 8 (Aug. 1985), pp. 44–48.

13. Van Veen, F., *Handbook of Stroboscopy*, General Radio Co., West Concord, Mass., 1966, p. 73.

14. Christopher, R. M. (ed), *ISA Directory of Instrumentation, 1985–1986*, Instrument Society of America, Research Triangle Park, N.C.

15. Bouche, R. R., *Accelerometers for Shock and Vibration Measurements*, Tech. Pap. 234, Endevco, Calif.

16. Baxter, R. D., J. J. Beckman, and H. A. Brown, Measurement Techniques, in *Shock and Vibration Handbook*, 2nd ed., C. M. Harris and C. E. Crede (eds.), McGraw-Hill Book Company, New York, 1976, p. 20–1.

17. Vezzeti, C. F., and P. S. Lederer, *An Experimental Technique for the Evaluation of Thermal Transient Effects on Piezoelectric Accelerometers*, NBS Technical Note 855, U.S. Department of Commerce, Washington, D.C., Jan. 1975.

18. Kaufman, A. B., Acoustic Sensitivity of Accelerometers, *Instrum. Control Syst.*, Vol. 32 (May 1959), pp. 720–721.

19. *Noise Suppression through Accelerometer Isolation*, Tech. Data A502, Endevco, Calif., Feb. 1973.

    Morrison, R., *Grounding and Shielding Techniques in Instruments*, John Wiley & Sons, Inc., New York, 1972.

20. Rasanen, G. K., and B. M. Wigle, Accelerometer Mounting and Data Integrity, *Sound Vib.*, Vol. 1 (Nov. 1967), pp. 8–15.

    Baxter, et al., Ref. 16, p. 20–6.

21. Ramboz, Ref. 5, Chap. 18.

22. Pennington, D., In-place Calibration of Piezoelectric Crystal Accelerometer Amplifier Systems, *Sixth ISA National Flight Test Instrumentation Symposium*, San Diego, Calif., 1960.

23. *Improving the Accuracy of Structural Response Measurements,* Application Note 2402, Hewlett-Packard, Palo Alto, Calif.

24. *Dynamic Calibrator, Model 612 and 613,* Document GALEC5, QUIZOTE Measurement Dynamics, Inc., Cincinnati, Ohio, 1983.

25. *Hammer Calibrator, Model 620,* Document GElTC9A, QUIZOTE Measurement Dynamics, Inc., Cincinnati, Ohio, Apr. 1983.

26. Cook, N. H., and E. Rabinowicz, *Physical Measurement and Analysis,* Addison-Wesley Publishing Company, Inc., Reading, Mass., 1963, p. 164.

27. King, J. D. of Southwest Research Institute, *Vibrating Diaphragm Pressure Transducer, Technology Utilization Report,* NASA SP-5020, National Aeronautics and Space Administration, Washington, D.C., Aug. 1966.

28. Chass, J., *The Variable MU Pressure Transducer,* Control Components Div. of International Resistance Co., Feb. 1963.

29. Whittier, R. M., Basic Advantages of the Anisotropic Etched, Transverse Gage Pressure Transducer, *Endevco Prod. Dev. News,* Vol. 16, No. 3 (1980).

30. Wolber, W. G., New Sensors for Automobile Engine Control, *Instrum. Technol.,* Vol. 25 (Aug. 1978), pp. 47–53.

31. Butler, R. I., and M. McWhirter, Current Practices in Shock and Vibration Sensors, *Instrum. Technol.,* Vol. 14 (Mar. 1967), pp. 41–48.

32. Kaminski, R. K., Measuring High Pressure Above 20,000 psig, *Instrum. Technol.,* Vol. 15 (Aug. 1968), pp. 59–62.

33. Newhall, D. H., Manganin High-Pressure Sensors, *Instrum. Control Syst.,* Vol. 35 (Nov. 1962), pp. 103–104.

34. Newhall, D. H., and L. H. Abbot, High Pressure Measurement, *Instrum. Control Syst.,* Vol. 34 (Feb. 1961), pp. 232–233.
Spain, I. L., and J. Paauwe, *High Pressure Technology,* Vol. 1, Marcel Dekker, Inc., New York, 1977.

35. Lawford, V. N., and I. Barton, Differential-Pressure Instruments, *Instrum. Technol.,* Vol. 21 (Dec. 1974), pp. 30–40.

36. *Pressure Transducer Handbook,* National Semiconductor Corp., Sunnyvale, Calif., 1977, Sec. 10.

37. Rogero, S., Measurement of High *Frequency Pressure Phenomena Associated with Rocket Motors,* Tech. Report 32-788, Jet Propulsion Laboratory, Aug. 1965.

38. Lally, R. W., Better Response from Pressure Transducers, *Instrum. Control Syst.,* Vol. 39 (Sept. 1966), pp. 103–105.

39. Reid, R. J., and E. M. Kops, Dynamic Response of Remote Pressure Pickups, *Instrum. Control Syst.,* Vol. 32 (Aug. 1956), pp. 1202–1204.

40. Hougen, J. O., O.R. Martin, and R. A. Walsh, Dynamics of Pneumatic Transmission Lines, *Control Eng.,* (Sept. 1963), pp. 114–117.

41. Shinskey, G., Pneumatic Transmission—How Far Can They Go?, *Instrum. Control Syst.,* Vol. 47 (June 1974), p. 83.
Franke, M. E., A. J. Malanowski, and P. S. Martin, Effects of Temperature, End-Conditions, Flow, and Branching on the Frequency Response of Pneumatic Lines, *J. Dyn. Syst., Meas. Control* (Mar. 1972), pp. 15–19.

Hougen et al., Ref. 40.

42. Thomson, T. B., *The Effect of Tubing on Dynamic Pressure Recording,* Tech. Report No. 61-3, Rocketdyne, Div. of North American Aviation, Inc., CAnoga Park, Calif., Feb. 1961.

Nyland, T. W., D. R. Englund, and R. C. Anderson, Frequency Response of Short Pressure Probes, *Instrum. Control Syst.,* Vol. 46 (Aug. 1973), pp. 27–29.

43. Schweppe, J. L., L. C. Eichberger, D. F. Muster, E. L. Michaels, and G. F. Paskusz, *Methods for the Dynamic Calibration of Pressure Transducers,* NBS Monograph 67, Natural Bureau of Standards, U.S. Department of Commerce, Washington, D.C., 1963.

Lederer, P. S., Performance-Testing Pressure Transducers, *Instrum. Control Syst.,* Vol. 40 (Sept. 1967), pp. 93–99.

44. Schweppe, et al., Ref. 43, p. 6.

45. *Notes on the Dynamic Behavior of Integrated Sensor (IS) Pressure Transducers,* Application Note KPS-AN 13, Kulite Semiconductor Products, Inc., Ridgefield, N.J.

8

Fluid-Flow Measurements

8-1. INTRODUCTION

Fluid flow touches every aspect of engineering, from industrial processing to research. Flow measurements and control are described in books and symposiums [1]. Surveys of the subject are published periodically in journals [2]. General-purpose flowmeters are often updated for high-tech applications [3].

Fluid flow measurement is a diverse field of study, which includes liquid and gas, compressible and incompressible, subsonic and supersonic, laminar and turbulent flow, steady state and pulsating, average flow and flow field, volumetric and mass flow, two-phase medium and slurry, closed and open channel. Flow measurements are influenced by a host of parameters, such as viscosity, temperature, pressure, and specific heat. Moreover, there are important variations within a class of flowmeters. In view of the diversity, this brief study cannot be comprehensive.

In this chapter we present a perspective of flowmeters by classifying flow measurements as "direct" and "indirect." The study is mostly for flow in closed channels. It is important to note that the direct method to be described is not the direct comparison or null-balance method presented in Chapter 3.

Flow is a transport phenomenon. This allows a "direct" measurement. Flow can also be measured by its effect. This is loosely called an "indirect" measurement. In other words, any variable in a flow-related equation can be utilized for a flow measurement. This approach is used in Secs. 7-6 and 7-8 for the discussion of force and pressure transducers. It is hoped that a perspective can be gained by illustrating the principles and physical laws employed, pointing out variations of the same principle, showing sufficient examples, and discussing some applications. For convenience, conversions of commonly used units to SI for flow measurements are shown in Table 8-1.

TABLE 8-1 Conversions to SI Units

To convert from:	To:[a]	Multiply by:
barrel (for petroleum)	m^3	1.589 873 E−1
centipoise	Pa·s	1.000 000 E−3
centistoke	m^2/s	1.000 000 E−6
foot	m	3.048 000 E−1
foot-pound-force	J	1.355 818 E+0
gallon (U.S. liquid)	m^3	3.785 412 E−3
pound-force	N	4.448 222 E+0
pound-force/in^2	Pa	6.894 757 E+3
pound-mass	kg	4.535 924 E−1

[a]m, meter; N, newton, Pa, pascal, J, joule, kg, kilogram; 1 N = 1 kg·m/s^2, 1 Pa = 1 N/m^2 = 1 kg/(m·s^2).

8-2. LAMINAR AND TURBULENT FLOW

Reynolds number, viscosity, laminar flow, and turbulent flow of incompressible fluids are examined briefly in this section. Turbulent flow occurs in most flowmeters.

Viscous force of a fluid is an internal frictional force. Let plate A shown in Fig. 8-1a move in a fluid with a velocity V parallel to a stationary plate B. Thin layers of fluid adhere to the plates. The flow profile is as shown in the figure. Absolute (dynamic) viscosity of the fluid is defined as

$$\text{absolute viscosity, } \mu = \frac{\text{shear stress}}{\text{rate of shear strain}} \tag{8-1}$$

$$\mu = \frac{\text{force/area}}{(\text{velocity } V)/(\text{distance } y)} \tag{8-2}$$

An *absolute* (or *dynamic*) *viscosity* of 1 poise is 0.1 Pa·s or 0.1 kg/(m·s). The centipoise, or 0.01 poise, is a common unit. Its conversion to SI units is shown in Table 8-1. *Kinematic viscosity* is absolute viscosity divided by density. The common unit is the centistoke, expressed in units of m^2/s. The stress/strain ratio in Eq. 8-1 resembles a shear modulus. A fluid in static equilibrium cannot sustain a shear stress as in a solid. Viscosity is due to the rate of shear strain between laminas of a fluid. The velocity gradient dV/dy is assumed constant. Viscosity is nonlinear for some fluids, such as greases.

The flow characteristics of fluids are described by the *Reynolds number* N_R, which is

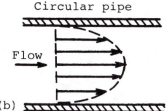

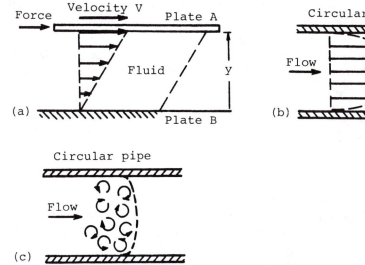

FIGURE 8-1 Laminar and turbulent flow in fluids. (a) Flow between parallel plates. (b) Laminar flow: axis-symmetric. (c) Turbulent flow.

$$\text{Reynolds number, } N_{\text{R}} = \frac{\text{inertia force}}{\text{viscous force}} \qquad (8\text{-}3)$$

It can be shown from dimensional analysis that inertia force is proportional to $D^2 \rho V^2$ and viscous force to $\mu V D$, where ρ is the fluid density in kg/m^3, V its velocity in m/s, D a characteristic dimension in meters (such as a pipe diameter), and μ the absolute viscosity in Pa · s. Thus N_{R} in SI units reduces to

$$N_{\text{R}} = \frac{\rho V D}{\mu} \qquad (8\text{-}4)$$

In English units, ρ is in lb$_{\text{m}}$/ft^3, V in ft/sec, D in feet, and μ in lb$_{\text{f}}$ · sec/ft^2. Since N_{R} is nondimensional, it must be written as $N_{\text{R}} = \rho V D / \mu g_c$, where $g_c = 32.17$ lb$_{\text{m}}$ · ft/lb$_{\text{f}}$ · sec^2 is a dimensional constant. Note that g_c in SI units is 1 kg · m/N · s^2 and it is not necessary in Eq. 8-4.

Laminar flow in a pipe, shown in Fig. 8-1b, occurs at low N_{R} when the viscous force dominates (see Eq. 8-3). Conditions for laminar flow are low density, low velocity, small diameter, and high viscosity, as shown in Eq. 8-4. The opposite is true for turbulent flow, as shown in Fig. 8-1c. This occurs at high N_{R} when the inertia force dominates. From experimental studies, the N_R for a transition from laminar to turbulent flow is from 2000 to 3000. The flow for this range is unstable.

Since N_R is based on dimensional analysis, relations obtained for one fluid are applicable for another fluid. For example, the drag coefficient of a circular disk versus Reynolds number is universally applicable for all fluids. This is the basis for target flowmeters described in a later section.

Flowmeters for viscous (laminar) flow at low N_R are governed by the Hagen–Poiseuille law.

$$Q = \frac{\pi r^4}{8\mu L}\Delta P \tag{8-5}$$

where Q is the volumetric flow rate, r the tube radius, μ the absolute viscosity, L the tube length, and ΔP the pressure drop. Flowmeters for turbulent flow at high N_R are governed by the square-root law. It can be shown from Bernoulli equation (Eq. 7-48) that the volumetric flow rate Q is proportional to the square root of the differential pressure ΔP across the flowmeter:

$$Q = C\sqrt{\Delta P} \tag{8-6}$$

where C is a constant. The equation is derived in Sec. 8-5. Many types of flowmeters are governed by the square-root law and characteristics of square-root-law flowmeters are examined in Sec. 8-5B.

8-3. DIRECT FLOW MEASUREMENTS

Flow is a transport phenomenon. Hence it can be measured directly. These methods can be classified as (1) weighing or volumetric, (2) positive displacement, (3) flow visualization, and (4) tracer technique. A tracer is a carrier in the sense that it moves with the flow and carries the flow information.

A. Weighing and Volumetric Methods

Flow measurement by a weighing or volumetric method removes the fluid from a system and integrates the flow rate over a specific period. The procedure is basic and conceptually simple, but it is not always convenient for field applications.

With proper precautions and procedures, this is the most accurate method. It relates a flow rate to the fundamental units of mass, length, and time. The uncertainty in measurement is of the order of 1 part in 10^3 [4]. It is often used for the steady-state calibration of other flowmeters, as shown schematically in Fig. 8-2.

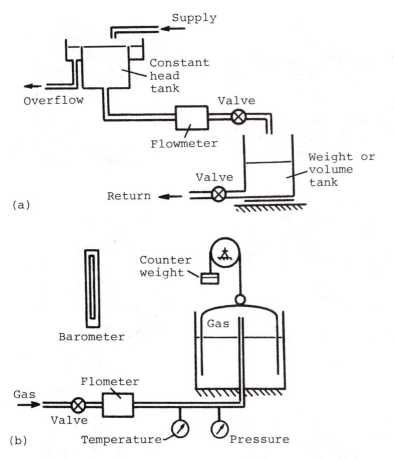

FIGURE 8-2 Calibration of flowmeters. (a) Calibration of liquid flowmeter. (b) Calibration of gas flowmeter. (From Ref. 5.)

B. Positive-Displacement Meters

A positive-displacement meter measures volumetric flow rate by dividing the flow into segments. If each segment has a precise volume, the number of segments counted is the total volume.

These flowmeters can be classified as reciprocating or rotary. Various constructions are used for both types. The reciprocating piston flowmeter in Fig. 8-3a is for liquids. The bellow type in Fig. 8-3b is for gases. As shown in the figure, chamber 1 is emptying to the outlet, chamber 2 is filling from the inlet, chamber 3 is empty, and chamber 4 is just filled. The gas flow is less pulsating by employing four chambers.

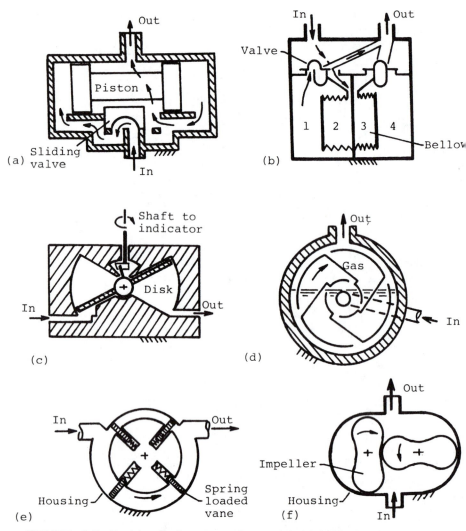

FIGURE 8-3 Positive-displacement flowmeters. (a) Reciprocating-piston meter. (b) Bellow-type meter (from Ref. 6). (c) Nutating-disk meter. (d) Sealed-drum meter. (e) Rotating-vane meter. (f) Lobed-impeller meter.

Four rotary flowmeters are shown in Fig. 8-3c to f. The nutating-disk meter is similar to a household water meter. The wet-type sealed-drum gas meter is superseded by the dry-type, because the liquid seal is subjected to freezing and evaporation. The rotary-vane meter is generally for liquids. The lobed-impeller meter is for liquid or gas. Many variations of the lobes are in use, including helical lobes, oval-shaped gears, and meters with more than two rotating parts.

The positive-displacement flowmeters above are essentially fluid motors. The pressure drop across the meter, typically 10 to 40 psi, is due to restrictions in the meter and friction of the moving parts. The accuracy is of the order of 1%, depending on the viscosity and the type of fluid. The performance of the meter will deteriorate with time because of wear and increasing leakages.

Positive-displacement meters are favored for start/stop short runs, medium- and high-viscosity fluids, and certain contaminated products. Their applications include tank truck metering, blending operations, refinery processing, and processes in crude oil production [7].

C. Flow Visualization

Flow visualization [8] utilizes the flow stream itself for a qualitative and/ or quantitative measurement of a flow field. Generally, the flow-related change in a property of fluid in the flow stream is examined, and the flow can be observed with minimum disturbance. A tracer is sometimes introduced in the fluid, assuming that the tracer follows the flow pattern and the flow pattern is least disturbed. The method is useful for research but rather inconvenient for most industrial applications.

Flow visualization for flow at low velocities [9] is well established. Smoke is commonly used as a tracer in airstreams. The smoke is generated by permitting air saturated with hydrochloric acid to come into contact with fumes of ammonia. Color dyes can be introduced in a fluid stream. Quantitative data for a three-dimensional flow can be obtained by using small particles of polystyrene suspended in the air and a stroboscopic light [10]. Oil can be injected in water with an atomizer. The oil droplets in suspension can be illuminated sharply by means of a thin sheet of light. Small pieces of aluminum and mica are suitable aids for flow visualization.

Schlieren, shadowgraph, and interferometry can be used for supersonic flows [11]. Schlieren and shadowgraph methods make use of the reflection of light due to variations in density in a compressible flow, similar to the passage of light through a glass prism. Interferometry makes use of changes in the speed of light passing through regions of varying density in a compressible flow. Quantitative data can be obtained from fringes (see Fig. 7-3) due to the interference between the affected and unaffected light beams.

D. Carrier Systems

The term *carrier systems* is coined to describe a large class of flowmeters suitable for industrial applications. It is a tracer technique. A carrier moves with the flow and carries the flow information. Similar to flow visualization, the method is direct because the fluid itself is used to convey the flow information. Typical examples are described in this section.

The term *carrier* or *tracer* is used in the broad sense. It may be injected externally into a flow stream, such as radioisotopes, or be inherent in the fluid, such as air bubbles. The method can be used to obtain an average steady-state flow velocity or to probe the local velocity in a fluid. For an average velocity measurement, the tracer may not reveal the true profile of the flow, and the accuracy of measurement is influenced by this uncertainty.

1. Electrolytic Tracers [12]

A transit time method is shown in Fig. 8-4a. A shot of salt solution is injected into the water in a pipe. The subsequent cloud of electrolyte in the water has higher conductivity than the water. Electric currents I_1 and I_2 are detected as the cloud of electrolyte passes the electrodes at A and B, separated by a distance L. The time difference or the transit time between I_1 and I_2 is Δt. The average velocity is $V = L/\Delta t$. The volumetric flow rate is the product of V and the cross-sectional area of the pipe.

The method is used for intermittent measurement of constant flow rates in large pipes. The accuracy is from 1 to 6%. The actual flow rate tends to be larger than that indicated, but the accuracy improves with higher Reynolds number because of greater turbulence. Error due to uncertainties in Δt can be minimized by using the Δt between the center of gravity of the current–time areas, as shown in the figure.

Alternatively, the volumetric flow rate can be determined by the dilution method, shown in Fig. 8-4b. A salt solution of concentration C_1 is introduced into the water at a steady and known flow rate q_1. The flow rate of the water q_2 is unknown. The concentration after mixing is C_2. Thus $q_2 = q_1(C_1/C_2)$. The accuracy of the method is from 0.03 to 0.06% [13].

2. Radioisotope Tracers

Radioisotope tracers can be substituted for the salt solution for both the transit time and the dilution method described above. Moreover, in using the dilution method, a sample of the fluid, either liquid or gas, can be

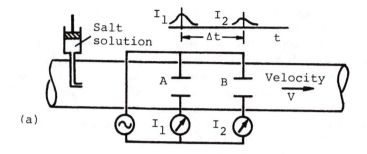

(a)

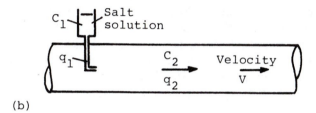

(b)

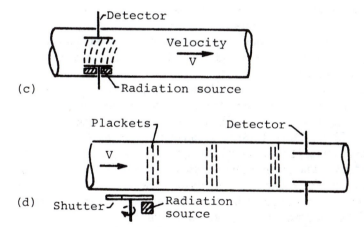

(c)

(d)

FIGURE 8-4 Tracer methods for flow measurement. (a) Transit-time method. (b) Dilution method. (c) Deflection method. (d) Signal modulation.

withdrawn down stream from the point of injection. A more precise reading can be obtained by counting the radiation off-line. The number of counts is inversely proportional to the average velocity of the flow. Krypton 85 and argon 41 have been used to measure gas flow rates [14].

The *deflection method* probes the flow in a gas stream at only one location, as shown in Fig. 8-4c. The radiation is from an alpha or a beta soruce. For zero flow, almost all the ions are collected at the detector from the continuous ionization. As the gas velocity increases, some ions are swept out by the flow before they are collected. The ions collected can be correlated with the gas velocity.

A *signal modulation method,* shown in Fig. 8-4d, consists of a radiation source, a shutter, and a detector. The shutter is a rotating slotted disk. It modulates the constant radiation source to produce a sequence of ion plackets in the flow stream. The technique eliminates the dependence on gas composition, temperature, pressure, viscosity, humidity, and the inherent instability of ion-current-detecting electronics [15].

3. Sonic Flowmeters [16]

Sonic or ultrasound flowmeters are applicable for pipe lines and open channels. The accuracy depends on the installation, operating conditions, and measuring technique. Accuracy of 0.02% is attainable [17]. A sales catalog [18] can be used as a guide, but the ultimate test of accuracy is by means of a field calibration [19].

The sonic method is generally considered a transit-time measurement. The flowmeter consists of a transmitter and a receiver; piezoelectric crystals are commonly used for both. The frequency of sonic signals is from 600 kHz to 10 MHz. Three basic techniques or their combinations are used: (1) transit time, (2) beam deflection, and (3) Doppler effect. Doppler effect describes the apparent frequency shift when a sonic wave is superposed on a flowing medium. An example is the increase in pitch of a train whistle as the train approaches an observer, followed by a decrease in pitch as the train passes and moves away from the observer.

a. Transit-time method.   Several methods using the transit-time technique and Doppler effect are described in this section. A sound burst or a train of sonic pulses is injected into a flowing fluid by means of a transmitter T and detected by a receiver R, as shown in Fig. 8-5a. When the sonic wave is injected downstream, the actual velocity of the wave is $(c + V)$, where c (= 5000 ft/sec for water) is the sonic velocity in the fluid at rest and V the flow velocity. For a distance L between T and R, the transit time is $t_1 = L/(c + V)$. Similarly, the transit time is $t_2 = L/(c - V)$ when the sonic wave is injected upstream. Assuming that $c \gg V$, the time difference Δt is

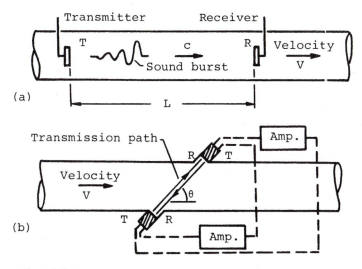

FIGURE 8-5 Ultrasound transit-time methods. (a) Basic transit-time method. (b) Nonintrusive transit-time method.

$$\Delta t = t_2 - t_1 = \frac{L}{c - V} - \frac{L}{c + V} = \frac{2LV}{c^2 - V^2}$$

$$\simeq \frac{2LV}{c^2} \tag{8-7}$$

An application of this method using four transducers is shown in Fig. 8-5b. If a transducer is used for both transmitting and receiving by time sharing, only two instead of the four transducers are required. The transducers are nonintrusive, and they can be clamped on the outside wall of a pipe. Note that this is a variation of the two-point transit-time method (see Fig. 8-4a) with the sonic wave superposed on the flow.

This technique is used successfully for large-diameter pipes, but the Δt is quite small and the uncertainty in its measurement can be large. For example, if $c = 5 \times 10^3$ ft/sec, $L = 6$ in., and $V = 1$ ft/sec, the Δt from Eq. 8-7 is 40×10^{-9} sec. When the leading edge of the pulse is used to indicate the pulse arrival time, error due to the overall pulse length or envelope shape has little effect on the measurement. The sonic velocity c, however, is not constant and is a function of fluid composition, temperature, and density.

The *phase difference* $\Delta \phi$ between the upstream and downstream signals can be used to measure the flow velocity V instead of Δt. When a sinusoidal signal of frequency f is used, the $\Delta \phi$ from Eq. 8-7 is

$$\Delta\phi = 2\pi f \,\Delta t = \frac{4\pi f L V}{c^2 - V^2} \tag{8-8}$$

This technique is subject to the same limitation as the transit time method, because the associated electronics use Δt for determining the phase angle.

A *frequency difference method* using two resonant circuits is widely used. A resonant circuit is created by using the pulses received to trigger the transmitting pulses in a feedback loop. The feedback paths for the two resonant circuits are shown as dashed lines in Fig. 8-5b. The resonant frequencies f_1 and f_2 and the frequency difference Δf are

$$f_1 = \frac{1}{t_1} = \frac{c + V}{L} \quad \text{and} \quad f_2 = \frac{1}{t_2} = \frac{c - V}{L} \tag{8-9}$$

$$\Delta f = \frac{1}{t_1} - \frac{1}{t_2} = \frac{2V}{L} \tag{8-10}$$

The method is independent of the sonic velocity c, but Δf tends to be small.

For the methods described above, if D is the internal diameter of the pipe and θ the angle as shown in Fig. 8-5b, then Δt, $\Delta\phi$, and Δf in Eqs. 8-7 to 8-10 become

$$\Delta t = \frac{2DV/\tan\theta}{c^2 - (V\cos\theta)^2} \tag{8-11}$$

$$\Delta\phi = \frac{2\pi f/\tan\theta}{c^2 - (V\cos\theta)^2} \tag{8-12}$$

$$\Delta f = \frac{V\sin 2\theta}{D} \tag{8-13}$$

b. **Deflection method.** In principle, the deflection method using ultrasound is similar to that using radiation shown in Fig. 8-4c. The transmitter and receiver are perpendicular to the pipe axis and separated by a distance L. The fluid flow with velocity V causes the sonic emission E to deflect by an amount X.

$$X = \frac{V}{c} L E \tag{8-14}$$

where c is the sonic velocity in the fluid at rest. Again, c is not constant. An additional uncertainty in the measurement is the strength of the emission E.

c. **Doppler shift method [20].** A clamp-on Doppler flowmeter with a transmitter T and a receiver R as a single unit is shown in Fig. 8-6,

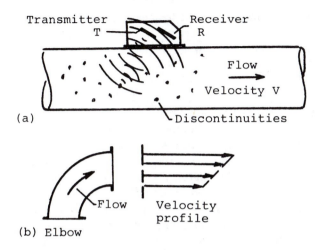

(a)

(b) Elbow

FIGURE 8-6 Effect of velocity profile on Doppler meter. (a) Doppler flowmeter. (b) Flow profile due to elbow.

although T and R may be separate units placed around the circumference of a pipe. T injects a continuous ultrasound beam at a frequency f_T (about 600 kHz) into the liquid. The beam is scattered by solid particles, bubbles, or any discontinuity in the liquid and is reflected back to R at a frequency f_R. The Doppler shift due to the velocity V of the liquid is the difference $\Delta f = f_T - f_R$.

$$\Delta f = f_T - f_R = \frac{2f_T V \cos \theta}{c} \tag{8-15}$$

where c is the sonic velocity in the liquid at rest, θ the angle as shown in Fig. 8-5b, and $c \gg V$.

The Doppler effect measures the point value at a discontinuity. Since discontinuities are dispersed in a liquid, the receiver detects multiple frequencies. Some averaging scheme is used to obtain an average velocity for the volumetric flow rate. Hence the flowmeter is sensitive to the velocity profile across a flow section, as illustrated in Fig. 8-6b. In other words, changing the location of a flowmeter could invalidate its calibration.

In addition to the transducer location, other uncertainties to be considered are concentration of sonic discontinuities, size of suspended particles, depth of penetration of the ultrasonic beam, and air pockets in a pipe [21]. Note that a Doppler flowmeter depends on discontinuities in a liquid for its operation, while a transit-time meter must work with clean liquids to minimize signal attenuation and dispersion.

Doppler flowmeters are commonly used to measure industrial waste and mineral slurries. They are used successfully in such industries as papermaking, pharmaceuticals, and sewage disposal.

4. Laser Velocimeter [22]

A laser velocimeter utilizes a frequency shift due to the scattering of a laser beam to measure local fluid velocities. It is suitable for measurements in liquids and gases. Both the Doppler flowmeter and the laser velocimeter employ the same principle; one is sonic and the other optical. Their operations can be compared by the similarity in properties of sound and light waves [23]. The laser velocimeter, however, is more complex.

Many optical configurations have been proposed [24], and three commonly used ones are described briefly below [25]. The one shown schematically in Fig. 8-7a is based on the Doppler frequency shift from light scattered by particles moving with the flow. The frequency shift is small, but the light can be mixed (heterodyned) with the unshifted light from the laser. The mixing produces a beat frequency, equal to the frequency difference Δf of the Doppler shift. The flow velocity is proportional to Δf, as shown in the figure.

The dual-scatter system shown in Fig. 8-7b is called the differential mode or the fringe method. Due to scattering, a set of fringes (see also Fig. 7-3d) is formed from the optical interference of the two beams intersecting at the probed volume. A sensor detects the passing fringes. The resulting signal is processed to give the flow velocity versus time at the output. The fringes can be observed over a wide viewing angle, since the differential Doppler frequency is independent of the direction of detection.

The two-focus system shown in Fig. 8-7c uses two beams from an optical beam splitter. The beams are brought into focus in the focal plane of a lens. The foci are separated by a short distance d_0. The transit time of particles across d_0 gives the flow velocity. The output shown is produced by a single particle.

Laser velocimetery has many advantages: (1) as the sensors are nonintrusive, this will be one of the more popular techniques for measuring flow fields and turbulence in turbomachinery; (2) the probed volume of the observation is very small, of the order of a cube 0.2 mm on a side; and (3) extremely high-frequency response is possible. The disadvantages are high cost and the need for tracer particles in the flow stream. The natural or seeded particles must be small enough to follow the local flow acceleration accurately. Ease of optical alignment, space limitation, and signal/noise ratio are the main considerations in applications.

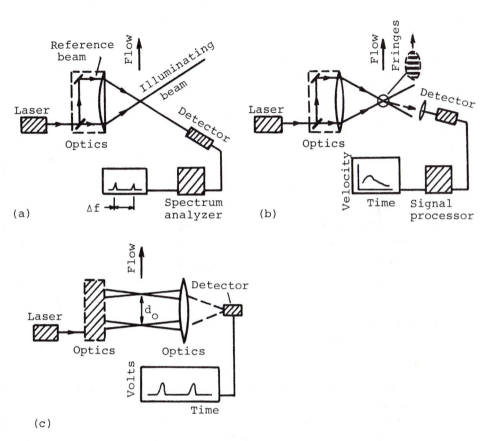

FIGURE 8-7 Laser-velocimetry systems. (a) Doppler anemometer, output shows Δf. (b) Dual-scatter system, output in velocity versus time. (c) Two-focus system, output from a single particle.

5. Magnetic Flowmeters [26]

The basis of magnetic flowmeters is Faraday's law of electromagnetic inductuion, which is the generator effect described in Chap. 5 (see Figs. 5-17 and 5-20). When a conductor moves with a transverse velocity $\dot{x}$ across a constant uniform magnetic field, as shown in Fig. 8-8a, the induced voltage v is proportional to $\dot{x}$ of the conductor.

The schematic of a magnetic flowmeter shown in Fig. 8-8b consists of a moving conducting liquid in a magnetic field. The liquid at the cross section forms the conductor. The pipe section at the flowmeter must be non-

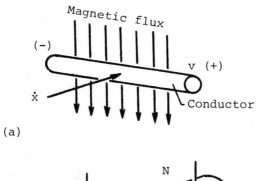

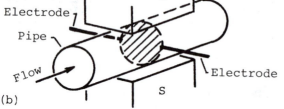

FIGURE 8-8   Principle of magnetic flowmeter. (a) Electromagnetic induction.
(b) Schematic of magnetic flowmeter.

magnetic to allow the magnetic flux to penetrate the pipe. The magnetic
flux from north to south is generated by two coils laid on the pipe wall.
Pulsed dc or ac can be used to power the coils. Electrodes are placed at
points of maximum potential difference for a voltage output. The conducting path in the meter is a function of the conductivity of the liquid
and the diameter of the electrodes.

The liquid must be conductive. The threshold is 10 μS/cm, where the
unit siemens (S) is 1/ohm. For tap water, the conductivity is 200 μS/cm.
Some liquids, such as gasoline, have low conductivity and are not appropriate for magnetic flowmeters. The conductivity of aqueous solutions is
given in the literature [27]. Note that electrical conductivity may be a
function of temperature, but it does not always increase with the solution
concentration for some liquids.

Magnetic flowmeters are suitable for numerous applications. They are
nonintrusive and have fast response. Forward and reverse flow can be
measured with equal accuracy. The flow rate may range from 10^{-3} mm/s
to high values. The lining for the pipe can be selected to accommodate the
type of liquid. The meters are insensitive to viscosity, density, and type of
flow conditions (turbulent or laminar) as long as the velocity profile is

symmetrical. The operating characteristics of magnetic flowmeters can be obtained from manufacturers [28].

6. Thermal Flowmeters

A thermal flowmeter uses temperature as its "tracer." Since heat transfer to a flow stream is proportional to its mass flow rate, thermal flowmeters are basically mass flowmeters. Many methods of heat injection and signal extraction are used. For this discussion, these flowmeters are classified as being of heated tube or immersion type.

Thermal flowmeters are particularly useful for precision measurement of gases at low flow rates [29] from vacuum to high pressure when other types of meters are not suitable. Examples of applications are the measurement of low-velocity air in buildings and the accurate control of tungsten hexafluoride in the vacuum processing of semiconductors [30]. Within their design ranges, these flowmeters are relatively insensitive to changes in density, pressure, fluid viscosity, and temperature. As a thermal device, however, the response time of the meters is generally not high, but can be as short as 20 ms [31].

a. **Heated-tube type.** The mass flowmeter shown in Fig. 8-9a consists of a heated capillary tube and two heat sinks. The heat sinks keep the ends of the tube at the same temperature. The temperature profile is symmetrical (solid line) for zero flow, as shown in the figure. The flow causes an asymmetrical temperature profile (dashed line), and $(T_2 - T_1)$ is a function of the mass flow rate. Note that this is another application of the deflection method (see Fig. 8-4c).

The laminar flowmeter shown in Fig. 8-9b consists of a sensor tube A and a reference tube B. A and B are exposed to an identical environment, but no flow occurs in B. Resistance thermal detectors RTDs can be used for both heating and temperature sensing. The flow is a function of the resistances of the RTDs, and the output is measured by means of a Wheatstone bridge. A range of 100:1 can be accomplished with this flowmeter. For low flow rates, only tube A is used. For high flow rates, tube A and the bypass tubes are used. Since the flow is laminar, the flow ratio between the sensor tube A and the bypass tubes is constant. Several flowmeters are based on variation of the same principle.

The boundary layer flowmeter shown in Fig. 8-9c is suitable for laminar and turbulent flow. It consists of a thin-walled pipe with temperature sensors, which are thermally insulated from each other. Assume that the thermal conductance of the pipe material is high and the heat loss to the surroundings is negligible. Heat is injected into the boundary layer of the fluid adjacent to the inner pipe wall, and the temperature profiles of the fluid at the sections are as shown in the figure. Heat transfer to the fluid

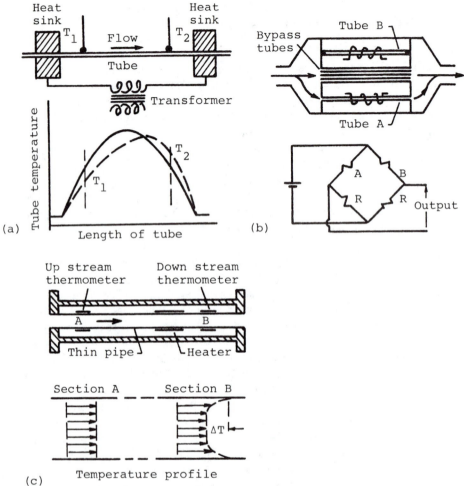

FIGURE 8-9 Heated-tube mass flowmeters. (a) Deflection method (Model ST/ STH Series Flowmeters, courtesy of Teledyne Hastings-Raydist, Hampton, Va.). (b) Laminar flowmeter (TMF Model 124, courtesy of Rosemont Engineering Co., Minneapolis, Minn.). (c) Boundary layer meter (from Ref. 32).

depends on the surface film coefficient h of the boundary layer, and h is a function of mass flow rate. For a given fluid composition and design configuration, we have

$$Q = C_1 \dot{m}^{0.8} \Delta T \qquad \text{for turbulent flow} \qquad (8\text{-}16)$$

$$Q = C_2 \dot{m}^{1/3} \Delta T \qquad \text{for laminar flow} \qquad (8\text{-}17)$$

where Q is the injected heat, C_1 and C_2 are constants, m is the mass flow rate, and ΔT is the temperature difference.

b. Immersion type. The immersion-type mass flowmeter shown in Fig. 8-10a consists of two resistance thermal detectors RTDs in a flow tube. The heated RTD is the mass flow sensor. The unheated one is for temperature compensation. The resistance of the sensors form opposite arms of a Wheatstone bridge. The two RTDs may be packaged in a single probe (e.g., Series 100, Datametrics, Watertown, Massachusetts; FMA 300 Series, Omega Engineering, Stamford, Connecticut).

The flowmeter in Fig. 8-10b consists of two noble-metal thermocouples

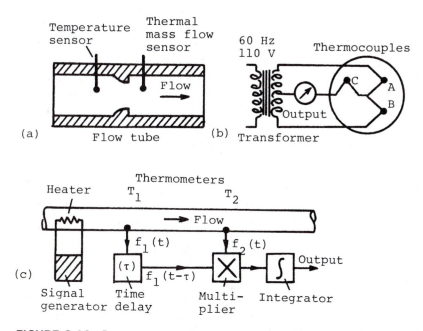

FIGURE 8-10 Immersion-type heat-transfer mass flowmeters. (a) Immersion type (Accu-Mass Series 370, courtesy of Sierra Instruments, Inc., Camel Valley, Calif.). (b) Heated thermocouples (Type K Flowtube, courtesy of Teledyne Hastings-Raydist, Hampton, VA.). (c) Cross-correlation of T_1 and T_2 (from Ref. 33).

A and B in a low-voltage ac bridge circuit. The couples are heated by an ac current. The flow causes a change in the temperature of A and B and therefore a change in the dc output from the thermocouples. An identical unheated thermocouple C is placed in the dc metering circuit for temperature compensation. Note that the flowmeters in Fig. 8-10a and b use the same principle, consisting of a temperature sensor for the flow measurement and another for temperature compensation.

A transit-time measurement using the cross-correlation technique is shown in Fig. 8-10c. The technique has general applications [34]. A train of heat pulses from a signal generator in a pseudo-random mode is injected into the field. Downstream are two fast-responding temperature sensors. The transit time detected by the sensors can be completely masked by noise. A cross-correlation function $R_{21}(\tau)$ measures the interdependency between $f_1(t)$ and $f_2(t)$ as a function of the parameter τ, which is the time shift of one function with respect to the other [35].

$$R_{21}(\tau) = \int_{-\infty}^{\infty} f_2(t)f_1(t - \tau)\, dt \tag{8-18}$$

As shown in the figure, (1) $f_1(t)$ is shifted (delayed) by the time τ to become $f_1(t - \tau)$, (2) $f_1(t - \tau)$ and $f_2(t)$ are multiplied, and (3) their product is integrated to yield an output. The process is repeated by varying the parameter τ. The correlated output peaks sharply when the delay time τ is equal to the transit time. The mathematical manipulations can be mechanized.

8-4. INDIRECT FLOW MEASUREMENTS

As stated in the introduction, methods of flow measurements are classified as direct and indirect in this chapter. The basis of direct methods is the transport phenomenon of the fluid. The fluid itself is utilized to convey the flow information as it moves from one point to another in a flow stream.

An indirect method examines the effect of flow in a given flow-measuring device, and the flow is measured by means of related physical quantities. In other words, an indirect method examines the flow and other physical quantities in a flow-related equation as prescribed by a physical law for the given problem. Any flow-related physical law can be utilized for a flow measurement.

The flowmeters described below are classified by the physical laws employed. For example, the square-root law, derived from Bernoulli's equation, is the basis for a large class of flowmeters, and the flow is measured by means of a differential pressure. This includes orifices, venturi meters, flow nozzles, flow tubes, pitot-static tubes, and variable-area flowmeters. Turbine flowmeters operate on the momentum principle. Hot-wire and hot-film anenometers utilize the rate of convective heat transfer. Only commonly used flowmeters are described. It is evident that this study cannot be exhaustive.

8-5. SQUARE-ROOT-LAW FLOWMETERS

The square-root-law flowmeter measures a fluid flow by means of a differential pressure, which is induced by placing an obstruction in the flow stream. This includes a large class of meters. They are generally designed for high Reynolds numbers.

The basis of the square-root law for incompressible flow is Bernoulli's equation. Characteristics of square-root-law flowmeters will be examined after the description of a few examples. Bernoulli's equation (Eq. 7-45) is modified to include the lost head h and to accommodate SI and English units.

$$\frac{P_1 - P_2}{\rho} g_c = \frac{V_2^2 - V_1^2}{2} + g(Z_2 - Z_1) + gh \qquad (8\text{-}19)$$

The equation shown has the dimensions $(m/s)^2$ or $(ft/sec)^2$.

P = absolute pressure	N/m^2 (or Pa)	lb_f/ft^2
ρ = density	kg/m^3	lb_m/ft^3
V = velocity	m/s	ft/sec
Z = elevation	m	ft
h = lost head	m	ft
g = acceleration due to gravity	$9.807\ m/s^2$	$32.17\ ft/sec^2$
g_c = dimensional constant	$1\ kg \cdot m/N \cdot s^2$	$32.17\ lb_m \cdot ft/lb_f \cdot sec^2$

From continuity of flow, the theoretical volumetric flow rate Q_t is

$$Q_t = A_{1f} V_{1f} = A_{2f} V_{2f} \qquad (8\text{-}20)$$

where the velocities V_{1f} and V_{2f} are measured at the fluid stream cross-sectional areas A_{1f} and A_{2f}, respectively. For an orifice, A_{2f} usually refers to the vena contracta or the minimum stream dimension. Neglecting the change in elevation and the lost head in Eq. 8-19, we get

$$Q_t = \frac{A_{2f}}{[1 - (A_{2f}/A_{1f})^2]^{1/2}} \left[\frac{2g_c(P_1 - P_2)}{\rho} \right]^{1/2} \qquad (8\text{-}21)$$

The theoretical volumetric flow rate Q_t is modified to obtain the actual flow rate Q_a.

$$Q_a = \frac{C_d A_2}{[1 - (A_2/A_1)^2]^{1/2}} \left[\frac{2g_c(P_1 - P_2)}{\rho} \right]^{1/2} \tag{8-22}$$

where A_1 is the internal cross-sectional area of a conduit, A_2 the flow area at the obstruction, and C_d the discharge coefficient. C_d = (actual flow rate)/(theoretical flow rate). Note that $1/[1 - (A_2/A_1)^2]^{1/2}$ is a velocity of approach coefficient. It accounts for the change in cross-sectional areas of the flow stream. The discharge coefficient C_d accounts for losses in the flow process and the geometry of the setup, such as the nonoptimum location of pressure taps. The value of C_d for a given setup is mainly a function of Reynolds number N_R. Hence the results from the calibration with one fluid can be used for another.

Experimental values of C_d have been extensively investigated [36]. If a flowmeter is constructed according to certain standard dimensions, including the pressure tap locations, the value of C_d is predictable to within 1% for $N_R > 10^5$. Although C_d is a function of N_R, it is relatively constant for certain design geometry of a flowmeter and $N_R > 10^4$. The values of C_d, however, are sensitive to upstream flow disturbances caused by elbows, tees, valves, and so on. The error introduced may be as much as 15%. A rule of thumb is that a minimum run of 20 to 30 pipe diameters or straightening vanes must be used to smooth out the flow disturbances ahead of a flowmeter.

A. Orifice and Venturi Flowmeters

1. Orifice Meter

The sharp-edged orifice is commonly used because of its simplicity, low cost, and standardization. A sharp-edged orifice is a flat piece of metal with a hole machined to precise dimensions. Note that the standard design of a sharp-edged orifice requires that the edge be sharp and the plate be sufficiently thin relative to the diameter. Wear from long usage or abrasion may round off the upstream edge of the orifice to cause a significant change in the value of its discharge coefficient C_d.

a. Incompressible flow. The flow pattern of a fluid in a concentric orifice is shown in Fig. 8-11a. The static pressure distribution and the permanent pressure loss (see Eq. 8-19) are as shown. Several schemes are used to locate the pressure taps, and the widely used flange taps are shown in the figure.

Three orifices are shown in Fig. 8-11b. The drain hole, when necessary, is for condensates in a gas flow, and the vent hole is for gases in a liquid flow. Eccentric and segment orifices are primary for liquids containing solids.

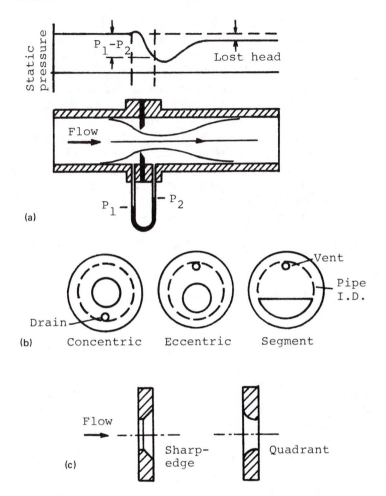

FIGURE 8-11 Orifices and flow patterns. (a) Flow pattern and pressure taps. (b) Orifices. (c) Profile of orifices.

The discharge coefficient C_d of an orifice is a function of Reynolds number N_R. The sharp-edged and quadrant orifices shown in Fig. 8-11c can be combined to give a constant C_d for N_R from 100 to 10^4 [37]. The C_d of a sharp-edged orifice drops from 0.7 to 0.6 for this range. The quadrant orifice is developed to measure liquid flow with low Reynolds number. Its C_d rises from 0.7 to 0.9 for the same range. Hence a constant C_d can be obtained by using the two orifices in series or in parallel.

b. Compressible flow. Compressible flow is a study in thermodynamics [38]. The isentropic flow equation for orifices of an ideal gas is derived in this section [39]. The idealized conditions are zero heat transfer, mechanical work, mechanical friction, change in elevation, unrestrained expansion, and fluid turbulence. The last assumption is more appropriate for a venturi-type meter, since turbulence is unavoidable in an orifice.

The idealized pressure–volume relation of a gas is

$$Pv^k = \text{constant} \tag{8-23}$$

where k is the ratio C_p/C_v of the specific heat, P the absolute pressure, and v the specific volume. For a one-dimensional compressible flow, the idealized energy balance equation is

$$\frac{dP}{\rho} g_c + V \, dV = 0 \tag{8-24}$$

where V is a velocity, $\rho = 1/v$ a density, and g_c a dimensional constant ($g_c = 1 \text{ kg} \cdot \text{m/N} \cdot \text{s}^2$ or $32.17 \text{ lb}_m \cdot \text{ft/lb}_f \cdot \text{sec}^2$). Substituting Eq. 8-23 into Eq. 8-24 and integrating, we obtain

$$\frac{V_2^2 - V_1^2}{2} = g_c \frac{k}{k-1} \left(\frac{P_1}{\rho_1} - \frac{P_2}{\rho_2} \right) \tag{8-25}$$

For continuity, the mass flow rate is $\dot{m} = \rho_1 A_1 V_1 = \rho_2 A_2 V_2$, where A_1 and A_2 are the areas of the flow sections. For circular sections, let the diameter ratio of the sections be $\beta = (A_2/A_1)^{1/2}$. Combining these with Eq. 8-25, we obtain the theoretical mass flow rate $\dot{m}_t$.

$$\dot{m}_t = \frac{A_2}{[1 - \beta^4 (P_2/P_1)^{2/k}]^{1/2}} \left\{ 2g_c \frac{k}{k-1} P_1 \rho_1 \left[\left(\frac{P_2}{P_1} \right)^{2/k} - \left(\frac{P_2}{P_1} \right)^{(k+1)/k} \right] \right\}^{1/2} \tag{8-26}$$

The equation is multiplied by an appropriate discharge coefficient C_d to obtain the actual flow. The same C_d can be used for liquid and gas flow having the same Reynolds number. Note that the simpler incompressible flow equation, (Eq. 8-22), is often used for compressible flow at low velocities. An investigation of the error due to the simplification is given as an exercise.

There is a lack of basic references for measuring the flow of compressible fluids. The properties of a fluid, such as pressure and temperature, are more variable for a compressible flow. Different types of meters for compressible flow can give different results in a field test with the meters on the same line [40].

2.   Venturi, Dall Flow Tube, and Flow Nozzle

The venturi meter shown in Fig. 8-12a has a smooth entrance and exit cone, like a convergent-divergent nozzle. It has a well-formed streamlined flow. The Dall flow tube in Fig. 8-12b is similar to a venturi. The flow nozzle in Fig. 8-12c does not have an exit cone. These meters can handle large flows at high Reynolds number N_R. They cost more than orifices, occupy more space, and are not easily removed from the line for inspection.

Due to the more complex geometry, these flowmeters are less standardized compared with sharp-edge orifices and are calibrated individually. The discharge coefficient C_d of a venturi and a flow nozzle ranges from 0.94 at $N_R = 10^4$ to 0.99 at $N_R = 10^6$. The C_d of a Dall tube is about 0.67, like that of an orifice.

Flow nozzles have less permanent lost head than do orifices (see Fig. 8-11). Dall flow tubes have the least head loss among these meters. The lost head is an energy cost in pumping. The energy cost of an orifice is about \$900 per year for an industrial setting [41]. The total cost, of course, must include many items, such as the initial cost, startup, and maintenance.

B.   Characteristics of Square-Root-Law Flowmeters

Characteristics of the square-root-law or "obstruction" type flowmeters are briefly examined. Their range is limited because the flow rate is proportional to the square root of the differential pressure for the flow measurement. If the readable range of $(P_1 - P_2)$ is 16:1, the range of the flow rate is only 4:1.

The meter is more sensitive and therefore more accurate for higher flow rates than for lower flow rates, since $(P_1 - P_2)$ is proportional to Q^2 and $\Delta P/\Delta Q$ is larger for higher flow rates. For consistent accuracy, the error should be a constant percentage of the flow. By the same token, the error in a total flow for the same meter is also a function of its flow rates. For example, the total flow over a time interval is the integral of flow rates. If the total flow is mainly from low flow rates, accuracy cannot be improved by integrating over a longer interval.

Error in flow rates for pulsating flow is somewhat controversial. An ex-

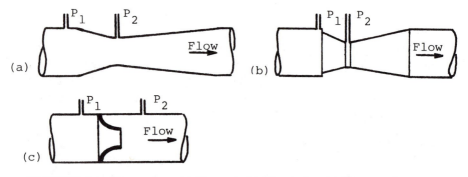

FIGURE 8-12   Flowmeters. (a) Venturi. (b) Flow tube. (c) Flow nozzle.

perimental study shows that the pulsating flow of incompressible fluids can be accurately metered with orifices if the time average of the square root of ΔP is determined [42]. It is erroneous to use the square root of the time average of ΔP. This is expressed in the inequality

$$\frac{1}{T}\int_0^T (\Delta P)^{1/2}\,dt \neq \left(\frac{1}{T}\int_0^T \Delta P\,dt\right)^{1/2} \tag{8-27}$$

In other words, the average should be based on the square-root law, but not on the average of the differential pressure. This seems self-evident, because the change in flow rate is larger for a $+\Delta P$ than for a $-\Delta P$. Hence an average differential pressure tends to give a larger flow rate.

The accuracy of an orifice meter also depends on its physical dimensions [43]. Assuming that there is no change in ΔP and fluid properties, it can be shown from Eq. 8-22 that

$$\frac{\Delta Q}{Q} = \frac{\Delta C_d}{C_d} + \frac{2}{1-\beta^4}\frac{\Delta d_2}{d_2} - \frac{2\beta^4}{1-\beta^4}\frac{\Delta d_1}{d_1} \tag{8-28}$$

where $\beta = d_2/d_1$ is a ratio of the diameter of the orifice and the pipe. The error $\Delta C_d/C_d$ may be due to changes in flow conditions or an erosion of the sharp edges of an orifice. The error $\Delta d_2/d_2$ due to the diameter of the orifice, however, may dominate the overall error in a measurement.

C.   Pitot-Static Tubes

A pitot-static tube measures the local velocity of a flow instead of an average flow as in flowmeters. The pitot or impact tube shown in Fig. 8-13a is a tube with one end bent at a right angle toward the flow direction. The flow stagnates at the impact tube. The difference between the stagnation and static pressure is used to indicate the local velocity of a free stream.

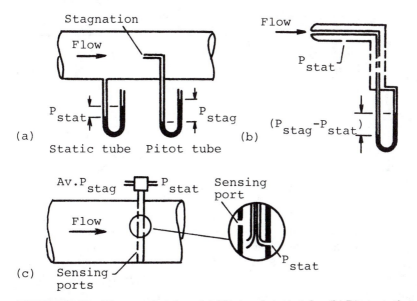

FIGURE 8-13 Pitot-static tubes. (a) Pitot and static tube. (b) Pitot-static tube. (c) Annubar.

The pitot-static tube as a single unit is shown in Fig. 8-13b. The annubar [41] shown in Fig. 8-13c measures the average stagnation pressure across a pipe section by means of a number of sensing ports. It measures the average flow rate across a pipe without the penalty of high lost head as in orifices.

1. Incompressible Flow [44]

The basis of the pitot-static tube is the square-root law. From Eq. 8-19, P_1 is the static pressure P_s of the undisturbed free stream at the velocity V_1. The fluid is brought to rest ($V_2 = 0$) isentropically at stagnation. Hence the stagnation pressure is the total pressure P_t ($= P_2$) due to P_s and the velocity V_1. From Eq. 8-19 we get

$$V = \left[\frac{2g_c}{\rho_{stat}} (P_t - P_s) \right]^{1/2} \tag{8-29}$$

The static pressure is usually more difficult to measure. The equation can be modified by a calibration constant if the static pressure indicated is not the true value. A correction is not necessary in a Prandtl pitot-static tube [45].

A pitot-static tube becomes insensitive at low velocities. A larger ΔP

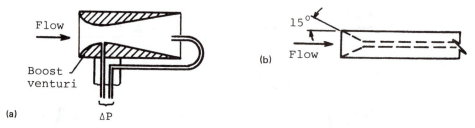

FIGURE 8-14 Configurations of pitot-static tubes. (a) Pitot-static tube. (b) Pitot tube.

signal can be obtained by means of a boost venturi, as shown in Fig. 8-14a, but the P_s is no longer that of the free stream. Error in the total pressure due to a misalignment depends on the geometry of the impact tube. The square-ended tube with a 15° internal bevel angle shown in Fig. 8-14b is capable of providing total pressure data with errors of less than 1% if the misalignment with the flow stream is less than 25°.

2. Compressible Flow

The free-stream velocity V for compressible flow is obtained by substituting Eq. 8-23 into Eq. 8-25. Letting $V_1 = V$, $V_2 = 0$, $P_1 = P_s$, $P_2 = P_t$, and simplifying, we get

$$V = \left\{ \frac{2g_c k}{k-1} \frac{P_s}{\rho_{\text{stat}}} \left[\left(\frac{P_t}{P_s} \right)^{(k-1)/k} - 1 \right] \right\}^{1/2} \tag{8-30}$$

In gas dynamics, Eq. 8-30 is usually expressed in terms of Mach number N_M of the free stream, where $N_M = V/c$ and $c = (kP/\rho)^{1/2}$ is the sonic velocity. Thus Eq. 8-30 can be rewritten as

$$P_t = P_s \left(1 + \frac{k-1}{2g_c} N_M^2 \right)^{k/(k-1)} \tag{8-31}$$

For supersonic flow ($N_M > 1$), a shock wave is formed ahead of the pitot tube. Velocity calculations for this case is beyond the intended scope of this book.

D. Variable-Area Flowmeters

A variable-area flowmeter may have a fairly wide working range. For the meters described above, such as a flow orifice, the area of the obstruction is fixed, and the differential pressure changes with flow rate. For a variable-area meter, the area of the obstruction changes with flow rate, but the differential pressure generally remains constant.

1. Flow Indicators

A flow indicator with a swinging vane is shown in Fig. 8-15a. A spring load (not shown) closes the meter for zero flow. An increasing flow forces the vane to rotate, thereby providing a larger area for the flow. The flow indicator shown in Fig. 8-15b changes the flow area between a taper pin and a fixed-size orifice in a spring-loaded piston. An increasing flow forces the piston to rise, thereby increasing the annular area between the pin and the orifice.

2. Rotameter

Rotameters are commonly used for medium flow rates and for visual flow indications. Their working range is 10:1 instead of 4:1 as described above. They can also be used for very small flow rates, of the order of 0.05 cm^3/min for liquids and 2 cm^3/min for gases [46]. The accuracy is about 2%.

The rotameter shown in Fig. 8-15c consists of a float and a vertical transparent tapered tube with graduations along its axial direction. The tube diameter is larger at the top than the bottom. Fluid enters from the bottom of the tube, passes upward through the annulus between the float and the tube, and exits through the top. For a given flow rate, the float

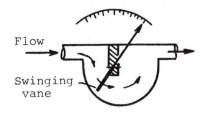

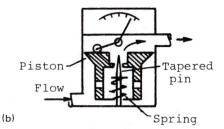

FIGURE 8-15 Variable-area flowmeters. (a) Swinging-vane meter (Series LL and LG, courtesy of Universal Flow Monitors, Hazel Park, Mich.). (b) Tapered-pin meter (Series W, courtesy of Universal Flow Monitors, Hazel Park, Mich.). (c) Rotameter.

assumes an equilibrium position y to indicate the flow rate. The float may be spherical or cylindrical. It may be shaped to obtain particular effects, such as to induce turbulence in the flow to minimize the effect of viscosity.

When the float is at equilibrium, the vertical forces due to differential pressure, viscosity, buoyancy, and gravity are balanced. A simplified model assumes that the downward force due to gravity minus buoyancy is equal to the upward force due to the differential pressure across the float. Thus a force balance on the float gives

$$A_F \, \Delta P = \text{Vol}_F(\rho_F - \rho) \, \frac{g}{g_c} \tag{8-32}$$

where

A_F = projected area of the float
ΔP = differential pressure across the float
Vol_F = volume of the float
ρ_F = density of the float
ρ = density of the fluid
g = acceleration due to gravity
g_c = dimensional constant (see Eq. 8-19)

The volumetric flow rate Q_a is obtained by substituting ΔP for $(P_1 - P_2)$ from Eq. 8-32 into Eq. 8-22:

$$Q_a = \frac{C_d(A_T - A_F)}{\{1 - [(A_T - A_F)/A_T]^2\}^{1/2}} \left[\frac{2 \, \text{Vol}_F(\rho_F - \rho)g}{\rho A_F} \right]^{1/2} \tag{8-33}$$

where C_d is the discharge coefficient, A_T the cross-sectional area of the tapered tube at the float, and $A_T - A_F$ the annular flow area. Hence $A_T = A_1$ and $A_T - A_F = A_2$.

Q_a is approximately linear with the float position y when the variation in C_d is small, the fluid density ρ is constant, and the tube is shaped that the annular area $A_T - A_F = Ky$, where K is a constant. Since $[(A_T - A_F)/A_T]^2$ is always much less than 1, we obtain the approximate equation

$$Q_a \simeq (\text{constant})y \tag{8-34}$$

If the density of the float to that of the fluid is 2:1, the same meter can be used to measure the mass flow rate of any fluid [47]. Mass flow rate is the product of Q_a and the fluid density ρ. Assume that $A_T - A_F \simeq Ky$ and $\{1 - [(A_T - A_F)/A_T]^2\}^{1/2} \simeq 1$. From Eq. 8-33, the mass flow rate is

$$\dot{m} = C_1 y[\rho(\rho_F - \rho)]^{1/2} \tag{8-35}$$

where C_1 is constant. The differential $d(\dot{m}/y)/d\rho$ gives

$$\frac{d(\dot{m}/y)}{d\rho} = \frac{C_2(\rho_F - 2\rho)}{[\rho(\rho_F - \rho)]^{1/2}} \tag{8-36}$$

where C_2 is constant. The condition for $\dot{m}/y$ independent of the fluid density ρ is

$$\rho_F = 2\rho \tag{8-37}$$

E. Drag-Force Flowmeters

The drag-force or target flowmeter shown in Fig. 8-16 consists of a drag body immersed in a flow stream and a flexure with strain gages to indicate the drag force due to the flow. The drag force on the immersed body is

$$F = CA\,\frac{\rho V^2}{2g_c} \tag{8-38}$$

where

F = drag force on the body
C = nondimensional drag coefficient
A = target area
ρ = fluid density
V = free-stream velocity
g_c = dimensional constant

The volumetric flow rate is the product of V and the free-stream cross-sectional area. For sufficient high Reynolds number and properly shaped body [48], C is fairly constant. The flowmeter is also governed by the square-root law, since the velocity V is proportional to the square root of the drag force F.

The meter is bidirectional if the drag body is symmetrical. The flow measured is independent of pressure and temperature when the value of C and the fluid density ρ are reasonably constant. The flexure-and-drag

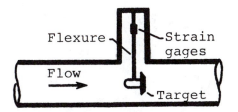

FIGURE 8-16 Drag-force (target) flowmeter.

body is a lightly damped second-order system. Hence one advantage of the meter is its high frequency response.

8-6. TURBINE FLOWMETERS

A turbine-type flowmeter works on the momentum principle. It consists of a free-running small turbine suspended in a pipe. The rotation speed of the turbine is proportional to the velocity of the fluid, similar to the familiar anemometer for wind velocity measurements. Since the flow area of the meter is fixed, the rotational speed is proportional to the volumetric flow rate.

The turbine flowmeter shown in Fig. 8-17a consists of a straight pipe, a free-running multibladed turbine wheel on bearings, the upstream and downstream flow straighteners (not shown), and a pickup outside the fluid passage. Generally, an installation requires a minimum of 10 diameters of straight pipe upstream and five downstream.

The pickup gives one voltage pulse for each blade passing the pickup.

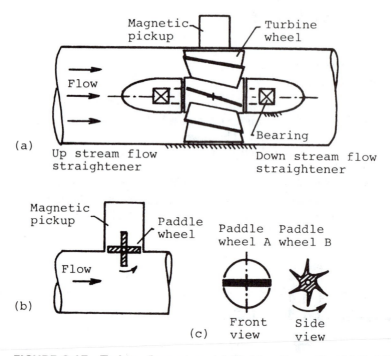

FIGURE 8-17 Turbine flowmeters. (a) Turbine flowmeter. (b) Paddle-wheel flowmeter. (c) Paddle wheels.

Thus the flow rate or the total flow is measured by counting. The pulse output is a convenience for digital control. A magnetic pickup is commonly used (see Fig. 5-20a). The magnetic drag on the turbine wheel will degrade the linearity of the meter at low flow rates, but this can be compensated by using a carrier type system.

Turbine flowmeters are commonly used. Since the operation is not based on the square-root law, the range is about 10:1 instead of 4:1. They are available in sizes from ⅛ to 8 in. in diameter, and are used for many type of fluids for a wide range of pressure and temperature. The effect of viscosity is secondary within the linear range. The pressure drop across a meter is about 3 to 10 psi. The accuracy is about 0.1% for liquids and 0.5% for gases with excellent repeatability. With a low-inertia rotor, the meter has good dynamic response, although it always reads high for pulsating gas flows [49].

In addition to the advantages above, these flowmeters are relatively maintenance free. A limitation is the bearing wear. Since the meter has a large range, an oversized flowmeter should be selected to prolong bearing life and to avoid damage to the meter due to overspeeding.

The paddle-wheel type shown in Fig. 8-17b is less expensive. The accuracy is about 2%, but the same transducer can be used for a limited range of pipe sizes. Two designs for paddle wheels, labeled A and B, are shown in Fig. 8-17c.

8-7. VORTEX-SHEDDING FLOWMETERS

Vortex-shedding and swirl flowmeters are described in this section. The flow of fluid around solid bodies is studied extensively [50]. Vortex-shedding flowmeters are based on the von Kármán effect. These meters were introduced in the late 1960s. They have gained wide acceptance, and their applications are studied actively [51].

Vortices or eddies shed alternately from a body immersed in a fluid at a steady flow, and the vortices travel downstream with the flow. The frequency of the shedding is given by the Strouhal number N_S, which is a function of Reynolds number N_R:

$$N_S = \frac{fD}{V} = \phi(N_R) \qquad (8\text{-}39)$$

where f is the shedding frequency, D a characteristic dimension of the shedding body, and V the fluid velocity. For the triangular bluff body or strut shown in Fig. 8-18a, N_S is reasonably constant for $10^4 < N_R < 10^6$. Within this range, the frequency of shedding is directly proportional to the fluid velocity and therefore the volumetric flow rate. Note that the

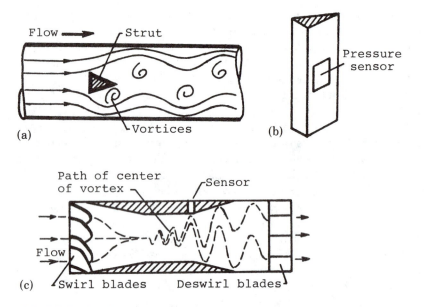

FIGURE 8-18 Flowmeters based on fluid oscillations. (a) Vortex shedding. (b) Triangular strut. (c) Swirlmeter.

output signal is the shedding frequency rather than the magnitude of the vortices.

Many schemes are used to count the shedding frequency. Since the vortices cause local pressure pulses to alternate on the shedder, the frequency signal can be sensed by a pressure transducer. The frequency signal can also be detected by means of the turbulence from the vortices, which are at regular intervals. An ultrasonic, laser Doppler, or a resistance thermal detector RTD-type sensor can all be employed for this purpose.

Many bluff bodies are used to produce vortex shedding. Two bluff bodies of radical different shapes could be identical vortex shedders if their Strouhal numbers were identical. In fact, the same calibration of a meter can be used for both liquid and gas. This "universal" design simplifies calibration, maintenance, and the interchanging of parts. A triangular strut with provision for pressure sensing is shown in Fig. 8-18b. In some meters, a second strut can be used for the hydrodynamic amplification of the vortex shedding (e.g., DV Series, Fisher Controls, St. Louis, Missouri).

Accuracy of vortex-shedding flowmeters is about 0.5% of reading with excellent repeatability. The working range is 20:1 and a linear range of 100:1 is attainable. Upstream and downstream straight piping are necessary for proper installation. The meter can be used for a wide range of fluids, but slurries or high-viscosity liquids are not recommended.

The main advantages of vortex-shedding flowmeters are no moving parts, good accuracy with long-term repeatability, large operating range, linear output with reasonable lost head, suitable for gas or liquid, and calibration not affected by viscosity, density, pressure, or temperature. It is cost-effective compared with many common flowmeters. The main sources of error are turbulent eddies carried into the meter, mechanical vibration, and sound waves in the fluid.

The swirlmeter [52] shown in Fig. 8-18c also uses oscillations of a fluid for velocity measurement, but employs the vortex precession principle instead of vortex shedding. It has no moving parts. The swirl traces a helical path through the flow. The pitch of the helix is fixed by the meter, but the rate at which the helix passes a point is proportional to the volumetric flow rate. The operation is analogous to the threads of a screw which travel in accord with the rate of rotation. The signal can be sensed by a thermister or a RTD-type transducer.

8-8. HOT-WIRE AND HOT-FILM ANEMOMETERS

Hot-wire anemometers have been used since the late nineteenth century. The hot-film type is more rugged and is a recent introduction. Constant-temperature and constant-current hot-wire anemometers are described in this section. The latter has largely been replaced by the constant-temperature type.

The instrument is based on the convective heat transfer from a heated sensor to its surrounding fluid. The heat loss depends on the fluid velocity, temperature, pressure, density, and thermal properties, as well as the temperature and physical parameters of the sensor. Since these are potential noise inputs in a measurement, they also show potential applications of the instrument (see Eq. 2-4). Literature on th subject is extensive [53]. A basic text by Lomas [54] is well worth reading.

Hot-wire anemometers are used mainly to measure rapid fluctuating velocities in gas and liquid. The instrument has excellent frequency response and an upper frequency limit of 400 kHz is commercially available. Among the flowmeters described in this chapter, only the laser Doppler velocimeter is comparable in this performance.

The probe of a hot-wire anemometer, shown in Fig. 8-19a, consists of a fine-wire sensor, the wire supports, a probe body, and electrical leads. The wire is about 1 mm (40×10^{-3} in.) in length and 5 μm (0.2×10^{-3} in.) in diameter. Tungsten, platinum, and platinum alloys are used for the wire. The sensor of a hot-film probe, shown in Fig. 8-19b, is

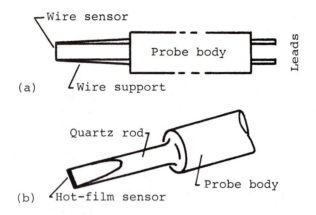

FIGURE 8-19 Probes for hot-wire and hot-film anemometers. (a) Hot-wire probe. (b) Wedge hot-film probe.

usually made of platinum deposited in a thin layer onto a substrate of fused quartz.

The energy balance of a heated wire at equilibrium is

$$\underbrace{I^2R}_{\substack{\text{power}\\\text{input}}} = \underbrace{hA(T - T_f)}_{\substack{\text{convective heat}\\\text{transfer}}} \tag{8-40}$$

where I is an electric current, R the wire reistance, h the heat transfer film coefficient, A the heat transfer area, and T and T_f are the wire and fluid temperatures. The heat transfer from a cylinder of infinite length is commonly expressed in King's law:

$$h = C_1' + C_2'\sqrt{N_R} \tag{8-41}$$

or

$$h = C_1 + C_2V^n \tag{8-42}$$

where the C's are parameters [55], N_R is the Reynolds number, V the fluid velocity, and $n \simeq 0.5$. The resistance R of a hot-wire sensor is

$$R = R_0[1 + \alpha(T - T_0)] \tag{8-43}$$

where R_0 is the resistance at the reference temperature T_0, and α the coefficient of resistivity. Thus the temperature of a wire can be measured by means of its resistance.

A. Constant-Temperature Anemometers

The simplified circuit of the constant-temperature anemometer shown in Fig. 8-20 consists of bridge circuit and a high-gain dc differential amplifier in a feedback loop. Let the bridge be initially balanced with zero excitation and fluid velocity V. The bridge is unbalanced by increasing R_4 and keeping $V = 0$. Thus the voltage across b-d is the unbalance input to the amplifier. This increases the excitation voltage to the bridge across a-c and the current I through the sensor. Subsequent increase in temperature and resistance R of the sensor reduces the unbalance voltage. The process is self-adjusting. An equilibrium is reached when the unbalance voltage becomes zero and the sensor is at a desired temperature. The differential temperature $(T - T_f)$ in Eq. 8-40 is typically maintained at about 450°F in air and 80°F in water. This gives the reference current I_0.

A fluid flow cools the sensor, decreases its resistance R, and unbalances the bridge. Since the feedback loop is self-adjusting, it automatically increases the excitation to the bridge to restore the balance. The velocity V of the fluid can be measured by means of the voltage across the terminals a and c at the bridge. Alternatively, a drop resistor can be inserted to measure the total current to the bridge as shown in the figure. Assuming $n \simeq 0.5$ in Eq. 8-40, it can be shown that

$$V = C_0 \left[\left(\frac{I}{I_0} \right)^2 - 1 \right]^2 \qquad (8\text{-}44)$$

where C_0 is a calibration constant and I_0 the current through the sensor to give the desired temperature for $V = 0$. An anemometer is calibrated by

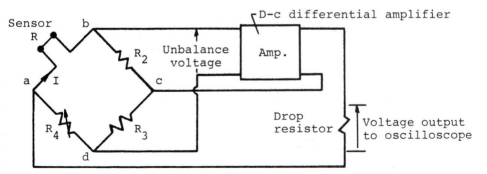

FIGURE 8-20 Constant-temperature anemometer with feedback.

exposing the probe to a known velocity in the fluid in which it is to be used.

B. Constant-Current Anemometers

Feedback is not used for constant-current anemometers. The circuits for the constant-current drive in Fig. 8-21a are the basic circuits presented in Sec. 3-5 (see Figs. 3-36b and 3-37b). A constant-current source can be replaced by a constant-voltage source with a high series resistance R_b ($R_b = 2$ kΩ and $R = 1$ Ω).

The frequency response of the anemometer in Fig. 8-21a is only about 160 Hz [56]. This can be deduced from the time constant of the sensor in an energy balance equation, similar to finding the time constant of a thermometer (see Prob. 4-2). The frequency response can be improved by means of series compensation as shown in Fig. 8-21b (see Fig. 4-30). The time constant of the sensor, however, is unknown, and the scheme for obtaining the correct compensation involves additional circuitry [57].

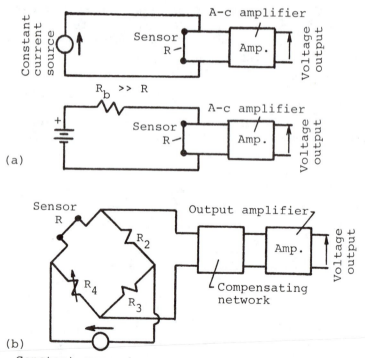

FIGURE 8-21 Constant-current anemometer circuits. (a) Velocity fluctuation measurement. (b) Anemometer with compensation.

Hot-wire instruments are sensitive and versatile. They are available in numerous configurations [58] for specific applications. However, they are delicate instruments and must be used properly. Breakage and burnout are not uncommon. An unexpected flow reversal in the vicinity of the probe may be interpreted as an decrease in velocity followed by an increase. Some conditions that may cause difficulties are misalignment, probe fouling, sensor aging, conducting liquids such as mercury [59], low-velocity measurements, air bubbles in the liquid, vortex shedding from parts of the probe, and hydrodynamic interference from another probe or a nearby object.

8-9. MASS FLOWMETERS

Mass flowmeters are used when a product is sold by weight or when the performance of a machine is based on mass measurements, such as a liquid-fueled rocket engine. The mass flow rate of a fluid can be obtained from the product of its density and volumetric flow rate. This is an indirect method and commonly used, but the density measured at one location is not always that at the flowmeter, such as in cryogenics [60]. Mass flow rate can also be measured directly. Examples of both methods are described in this section.

A. Density Measurements

Density measurements as an intermediate step for finding mass flow rates are described in this section. Some instruments must be compensated for effects of flow, pressure, temperature, or viscosity.

The turbine–orifice combination shown in Fig. 8-22a measures the fluid velocity V and the ΔP across the orifice. Since V is determined, the density ρ can be deduced from Eq. 8-22. The submerged plummet in Fig. 8-22b is a "hydrometer." A change in density changes the effective weight of the plummet and chain. The jet and the coaxial receiver shown in Fig. 8-22c are separated by a fixed distance. Clean air is supplied to the jet. The pressure detected at the output is a measure of the gas density between the jet and the receiver. The density of a flowing liquid is measured as shown in Fig. 8-22d, in which the resonant frequency of the liquid-filled pipe is sustained by means of feedback. A density change in the liquid causes a change in resonant frequency. Similarly, a vibrating plate is used in the densitometer shown in Fig. 8-22e. The resonant frequency of a plate immersed in a confined volume changes with the density of the surrounding fluid. The capacitance between the coaxial plates shown in Fig. 8-22f changes with the dielectric constant of the fluid, which is a function of its density. Dielectric densitometers are used for flowmeters in cryogenics.

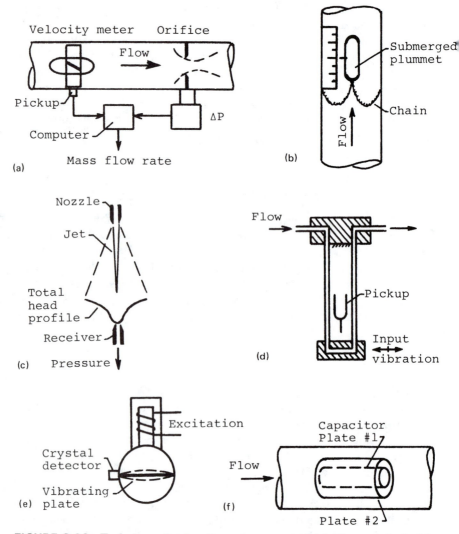

FIGURE 8-22 Techniques for density measurement. (a) Indirect method. (b) "Hydrometer." (c) Jet pressure (Model 601D, Turbojet Densitometer, courtesy of Fluid Dynamic Devices Ltd., Mississauga, Canada). (d) Resonant frequency. (e) Vibrating plate. (f) Dielectric "constant."

The examples above show that any density-related equation can be used for a density measurement. Some equations feasible for density measurements are

$$I = I_0 e^{-\mu d} \tag{8-45}$$

$$c = \sqrt{\frac{kP}{\rho}} \tag{8-46}$$

$$\rho = \frac{P}{RT} \tag{8-47}$$

The nuclear-radiation absorption coefficient μ of a material is proportional to its density. It is feasible to use radiation attenuation in Eq. 8-45 for density measurement, where the I's are radiation intensities and d is the separation between the radiation source and detector. The density ρ of a gas in Eq. 8-46 is related to the sonic velocity c, pressure P, and specific heat ratio k ($= C_p/C_v$). The gas law in Eq. 8-47 relates the density ρ, absolute pressure P, absolute temperature T, and gas constant R.

B. Direct Mass Flowmeters

A true mass flowmeter should be independent of other physical variables of the fluid, the type of flow, and ambient conditions. The simplest method is to determine the mass flow rate by weighing (see Fig. 8-2), but this is not always convenient in field applications. Thermal flowmeters (see Figs. 8-9 and 8-10) are basically mass flowmeters, provided that the specific heat remains constant. Examples of direct mass flowmeters are described in this section.

Angular momentum principle is used in the direct mass flowmeter shown in Fig. 8-23a. It consists of an impeller driven at a constant angular velocity ω, a stationary disk to decouple the viscosity effect of the fluid, a turbine for the momentum transfer, and a torsional spring to restrain the turbine from free rotation. The impeller gives a swirl or an angular momentum to the flowing fluid. The swirl is removed by the turbine. From Newton's second law, the torque T on the turbine is

$$T = \frac{d}{dt}(J\omega) \tag{8-48}$$

or

$$T = \omega k^2 \frac{dm}{dt} \tag{8-49}$$

where J is the mass moment of inertia of the fluid of mass m, ω a constant

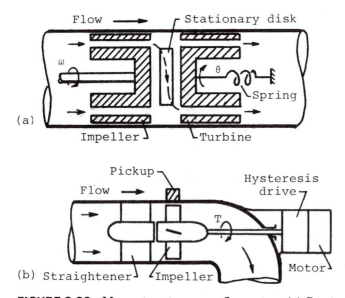

FIGURE 8-23 Momentum-type mass flowmeters. (a) Constant-ω impeller/turbine meter (courtesy of General Electric Aerospace Div., Wilmington, Mass.). (b) Constant-torque impeller meter.

angular velocity of the motor, and k the radius of gyration of the fluid mass m. For constant ω and k, the torque T indicated by the spring is proportional to the mass flow rate dm/dt. The range of the meter is 10:1 for constant flow rates. Under unsteady flow conditions, the viscous force in the system may become large.

The mass flowmeter shown in Fig. 8-23b works on the same principle. It consists of an impeller, a motor, and a hysteresis drive (similar to a slip clutch) to give a constant torque T. The constant T can also be maintained by means of a servo system. For T and k constant in Eq. 8-49, the mass flow rate dm/dt is proportional to $1/\omega$. If a magnetic pickup is used to measure the speed, the time interval Δt between pulses from the pickup is inversely proportional to ω. Hence dm/dt is linear with Δt. The meter has high resolution at low flow rates, but the resolution decreases at higher flow rates.

Variations of the momentum principle are used in other mass flowmeters. The design shown in Fig. 8-24a consists of two free-running rotors coupled through a torsional spring. Each rotor has a different pitch angle for its blades. The spring allows the rotors to rotate at the same speed but out of phase with one another. The mass flow rate is proportional to the

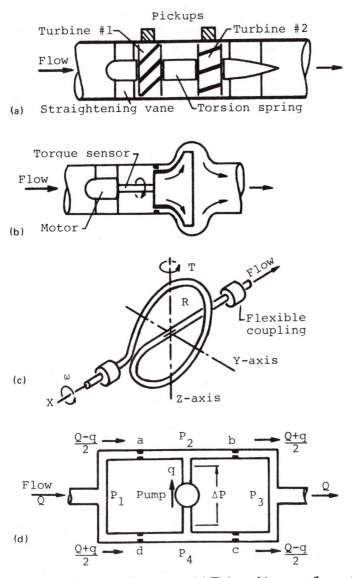

FIGURE 8-24   Mass flowmeters. (a) Twin-turbine mass flowmeter. (b) Coriolis mass flowmeter (from Ref. 61). (c) Gyroscopic mass flowmeter. (d) Bridge-circuit mass flowmeter (courtesy of Flo-Tron, Inc., Paterson, N.J.).

relative angular displacement between the rotors. The "Coriolis" meter shown in Fig. 8-24b measures mass flow rate by means of the torque required to maintain a constant angular velocity in the fluid while it moves radially. The fluid in the gyroscopic mass flowmeter shown in Fig. 8-24c is forced through a 360° loop while the loop is rotating about the X axis at a constant angular velocity. The precession moment produced about the Z axis is proportional to the mass flow rate. This is measured at the flexible couplings shown in the figure.

A differential pressure bridge as a mass flowmeter is shown in Fig. 8-24d. It consists of a bridge with four identical orifices and a positive-displacement pump at a constant flow rate q. Liquid at a volumetric rate Q is supplied to the meter. Assume that $q < Q$. The flow through the orifices are as shown in the figure. The flow and pressure drop across orifice a and orifice d are

$$\frac{Q - q}{2} = C \sqrt{\frac{P_1 - P_2}{\rho}} \quad \text{and} \quad \frac{Q + q}{2} = C \sqrt{\frac{P_1 - P_4}{\rho}} \qquad (8\text{-}50)$$

where C's are constants. Combining the equations and simplifying, the mass flow rate ρQ is

$$\rho Q = \frac{C}{q}(P_2 - P_4) = k(P_2 - P_4) \qquad (8\text{-}51)$$

where $k = C/q$ is a constant. It can be shown that the mass flow rate ρQ is proportional to $(P_1 - P_3)$ for $q > Q$. The operation of the meter is independent of pressure, temperature, and density of the liquid, but the density must remain unchanged within the meter. Hence the meter is suitable only for liquids. Its range is from 30:1 to 100:1 with accuracy of 0.5% of reading. These flowmeters are available from 0.1 to 50,000 lb_m/hr.

PROBLEMS

8-1.   Water flows in a 4-in.-diameter pipe with an average velocity of 15 ft/sec. The temperature is $T = 80°F$, density $\rho = 62.2$ lb_m/ft^3, and viscosity $\mu = 1.77 \times 10^{-5}$ lb_f-sec/ft^2. (a) Calculate the Reynolds number. (b) Convert the data into SI units and verify the answer.

8-2.   The on-line measurement of the viscosity of a fuel oil is as shown in Fig. 7-33e. The apparatus consists of a temperature-controlled tubing ($L = 25$ cm and $D = 4$ mm) and a constant-delivery pump. Calculate the absolute viscosity if the average velocity of the flow is $V = 1.5$ m/s, the differential pressure across the tubing ($P_1 - P_2$) $= 2.5$ kPa, temperature $T = 40°C$, the specific gravity is 0.90, and the density of water is 992 kg/m^3.

8-3. A model study is proposed for airflow in a 1.0-m-diameter duct with a gage pressure of 10 kPa and an average velocity of 10 m/s. Select one of the models below and give your reasons for the selection. (a) Airflow in a 10-cm-diameter pipe. (b) Water flow in a 6-cm-diameter pipe. (c) Oil flow in a 6-cm diameter pipe.

Data (at atmospheric pressure):

	Absolute viscosity (Pa · s)	Kinematic viscosity (m²/s)
Air	1.78×10^{-5}	1.49×10^{-5}
Water	1.08×10^{-3}	1.08×10^{-6}
Oil	3.75×10^{-3}	4.38×10^{-6}

8-4. The drag force F on a submerged body is a function of the fluid density ρ, absolute viscosity μ, velocity V, and a characteristic dimension D. Use dimensional analysis (a) to find the dimensionless ratio of F and μ, and (b) to show that the force coefficient is a function of Reynolds number N_R.

8-5. The accuracy of a lobed-impeller meter (see Fig. 8-3f) for volumetric airflow from 1000 to 4000 ft³/min is ±1%. If the air pressure is (20.0 ± 0.2) psia and temperature is $(120 \pm 2)°F$, calculate the uncertainty in the mass flow rate.

8-6. Show a sketch of the ASME recommended proportions for each of the obstruction meters and give reference(s) for the information: (a) sharp-edged orifice; (b) venturi tube; (c) long-radius flow nozzle.

8-7. Show a plot of the discharge coefficient C_d versus Reynolds number N_R for each of the obstruction meters and give reference(s) for the information: (a) sharp-edged orifice; (b) venturi tube; (c) long-radius flow nozzle.

8-8. A sharp-edged concentric orifice is used to measure the flow of water in a 10-cm-diameter pipe. The diameter ratio of orifice to pipe is 0.40 and the differential pressure is 100 kPa. (a) Calculate the flow in gallons per minute. Data: Density $\rho = 997$ kg/m³, viscosity $\mu = 8.94 \times 10^{-4}$ Pa · s, $C_d = 0.61$, 1 gallon $= 3.786 \times 10^{-3}$ m³. (b) Convert the data to SI units and verify the calculations.

8-9. A venturi for metering fuel oil is calibrated as shown in Fig. 8-2a. Pipe diameter $D_1 = 4$ in. and $D_2{:}D_1 = 0.5{:}1$. The flow rate is 500 gal/min. (a) Calculate its discharge coefficient C_d. Data: oil specific gravity $= 0.86$, $\mu = 6.02$ lb$_f$ · sec/ft², water density $= 62.4$ lb$_m$/ft³, $\Delta P = 15$ psi. (b) Convert the data into SI units and verify the calculations.

8-10. The flow of a heavy oil through a 20-cm-diameter pipe at 3000 barrels per hour is measured by means of a sharp-edged concentric orifice. (a) Calculate the pressure drop across the orifice and estimate the power requirement. Data: temperature $T = 40°C$, density $\rho = 892$ kg/m^3, viscosity $\mu = 0.0521$ Pa $\cdot$ s, $C_d = 0.62$, diameter ratio (orifice):(pipe) = 0.5:1. (b) Repeat part (a) for a ventuir tube with $C_d = 0.95$.

8-11. (a) Derive Eq. 8-28, showing the error in flow rate $\Delta Q/Q$ due to $\Delta C_d/C_d$, $\Delta d_2/d_2$, and $\Delta d_1/d_1$. (b) If the uncertainty in C_d is $\pm 0.2\%$, that in d_2 is $\pm 0.15\%$ and in d_1 is $\pm 0.25\%$, plot $(\Delta Q/Q)_{rss}$ versus β for $0.3 < \beta < 0.8$.

8-12. A sinusoidal pulsating flow $Q_p \sin \omega t$ is superposed on the average flow Q_{av} of an obstruction-type flowmeter, where $Q_{av} > Q_p$. Verify the inequality in Eq. 8-27, where ΔP is the differential pressure across the obstruction.

8-13. (a) Select a venturi flowmeter and a differential pressure transducer for metering water from 100 to 200 gal/min. Data: throat Reynolds number $N_R \geqslant 2 \times 10^5$, $d_2/d_1 = \beta = 0.5$, $C_d = 0.975$, water $T = 70°F$, density $\rho = 62.3$ lb$_m$/ft^3, absolute viscosity $\mu = 2.02$ lb$_f \cdot$ sec/ft^2, and 1 gallon = 0.1337 ft^3. (b) Estimate the root-sum-square error $(\Delta Q/Q)_{rss}$ for the flow rates if the tolerance in d_1 is ± 0.003 in., in d_2 is ± 0.001 in., the uncertainty in C_d is ± 0.003, and in $(P_1 - P_2)$ is $\pm 1\%$ full scale. (c) Estimate the $(\Delta Q/Q)_{rss}$ for the flow rate of 50 gal/min.

8-14. Show the details in deriving Eq. 8-26 for compressible flow of an obstruction-type flowmeter.

8-15. (a) Calculate the mass flow rate of air in a 5-in.-diameter duct with a 3-in. sharp-edged orifice. Data: Inlet pressure $P_1 = 20$ psia at 80°F, differential pressure $(P_1 - P_2) = 15$ in. of 0.84 specific gravity oil, density of water $\rho_w = 62.3$ lb$_m$/ft^3, density of air (at atmospheric pressure of 14.696 psia) $\rho = 0.0735$ lb$_m$/ft^3, absolute viscosity of water $\mu = 0.385 \times 10^{-6}$ lb$_f \cdot$ sec/ft^2, and the discharge coefficient $C_d = 0.62$. (b) Find the percentage error if the calculation is based on the incompressible flow equation (Eq. 8-22).

8-16. It is desired to calculate the mass flow rate for compressible flow (Eq. 8-26) by means of the simplier incompressible flow equation (Eq. 8-22). Plot the percentage error versus pressure ratio P_2/P_1 with the diameter ratio β as a parameter. Use $0.99 > P_2/P_1 > 0.80$ and $\beta = 0.30, 0.50, 0.6$, and 0.70.

8-17. A pitot-static tube is used to measure the air velocity V in a 15-cm-diameter duct. Determine V if the air temperature is $T = 30°C$, specific heat ratio $k = 1.4$, density (at atmospheric pressure) $\rho = 1.14$ kg/m^3, static pressure $P_s = 150$ kPa, differential pressure

$(P_t - P_s)$ = 12 cm of specific gravity 0.82 oil, and water density ρ_w = 996 kg/m^3.

8-18. (a) A pitot-static tube is used to measure the velocity V of an aircraft. Calculate V in mph if the air temperature is $T = 40°$F, specific heat ratio $k = 1.4$, density (at atmospheric pressure) $\rho = 0.0794$ lb$_m$/ft^3, static pressure $P_s = 13$ psia, differential pressure $(P_t - P_s) = 20$ in. of water, water density $\rho_w = 62.42$ lblb$_m$/ft^3. (b) Calculate the error in V if the uncertainty in $(P_t - P_s)$ is ± 1 in. of water.

8-19. A rotameter with a steel float is designed to meter the volumetric flow rate Q of water at 5°C (41°F). (a) Estimate the error in Q if the water is at 50°C (122°F). Data: water density is ρ (5°C) = 1000 kg/m^3, ρ (50°C) = 988.1 kg/m^3, and steel density $\rho_F = 0.382$ lb$_m$/in^3. (b) Estimate the error in Q due to the temperature change in the steel float. The linear expansion of steel is $+13.6 \times 10^{-6}/$°C.

8-20. The velocity profile for the turbulent flow of air in a smooth pipe of diameter d $(= 2R)$ is

$$V_r = V_c(1 - r/R)^{1/n}$$

where r is a radius from the pipe center, V_r its velocity, V_c the velocity at the center, and $n = 6$, 8, or 10, depending on the Reynolds number. (a) For $n = 6$, find the radius at which the average velocity V_{av} can be probed by means of a single measurement with a pitot-static tube. (b) Repeat part (a) for $n = 8$ and $n = 10$. (c) Discuss the possible disadvantage of this method of measurement.

8-21. The velocity profile of air in a circular duct is asymmetrical. To obtain an average velocity, the flow area is divided into three equal-area annuli and a pitot-static tube is used to probe the center of area of each annulus. (a) Calculate the probe locations. (b) If the true velocity profile is $V_r = V_c(1 - r/R)^{1/8}$ (see Prob. 8-20), calculate the error in the measurement. (c) Repeat the problem using an annubar (see Fig. 8-13c) that divides the flow area into two equal areas.

8-22. Typical review questions: With the aid of a sketch, briefly discuss each of the following:
(a) Positive-displacement flowmeters.
(b) Techniques using ultrasound for flow measurements.
(c) Principles used in magnetic flowmeters.
(d) Characteristics of obstruction-type flow meters.
(e) Vortex shedding flowmeters.
(f) Constant-temperature anemometers.

REFERENCES

1. Miller, R. W. (ed.), *Flow Measurement Engineering Handbook,* McGraw-Hill Book Company, New York, 1983.
   Dowdell, R. B., *Flow Its Measurement & Control in Science & Engineering,* Instrument Society of America, Pittsburgh, Pa., 1974.
   IMEKO Conference, Elsevier Science Publishing Co., Inc., New York, 1984.
2. Lomas, D. J., Selecting the Right Flowmeter, Part I, *Instrum. Technol.,* Vol. 24, No. 6 (1977), pp. 71–77.
   Flowmeter Survey, *Instrum. Control Syst.,* Vol. 42 (Mar. 1969), pp. 115–130; (July 1969), pp. 100–102.
3. Krigman, A., Flow Measurement: Some Recent Progress, *InTech,* Vol. 30, No. 4 (1983), pp. 9–13.
   Krigman, A., Flow Measurement: A State of Flux, *InTech,* Vol. 31, No. 10 (1984), pp. 9–13.
   Walters, S., New Instrumentation for Advanced Turbine Research, *Mech. Eng.,* Vol. 106, (Feb. 1984), pp. 43–51.
   Mattingly, G. E., Improving Flow Measurement Performance: Research Techniques and Prospects, *InTech,* Vol. 32, No. 1 (1985), pp. 57–65.
4. Galley, R. L., Aerospace Flow Metrology, *Instrum. Control Syst.,* Vol. 39, No. 12 (1966), pp. 113–117.
5. *Specification, Installation, and Calibration of Turbine Flowmeters,* DR31.1, Instrument Society of America, Pittsburgh, Pa., 1972.
6. Walker, R. K., Displacement Meters, *Instrum. Control Syst.,* Vol. 39 (Oct. 1966), pp. 141–144.
7. Bloser, B. L., Positive Displacement Liquid Meters, *Instrum. Control Syst.,* (Sept.–Oct. 1977), pp. 80–83.
   Hendrix, A. R., Positive Displacement Flowmeters: High Performance—with Little Care, *InTech,* Vol. 29 (Dec. 1982). pp. 47–49.
8. VanDyke, M., *An Album of Fluid Motion,* Parabolic Press, Stanford, Calif., 1982.
9. Prandtl, L., and O. G. Tietjens, *Applied Hydro- and Aeromechanics,* McGraw-Hill Book Company, New York, 1934, pp. 265–274.
10. Allen, M., and A. J. yerman, Visualizing Three-Dimensional Flow, *Instrum. Control Syst.,* Vol. 39 (Mar. 1966), pp. 93–95.
11. Wu, J. H., and J. H. Lee, A Simple Schlieren System, *ISA J.,* Vol. 13 (April 1966), pp. 56–58.
12. Lion, K. S., *Instrumentation in Scientific Research,* McGraw-Hill Book Company, New York, 1959, p. 119.
13. Lion, Ref. 12, p. 120.
14. Rhodes, D. F., Measuring Flow with Radiotracers, *Instrum. Technol.,* Vol. 22, (Oct. 1975), pp. 43–48.
15. Fishmann, J. B., Using Radioactivity in the Natural Gas Industry, *Instrum. Control Syst.,* Vol. 40, No. 11 (1967), pp. 111–113.
16. MacKenzie, K. V., A Decade of Experience with Velocimeters, in *Under*

Water Sound, V. M. Albers (ed.), Dowden, Hutchinson & Ross, Inc., Stroudsburg, Pa., 1972, Sec. 12.

Urick, R. J., *Principles of Underwater Sound,* McGraw-Hill Book Company, New York, 1983.

17. Brown, A. E., and G. W. Allen, Ultrasonic Flow Measurement, *Instrum. Control Syst.,* Vol. 40, No. 3 (1967), pp. 130–134.

Liptak, B. G., Flow Metering Accuracy, *Instrum. Technol.,* Vol. 18, (July 1971), pp. 35–38.

Lowel, F. C., Jr., Acoustic Flowmeters for Pipelines, *Mech. Eng.,* Vol. 101, (Oct. 1979). pp. 29–35.

18. *Flow and Level Measurement Handbook and Encyclopedia,* Omega Engineering, Inc., Stamford, Conn., 1986.

19. Liptak, B. G., and R. K. Kaminski, Ultrasonic Instruments for Level and Flow, *Instrum. Technol.,* Vol. 21 (Sept. 1974), p. 59.

20. Sanderson, M. L., A Review of State of the Art, in *Advances in Flow Measurement Technology,* BHRA Fluid Engineering, Cranfield, Bedford, U.K., 1981, pap. G1.

Schmidt, T. R., What You Should Know About Clamp-on Ultrasonic Flowmeters, *InTech,* Vol. 28, (May 1981), pp. 59–62.

21. Waller, J. M., Guidelines for Applying Doppler Acoustic Flowmeters, *InTech,* Vol. 27, (Oct. 1980), pp. 55–57.

22. Kock, W. E., *Engineering Applications of Lasers and Holography,* Plenum Press, New York, 1975.

Durst, F., *Principles and Practice of Laser-Doppler Anemometry,* Academic Press, Inc., New York, 1976.

Borrego, C., Simultaneous Measurements of Hot-Wire Anemometer and Laser Doppler Velocimeter in a Turbing Flow, in *Advances in Flow Measurement Techniques,* BHRA Fluid Engineering, Cranfield, Bedford, U.K., 1981, pap. E3.

Drain, L. E., *Laser Doppler Techniques,* John Wiley & Sons, Inc., New York, 1980.

Augus, J. C., D. L. Morrow, J. W. Dunning Jr., and M. J. French, Motion Measurement by Laser Doppler Techniques, *Ind. Eng. Chem.,* Vol. 61, No. 2 (1969), pp. 8–20.

23. Kock, W. E., *Sound Waves and Light Waves,* Doubleday & Company, Inc., New York, 1965.

24. Durrani, T. S., and C. A. Greated, *Laser Systems in Flow Measurements,* Plenum Press, New York, 1977, Chap. 2.

25. Walters, S., New Instrumentation for Advanced Turbine Research, *Mech. Eng.,* Vol. 105 (Feb. 1983), pp. 41–51.

26. Lion, Ref. 12, pp. 120–123.

27. Ref. 18, pp. H-19–20.

28. Christopher, R. M. (ed.), *ISA Directory of Instrumentation 1985–1986,* Instrument Society of America, Research Triangle Park, N.C.

29. Laub, J. H., Read Mass Flow Directly with Thermal Flowmeters, *Control Eng.* Vol. 13 (Apr. 1966), pp. 69–72.

30. Benson, J. M., Thermal Flow Sensors, *Instrum. Technol.*, Vol. 18, (July 1971), pp. 39–43.
31. Laub, J. H., Ref. 29, p. 72.
32. Laub, J. H., Measuring Flow with the Boundary Layer Flowmeter, *Control Eng.* (March 1957), p. 112.
33. Sydenham, P. H., *Transducers in Measurement and Control,* Instrument Society of America, Research Triangle Park, N.C., 1980, p. 62.
34. Beck, M. S., Correlation in Instruments: Cross-Correlation Flowmeters, *J. Phys.* Vol. E41, No. 1 (1981), pp. 7–19.
35. Hsu, H. P., *Fourier Analysis,* Simon and Schuster, New York, 1970. p. 94.
36. *Fluid Meters: Their Theory and Applications,* 6th ed., American Society of Mechanical Engineers, New York, 1971.
    Miller, Ref. 1.
37. Bloom, G., Errorless Orifices, *Prod. Eng.* (Oct. 1965), pp. 61–64.
38. Shapiro, A. H., *The Dynamics and Thermodynamics of Compressible Fluid Flow,* Vol. 1, The Ronald Press Company, New York, 1953.
39. Benedict, R. P., *Fundamentals of Temperature, Pressure, and Flow Measurements,* 2nd ed., John Wiley & Sons, 1977, p. 416.
40. Evans, H. J., Measurement of Compressible Fluids, *Instrum. Control Syst.*, Vol. 47 (Feb. 1977), pp. 61–64.
41. Alston, N. A., Flow Measurement Energy Cost, *Meas. Control.*, Vol. 11 (Sept.–Oct. 1977), pp. 97–99.
42. Sauser, H. J., Jr., P. D. Smith, and L. V. Field, Metering Pulsating Flow in Orifice Installations, *Instrum. Technol.*, Vol. 14, (Mar. 1967), pp. 41–44.
43. Kinpis, L., Simplified Flow Instrument Specification by Standardizing Primary Element Sizes, *InTech,* Vol. 27, (June 1980), pp. 40–42.
44. Folson, R. G., Review of the Pitot Tube, *Trans. ASME,* (Oct. 1956), p. 1447.
45. Prandtl and Tietjens, Ref. 9, p. 230.
46. Blechman, S., Techniques for Measuring Low Flows, *Instrum. Control Syst.*, Vol. 36 (Oct. 1963), pp. 82–85.
47. Buzzard, W., Measure Mass Flow with a Rotameter, *Control Eng.* (Mar. 1966), pp. 96–98.
48. Binder, R. C., *Fluid Mechanics,* 3rd ed., Prentice-Hall, Inc., Englewood Cliffs, N.J., 1955, pp. 179–184.
49. Haalman, A., Pulsating Errors in Turbine Flowmeters, *Control Eng.* (May 1965), pp. 89–91.
50. *Flow Around Bluff Bodies: Numerical and Experimental Approaches,* von Kármán Institute of Fluid Dynamics, Lecture Series 1984-06, Rhode Saint Genese, Belgium, May 1984.
51. Krigman, Ref. 3.
    Mattingly, Ref. 3.
52. Woodring, E. D., Magnetic Turbine Flowmeters, *Instrum. Control Syst.*, Vol. 42 (June 1969), pp. 133–135.

Rodely, A. E., The Swirl Flowmeter, *Instrum. Control Syst.,* Vol. 41 (Mar. 1968), pp. 109–111.

53. Freymuth, P., A Bibliography of Thermal Anemometer, *TSI Q.,* TSI Inc., St. Paul, Minn., Vol. 4, No. 4 (Nov.–Dec. 1978); Vol. 5, No. 1 (Feb.–Mar. 1979); Vol. 5, No. 4 (Nov.–Dec. 1979).

54. Lomas, C. G., *Fundamentals of Hot Wire Anemometry,* Cambridge University Press, New York, 1986.

55. Smol'yakov, A. V., and V. M. Tkachenko, *The Measurement of Turbulent Fluctuations,* Springer-Verlag, New York, 1980, p. 68.

56. Doebelin, E. O., *Measurement Systems, Application and Design,* McGraw-Hill Book Company, New York, 1983, pp. 508–510.

57. Doebelin, Ref. 56, p. 512.

58. *DISA Probe Manual,* DISA Electronik A/S, DK 2730, Herler, Denmark.

59. Hoff, M., Hot-Film Anemometry in Liquid Mercury, *Instrum. Control Syst.,* Vol. 42 (Mar. 1969), pp. 83–86.

60. Alspach, W. J., C. E. Miller, and T. M. Flynn, Flow Measurement, Part 2: Mass Flowmeters in Cryogenics, *Mech. Eng.,* Vol. 89 (May 1967), pp. 105–113.

61. Li, Y. T., and S. Y. Lee, A Fast Response True Mass Rate Flowmeter, *ASME Trans.,* Vol. 75 (1953), p. 835.

9

Temperature Measurements

9-1. INTRODUCTION

The objective of this chapter is to give an overview of modern industrial temperature sensors. The emphasis is on thermocouples and resistance devices, since they are used for more than 80% of temperature measurements in industry [1]. Pyrometers and integrated circuits are gaining popularity. Some sensors for temperature monitoring are mentioned for completeness.

Temperature is measured by its effect. Any object that has a property influenced by temperature is potentially a thermometer. The effect may be a change in (1) the physical/chemical states of an object, (2) dimensions, (3) electrical properties, (4) radiation properties, and (5) others [2]. The topics in the chapter are also grouped by the type of changes.

The temperature of an object can also be related to the mean kinetic energy of its molecules, but the kinetic energies are not measurable at present. The thermodynamic temperature scale based on the Carnot cycle is independent of material properties, but it is not practical and gives only the ratio of temperatures. The temperature scale based on ideal gas law is identical to the thermodynamic scale [3], but its implementation is not practical except in a standards laboratory [4].

The four fundamental units in measurement are mass, length, time, and temperature. The first three are extensive quantities. For example, the addition of two bodies of equal mass gives twice the mass. Temperature is an intensive quantity. When two bodies of the same temperature are brought together, they are in thermal equilibrium, and no change in temperature is observed. This is the basis for using fixed points in the International Practical Temperature Scale (IPTS) for calibration. Instrumentation and temperature measurement practices are covered by many standards [5]. For the foreseeable future, the IPTS based on property of materials is the ultimate standard temperature scale.

9-2. INTERNATIONAL PRACTICAL TEMPERATURE SCALE [6]

The *International Practical Temperature Scale* (IPTS) is based on changes in the physical state of substantances, such as from a solid state to a liquid state. It is set up by agreement to conform as closely as practical with the thermodynamic scale. Several revisions were made since its acceptance in 1927 [7]. The IPTS-68 is based on six primary and many secondary reproducible equilibrium temperatures (fixed points), to which numerical values are assigned. The primary and secondary fixed points are shown in Table 9-1, in which the asterisks (*) denote the primary fixed points. A freezing point (fp) is an equilibrium temperature for the liquid-solid phase of a substance, a boiling point (bp) for the liquid–vapor phase, and a triple point (tp) for the solid–liquid–vapor phase. All primary fixed points are in degrees Celsius, and are at the pressure of 1 standard atmosphere except the triple point of water.

The IPTS-68 is divided into four temperature ranges. Instruments, equations, and precise procedures are specified for the interpolation in each of the ranges. Range 1 is from 13.81 K (tp hydrogen) to 273.15 K (fp water). Range 2 is from 0°C (fp water) to 630.74°C (fp antimony). The interpolating instrument for ranges 1 and 2 is a platinum resistance ther-

TABLE 9-1 Fixed-Point Values; IPTS-68[a,b]

Fixed points	°C	°F	K
tp hydrogen	−259.34	−434.81	13.81
bp hydrogen	−252.87	−432.17	20.28
bp neon	−246.048	−410.89	27.102
tp oxygen	−218.789	−361.820	54.361
*bp oxygen	−182.962	−297.332	90.188
*tp water	0.01	32.02	273.16
*bp water	100.00	212.00	373.15
*fp zinc	419.58	787.24	692.73
*fp silver	961.93	1763.47	1235.08
*fp gold	1064.43	1947.97	1337.58
fp tin	231.9681	449.5426	505.1181
fp lead	327.502	621.504	606.652
bp sulfur	444.674	832.413	717.824
fp antimony	630.74	1167.33	903.89
fp aluminum	660.37	1220.67	933.52

[a]tp, triple point; bp, boiling point; fp, freezing point.
[b]Asterisk denotes primary fixed points.

mometer. Range 3 is from 630.74 to 1064.43°C (fp gold). A platinum/10% rhodium and platinum thermocouple is used for the interpolation. Range 4 is above the gold point, and the IPTS-68 is defined by Planck's radiation formula with an optical pyrometer. If extreme high accuracy is not required, chemicals are available as secondary temperature standards for the 50 to 300°C range, certified to ±0.05% (e.g., Fisher Scientific Co., Pittsburgh, Pennsylvania; Omega Engineering, Inc., Stamford, Connecticut).

9-3. EXPANSION AND FILLED THERMOMETERS

Expansion thermometers are based on dimensional changes, such as the increase in length of metals with temperature. The change in length is quite small. Generally, a differential or some scheme is used to magnify the change. A *filled thermometer* works on the expansion principle if it is completely filled with fluid, and works on the pressure principle if filled with a gas or partially filled with a volatile liquid [8]. Expansion and filled thermometers are described briefly in this section.

A *bimetallic thermometer* measures temperature by means of the differential thermal expansion of two metals. The bimetallic strip, shown in Fig. 9-1a, consists of a bonded composite of two metals. One of the metals is usually a copper alloy and the other Invar, a nickel steel with low thermal expansion coefficient. A temperature change will cause the bimetallic strip to bow, as shown in the figure.

Many types of bimetallic thermometers are commonly used. The bimetallic spiral with a mercury switch at its free end, shown in Fig. 9-1b, is for temperature sensing in a home thermostat. The bimetallic helix, shown in Fig. 9-1c, is placed in a thermometer well to serve as a mercury-in-glass thermometer. In this case, the output is the rotation of a pointer. The bimetallic probe, shown in Fig. 9-1d, can be placed in the duct of a jet engine to measure the average gas temperature [9]. Accuracy of bimetallic thermometers is from ±2 to ±5% and the upper temperature limit is about 500°F.

The familiar *mercury-in-glass thermometer* is an example of a filled thermometer that works on the expansion principle. It consists of a thin-walled glass chamber (bulb) filled with mercury, a uniform capillary in a glass stem with a scale, and an expansion chamber above the capillary for protection. The volume coefficient of expansion of mercury is about eight times that of glass. Due to the difference in coefficient between mercury and glass, the mercury rises up the capillary in the stem to indicate temperature. Other liquids are also used, because mercury freezes at −38.9°C.

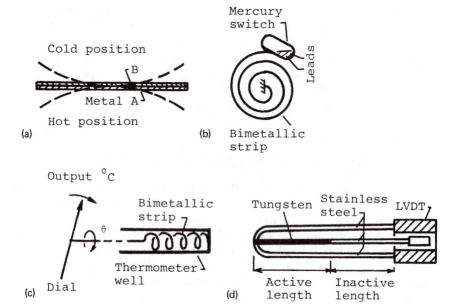

FIGURE 9-1 Bimetallic thermometers. (a) Bimetallic strip. (b) Spiral. (c) Helix. (d) Differential.

If a thermometer is for partial immersion, an immersion ring is etched on the glass stem to indicate the correct depth of immersion. A total immersion type is preferred, because of the uncertainty of the exposed mercury thread to ambient temperature. When a total immersion thermometer is used in a partial immersion mode, the mercury thread above the liquid surface is not at the temperature of the bath. The temperature correction [10] is

$$C_s = kn(T_B - T) \qquad (9\text{-}1)$$

where

   C_s = temperature correction
   k = correction factor (for mercury thermometers, $k = 0.00016$ for Celsius scale, and $k = 0.00009$ for Fahrenheit scale)
   n = number of degrees between the surface of the bath and the end of the mercury thread in the capillary
   T_B = indicated bulb temperature
   T = average temperature of the emergent mercury column, measured by means of another thermometer attached to the stem

Certified mercury-in-glass thermometers are widely used as working standards. For the total immersion type, the maximum accuracy is ±0.01°C from 0 to 150°C, and ±1°C from 300 to 500°C. Error due to viscous flow of glass under stress is about 0.2°F, due to hysteresis effect in heating and cooling is about 0.01°F per 10°F, and due to external pressure is 0.2°F per atmosphere [11]. The ice point must be checked or the thermometer calibrated [12] if an accuracy of ±0.2°F or better is required.

The *constant-volume gas thermometer,* shown in Fig. 9-2a, is an example of a filled system that works on the pressure principle. It follows the gas law and has a pressure gage for its readout. The bimetallic strip shown in the figure is for ambient temperature compensation. The bulb of a gas-filled thermometer tends to be large, but the averaging effect of a large bulb may be advantageous for some applications.

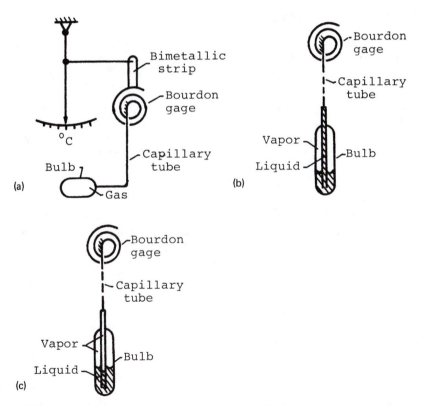

FIGURE 9-2 Fluid-filled thermometers. (a) Constant-volume gas thermometer. (b) Bulb temperature > ambient temperature. (c) Bulb temperature < ambient temperature.

A *vapor-pressure thermometer* has a larger pressure change with temperature than that of a gas thermometer. The bulb can be smaller, and its operation is not influenced by dimensional changes of the bulb. The scheme shown in Fig. 9-2b is used when the bulb temperature is higher than the ambient temperature, and that in Fig. 9-2c used when the bulb temperature is lower than ambient. The discussion of these arrangements is left as an exercise.

Filled thermometers are used mostly for monitoring, although the pressure output can be utilized for pneumatic control. Their dynamic response is slow, and the time constant is of the order of 5 to 10 s. The trend is toward other types of sensors in order to obtain multipoint measurements, electric signal transmission, multiplexing, data logging, and computer control.

9-4. THERMOCOUPLES

The thermocouple, or thermoelectric thermometer, is probably the most versatile and inexpensive temperature sensor. It is applicable for almost the entire temperature range, and is used for 50% of temperature measurements in industry [1]. Numerous metals, alloys, and even refractory materials can be paired as thermocouples [13]. Thermoelectric effects of materials, thermocouple "laws," the gradient approach, construction, and practical thermocouple measurements are described in this section.

A. Thermoelectric Effects

A thermocouple, shown in Fig. 9-3a, consists of two dissimilar metallic wires A and B. It measures the differential junction temperatures T_1 and T_2. The electric current or voltage measured by the meter in the circuit is a function of $(T_1 - T_2)$. Generally, (1) the circuit is as shown in Fig. 9-3b, with the "hot" junction at T_{Hot} and the "cold" junction at T_{Ref}, the reference temperature of an electrically insulated isothermal block; and (2) the voltage is measured under zero-current condition by means of a thermocouple potentiometer (see Fig. 3-1b) or a sensitive high-input-impedance voltmeter. In other words, the thermal electromotove force (EMF) is measured with an open circuit.

There are three effects in a thermocouple circuit: the Seebeck, Peltier, and Thomson effects. The Seebeck effect describes the open-circuit voltage developed in a thermocouple circuit. The Peltier effect relates the reversible heating and cooling that usually occurs when an electric current crosses a junction between two dissimilar metals. The Thomson

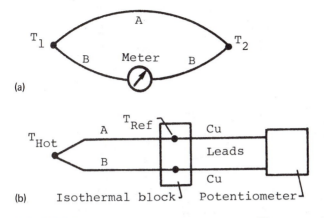

(a)

(b)

FIGURE 9-3 Basic thermocouple circuits. (a) Thermocouple circuit. (b) Circuit with potentiometer readout.

effect relates the reversible heating and cooling in a homogeneous conductor, subjected to a thermal gradient and a current flow. The physical effects can be explained from a macroscopic viewpoint by irreversible thermodynamics [14]. Generally, no current is drawn from a thermocouple circuit in a temperature measurement, and the Seebeck effect is sufficient to explain the circuit behavior.

The voltage–temperature $(E-T)$ characteristics of materials that are usually paired as thermocouples are shown in Fig. 9-4a. The slope S_A of the $E-T$ curve of a material A is its voltage–temperature sensitivity, or the *Seebeck coefficient,* commonly called the *thermoelectric power*. S_A is a material property of A, but it can not be determined alone because it takes two dissimilar materials to produce a Seebeck EMF. Traditionally, pure platinum is used as the reference material. Hence S_A is not an absolute coefficient. It should be written as $S_{AP} = S_A - S_P$, referring to platinum as the datum.

The Seebeck EMF E_{AB} from a pair of thermocouple wires $A-B$ is illustrated in Fig. 9-4a. Let S_A be the sensitivity of material A (chromel). S_A is positive, because the $E-T$ curve of A has a positive slope relative to platinum. The Seebeck EMF of material A is E_{AP} at a given temperature. Similarly, S_B is the sensitivity of material B (constantan). S_B is negative, and its Seebeck EMF is E_{PB}. Evidently, the Seebeck EMF of thermocouple $A-B$ is $E_{AB} = E_{AP} + E_{PB}$. The Seebeck coefficient of thermocouple $A-B$ is $S_{AB} = S_{AP} - S_{PB}$.

If the thermal EMF of each material realtive to platinum is known, as shown in Fig. 9-4a, the EMF from the combination of two materials is the

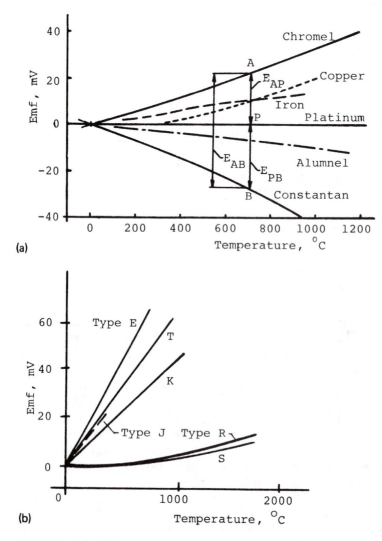

(a)

(b)

FIGURE 9-4 Voltage–temperature characteristics of thermoelements and thermocouples. (a) EMF of thermal elements versus platinum. (b) EMF of thermocouples.

algebraic sum of their EMFs. In other words, it is only necessary to determine the voltage–temperature characteristics of each material with respect to a reference separately. Thermocouples can then be selected from pairs of these materials without a recalibration.

A number of common thermocouples have been given designations by the Instrument Society of America ISA. Types E, J, K, and T are *base-metal thermocouples*. They are useful up to about 1000°C. Types S, R, and B are *noble-metal thermocouples*. They are useful up to about 2000°C.

Type E: chromel vs. copper–nickel alloy (constantan)
Type J: iron vs. copper–nickel alloy (constantan)
Type K: chromel vs. nickel–aluminum alloy (alumel)
Type T: copper vs. copper–nickel alloy (constantan)
Type S: platinum/10% rhodium vs. platinum
Type R: platinum/13% rhodium vs. platinum
Type B: platinum/30% rhodium vs. platinum/6% rhodium alloy

It is important to note that the names for the alloys above identify only the voltage–temperature characteristics of the alloys, but not the exact chemical composition. For example, the constantan for type J thermocouple is not thermoelectrically interchangeable with the constantan for type T.

Typical voltage–temperature characteristics of thermocouples are shown in Fig. 9-4b. Reference tables for thermocouples and the individual materials (thermoelements) relative to platinum are given in the NBS monograph 125 [15]. The Seebeck coefficients are also tabulated because they are not constants; that is, the Seebeck EMF is not linear with temperature. A scheme is used to designate the polarity of thermocouple wires. If wire A is positive relative to wire B, A is positive relative to B at the reference junction, when $T_{\text{Hot}} > T_{\text{Ref}}$ (see Fig. 9-3b). The first-named material of a thermocouple is always positive, and the second-named material is negative. For example, iron is positive and constantan negative for an iron–constantan thermocouple. Color codes are used to identify the type of couples and the polarity of thermocouple wires; however, red, or red with a color trace, is always negative.

B. Thermoelectric Laws

Referring to Fig. 9-5, the behavior of thermocouple of homogeneous materials can be described by thermoelectric laws [16].

1. Intermediate Temperature

The thermal EMF E_{AB} of a thermocouple A–B due to temperatures T_1 and T_2 at the junctions is not affected by any intermediate temperature T_3 in the circuit, as shown in Fig. 9-5a. This allows thermocouple wires to be exposed to unknown and varying temperatures of the environment.

2. Intermediate Metal

The thermal EMF E_{AB} of the circuit shown in Fig. 9-5b is not affected by any intermediate material C, provided that the junctions are at the same temperature, T_3. This permits the insertion of a material at any intermediate point in the circuit, such as shown in Fig. 9-5c. This becomes the basic thermocouple circuit shown in Fig. 9-3b, in which T_2 is at T_{Ref} of the isothermal block. Another implication is that soldering, welding, or other methods can be used to form thermocouple junctions, since the solder is an intermediate material. Thermocouple wires can be welded directly to a metal to measure its temperature, because the metal is an intermediate material. In fact, a weld may be unnecessary if good contacts can be made between the thermocouple wires and the metal.

3. Successive Metals

From Fig. 9-5d, if the thermal EMF for thermocouple A–C of materials A and C is E_{AC} and that for C–B is E_{CB}, the EMF for the thermocouple A–B is $E_{AB} = E_{AC} + E_{CB}$. This allows the pairing of materials to form thermocouples.

4. Successive Temperature

Referring to Fig. 9-5e, if E_{21} is the thermal EMF due to the temperatures (T_2, T_1) and E_{32} due to (T_3, T_2), the thermal EMF due to (T_3, T_1) is $E_{31} = E_{32} + E_{21}$. This permits the use of standard tables referenced at $0°C$ when the actual T_{Ref} is at another temperature. For example, let $T_3 =$ unknown temperature being measured, $T_2 = 20°C =$ actual T_{Ref}, and $T_1 = 0°C =$ standard reference temperature. From reference tables, $E_{21} = 1.019$ mV. Let the measured thermal EMF be $E_{32} = 6.438$ mV. Thus $E_{31} = E_{32} + E_{21} = 6.438 + 1.019 = 7.457$ mV. The unknown T_3 from tables referenced at $0°C$ is $140°C$.

C. Gradient Approach to Thermocouple Circuitry

The gradient approach to thermocouple circuitry by Moffat [17] is an invaluable supplement to the thermocouple laws, which assume homogeneous materials. Yet inhomogeneity is often inevitable due to environmental contamination, cold work, and selective oxidation at high temperatures [18]. Note that a thermocouple always gives an output signal, regardless of its validity or the condition of the installation, unless a wire is broken. The gradient approach gives a systematic method of dealing with inhomogeneity, to troubleshoot, and to understand the behavior of more complex thermocouple circuits.

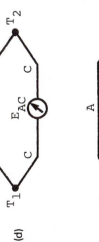

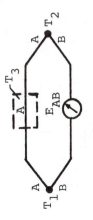
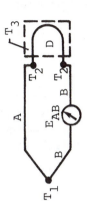
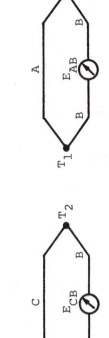
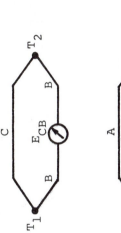

FIGURE 9-5 Circuits illustrating thermoelectric laws. (a) Homogeneous circuit. (b), (c) Intermediate metal. (d) Three metals. (e) Successive temperature.

The gradient approach states that the net thermal EMF E_{net} in a thermocouple, consisting of materials A and B, is due to the thermal gradient along the wires.

$$E_{net} = \int_0^L S_A \frac{dT}{dx}\, dx + \int_L^0 S_B \frac{dT}{dx}\, dx \tag{9-2}$$

where S_A and S_B are the total thermoelectric power of materials A and B, T = temperature, x = distance along the wires, and L = length of wire. It can be shown that Eq. 9-2 does not invalidate the thermoelectric effects described above. If S_A and S_B are not functions of position, then E_{net} is a function of the junction temperatures only.

$$E_{net} = \int_{T_0}^{T_L} S_A\, dT + \int_{T_L}^{T_0} S_B\, dT \tag{9-3}$$

$$E_{net} = \int_{T_0}^{T_L} (S_A - S_B)\, dT \tag{9-4}$$

Let us illustrate the method with a simple example. The steps to find E_{net} for the thermocouple of materials A and B, shown in Fig. 9-6a, are as follows:

1. Identify the points of interest (1,2,3) and assign nominal temperatures (T_{Ref}, T_{Hot}) to each point.

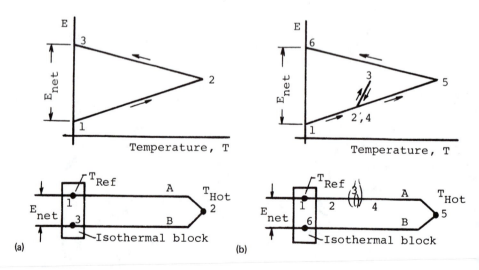

FIGURE 9-6   Gradient approach to find E_{net}. (a) E_{net} from thermocouple AB. (b) Intermediate temperature.

2. Obtain the EMF versus temperature $E-T$ calibration curves for the materials (see Fig. 9-4a).
3. Starting from T_{Ref} at point 1, construct a curve in the $E-T$ plot with the slope S_A of metal A until T_{Hot} at point 2, where A contacts B.
4. Similarly, starting from point 2, construct a curve with the slope S_B back to T_{Ref} at point 3.
5. The E_{net} for the thermocouple is from the $E-T$ plot. Note that the graphical construction simply performs the line integration in Eq. 9-3.

Using the gradient approach, the $E-T$ plot to verify the "law of intermediate temperature" is shown in Fig. 9-6b. The points of interest are from 1 to 6. Following the procedure above, the steps for obtaining E_{net} from the $E-T$ plot are self-evident.

The decalibration of a thermocouple and the difficulties in its recalibration are well known. The thermocouple, shown in Fig. 9-7a, is used to measure the temperature of a metal bath. Since steep thermal gradients exist in the shaded section, this section will produce virtually the entire output EMF. Now, let the decalibration of the thermocouple wires, shown in Fig. 9-7b, occur in the sections between points 2–3 and 5–6, and the thermocouple is being recalibrated. The output E is from the recalibration with section 3–4–5 at the same temperature. The output E^* is from the recalibration of the same thermocouple with section 2–3–4–5–6 at the same temperature, as shown in Fig. 9-7c. The EMF E^* is generated by the unaffected wires, because all the affected wires are at a uniform temperature, that is, by the good thermocouple wires that have not been decalibrated. The result of the recalibrations is not predicted by the thermocouple laws above. The gradient approach is a useful operational method for troubleshooting.

D. Reference Junctions

The thermal EMF E_{net} of a thermocouple is a function of $(T_{Hot} - T_{Ref})$. Hence the T_{Ref} of the reference junction must be known accurately. A $\pm 1°C$ error in T_{Ref} will result in a $\pm 1°C$ error in the measured temperature.

If the T_{Ref} is at $0°C$, reference tables can be used to convert the measured EMF directly into temperature. An ice bath, shown schematically in Fig. 9-8a, is the simplest method to keep the T_{Ref} at $0°C$. Details of an ice bath are described in the literature [19]. However, errors of more than $\pm 1°C$ are often found in a poorly made ice bath [20]. An ice-bath is generally restricted to laboratory use, since it is awkward for a mobile test.

When the T_{Ref} is at a fixed temperature, software compensation can be

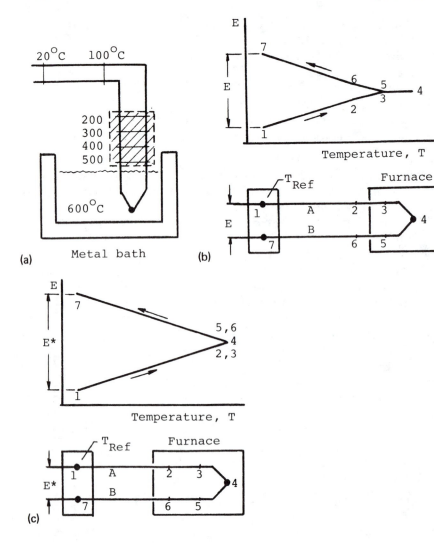

FIGURE 9-7 Recalibration of decalibrated thermocouple. (a) Thermal gradient. (b), (c) Recalibration.

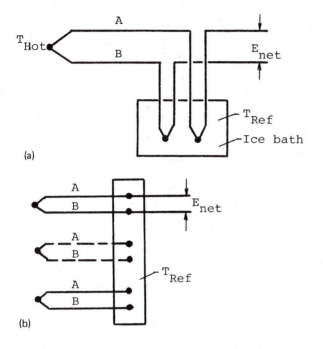

(a)

(b)

FIGURE 9-8 Thermocouple reference junctions. (a) $T_{Ref} = 0°C$. (b) $T_{Ref} > 0°C$.

used in a data acquisition system to convert the thermocouple output to temperature. For example, when the T_{Ref} is at $0°C$, the conversion from the output x mV to $T°C$ is expedited using a power series polynomial [21] such as

$$T = a_0 + a_1 x + a_2 x^2 + a_3 x^3 + \cdots + a_n x^n \tag{9-5}$$

where the a's are coefficients unique to each type of thermocouple. It is expedient to compute the temperature, because to store the lookup-table values in a computer could consume an inordinate amount of memory. Similarly, the temperature can be computed when the T_{Ref} is at a known elevated temperature instead of $0°C$. A large isothermal block, as shown in Fig. 9-8b, can be used to maintain a fixed T_{Ref} for many themocouples.

Many schemes are used for hardware reference junction compensation [22]. Essentially, a compensating voltage E_{Ref} is added to the thermocouple output to simulate a constant T_{Ref} at $0°C$. The scheme shown in Fig. 9-9a consists of a bridge circuit in which R_4 is temperature sensitive. Both R_4 and T_{Ref} are allowed to drift with the ambient temperature. The voltage change across R_4 is added to the thermocouple output for the com-

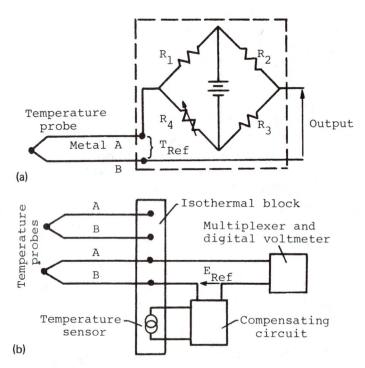

(a)

(b)

FIGURE 9-9 Compensation for drifts in reference junction. (a) Bridge circuit. (b) Multiple input.

pensation. The scheme shown in Fig. 9-9b is widely used in data acquisition systems. The T_{Ref} at the isothermal block is allowed to drift with the ambient temperature. A sensor at the isothermal block measures the T_{Ref} and actuates a circuit to yield a compensating voltage E_{Ref} for the thermocouples.

E. Thermocouple Probes

General- and special-purpose thermocouple probes are available for all temperature ranges [23]. General-purpose thermocouples are fabricated with a soldered, welded, or butt joint, as shown in Fig. 9-10a to c, in which A and B are thermocouple wires. Details for fabrication are given in the literature [24].

Three special-purpose thermocouples are illustrated in Figs. 9-10d to f. The intrinsic couple in Fig. 9-10d has an extremely fast response [25]. The couple is made from 0.001-in. wires, flattened to 200-μin. ribbons, and

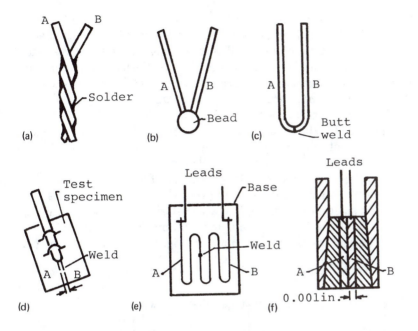

FIGURE 9-10 Fabrication of general- and special-purpose thermocouple probes. (a) Soldered junction. (b) Beaded junction. (c) Butt-welded junction. (d) Intrinsic welded. (e) Filament sensor (courtesy of BLH Electronics, Waltham, Mass.). (f) Renewable junction (Model P, courtesy of Nanmac Corp., Indian Head, Md.).

spot welded to the surface of a test specimen. The metal between the wires is an intermediate metal in the circuit. The time constant of this thermocouple is about 20 μs. The filament thermocouple shown in Fig. 9-10e is on an insulating base, similar to a strain gage. It can be attached to an irregular surface. The renewable junction thermocouple, shown in Fig. 9-10f is for extremely abrasive environments, such as the cone of a rocket engine. The sensor consists of 0.001-in. thermocouple ribbons insulated by 0.0002-in. mica. The abrasive action destroys the junctions of the thermocouple, but as fast as the old junctions are destroyed, new junctions are formed continuously by the abrasive action itself. In the same vein, a fine-wire platinum thermocouple (not shown) that destroys itself after one reading is available for temperature sensing of hot, molten metals with a 0.25% accuracy (e.g., Leeds and Northrup, North Wales, Pennsylvania). The thermocouple is destroyed, but it has performed the task and costs less than a dollar.

Thermocouples for industrial applications are usually protected by sheathing, as shown in Fig. 9-11a to c, in which A and B are the thermocouple wires. The thermocouple may be insulated from the sheath, grounded to the sheath, or end welded to the sheath. For faster response, the thermocouple junction may protrude beyond the sheath (not shown).

Many special probes are commercially available. The probe shown in Fig. 9-11d is for sensing the surface temperature of a moving treadline of nylon or similar textile material [26]. The sensor consists of a cross arrangement with a thermocouple junction. Half of the cross is heated by an electric current, and the system is initially balanced. It is unbalanced when a treadline touches the junction. Temperature is measured by the current required to rebalance the system. Another example is a thermocouple in the form of a leave spring (not shown) to measure the temperature of a smooth-moving surface with velocity up to 300 ft/min (e.g., Series 68000, Omega Engineering, Inc., Stamford, Connecticut). The simple total-temperature probe shown in Fig. 9-11e consists of a ther-

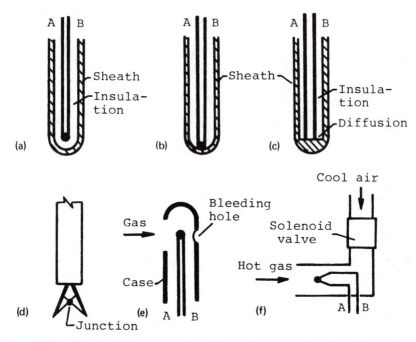

FIGURE 9-11 Industrial general- and special-purpose thermocouple probes. (a) Insulated junction. (b) Grounded junction. (c) End-welded probe. (d) Filament probe. (e) Total temperature (from Ref. 27). (f) Pulse-cooled sensor (from Ref. 28).

mocouple in a cavity. It is used when maximum economy, small size, and moderate accuracy are required.

The pulse-cooled probe shown in Fig. 9-11f extends the range of a chromel–alumel thermocouple (melting point 2550°F) to 7000°F. Normally, the thermocouple is air cooled. The solenoid valve shuts off the normal cooling air in order to measure the hot gas temperature T_{gas}. Thus a step temperature input is applied to the thermocouple, and it heats up as a first-order instrument (see Sec. 4-4C and Eq. 4-50):

$$\tau \frac{dT}{dt} + T = T_{gas} \tag{9-6}$$

where τ is the time constant of the thermocouple, T the instantaneous temperature, and dT/dt the time rate of change of T. Both τ and dT/dt are measured with additional circuitry and a computer. Thus T_{gas} can be estimated from the sum $[\tau \, dT/dt + T]$. The air is turned on again before the thermocouple becomes overheated.

F. Practical Thermocouple Measurements

The overall error in a thermocouple measurement arises from many sources, from materials tolerances of wires, through all stages of the measurement, to data acquisition and noise documentation (see Secs. 3-7 and 3-8). Noise consideration is particularly important for thermocouple measurements, because of the low-level signal. For example, the Seebeck coefficient of base-metal thermocouples is of the order of 40 μV/°C and that of noble-metal couples 7 μV/°C. Hence the detection of 0.1°C requires resolutions of 4 μV and 0.7 μV, respectively, for the overall system. Instead of a systematic discussion of errors, some common problems in thermocouple measurements are described in this section.

1. Thermocouple Wires

The quality of thermocouple wires must conform to standard reference tables. The industry-accepted ANSI Standard specify the allowable deviations for a temperature range [29]. The limits are materials tolerances for thermocouple wires only, that is, the possible error inherent of new thermocouples, prior to an exposure to adverse temperature or operating conditions. The error limit for special-grade wires is one-half that of the standard grade.

For example, the ASTM wire error [30] for a 8 AWG type J standard grade thermocouple is ±2.2°C from 0 to 277°C and ±0.75% from 277 to 760°C, as shown in Fig. 9-12. The possible error at 760°C is ±[2.2 + 0.0075(760 − 277)] = ±5.8°C, and the thermocouple still meets the guar-

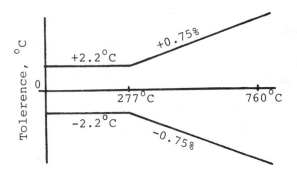

FIGURE 9-12   ASTM Standard for No. 8 AWG Iron–constantan thermocouple.

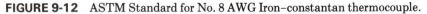

antee. Again, this is an "acceptable" error for a new thermocouple, not necessarily its accuracy in service. Evidently, a thermocouple should be calibrated individually when a higher accuracy is required.

Less expensive extension wires for thermocouples have the same error limit, but their applications are restricted to lower temperature ranges. Extension-grade wires for base-metal thermocouples are of the same material as the thermocouples. Extension-grade wires for noble-metal thermocouples are made from less expensive proprietary alloys, to match the relative Seebeck coefficient of the thermocouples. Since extension wires are relegated to much lower temperature ranges, their thermal gradient is small and their contribution to the overall error is minimal.

2. Practical Thermocouple Measurements

Some errors pertaining to thermocouple operations are described. Topics such as decalibration and reference junctions are not repeated here. Note that long-term stability and repeatability are often as important as absolute accuracy for some applications [31].

a. Thermal Shunting. A thermocouple is designed to measure the temperature of its "hot" junction. Thermal shunting is due mostly to the heat transfer at the hot junction. The thermocouple and its sheath may change the temperature of the hot junction and/or the local temperature at the point of measurement. In other words, thermal shunting is a thermal "loading" problem. Loading can occur under both steady-state and dynamic conditions.

A method to minimize thermal shunting in a surface temperature measurement is to place the leads from the hot junction along an isotherm, as shown in Fig. 9-13a. An elaborate system to avoid thermal shunting for a surface temperature measurement is shown in Fig. 9-13b.

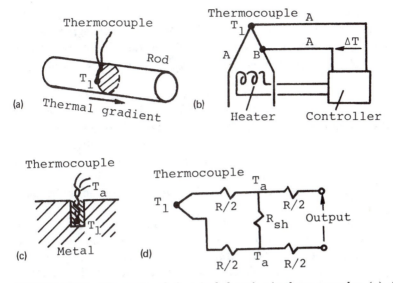

FIGURE 9-13 Thermal and electrical shunting in thermocouples. (a), (b) Surface temperature measurement [(b) from Ref. 32]. (c), (d) Electrical shunts.

The thermocouple A–B is used to measure the surface temperature T_1. A portion of B is used as a differential thermocouple. A ΔT is detected if there is a thermal gradient along B, that is, if there is heat transfer to or from T_1. The signal ΔT is fed back to regulate a heater to render ΔT to zero. Thus the thermocouple wires are long an isotherm.

b. Electrical shunting. Electrical shunting may be due to a local short circuit or a deterioration of insulation. A short circuit is shown in Fig. 9-13c. The wires are shorted at T_a. Thus T_a become the temperature of the hot junction instead of T_1. A short between lead wires may be due to a mechanical damage. Moisture is often the culprit in lowering the resistance between leads. Moreover, the voltaic effect, due to moisture and the dye for wire insulation, or moisture and traces of flux from soldering, may be of the order of millivolts. The error follows no particular pattern [33].

The distributed shunt resistance in a thermocouple circuit is generally caused by a deterioration of insulation. The resistivity of common oxide insulations decreases exponentially with increasing temperature. Hence shunting is more prominent at high temperatures, such as in the insulation between the thermocouple wires in a sheath. It is a good practice to avoid using thermocouples near their maximum rated temperatures. The schematic of an electrical shunt is shown in Fig. 9-13d, where R_{sh} is the shunt resistance and R that of the thermocouple. Errors for different degree of shunting can be estimated [34]. Shunting can only degrade a temperature measurement. If $R_{sh} \ll R$, the T_a at the shunt becomes a virtual junction in the circuit, and the measured temperature is dictated by T_a.

c. Series–parallel circuits. The series circuit (thermopile) shown in Fig. 9-14a is often used to increase the voltage output for greater sensitivity. All junctions must be electrically insulated from one another. A series circuit is analogous to batteries in series. The total voltage output is the sum of the individual voltages. If the hot junctions are at T_a, T_b, and T_c and the cold junctions at T_{Ref}, an arithmatic mean of the hot-junction temperatures can readily be obtained. This is the preferred method for temperature averaging.

The parallel circuit shown in Fig. 9-14b is also used for temperature averaging. The circuit is analogous to batteries in parallel, but current may flow among the thermocouples. The parallel circuit for averaging should be used with caution [35] since there is an undetermined current flow in the circuit and the voltage output is not entirely from Seebeck effect.

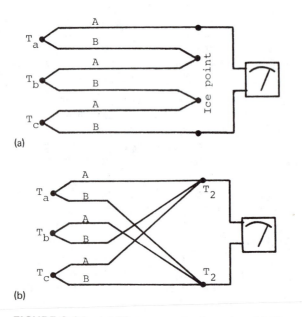

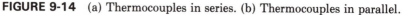

FIGURE 9-14 (a) Thermocouples in series. (b) Thermocouples in parallel.

3. Data Transmission

Line-related noise, or error due to data transmission, is described briefly. The low-level dc signal from a thermocouple can be completely masked by noise pickups. A guard input [36] is generally provided at the digital voltmeter (DVM) for noise reduction.

A high common-mode voltage V_{CM} (see Sec. 3-1C) may occur in a thermocouple circuit. For example, assume that a grounded-junction thermocouple is used to measure the temperature of molten metal in an electric furnace. The thermocouple is placed in the molten metal and a 240 V_{rms} supply is used to heat the metal. Thus the V_{CM} at the thermocouple junction is on the order to 100 V.

As shown in Fig. 9-15a, the V_{CM} causes a parasitic current I in the ground loop, which consists of the V_{CM}, the R_s of the LO lead, and the stray capacitance from the LO terminal of the DVM to ground. The IR_s drop in the LO lead is a signal between the HI and LO terminals at the DVM. This is

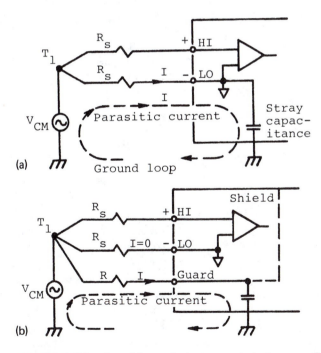

FIGURE 9-15 Guard input in instruments to increase CMR. (a) Error due to parasitic current. (b) Passive guard.

interpreted as a signal from the thermocouple. Note that the HI terminal represents a higher input impedance than the LO terminal, because the power supply in the instrument is referenced to the LO terminal. Hence the ground loop flows through the LO lead.

The guard, shown in Fig. 9-15b, is physically a metal box surrounding the entire voltmeter circuit. It is connected to the shield of the thermocouple wires to shunt the parasitic current I. The guard input provides a low-impedance path for the V_{CM}. If R_s is 1 kΩ, guarding can yield a 40-dB common-mode rejection [37]. Many schemes, including active electronic components, are used to minimize line-related noise in instruments [38].

Other noise reduction techniques described in this book are filtering/shielding and analog-to-digital (A/D) conversion. The power-line-related noise and its harmonics are virtually eliminated when the integration during the A/D conversion is an integer multiple of the power-line frequency (see Sec. 6-7E).

9-5. RESISTANCE TEMPERATURE DETECTORS

Resistance temperature detectors (RTDs) are simply resistive elements. The resistance of metals increases with temperature. This common property is used for temperature sensing. Characteristics and applications of metallic RTDs are described in this section.

A. Materials and Construction

Common materials for RTDs are platinum (from -260 to $1000°C$), copper (from -200 to $260°C$), nickel and Balco (70% Ni/30% Fe) (from -100 to $230°C$), and tungsten (from -100 to $2500°C$). The temperatures in parentheses are approximate figures. Temperatures recommended by manufacturers may differ considerably, depending on the construction of the sensor, materials for the capsule and sheathing, and the duty cycle in application. For example, a sensor can be exposed to a higher-than-steady-state temperature for a short duration without damage.

The sensing element can be broadly classified as wire-wound or film type. The latter is a more recent introduction. The wire-wound RTD, illustrated in Fig. 9-16a, is an early design to obtain strain-free wires. Helical coils of annealed platinum wire are loosely supported on a crossed mica web. The assembly is encapsulated in a glass tube for protection. This RTD is fragile and the thermal coupling between the sensing element and the point of measurement is poor.

The partially supported sensing element, shown in Fig. 9-16b, is more

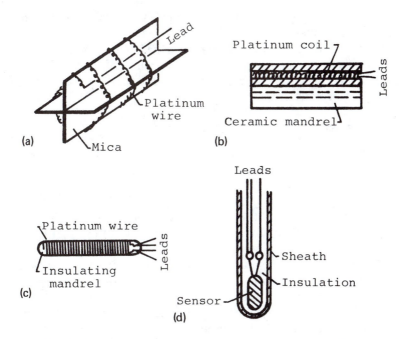

FIGURE 9-16 Construction of RTD sensors and probe. (a) Platinum RTD. (b) Partially supported. (c) Fully supported. (d) RTD probe.

suitable for industrial applications. It consists of small coils of wire inserted into the axial holes of an insulating mandrel. Adhesive is introduced into the holes and the assembly is fired. Part of each turn of the coils is thus sintered to the mandrel, but the remainder of each turn is free.

A fully supported and less expensive element is shown in Fig. 9-16c. The wire for sensing is wound on an insulating mandrel and then coated with an insulation. A fully supported element is more rugged and can survive shocks of 100g. It must be carefully designed to minimize a strain-related resistance change.

A RTD probe usually consists of an encapsulated sensing element in a protective sheath, as shown in Fig. 9-16d. General-purpose and special probes are commercially available [39]. Some are designed to operate in fluids, and others for the surface temperature measurement of solids. The time constant of wire-wound RTDs ranges from 0.5 to 10 s, depending on the heat transfer rate of the service condition.

Metallic film RTDs (not shown) are made by depositing a metal film on a substrate and then encapsulated. The film type is usually smaller than the wire-wound type and has a shorter time constant. Bondable film of nickel, Balco, or copper foil can be handled like strain gages (e.g., TG Series, Measurement Group, Inc., Raleigh, North Carolina). These RTDs are used for surface temperature measurements from −195 to 250°C. Due to the low thermal mass and large bonded area, the time lag of bondable RTDs is almost negligible.

B. Characteristics and Standards

The resistance–temperature $(R-T)$ characteristics of metals can be expressed as a polynomial:

$$R = R_0(1 + a_1 T + a_2 T^2 + \cdots + a_n T^n) \tag{9-7}$$

where the a's are constants, R the resistance at a temperature T, and R_0 the resistance at base temperature T_0. A polynomial interpretation is used in the IPTS-68 to define temperatures from −190 to 660°C between fixed points by means of a platinum RTD [40]. Platinum and copper have almost linear $E-T$ characteristics for a reasonable temperature range, as shown in Fig. 9-17. Nickel and Belco (70% Ni/30% Fe) are nonlinear, but Belco bondable sensors are easy to fabricate and the alloy has 2.4 times the resistivity of pure nickel. The result is lower-cost sensors and the ability to make higher-resistance sensors in smaller sizes [41].

When a material is almost linear over a temperature range, the first approximation from Eq. 9-7 can be used to determine temperature from a measured resistance.

$$R = R_0(1 + \alpha T) \tag{9-8}$$

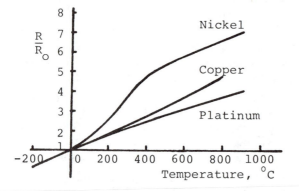

FIGURE 9-17 Resistance–temperature characteristics of metals.

From Eq. 9-8, the α or average value of the temperature coefficient of resistance between 0 and 100°C is

$$\alpha = \frac{R(100°C) - R(0°C)}{100 \times R(0°C)} \tag{9-9}$$

where α has dimensions of $(\Delta R/R)/°C$. Since α is a linearized value, the base temperature for the linearization, commonly at 0°C, must be specified [42] (see Fig. 2-11). An equivalent expression for Eq. 9-9 is the resistance ratio:

$$\text{resistance ratio} = \frac{R(100°C)}{R(0°C)} \tag{9-10}$$

For example, if $\alpha = 0.003850$ for a 100-Ω platinum RTD, $R(100°C):R(0°C) = 138.50:100 = 1.3850:1$.

Over a limited temperature range, an unknown temperature T can be calculated conveniently from the measured resistance R by means of Eq. 9-8.

$$T = \frac{R/R_0 - 1}{\alpha} \tag{9-11}$$

For example, if $R = 132.5$ Ω for a 100-Ω RTD and $\alpha = 0.00385$, the temperature is $T = (132.5/100 - 1)/0.00385 = 84.42°C$.

Unfortunately, Eq. 9-11 must be used with caution, because there are several standards of α for platinum RTDs [43]. A manufacturer may use its own standard or that of several specifying agencies: for example, $\alpha = 0.00392$ (US-MIL-T-24388) and $\alpha = 0.003850$ (German-DIM43760). The difference between these two standards is 1.9%. Higher-purity platinum is required for the U.S.standard, but the trend is toward the Europen standard (DIM). Standards in some countries may differ. The British standard includes grade I and grade II. The nominal resistance of platinum RTDs at 0°C is 100 Ω, but the American Scientific Apparatus Manufacturers Association standard is 98.129 Ω [43].

The overall resistance of RTDs is not standardized. Platinum RTDs with resistances of 10, 50, 100, 200, 1000, and 2000 Ω are available for different applications. Many "standard" values are found in sales literature, such as 10 Ω for copper RTDs ($T_{Ref} = 25°C$), 120 Ω for nickel RTDs ($T_{Ref} = 0°C$), and 670 Ω for nickel–iron RTDs ($T_{Ref} = 25°C$). The resistance of thin film RTDs can be thousands of ohms.

C. Circuits

Since a RTD is a resistive element, the basic circuit for its measurement is a bridge or an ohmmeter. A precision bridge is necessary for the resistance measurement. For example, the sensitivity of a 100-Ω platinum RTD is 0.385 Ω/°C, and a modest 0.1°C accuracy would require a resolution of 0.0385 Ω. This is a small value, although a high-precision Mueller bridge with an uncertainty not exceeding 3 $\mu\Omega$ is possible [44]. On the other hand, a reasonable output is obtainable by measuring the voltage across an RTD. A current of 1 mA through a 100-Ω RTD gives an output of 100 mV, which is a large signal compared with that from a thermocouple. The associated I^2R heating is examined later.

The null-balance bridge, shown in Fig. 9-18a, is a common technique for resistance measurements, where R is the resistance of the RTD and the DVM is used as a null detector (see Sec. 3-4). The circuit is modified, as shown in Fig. 9-18b, for three reasons.

1. R must be physically separated from the bridge by extension wires to avoid subjecting the balancing resistors to the temperature of the RTD.
2. The bridge has three lead wires. The source leads A and B have no effect on the bridge balance, and the third lead C is a sensing lead and carries no current.
3. A potentiometer is used to balance the bridge. This avoids the uncertainty of the contact resistance for balancing at R_4, as shown in Fig. 9-18a.

When the bridge is used in an unbalanced mode, the voltage output V_o is not linear with the resistance in the RTD. This nonlinearity of an unbalanced bridge is acceptable for metallic strain gages, because $\Delta R/R$ is small in strain measurements (see Eq. 3-10), but the $\Delta R/R$ is large for a RTD. For example, if $\alpha = 0.00385$, the $\Delta R/R$ for a $\Delta T = 100$°C is 38.5%. If the bridge is initially balanced, the resistance R from Fig. 9-18a is

$$R = \frac{V_i + 2V_o}{V_i - 2V_o} R_2 \tag{9-12}$$

If the resistance R_L of the leads must be considered, the R of the RTD from the circuit in Fig. 9-18b is

$$R = \frac{V_i + 2V_o}{V_i - 2V_o} R_2 + \frac{4V_o}{V_i - 2V_o} R_L \tag{9-13}$$

The resistance R of a RTD can be measured with a constant-voltage or constant-current drive (see Sec. 3-5). The constant-voltage drive, shown in Fig. 9-18c, is not recommended, because the resistance R_L of the leads may cause a large error. For example, if $R_L = 2\Omega$ and the sensitivity of the RTD is 0.385 Ω/°C, the output is biased by an error of 2/0.385 = 5.2°C.

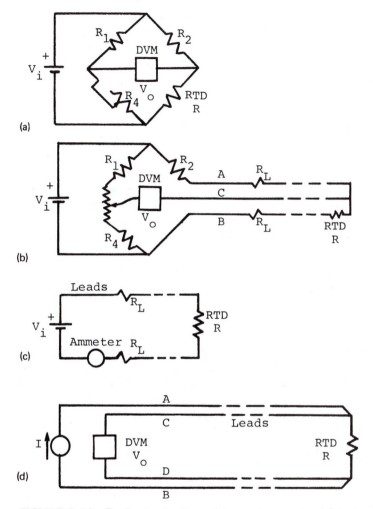

(a)

(b)

(c)

(d)

FIGURE 9-18 Basic circuits for resistance–temperature detectors. (a) Bridge circuit. (b) Bridge circuit with extended leads. (c) Constant-voltage drive. (d) Constant-current drive.

A four-wire ohmmeter circuit with a constant-current drive, shown in Fig. 9-18d (see also Fig. 3-51), is preferred, because the constant current is independent of the resistance of the source leads A and B. The input impedance of the DVM is of the order of 100 MΩ; therefore, the resistance in the signal leads C and D has no effect on the voltage signal at the DVM.

D. Sources of Error

Both the thermocouple and the RTD are susceptible to errors from materials tolerances, electrical shunting, and thermal shunting. Materials tolerances for RTDs are similar to that for thermocouples (see Fig. 9-12). If the material standard is $\pm 0.3°$C at $0°$C and up to $\pm 3°$C at $600°$C, the readings from two RTDs at $600°$C may differ by as much as $6°$C. Thermal shunting may be more a problem with RTDs because the physical bulk of the RTD is greater than that of the thermocouple.

The RTD is also susceptible to errors in lead resistances, self-heating, and thermoelectric effects. The effect of lead resistance is minimized by using larger lead wires or a four-wire ohmmeter circuit as shown in Fig. 9-18d.

The *self-heating effect* is the I^2R heating from a resistance measurement. It is stated as a self-heating error in $°$C/mW for a specific environment, such as in still air, air at 1 m/s, or water at $70°$F and 3 ft/s. It is also called a dissipating constant, expressed in mW/$°$C. Self-heating is not problematic for most applications. For example, the I^2R due to 1 mA through a 100-Ω RTD is 0.1 mW. If the self-heating error is $0.1°$C/mW, the heating due to 1 mA is $0.01°$C.

The measuring current I for a RTD ranges from 2 to 20 mA. A method to reduce self-heating and to increase sensitivity is to use short current pulses for the measurement. It takes several seconds to minutes for the self-heating to reach its final value. If the measuring current is in pulses of millisecond duration, the current can be increased by 10 to 100 times. Most pitfalls with platinum RTD applications result from using probes exceeding their specifications, particularly for temperatures greater than $550°$C [45].

The thermoelectric effect is reduced by placing junctions of dissimilar metals closed to one another and by avoiding steep thermal gradients near these junctions. Alternatively, an ac can be used for the resistance measurement, since a thermal EMF is not detectable by ac instruments.

A mistake, peculiar to RTD measurements, is the inadvertent mixing of RTDs and readout or control devices [46]. As noted in Sec. 9-5B, the difference in α between the U.S. and DIM standard is 1.9%. A readout device for one RTD cannot be used for another RTD with a different specification.

9-6. THERMISTORS

Both the RTD and the thermistor are simply resistive elements. Thermistors are used for temperature sensing as well as components in electronics and control systems [47]. The thermocouple, RTD, and thermistor as temperature sensors are compared briefly in this section.

The temperature ranges for the three type of thermometers are compared qualitatively in Fig. 9-19. The thermocouple has the widest temperature range but the lowest sensitivity. The RTD and thermistor do not require reference junctions. The RTD is more limited in temperature range but has higher sensitivity than the thermocouple. The advantages of RTDs are long-term stability and the capability for high precision. The thermistor has the least temperature range, typically from -100 to $150°C$, but its sensitivity is 10 times that of the RTD. The resistivity of metals increases with temperature; therefore, RTDs have positive temperature coefficients. Thermistors are available with positive or negative coefficients. For temperature sensing, thermistors with negative resistance–temperature coefficients are used almost exclusively.

Thermistors are available commercially in the form of beads, rods, flakes, and so on. A thermistor probe can be made very small, and its time constant comparable with that of a thermocouple, but much shorter than

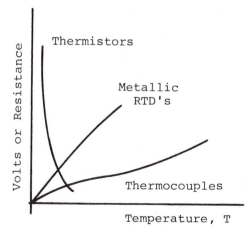

FIGURE 9-19 Characteristics of temperature transducers.

that of a RTD. Within its narrow temperature range, the thermistor compares favorably with the RTD in stability, repeatability, and interchangeability, but is far superior in response time and sensitivity.

Materials tolerances for thermocouples, RTDs, and thermistors are specified in like manner. Tolerance data must be considered when speaking of errors or interchangeability. Thermistors with a 0.2°C interchangeability are available. Data for a typical thermistor are shown in Fig. 9-20. A tolerance is the possible error for a new probe, prior to exposure to service conditions, although the error represents the "worst case," or limits, rather than typical error.

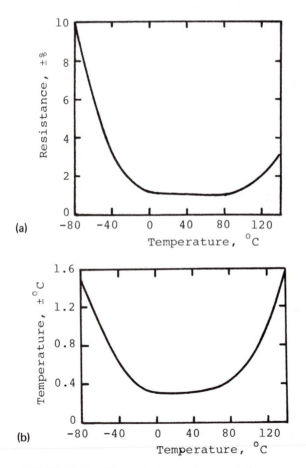

(a)

(b)

FIGURE 9-20 Characteristics of typical thermistor. (a) Resistance tolerance versus temperature. (b) Temperature tolerance versus temperature. (From Ref. 48.)

The negative resistance–temperature relationship of a thermistor can be expressed as [49]

$$R = R_0 e^{\beta(1/T - 1/T_0)} \tag{9-14}$$

where R is the resistance in ohms at the temperature T, T the absolute temperature K, R_0 the resistance at T_0, and $3000 < \beta < 5000$ is a material constant. Generally, T_0 is at 980 K (25°C). The value of R_0 ranges from 100 Ω to 1 MΩ, and a typical value is $R_0 = 5$ kΩ. The value of R for a thermistor may change from 3700 kΩ at -80°C to 100Ω at 145°C.

The sensitivity S of a thermistor from Eq. 9-14 is

$$S = \frac{d(R/R)}{dT} = \frac{-\beta + T\, d\beta/dT}{T^2} \tag{9-15}$$

$$S = \frac{d(R/R)}{dT} \simeq -\frac{\beta}{T^2} \tag{9-16}$$

$$\frac{dR}{dT} \simeq -\frac{\beta}{T^2} R \tag{9-17}$$

From Eq. 9-16, if $\beta = 4000$, $T = T_0 = 298$ K, and $R_0 = 5$ kΩ, the sensitivity is 0.045, or $\Delta R/R = 4.5\%$ per °C. For a platinum RTD, if the sensitivity from Eq. 9-8 is 0.00385, the $\Delta R/R$ is 0.38% per °C. Hence the thermistor has 10 times the sensitivity of a RTD.

An empirical expression is used to convert the measured resistance R of a thermistor into temperature T [48].

$$\frac{1}{T} = A + B \ln R + C(\ln R)^3 \tag{9-18}$$

where R is in ohms, T in kelvin, and A, B, and C are curve-fitting constants. Their values are obtained by substituting three pairs of values of (R, T) about the operating range into Eq. 9-18 to obtain three simultaneous equations. The equations are then solved simultaneously to give the constants. The output of a thermistor is nonlinear, but the computation of T from a measured value R is not a problem in data acquisition.

Due to the high resistance and high sensitivity, the effects of lead-wire resistance and self-heating are negligible for thermistors. For example, if $R_0 = 5$ kΩ and $\Delta R/R = 4.5\%$ per °C, the resistance change is $\Delta R = 225$ Ω/°C. This is much larger than the resistance of 500 ft of No. 18 AWG copper extension wire (6.5 Ω/1000 ft). The self-heating effect is also small. If $R_0 = 5$ kΩ and the rated dissipation constant is 4 mW/°C, an I^2R input of 0.04 mW relates only to 0.01°C of self-heating, but gives an output of 0.45 V.

All the circuits in Fig. 9-18 can be used for thermistors, because of neglectable lead-wire and self-heating errors. Using the bridge in Fig.

9-18a in an unbalanced mode with equal resistors and initial $V_0 = 0$, the measured resistance R is

$$R = R_0 \frac{1 + 2V_0/V_i}{1 - 2V_0/V_i} \tag{9-19}$$

The resistance R can be used to calculate the temperature T from Eq. 9-18, or to obtain T from tables. For the four-wire ohmmeter circuit in Fig. 9-18d, the current I is constant and R at the measured temperature T is

$$R = R_0 \frac{V}{V_0} \tag{9-20}$$

where R_0 and V_0 are the respective initial values at T_0, and R and V are the respective values at the temperature T.

9-7. PYROMETERS: PRINCIPLES [50]

A *pyrometer,* or *radiation thermometer,* is a noncontact instrument that measures the electromagnetic radiation emitted from a body and infers its temperature from the detected radiation. All objects emit radiation by virtue of their temperature. Principles employed for pyrometry are Planck's law and the Stefan–Boltzmann law. It will be shown in Sec. 9-8 that pyrometers can generally be classified as narrowband and broadband. Planck's law is used for the narrowband and the Stefan–Boltzmann law for the broadband or total radiation pyrometer.

Pyrometers are applicable for wide temperature ranges and are not restricted to high temperatures. The electromagnetic radiation spectrum is shown in Fig. 9-21. A portion of the spectrum can be used for pyrometry, and the spectrum normally used is from 0.3 to 40 μm. The visible spectrum, shown crosshatched, occupies only a very limited band, from 0.35 μm (blue-violet) to 0.78 μm (red). In the visible spectrum, a piece of steel in a furnace appears red at about 600°C, and it changes from a dull red, through orange and yellow, to white at about 1600°C. The infrared spectrum is not visible, but infrared pyrometers are common. A commercial unit can be used to −100°C (e.g., Model ST, Barber-Colman, Loves Park, Illinois). The frequency of radiation for other applications is also shown in the figure. As radiation propogates at the speed of light at 3×10^8 m/s, the product of frequency and wavelength is 3×10^8 m/s.

A. Physical Laws

The distribution of radiant power intensity from a blackbody at a given temperature for varying wavelengths is described by *Planck's law* in Eq. 9-21. The concept of a blackbody will be described presently.

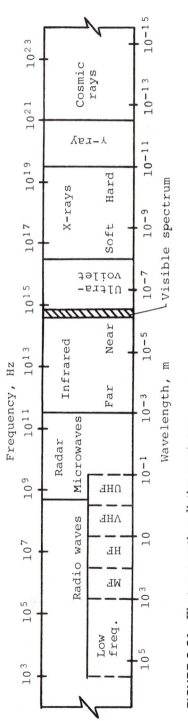

FIGURE 9-21 Electromagnetic radiation spectrum.

$$W_b(\lambda, T) = \frac{2\pi c^2 h}{\lambda^5 [e^{(hc/k\lambda T)} - 1]} = \frac{C_1}{\lambda^5 [e^{(C_2/\lambda T)} - 1]} \qquad (9\text{-}21)$$

where

$W_b(\lambda, T)$ = radiant power intensity from a blackbody in W/m^3
λ = wavelength in m
T = absolute temperature in kelvin
h = Planck's constant = 6.625×10^{-34} J·s
c = speed of light = 3×10^8 m/s
k = Boltzmann's constant = 1.380×10^{-23} J/K
$C_1 = 2\pi c^2 h = 3.74 \times 10^{-16}$ W·m^2
$C_2 = hc/k = 1.44 \times 10^{-2}$ m·K

The distribution of $W_b(\lambda, T)$ versus λ is shown in Fig. 9-22a. $W_b(\lambda, T)$ is the intensity of radiant energy emitted from a blackbody at temperature

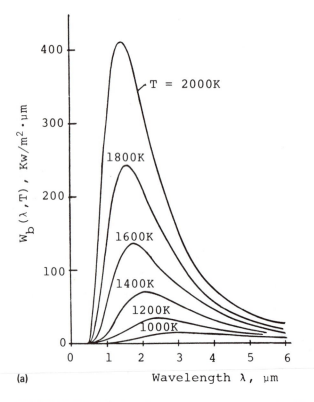

(a)

FIGURE 9-22 Blackbody radiation power intensity $W_b(\lambda, T)$.

T and wavelength λ, per unit time, per unit wavelength interval, per unit area, per unit solid angle, that is, the radiation from a flat surface onto a hemisphere. It is also called the hemispherical spectral radiant intensity.

Wien's radiation law is often used instead of Planck's law:

$$W_b(\lambda, T) = \frac{C_1}{\lambda^5 e^{(C_2/\lambda T)}} \tag{9-22}$$

The difference between the two equations is the -1 term in the demominator. The -1 can be omitted when the exponential term is large compared with unity. For example, for $T = 2500$ K and $\lambda = 1$ µm, values of $W_b(\lambda, T)$ from the two equations differ only by 0.3%. It may be of interest to note that historically Wien's law preceded Planck's law. The wavelengths λ_p at peak intensities for higher temperatures skew toward shorter wavelengths, as shown in Fig. 9-22b. This is described by *Wien's displacement law:*

$$\lambda_p T = 2891 \times 10^{-6} \tag{9-23}$$

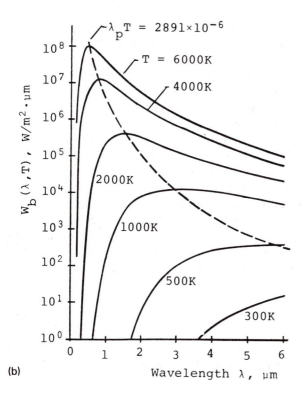

(b)

The area under each of the curves in Fig. 9-22 is the total power $W_b(T)$ emitting from a blackbody at a given temperature T. This is described by the *Stefan–Boltzmann law* in Eq. 9-25.

$$W_b(T) = \int_\lambda W_b(\lambda, T) \, d\lambda \tag{9-24}$$

$$W_b(T) = \sigma T^4 \tag{9-25}$$

where $\sigma = 5.67 \times 10^{-8}$ W/m$^2 \cdot$K is the Stefan–Boltzmann constant. The total power emitting from a blackbody is proportional to the fourth power of its absolute temperature.

B. Emittance

The concepts of blackbody and emittance are central to the study of pyrometry. Let us first describe a blackbody. Imagine that radiation is beamed at a small hole on the surface of a hollow sphere. All the energy entered the hole is being absorbed because there is no escape. Hence the area of the hole receiving the radiation is a perfect absorbing surface, or *blackbody. Black* here refers to the ability to absorb radiant energy rather than color.

Conversely, radiation emitting from the hole on the surface of the hollow sphere is also radiation from a blackbody. From Kirchhoff's law, the absorptivity of a material is equal to its emissivity, and a perfect absorber is also a perfect emitter. The emittance of a blackbody is unity; that is, a blackbody would emit the maximum amount of thermal radiation possible for a given temperature. If the sphere is at a constant temperature, radiation emitting from the hole is radiation from a blackbody, as described by Planck's law in Eq. 9-21. In practice, the blackbody behavior can be approximated by means of a blackened conical cavity of about 15° at a constant temperature. A blackbody is useful for the calibration of pyrometers and as a thermal reference in instruments.

Emittance describes the relative amount of thermal energy radiating from a body. It is expressed as a ratio of the actual radiation W_a versus the blackbody radiation W_b at the same temperature. The emittance of an actual body is always less than unity, and $W_a < W_b$. From Eq. 9-21, the spectral emittance ε_λ of a body at a constant temperature T is defined as

$$\varepsilon_\lambda = \frac{W_a(\lambda, T_m)}{W_b(\lambda, T)} \leqslant 1 \tag{9-26}$$

where T is the target temperature, or the true temperature, and T_m is the measured temperature. In other words, if the target were a blackbody, the pyrometer would give the true temperature T. Since $W_a(\lambda, T_m) < W_b(\lambda, T)$, the measured temperature is T_m. Evidently, $\varepsilon_\lambda \leqslant 1$ and $T_m \leqslant T$, because the greater the amount of energy emitting from the target, the higher the temperature indicated by a pyrometer. The integration of Eq. 9-26 gives the total emittance ε due to all wavelengths:

$$\varepsilon = \frac{W_a(T_m)}{W_b(T)} \leqslant 1 \qquad (9\text{-}27)$$

A gray body is a body whose emittance is independent of wavelength; that is, ε_λ = constant for all λ at a given T, or $\varepsilon_\lambda = \varepsilon$. Hence the actual $W_a(\lambda, T_m)$ versus λ curve has exactly the same shape as that for $W_b(\lambda, T)$ shown in Fig. 9-22.

Emittance rather than *emissivity* is used in this discussion. Like emittance, emissivity is a ratio, but emissivity is a material property in the sense that it is defined only for highly polished surfaces or controlled conditions [51]. Emittance describes an actual condition. It depends on many variables, such as size and shape of an object, surface roughness and oxidation, angle of viewing, temperature, wavelength, and so on. There is no simple relationship between the variables [52].

It will be shown in the next section that the importance of using accurate values of emittance in pyrometry cannot be overstated.

1. Due to the difference in emittance, two targets at the same temperature may emit significantly different amount of energies. Consequently, the same pyrometer may give different readings for two bodies of identical temperature.
2. Due to a change in emittance, such as oxidation or even angle of viewing, the same pyrometer may give different readings for the same body at the same temperature.

C. Measurement Uncertainty

The emittance of an object must be known in a temperature measurement. An inherent error in pyrometry is due to the uncertainty in the value of emittance. The error relating to Planck's law and the Stefan–Boltzmann law is described in this section.

Consider a *narrowband pyrometer* employing Planck's law for a temperature measurement at a known wavelength λ. From Eq. 9-26, T_m relates to the actual radiation $W_a(\lambda, T_m)$ from the target, but the target temperature T relates to $W_b(\lambda, T)$. Writing $W_b(\lambda, T)$ explicitly and substituting into Eq. 9-26, we get

$$W_a(\lambda, T_m) = \frac{\varepsilon_\lambda C_1}{\lambda^5 [e^{(C_2/\lambda T)} - 1]} \qquad (9\text{-}28)$$

Although $W_a(\lambda, T_m)$ is known, there are two unknowns in one equation in Eq. 9-28, ε_λ and T. The procedure in using a pyrometer is to assume a reasonable value for the emittance ε_λ, adjust the pyrometer accordingly, and then proceed with the measurement to deduce the temperature T.

The error in temperature due to the uncertainty in emittance ε_λ can be estimated from Eq. 9-26. Using Wein's law to simplify the calculation, defining

$$W_a(\lambda, T_m) = \frac{C_1}{\lambda^5 e^{(C_2/\lambda T_m)}} \quad \text{and} \quad W_b(\lambda, T) = \frac{C_1}{\lambda^5 e^{(C_2/\lambda T)}} \quad (9\text{-}29)$$

substituting Eq. 9-29 into Eq. 9-26, and simplifying, we obtain

$$T = \frac{T_m}{1 + (\lambda T_m/C_2)(\ln \varepsilon_\lambda)} \quad (9\text{-}30)$$

where T is the true temperature of the emitting body, T_m the measured temperature, and ε_λ the assumed emittance. The error dT/T from Eq. 9-30 is

$$\frac{dT}{T} = \frac{\lambda T}{C_2} \frac{d\varepsilon_\lambda}{\varepsilon_\lambda} \quad (9\text{-}31)$$

The error in T is not greatly influenced by the uncertainty in ε_λ for this type of pyrometer. For example, if $\lambda = 0.65$ μm and $T = 2000$ K, a 10% error in ε_λ results only in a 0.90% error in T.

Consider a *broadband pyrometer* using the Stefan–Boltzmann law for a temperature measurement. Substituting Eq. 9-25 into Eq. 9-27 and simplifying, we get

$$T = T_m \varepsilon^{-1/4} \quad (9\text{-}32)$$

$$\frac{dT}{T} = -\frac{1}{4} \frac{d\varepsilon}{\varepsilon} \quad (9\text{-}33)$$

where T is the true target temperature, T_m the measured temperature, and ε the total emittance. The uncertainty due to the value of emittance is greater for this instrument than that using Planck's law above. For example, a 10% error in ε results in a 2.5% error in T.

It should be restated that the uncertainty due to the value of emittance is inherent in pyrometry. Emittance, however, is a complex relation of many variables and not a simple constant. No single emittance value can be used for all types of pyrometers, for widely different temperatures, or for the same pyrometer under different conditions. For example, the spectral emittance ε_λ of tungsten increases with increasing temperature and decreasing wavelength, but existing theory is insufficient to predict this behavior accurately [52].

Finally, the calibration for emittance uncertainty depends largely on past experience.

1. It is proper to use relaible published data when similar conditions exist. The error can then be estimated from Eq. 9-31 or 9-33.
2. It may be possible to create a blackbody condition at the target.
3. Preliminary tests can be performed to obtain typical emittance values for various target conditions. For example, pyrometer readings can be made for targets of known temperatures, obtained by means of contact thermometers such as thermocouples. This is creating one's own lookup table.
4. The effect of emittance may be minimized by using a ratio pyrometer. This is discussed in the next section.

9-8. PYROMETERS

The theory of pyrometry described above is not new. In response to current demands [53], however, design efforts have led to more convenience, improved sensitivity, new materials, faster response, specialized applications, and (unfortunately) greater variations in detail. In this section we describe applications, general construction of narrowband and broadband pyrometers, and components common to the instrument.

Pyrometers are noncontact radiation instruments for measuring the average temperature of objects. The noncontact, radiation, and averaging features are necessary for some problems. These features also dictate the areas of application.

A noncontacting instrument is used when a contacting type, such as a thermocouple, is not practical. For example, pyrometers are used for temperature measurements of (1) delicate surfaces, such as paper and plastic sheets; (2) fast-moving objects, such as ingots in a steel mill; and (3) one-time applications, such as determining the temperature of high-tension power lines and energy audit of buildings [54]. Furthermore, a thermocouple must assume the high temperature of the object, but the sensor of a pyrometer is at a low temperature.

Temperature measurement by radiation is a convenience under hostile environmental conditions, such as in the presence of strong chemicals, high pressures, and extreme temperatures. The temperature of an object under strong electrical interference, as in an induction heating furnace, can be measured routinely by means of pyrometers. Fiber optics can be used for the signal transmission of high temperatures, such as inside the catalytic converter of an automobile (e.g., Model TM-2, Vanzetti Systems, Stoughton, Massachusetts). It would be difficult to measure the temperature of an erupting volcano by any other means.

The averaging effect has many advantages for some problems, such as averaging the surface temperature of a wall or that of an ocean [55]. The target size can be very large or extremely small. For example, an infrared microscope is capable of focusing on an area with a diameter as small as 0.0003 in. (e.g., Barnes Engineering Co., Stamford, Connecticut). For most applications, the target size of infrared pyrometers ranges from 0.10 in. at 2 in. to 2 in. at 10 in. (e.g., Omega Engineering, Inc., Stamford, Connecticut).

To satisfy diverse requirements, pyrometers are available with numerous features. Tables of instruments, specifications, and manufacturers are published periodically in trade journals [53] and buyer's guides. Some manufacturers have a "complete" line of products, for applications from aircraft engines to health care [56]. Recent advances are largely associated with infrared and spectral-selective devices, that is, pyrometers to operate in a selected spectrum, optimized for a given task.

A. Brightness Pyrometers

The *brightness pyrometer,* also known as the *disappearing filament* or *optical pyrometer,* is the most accurate instrument for high-temperature measurements above 700°C. It is employed to realize the IPTS-68 above the gold point at 1063°C. A classical form of the instrument consists of the optics, an absorption filter, a tungsten lamp of variable brightness, a red filter, and an eyepiece, as shown in Fig. 9-23a. The absorption filter is for viewing targets above 1300°C.

The pyrometer operates with an extremely narrow band. The red filter has a sharp cutoff at $\lambda < 0.63$ µm. The human eye provides the cutoff at the longer wavelengths of the visible spectrum. Thus the effective wavelength of the instrument is 0.653 µm. The test object and the lamp filament are viewed simultaneously. Hence the brightness of the target and that of the lamp filament are compared under monochromatic conditions.

The brightness of the lamp filament is adjusted manually until the filament seems to disappear, as shown in Fig. 9-23b, that is, when the test object and the filament are at the same temperature. The unknown target temperature is T and that of the lamp filament is T_m in Eq. 9-30. The spectral emissivities of target materials at 0.65 µm are given in the literature [51]. It will be shown that the manual adjustment can be automated.

The *ratio* or *two-color pyrometer* is an attempt to minimize the emittance uncertainty in pyrometry. The instrument is essentially a combination of two optical pyrometers, using two separate filters in order to operate at two separate wavelengths λ_1 and λ_2. The actual radiation

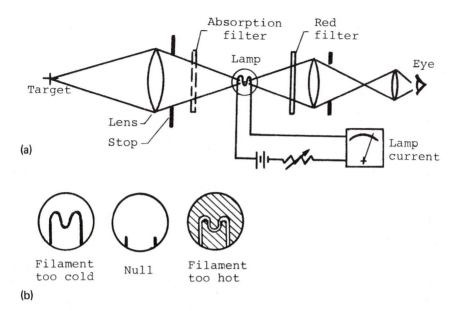

FIGURE 9-23 Schematic of a brightness (optical) pyrometer. (a) Brightness (optical) pyrometer (courtesy of Leeds and Northrup Co., North Wales, Pa.). (b) Disappearing filament.

received at λ_1 is $W_a(\lambda_1, T_{m1})$, as shown in Eq. 9-26, and that at λ_2 is $W_a(\lambda_2, T_{m2})$. The ratio R of the radiations at λ_1 and λ_2 is

$$R = \frac{\varepsilon_{\lambda 1}}{\varepsilon_{\lambda 2}} \frac{\lambda_2^5}{\lambda_1^5} \frac{e^{(C_2/\lambda_2 T)} - 1}{e^{(C_2/\lambda_1 T)} - 1} \tag{9-34}$$

where R is a known measured value. The ratio pyrometer is free of emittance error if $\varepsilon_{\lambda 1} = \varepsilon_{\lambda 2}$, since the true temperature T of the target can be calculated directly from Eq. 9-34.

If the values of $\varepsilon_{\lambda 1}$ and $\varepsilon_{\lambda 2}$ are unequal, the target temperature can be estimated by two methods. (1) The error is negligible if a narrow spectrum is chosen such that ε_λ varies slowly with λ, that is, if the ratio $\varepsilon_{\lambda 1}/\varepsilon_{\lambda 2}$ is approximately unity. (2) The ratio is assumed unity to obtain an approximation. Then the value of the ratio is estimated to provide a correction. First, if $\exp(C_2/\lambda T) \gg 1$ and the ratio $\varepsilon_{\lambda 1}/\varepsilon_{\lambda 2}$ is assumed to be unity, the approximated temperature T_R from Eq. 9-34 is

$$T_R = \frac{C_2(1/\lambda_2 - 1/\lambda_1)}{\ln [R(\lambda_1/\lambda_2)^5]} \tag{9-35}$$

Now, the ratio of the unequal emittances $\varepsilon_{\lambda 1}/\varepsilon_{\lambda 2}$ is estimated. Substituting Eq. 9-35 into Eq. 9-34 and simplifying, the estimated temperature T is

$$T = \frac{T_R}{1 + [\lambda_2 \lambda_1/(\lambda_2 - \lambda_1)](T_R/C_2) \ln(\varepsilon_{\lambda 1}/\varepsilon_{\lambda 2})} \tag{9-36}$$

B. Wideband and Selected-Band Pyrometers [57]

The *wideband* or *total-radiation pyrometer* operates on the Stefan–Boltzman law. The instrument consists of the optics, a detector, the housing, and sighting lenses, as shown in Fig. 9-24. The optics concentrate the radiation from a target onto a small detector. Thermal or photon detectors are used. The housing may be thermostatically control for greater thermal stability.

A *thermal detector* senses the radiation and infers the target temperature from its own temperature. Thermistors and RTDs are thermal detectors and are called *bolometers*. Since a thermal detector depends on its own temperature to infer the detected radiation, its time constant is fairly long, from 5 ms to 0.5 s. A pyrometer with a thermal detector is essentially a first-order instrument. A simplified analysis of the pyrometer shown in the figure with a thermopile detector gives

$$C(T_1 - T_2) = \varepsilon \sigma T^4 \tag{9-37}$$

where T is the target temperature, $T_{1,2}$ are those of the thermopile, C is a constnat, σ the Stefan–Boltzmann constant, and ε the total emittance of the target. Another type of thermal detector is pyroelectrics. These have high response speeds, because the electrical output depends on the time rate of change in temperature rather than on the detector temperature itself.

A *photon detector* generates a voltage as a function of its detected photon flux. The flux frees electrons in the detector to produce a measur-

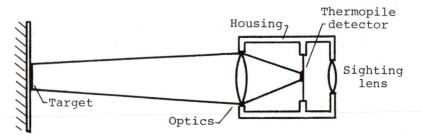

FIGURE 9-24 Schematic of a total radiation pyrometer.

able electrical output. As the reaction occurs at an atomic level, photon detectors have very high response speeds. In fact, photon detectors can be used for infrared scanning, that is, to display the dynamic temperature distribution of a small target on a TV monitor. The spectral response of photon detectors is not constant compared with that of thermal detectors. its output voltage also varies approximately as T^3, instead of to the fourth power as in the Stefan–Boltzmann law.

The disadvantage of a truly wideband pyrometer is the uncertainty in the value of emittance. The spectral emittance of most materials is not constant over the entire bandwidth; therefore, the total emittance is also temperature dependent. Most pyrometers include an emittance adjustment, from 0.2 to 1.0.

The study of pyrometry is further complicated by the fact that radiant energy can be partially absorbed, reflected, and transmitted. This is expressed as

$$\varepsilon = 1 - r - t \qquad (9\text{-}38)$$

where ε is the emittance, r the reflectance, and t the transmittance. The transmittance of a solid body is $t = 0$. The values of ε, r, and t are not constants. For example, soda-lime glass is transparent to solar radiation but opaque to infrared; that is, a material can be opaque or transparent, depending on the wavelength.

Recent advances have been largely associated with spectrally selective or narrowband pyrometers in the infrared spectrum. A selected-band pyrometer dictates the spectrum with which to operate. It may be designed (1) to match or to optimize the spectral properties of its components, or (2) to select the most advantageous spectrum for the task. For example, in mapping the temperature of the earth's surface from space, the main barriers are CO_2 and water vapor in the atmosphere. The mapping can be performed by using the 8- to 13-μm "window" in the infrared spectrum, because CO_2 and water vapor are largely transparent for this spectrum and for the temperature range considered. The other side of the coin is that filters can be used to pass the 15-μm band for which CO_2 is a strong emitter. Thus air temperatures miles ahead of an aircraft can be measured.

C. Components

Some components common to pyrometers are described in this section [58]. The basic items are the optics, filters, detectors, optical choppers, and black bodies for thermal reference. Electronics are omitted in this discussion. The sighting path between the target and the pyrometer is also an integral part of the measuring system, since the radiation may be partially absorbed, reflected, or transmitted in the transmission path, as shown in Eq. 9-38.

The sighting path must be transparent to the radiation. Any interference due to dust, flame, fumes, moisture, or background radiation is noise in the signal to the pyrometer. A *purging tube* with clean nonabsorbing gas is used when the interference cannot be filtered out optically. The gas, or the medium for the signal transmission, need only be transparent for the operating spectrum of the pyrometer. For example, fiber optics can be used instead of a purging tube for signal transmission at high temperatures.

Optics, filters, and detectors for pyrometers have different spectral sensitivities. In other words, the sensitivity versus wavelength plot of a material may peak sharply, or remain fairly flat with cutoffs at both ends of a given spectrum, like the frequency response characteristics of an instrument (see Fig. 4-22). The characteristics of the components should be optimized in order to work as an integral unit for the intended application. It is difficult to give details on spectral sensitivity of materials. For example, glass is opaque for wavelengths greater than 2.5 μm. Hence glass lenses are not suitable for most infrared detectors, but glass can be used for a lead sulfide detector, which has a spectral sensitivity peak at 2 μm. Furthermore, light of different wavelengths is focused differently; the result is poor target definition or background noise unless the optical system is achromatic, such as focusing mirrors or compound lenses. Visible and infrared radiation follow the same optical laws, but some infrared lenses, such as arsenic trisulfide, are opaque to the visible spectrum.

An *optical chopper* (see Fig. 3-54) can be used (1) to convert the dc input to a detector to ac, (2) to allow radiation of different wavelengths to reach the detector alternately, or (3) to compare the radiation from a target with that of a reference. For a dc-to-ac conversion, the chopper simply interrupts the radiation from the target to the detector. A circuit with chopped radiation to a thermistor detector is shown in Fig. 9-25a. This is the ballast circuit for strain gages (see Probs. 1-3 to 1-5). The ac output can then amplified with an ac amplifier. This circuit is less susceptible to the slowly varying ambient temperature. Another scheme uses two detectors in a bridge circuit, as shown in Fig. 9-25b (see Fig. 3-33). The detector D_1 is exposed to the chopped radiation, but D_2 is shielded from radiation and it is at the controlled temperature of the housing.

The pyrometer with an optical chopper, shown in Fig. 9-26a, is suitable for comparing the radiation from a target with that of an internal blackbody [59]. If the optical chopper has a mirror-finished surface, radiation from a built-in blackbody (not shown) can be reflected to the detector.

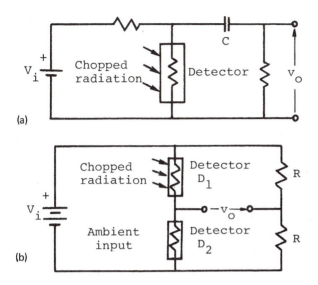

FIGURE 9-25 Circuits for chopped radiation detection. (a) Ballasat circuit. (b) Bridge circuit.

Thus the detector receives radiation from the target and a blackbody alternately. If an internal blackbody is not used and the chopper has a blackened surface, the detector receives radiation alternately from the target and the chopper alternately. Thus the controlled temperature of the housing becomes the thermal datum. The beam splitter simply allows direct sighting of the target.

The output from the detector is a train of square pulses. Its amplitude relates to the difference in radiation received between the target and the blackbody. This is an amplitude-modulation process (see Sec. 5-4A). The chopper modulates the input to the detector. A magnetic pickup can be placed at the chopper to obtain a synchronous signal for the phase-sensitive demodulation (see Figs. 5-10 to 5-12). The output is an analog signal corresponding to the temperature detected by the target. If the chopping frequency is 180 Hz, the time constant of the system is about 8 ms. Alternatively, the temperature of the blackbody can be adjusted by means of feedback for the system to operate in a null-balance mode (e.g., Barnes Engineering Co., Stamford, Connecticut), but the higher accuracy is obtained at the expense of speed of response (see Sec. 3-2).

Similarly, the automatic optical pyrometer, shown in Fig. 9-26b, employs an optical chopper to modulate the radiation from the target and the standard lamp to the photomultiplier tube. The chopper is at a fixed

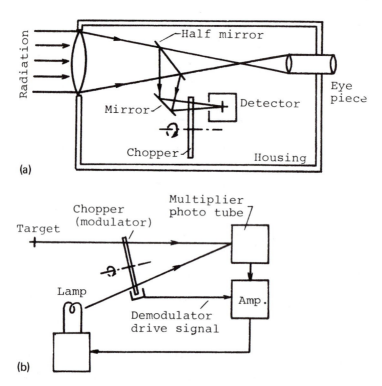

(a)

(b)

FIGURE 9-26 Pyrometers with optical chopper. (a) Optical head with internal reference. (b) Automatic brightness pyrometer (Model 8641—Mark 1, courtesy of Leeds and Northrup Co., North Wales, Pa.).

frequency. The standard lamp is used as an internal thermal reference instead of a blackbody. To operate in the null-balance mode, the brightness of the lamp is adjusted by means feedback until a null intensity is sensed by the photomultiplier. To operate in the unbalanced mode, the brightness of the lamp is unchanged, and the difference in brightness gives the output.

9-9. MISCELLANEOUS TEMPERATURE SENSORS

Several temperature sensors for industrial applications are described briefly in this section. Many new sensors, such as the Johnson noise thermometer [60], are made possible by advances in electronics in recent years, but they are used mostly in the laboratory.

The *integrated-circuit* (IC) *thermometer* is a semiconductor-based temperature sensor, developed in the electronic industry. It is widely accepted and has a market share of about 5%. Only an unregulated power supply is required for the IC. The sensor has good linearity and an analog output of 1 µA/K at 25°C, but has a very limited range, from −50 to 150°C. It has good sensitivity, however, and is often employed for the reference-junction compensation of thermocouples (see Fig. 9-8d).

The external circuitry for IC thermometers is very simple, as shown in Fig. 9-27, where V_o is the output voltage across the trim potentiometer R_s, V_i the power supplied, R_L the resistance of the leads. The V_o can be trimmed to 1 mV/K. This type of sensor does not have some of the disadvantages of RTDs and thermocouples, such as low-level outputs and reference-junction compensation. The constant-current nature of the output makes the sensor ideal for remote temperature sensing because the effect of R_L is negligible.

A class of sensors for temperature monitoring is widely used in industry (e.g., Williams Wahl Corp., Los Angeles, California; Omega Engineering Co., Stamford, Connecticut). They are simple, convenient, and very inexpensive. The monitoring is in incremental steps, and the sensors usually include a set of items for a range of temperature, with each item of the set to indicate a fixed value.

An example is the self-adhering temperature-sensitive label, shown in Fig. 9-28. Its range is from 100 to 500°F. The dots in a label represent a thermometer scale, usually for 10°F increments. Each dot changes color at a fixed temperature. Labels with one or several dots can be attached to any surface for temperature monitoring, such as bearing housings or critical parts of an electronic circuit board.

Another example is thermal crayons, with a range from 125 to 800°F.

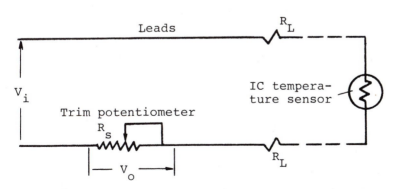

FIGURE 9-27 Circuit for integrated-circuit IC thermometers.

FIGURE 9-28 Temperature-sensitive label (dots represent thermometer scale).

These can be used on the surfaces of many materials. A crayon mark appears as a liquid smear at its rated temperature. Similarly, thermal sticks can be used to mark the surface of metals in a heat-treating plant. The stick leaves a dry, chalky mark on the metal surface; then the mark changes to a liquid smear at its rated temperature. The accuracy is ±1% and the upper temperature limit is 2500°F. A variation of the thermal stick is the thermal pellets for the same temperature range. The first sign of melting is the rated temperature. An example of its application is in the study of cracks in the cylinder head of a large diesel engine. It is difficult to install many thermocouples in the area of interest. Small pellets can be embedded in surface. The indicated temperatures can then be used to map isotherms in the cylinder head. Pyrometric cones for the ceramic industry are a similar product.

PROBLEMS

9-1. (a) A total-immersion mercury-in-glass thermometer is partially immersed in a hot-water bath. The indicated temperature is $T_B = 95°C$. The water surface is at the 45°C mark on the thermometer. A small thermometer attached to the middle portion of the exposed mercury thread reads $T = 35°C$. Find the corrected temperature T_B of the bath. (b) The thermometer is used in an oil bath with $T_B = 215.6°C$. The oil surface is at the 86°C mark, and $T = 70°C$. Find the corrected temperature T_B of the oil bath.

9-2. It is stated in the text that the pressure thermometer shown in Fig. 9-2b is for the bulb temperature greater than ambient, and that in Fig. 9-2c for less than ambient. (a) Justify the statements. (b) What would happen, if the bulb is at the average of a fluctuating ambient temperature?

9-3. Referring to Fig. 9-4, (a) the thermal EMF of a thermocouple consisting of materials $A–B$ is $E_{AB} = E_{AP} + E_{PB}$, and (b) the sensitivity is $S_{AB} = S_{AP} - S_{PB}$. Use the NBS Monograph 125, or equivalent, for the iron–constantan (type J) thermocouple to verify the statements for the temperatures of −200, 200, 400, and 600°C.

9-4. The output of a chromel–alumel (type K) thermocouple at T_1 is 29.1672 mV referenced 20°C. Find T_1°C.

9-5. (a) Chromel-alumel (type K) wires are used in the thermopile shown in Fig. P9-1a. Find the output E_o in mV. (b) Type K thermocouple wires are used in a thermopile shown in Fig. P9-1b. If $E_o = 28.0602$ mV, find the average temperature of the hot junctions. Conversely, calculate E_o from the temperatures shown in the figure.

9-6. A thermocouple circuit with copper extension wires is shown in Fig. P9-2. Apply Eq. 9-3 to verify that the extension wires have no effect on the temperature measurement. This verifies the law of intermediate metal.

9-7. The temperature 1000°C is measured with a chromel–alumel (type K) thermocouple, as shown in Fig. P9-3. A ±1% materials tolerance is probable for the less expensive type K extension wire. Find the tolerance in the measurement.

9-8. Referring to Fig. P9-3, a chromel–alumel (type K) thermocouple is used to measure T_{Hot}, but a (copper–constantan) type T extension wire is inadvertently used with the correct polarity instead of the type K. Find the error in the type K scale.

9-9. A thermocouple is shown in Fig. P9-4, where A' is a portion of material A that is decalibrated, but material B is not affected. Use Eq. 9-3 to find an expression for the error in E_o.

9-10. A temperature-controlled double oven is used to simualte the ice point for a chromel–alumel (type K) thermocouple, as shown in Fig. P9-5, where chromel is material A and alumel is material B. (a) Find the temperature of oven 1 if oven 2 is at 60°C. (b) Find the temperature of oven 2 if oven 1 is at 140°C.

9-11. Derive Eqs. 9-12, 9-13, and 9-17 for resistive-type thermometers.

9-12. The resistance of a metallic wire is $R = \rho L/A$, where ρ is the resistivity, L the length, and A the cross-sectional area. (a) If ρ for platimum is $9.83 \times 10^{-6}\ \Omega \cdot$ cm at 0°C, find the length of 0.05-mm ($= 0.002$ in.)-diameter wire for a 100-Ω platinum RTD. (b) If ρ for copper is $1.53 \times 10^{-6}\ \Omega \cdot$ cm at 20°C, find the length of 0.05-mm-diameter wire for a 25-Ω copper RTD.

9-13. A platinum RTD with $\alpha = 0.00392$ is used with an instrument designed for $\alpha = 0.00385$. Estimate the error at 100°C.

9-14. The resistance R of a thermistor is described by Eq. 9-14. If $R_0 = 5$ kΩ at $T_0 = 25$°C and $\beta = 4000$ K, calculate R for the temperatures -50, 0, 50, and 100°C.

9-15. The resistance–temperature characteristics of thermistors are described by Eq. 9-18. Three pairs of values of (T, R) are required to evaluate the curve-fitting constants (A, B, C) in the equation. Evaluate A, B, and C for each set of the data below, and check the calculations by substituting the given value of R in your equation to verify the corresponding temperature T. (Data for Probs. 9-15 and 9-16 are from a catalog of the Yellow Springs Instrument Co.) (a) $(-50°C, 335.3 \text{ k}\Omega)$, $(0°C, 16.33 \text{ k}\Omega)$, $(120°C, 194.7 \ \Omega)$. Use the value of $R = 1244 \ \Omega$ to verify that T is $60°C$. (b) $(-40°C, 884.6 \text{ k}\Omega)$, $(0°C, 94.98 \text{ k}\Omega)$, $(80°C, 4843 \ \Omega)$. Use the value of $R = 10.97$ kΩ to verify that T is $50°C$. (c) $(0°C, 333.1 \text{ k}\Omega)$, $(50°C, 34.78 \text{ k}\Omega)$, $(100°C, 6005 \ \Omega)$. Use the value of $R = 22.77$ kΩ to verify that T is $61°C$.

9-16. A thermistor is used in the bridge circuit shown in Fig. 9-18a. The bridge is initially balanced with equal resistors at $T_0 = 25°C$, $V_i = 10$ V, and an initial $V_o = 0$. The bridge is used in an unbalanced mode for temperature measurements, and the outputs are $V_o = 4$, 2, -2, and -4 V, respectively. (a) If $R_0 = 5$ kΩ at $25°C$ for the thermistor in Prob. 9-15a, calculate the temperatures for the V_o's enumerated above. (b) Repeat part (a) for the thermistor in Prob. 9-15b if $R_0 = 30$ kΩ. (c) Repeat part (a) for the thermistor in Prob. 9-15c if $R_0 = 100$ kΩ.

9-17. Derive Eqs. 9-31, 9-33, and 9-36 for radiation thermometers.

9-18. The measured temperature from a brightness pyrometer using $\lambda = 0.65$ µm is $T_m = 1100°C$. The target is copper oxide, and the value of its spectral emittance ε_λ is from 0.60 to 0.80 with the probable value of 0.70. Find the range of the true temperature.

9-19. The measured temperature from a total radiation pyrometer is $675°C$. It is estimated that the total emittance ε of the target is between 0.75 to 0.93. Estimate the uncertainty in the temperature measurement.

9-20. A ratio pyrometer operates with $\lambda_1 = 2.04$ µm and $\lambda_2 = 2.64$ µm. Calculate the indicated temperature T_R for $R = 0.9$. If the ratio of the spectral emittance is 0.9, estimate the target temperature.

9-21. The target temperature T from a brightness pyrometer is expressed in terms of T_m, ε_λ, and λ in Eq. 9-30, but λ is normally a fixed value. Define an error $\Delta T = T - T_m$. For $\lambda = 0.65$ µm, plot ΔT versus T_m with ε_λ as a parameter, for the values of T_m from $800°C$ to $2500°C$ and $\varepsilon_\lambda = 0.9$, 0.7, 0.5, and 0.3, respectively.

9-22. From Eq. 9-32 for a total radiation pyrometer, define an error $\Delta T = T - T_m$. Plot T versus T_m with ε as a parameter, for the values of T_m from 800 to $2500°C$ and $\varepsilon = 0.9$, 0.7, 0.5, and 0.3, respectively.

9-23. From Eq. 9-31 for a brightness pyrometer, plot the error dT/T versus T with $d\varepsilon_\lambda/\varepsilon_\lambda$ as a parameter. Use T from 1000 to 4000°C and the values of $d\varepsilon_\lambda/\varepsilon_\lambda = 0.1, 0.2, 0.3, 0.4,$ and 0.5.

9-24. Repeat Prob. 9-23 for the total radiation pyrometer, using Eq. 9-23 and total emittance instead of spectral emittance.

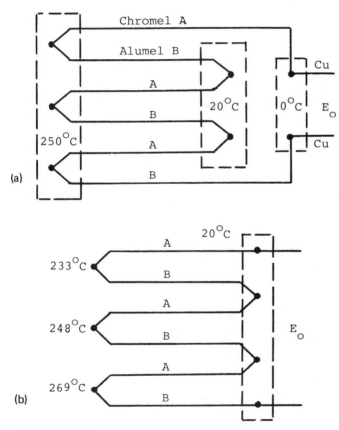

FIGURE P9-1

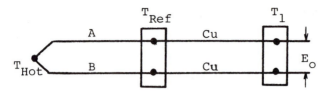

FIGURE P9-2

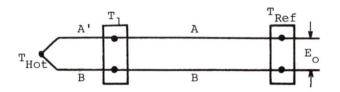

FIGURE P9-3

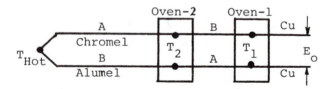

FIGURE P9-4

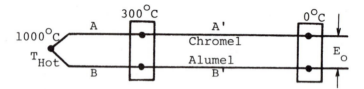

FIGURE P9-5

REFERENCES

1. Ball, K., Thermocouples and RTD's: A Controversy Continues, *InTech,* Vol. 33, No. 8 (1986), pp. 43–45.
2. Lion, K. S., *Instrumentation in Scientific Research,* McGraw-Hill Book Company, New York, 1959, Chap. 2.
   Plumb, A. A. (ed.), *Temperature,* Vol. 4, Pts. 1, 2, and 3, Instrument Society of America, Pittsburgh, Pa., 1972.
3. Sears, F. W., *Thermodynamics, Kinetic Theory, and Statistical Mechanics,* Addison Wesley Publishing Company, Inc., Reading, Mass., 1950, p. 116.
4. Eastermann, I. (ed.), *Methods of Experimental Physics,* Vol. 1, Academic Press Inc., New York, 1959, p. 240.
5. Krigman, A., What's Available in Temperature Standards, *InTech,* Vol. 30, No. 6 (1983), pp. 7–10.
   Process Instrumentation Technology, ANSI/ISA S51.1-1979, American National Standards Institute, New York, Dec. 1979.
6. Benedict, R. P., International Practical Temperature Scale of 1968, *Instrum. Control Syst.,* Vol. 42 (Oct. 1969), pp. 85–89.
7. *Evolution of the International Practical Temperature Scale of 1968,* ASTM STP 565, American Society for Testing and Materials, Philadelphia, Pa., 1974.
8. Rhodes, T. J., *Industrial Instruments for Measurement and Control,* McGraw-Hill Book Company, New York, 1941, Chap. 3.
9. Dittrich, R. T., Probe Averages Temperature Profiles, *Instrum. Technol.,* Vol. 15 (May 1968), p. 60.
10. Swindells, F., *Calibration of Liquid-in-Glass Thermometers,* NBS Circular 600, National Bureau of Standards, Washington, D.C., 1959.
11. Baker, H. D., E. A. Ryder, and N. H. Baker, *Temperature Measurement in Engineering,* Vol. 1, Omega Press, Stamford, Conn., 1975, p. 10.
12. Discussion of Apparatus for Verification & Calibration of Liquid-in-Glass Thermometers, *1980 Annual Book of ASTM Standards,* Part 44, American Society for Testing and Materials, Philadelphia, Pa., pp. 584–589.
13. Kinzie, P. A., *Thermocouple Temperature Measurements,* John Wiley & Sons, Inc., New York, 1973.
    Zysk, E. D., and A. R. Robertson, Nonmetallic Thermocouples, *Instrum. Technol.,* Vol. 19 (May 1972), pp. 42–45.
14. Benedict, R. P., *Fundamentals of Temperature, Pressure, and Flow Measurements,* 2nd ed., John Wiley & Sons, Inc., New York, 1977, pp. 73–86.
15. *Thermocouple Reference Tables,* NBS Monograph 125, Omega Press, Stamford, Conn., 1975.
16. Finch, D. I., General Principles of Thermoelectric Thermometry, in *Temperature,* C. M. Herzfold (ed.), Vol. 3, Pt. 2, Reinhold Publishing Corporation, New York, 1962.
    Manual on the Use of Thermocouples in Temperature Measurement, ASTM Spec. 407B, American Society for Testing and Materials, Philadelphia, 1981.

17. Moffat, R. J., The Gradient Approach to Thermocouple Circuitry, in *Temperature,* C. M. Herzfold (ed.), Vol. 3, Pt. 2, Reinhold Publishing Corporation, New York, 1962, pp. 33–38.
18. Keyser, D. R., How Accurate Are Thermocouples Anyway? *Instrum. Control Syst.,* Vol. 47, No. 2 (1974), pp. 51–54.
19. *Manual on the Use of Thermocouples in Temperature Measurements,* ASTM STP 407A, American Society for Testing and Materials, Philadelphia, 1979, p. 100.
20. Feldman, C. L., Automatic Ice-Point Thermocouple Reference Junction, *Instrum. Control Syst.,* Vol. 38, No. 1 (1965), p. 101.
21. Lamoureux, R. T., Millivolts to Temperature, *Instrum. Control Syst.,* Vol. 48, No. 1 (1975), pp. 43–44.
22. *Thermocouple Reference Junctions,* Consolidated Ohmic Devices, Inc., Carle Place, N.Y.
23. *1986 Temperature Measurement Handbook and Encyclopedia,* Omega Engineering, Inc., Stamford, Conn., Secs. A and D.
24. Baker et al., Ref. 11, pp. 47–55.
25. Dittbenner, G. R., Intrinsic Thermocouple for Fast Response, *Instrum. Control Syst.,* Vol. 39, No. 12 (1966), pp. 85–86.
26. Benson, J. M., and R. Horne, Surface Temperature of Thin Sheets and Filaments, *Instrum. Control Syst.,* Vol. 35, No. 10 (1962), pp. 115–117.
27. Werner, F. D., Total-Temperature Measurement, *Instrum. Control Syst.,* Vol. 33, No. 5 (1960), pp. 798–801.
28. Wormser, A. F., and R. A. Pfuntner, Pulse Technique Extends Range of Chromel-Alumel to 7000°F, *Instrum. Control Syst.,* Vol. 37, No. 5 (1964), pp. 101–103.
29. *American National Standard for Temperature Measuring Thermocouples,* ANSI MC96.1, American National Standards Institute and Instrument Society of America, 1975.
30. *Manual on the Use of Thermocouple in Temperature Measurements,* ASTM Spec. Pub. 470A, Omega Press, Stamford, Conn., 1974.
31. MacKenzie, D. M., and W. E. Kehret, Review of Temperature Measurement Techniques, Part I, *Instrum. Technol.,* Vol. 23, No. 9 (1976), pp. 43–48.
32. Renet, C., Try Counteraction to Measure Surface Temperature, *Instrum. Control Syst.,* Vol. 48, No. 6 (1975), pp. 33–35.
33. Brooks, E. J., and E. W., O'Neal, Keep Your Thermocouple Dry, *ISA J.* (Mar. 1965), pp. 94–95.
34. Baker et al., Ref. 11, pp. 56–57.
35. Kilpatrick, P. W., Accuracy of Thermocouples in Parallel, *Instrum. Autom.,* Vol. 30, No. 9 (1957), pp. 1706–1709.
36. Morrison R., *Grounding and Shielding Techniques in Instruments,* John Wiley & Sons, Inc., New York, 1972.
37. Epstein, J. S., and T. J. Heger, Versatile Instrument Makes High Performance Transducer-Based Measurements, *Hewlett Packard J.* (July 1981), pp. 9–15.

38. MacKenzie, D. M., and W. E. Kehret, Review of Temperature Measurement Techniques, Part II, *Instrum. Technol.,* Vol. 23, No. 11 (1976), pp. 49–54.

39. Ref. 23, Sec. E.

40. Benedict, Ref. 14, p. 29.

41. Bondable Resistance Temperature Sensors and Associated Circuitry, TN-506-1, Measurements Group, Inc., Raleigh, N.C.

42. Schurr, K., Understanding and Applying TCR, *Instrum. Control Syst.,* Vol. 48, No. 9 (1976), pp. 37–38.

43. Kerlin, T. W., and R. L. Shepard, *Industrial Temperature Measurements,* Instrument Society of America, Resarch Triangle Park, N.C., 1982, p. 157.

44. Mueller, E. F., Precision Resistance Thermometry, in *Temperature,* C. M. Herzfeld Vol. 1, Reinhold Publishing Corporation, New York, 1941, p. 162.

45. Trietley, H. L., Avoiding Error Sources in Platinum Resistance Temperature Measurements, *InTech,* Vol. 29 (Feb. 1982), pp. 57–60.

46. *Minco Newsletter,* Minco Products, Inc., Minneapolis, Minn., Sept. 1972.

47. *Handbook of Thermistor Applications,* Victory Engineering Corp., Springfield, N.J., 1963.

48. *YSI,* Precision Thermistor Products, Yellows Spring Instrument Co. Inc., Yellow Spring, Ohio.

49. *Thermistor Manual,* EMC-6B 1983, Fenwal Electronics, Clifton, N.J.

50. Benedict, Ref. 14, pp. 130–139.
    Kostkowski, H. J., and R. D. Lee, *Theory and Methods of Optical Pyrometer,* US NBS 41, U.S. Government Printing Office, Washington, D.C., Mar. 1962.

51. *ASME Performance Test Code,* PTC 19-3, American Society for Tsting and Materials, Philadelphia, 1974.

52. *Radiation Pyrometry,* Bulletin No. 31565, Milletron, Inc., Irwin, Pa.

53. Krigman, A., Guide to Selecting Non-contact Temperature Instruments, *InTech,* Vol. 30, No. 6 (1983), pp. 23–30.

54. Baur, P. S., Bringing Thermography Inhouse, *InTech,* Vol. 33, No. 8 (1986), pp. 9–21.
    How to Measure BTU-Energy Loss for Commercial and Residential Buildings, Energy Report No. E-277, William Wahl Corp., Los Angeles, Calif.

55. Jones, E. W., Space Application of IR Instruments, *Instrum. Control Syst.,* Vol. 41, No. 4 (1968), pp. 105–1110.

56. *Mikron Infrared Thermometers,* Application Note No. 7, Mikron Instrument Co., Wyckoff, N. J.

57. Doeblin, E. O., *Measurement Systems—Application and Design,* 3rd ed., McGraw-Hill Book Company, New York, 1983, pp. 618–650.
    Magison, E. C., Radiation and Optical Pyrometers, in *Process Instruments and Controls Handbook,* 2nd ed., D. M. Considine (ed.), McGraw-Hill Book Company, New York, 1974, pp. 2-71–2-100.

58. Tenney, A. S., III, Industrial Radiation Thermometery, *Mech. Eng.,* Vol. 108, No. 10 (1986), pp. 36–41.

59. Leftwich, R. F., Too Hot to Touch, *Instrum. Control Syst.,* Vol. 47, No. 10 (1974), pp. 51–54.

60. Kerlin and Shepard, Ref. 43, Chap. 14.

10
Laboratory Experiments

10-1. INTRODUCTION

The rationale for the selection of experiments in this chapter is discussed in this introduction. Readers may follow the chapter more easily with a few words of explanation. Note that a prefix E is used for all sections and figures pertaining to an experiment. For example, Sec. E1-5 denotes Section 5 in Experiment 1.

The laboratory, measurement, and instrumentation are essential parts in an engineering curriculum. From past experience it was difficult to teach a laboratory course, other than routine testing, without some background in measurement and instrumentation, and equally difficult to teach a meaningful course in measurement without an associated laboratory. The measurements laboratory has been taught for many years as the first laboratory course in the department. It consists of about two-thirds hands-on and one-third demonstrations. It is designed as a foundation for two follow-up required laboratory courses, which include performance tests of machinery, fluids, and digital data acquisition.

To cover the scope of a broad study, some topics in the measurements laboratory are examined in detail while others are demonstrations. It is only through a detailed study that students get a feeling for the magnitude of the variables involved and acquires a skill in experimentation. The guideposts for the selection of experiments are (1) simple experiments to complement the principles discussed in prior chapters, (2) a fairly thorough study in some areas of measurement, and (3) practical applications to give students a sense of reality.

As a first laboratory course, the experiments must be simple and basic to avoid being ensnared in details. For example, the study of transducers and measuring systems in Expts. 2 and 3 is a survey of the topics, and it depends mainly on demonstrations. The same can be said for signal conditioning in Expt. 7. However, the time and frequency response in Expts. 5 and 6 by means of model study are simple, basic, and in detail. The model study is justified as discussed in Chapter 1, because the expense and time in having multiple setups of real systems for these experiments would be prohibitive.

The output of transducers are mostly electrical, and the basic quantities measured are voltages, currents, resistances, and frequencies or digital pulses. The measurement of analog quantities are covered in fair detail in Expts. 1 and 7. Some students do not get excited in handling electrical quantities, but the instruments employed are basic and common in any modern laboratory.

To satisfy the students' desire for reality, the experiments on strain measurements are treated in detail in Expt. 4. This is cost-effective because the expenses in instrumentation and multiple setups are reasonable. A modest laboratory budget could afford students the opportunity to apply strain gages themselves in some setups.

The experiments are written for (1) semi-self-pacing, and (2) semi-self-contained. The former will alleviate the instructor's time and congestion in the laboratory. The latter relieves the burden of referring back and forth in the text, and allows the experiments to be taught as an independent laboratory course with the text material as reference.

Students should be sympathetic to the fact that it is difficult to sequence the experiments exactly with the concurrent lecture course. It is strongly urged that students prepare the data sheet before conducting an experiment, and the original data must be a part of the laboratory report.

The hours suggested for each experiment are listed below. The schedule for some may be tight, but the contents and the sequence of experiments can be altered to suit one's desire. The experiments should not be rushed for coverage.

Finally, it is suggested that all equipment, such as oscilloscopes and strain indicators, should be of industrial quality to be realistic. It may be more satisfactory in the long run. From past experience, some of the commercially available educational equipment does not impress students and is not always reliable.

Suggested Laboratory Hours for Experiments

Experiment	Lab hours
1. Voltage Measurements	6
2. Transducers and Physical Laws	1½
3. Measuring Systems	1½
4. Strain Gage Applications	9
5. Time Response of Instruments	3
6. Frequency Response of Instruments	3
7. Signal Conditioning	3

10-2. EXPERIMENT 1. VOLTAGE MEASUREMENTS: OSCILLOSCOPES, MULTIMETERS, AND DIGITAL MULTIMETERS

E1-1. Objectives

1. To familiarize with typical general-purpose laboratory instruments, and their capabilities and limitations
2. To study steady-state and transient voltage measurements
3. To examine loading in measurements

E1-2. Introduction

The output of transducers for mechanical measurements is generally electrical. This is expressed as voltages, currents, resistances, frequencies, or digital pulses. An oscilloscope, a digital multimeter, and a multimeter with a d'Arsonval movement are used for voltage sensing in this experiment. An oscillator is used as a voltage source to simulate a transducer.

The emphasis of the study is on the efficient use of instruments rather than on their electronics per se. Simplified functional block diagrams of these instruments are described below, because some insight on how the functions are performed are necessary. Specifications in the user's manuals should be consulted as a part of this study. Since the oscilloscope is used extensively, the operation of a general-purpose oscilloscope is examined in detail.

To a greater or lesser extent, loading is inherent in a measurement. Loading in voltage measurements is examined. Note that a valid measurement does not necessarily include "all" the information of a measurand; that is, the answer may depend on the objective of the measurement. To this end, an oscilloscope gives only the time history of its input voltage, a digital multimeter gives only the rms value, and a multimeter gives a rms voltage only by virtue of a scale adjustment.

E1-3. Description of Equipment

A. Oscilloscope

The simplified block diagram of a single-channel oscilloscope is shown in Fig. E1-1a. A trace is displayed on the screen of the cathode-ray tube. The vertical deflection of the trace is proportional to the input voltage $v(t)$, and the horizontal deflection is linear with time. Thus the trace gives the time history of an input voltage (see Sec. 6-4A). For most applications, an oscilloscope is simply a voltmeter with a built-in time base, or a time sweep.

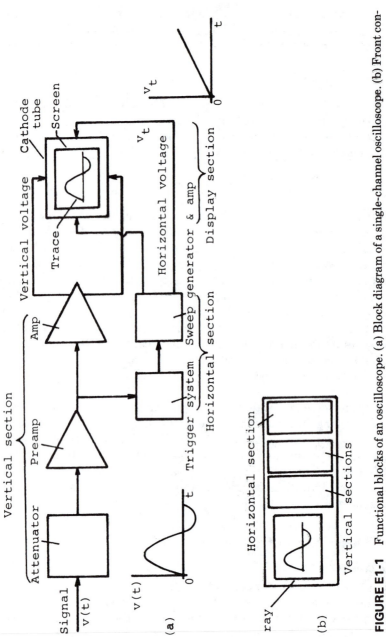

FIGURE E1-1 Functional blocks of an oscilloscope. (a) Block diagram of a single-channel oscilloscope. (b) Front controls of a dual-channel oscilloscope.

The block diagram has three main sections.

1. A voltage or input signal $v(t)$ is applied to the *vertical section*. This is conditioned by means of an attenuator adjusted to yield a convenient volts/division for the display. The signal is applied to the preamplifier, the amplifier, and then to the cathode-ray tube to give the vertical deflection.
2. The *horizontal section* controls the left-to-right deflection by means of a ramp voltage. The time sweep has a uniform time scale, because the ramp voltage v_t is linear with time, as shown in the figure.
3. The *display section* consists of the cathode-ray tube, which combines the vertical and horizontal deflections to give a trace on the screen (see Fig. 6-41). The schematic of the front panel of a dual-channel oscilloscope is shown in Fig. E1-1b.

When the input voltage $v(t)$ is periodic, as shown in Fig. E1-2, a consistent *trigger point* in $v(t)$ can be selected to initiate the ramp voltage v_t. A stable trace is shown because of the repeated display of the waveform, the persistence of the phospor on the screen, and the persistence of our vision.

B. Multimeter with D'Arsonval Movement

A multimeter (d'Arsonval movement) as a *dc voltmeter* is shown in Fig. E1-3a. It consists of a moving-coil microammeter with internal resistance R_m in series with a high resistance R_s. The meter without the R_s would have a very limited voltage range. From Ohm's law, a 200-μA meter with only its internal R_m of 1 kΩ has a range of 0 to 200 mV. The range is extended by inserting a R_s in series with the meter. For example, if R_s is 49 kΩ, the range is extended from 0–200 mV to 0–10 V.

A multimeter as an *ac voltmeter* is shown in Fig. E1-3b. The microammeter with a series R_s becomes a voltmeter as before. The ac voltage is rectified by means of diodes. Since the microammeter is a dc meter, the average value V_{av} of the rectified voltage is indicated by the meter as an ac voltage.

The V_{av} is changed to read V_{rms} (root mean square) by means of a scale adjustment. For a sinusoidal voltage, the rms value is $V_{rms} = 0.707\ V_p$, and $V_{av} = 0.637\ V_p$ for full-wave rectification, where V_p is the peak voltage. The scale adjustment is $0.707/0.637 = 1.11$. This scale factor is correct only if the waveform is sinusoidal; that is, a multimeter does not give the true V_{rms} for a periodic but nonsinusoidal voltage.

The input resistance R_i of a multimeter due to R_s and R_m is shown in Fig. E1-3. The value of R_i is changed by means of R_s to accommodate the input voltage. For example, if the specification of a multimeter is 5 kΩ/

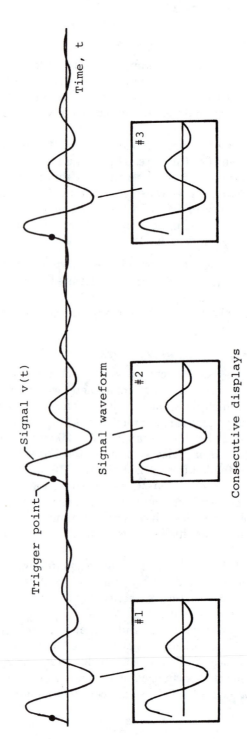

FIGURE E1-2 Displays of a repetitive pattern.

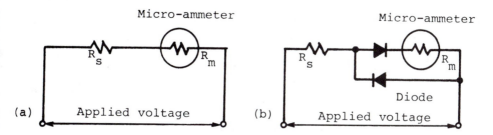

FIGURE E1-3   Schematic of a voltmeter. (a) Dc voltmeter. (b) Ac voltmeter.

V_{ac}, the R_i is 25 kΩ (= 5 × 5 kΩ) for the range 0 to 5 V. Similarly, R_i is 50 kΩ for the range 0 to 10 V.

C.   Digital Multimeter

The simplified block diagram of a digital multimeter for voltage measurements is shown in Fig. E1-4. The input voltage is preconditioned by means of an attenuator and then converted to a rms voltage. The analog-to-digital (A/D) converter (see Sec. 6-7E) changes the analog output from the rms converter to digital. Its value is then shown in the liquid-crystal display (LCD). A digital multimeter with a rms converter gives the true rms value for a sinusoidal as well as a nonsinusoidal periodic signal.

D.   Oscillator

As oscillator is used as a voltage source to simulate the signal from a transducer. The output waveforms of an oscillator are the sine wave, sawtooth wave, and square wave, as shown in Fig. E1-5a. The frequency of the output voltage can be changed over a wide range, and the amplitude V_p of the waveforms is essentially constant and independent of frequencies. A built-in dc bias, or offset, can be added to these waveforms, as shown in Fig. E1-5b.

E.   Equipment Required

General-purpose laboratory equipment are used for the experiments. This includes the oscillator, oscilloscope, multimeter, digital multimeter, capacitance and resistance decade boxes, and an assortment of resistors.

E1-4.   Voltage Measurements: Experiment 1-1

The objective of this experiment is to familiarize the student with basic laboratory instruments and to verify the relation between peak, rms, and average values for different waveforms.

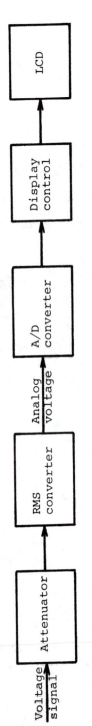

FIGURE E1-4 Functional blocks of a digital voltmeter.

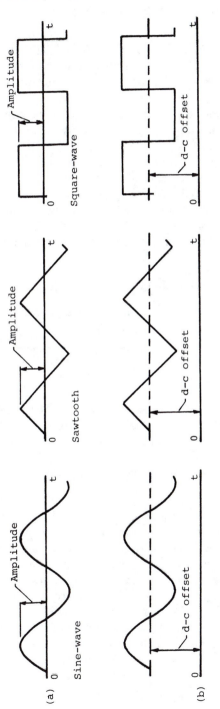

FIGURE E1-5 Waveforms of an oscillator. (a) Periodic waveforms. (b) Waveforms with dc offset.

A. Procedure

1. Setup the equipment as shown in Fig. E1-6a. Set the oscillator at 1 kHz, 6 V_{p-p} (do not use the offset unless otherwise stated). Use the oscilloscope as the sensor. The instructor initializes the oscilloscope.
2. Vary the frequency of the oscillator over a wide range to show that the amplitude of the sine wave, sawtooth wave, and square wave are independent of frequency.
3. Familiarize yourself with the controls of the vertical channels and the horizontal channel of the oscilloscope.
4. Set the oscillator at 1-kHz 5-V_{rms} since wave. Connect the oscillator to the multimeter, digital meter, and the oscilloscope one at a time.
5. Measure the voltages for different waveforms and enter the values in the table.

Waveforms	Oscilloscope (V_{p-p})	Multimeter (V_{rms})	Digital Meter (V_{rms})
Sine wave			
Sawtooth wave			
Square wave			

B. Discussion

Derive the theoretical V_{rms} values from V_{p-p} for the voltages in the table and compare with the measured values. Is it possible to find the rms value of a nonperiodic signal?

E1-5. Input/Output Impedances: Experiment 1-2

A. Introduction

The objective of this experiment is (1) to measure the output impedance of an oscillator, and (2) the input impedance of a readout device. It will be shown in the next section that input and output impedances are important parameters in specifications. The components in Fig. E1-6a are modeled as shown in Fig. E1-6b. The model of the oscillator is its Thévenin's equivalent, consisting of an ideal voltage source V_{eq} in series with an output impedance R_o, and R_o is generally 50 or 600 Ω. The model of the oscilloscope is a resistive element with an impedance R_i, and R_i is from 1 to 20 MΩ. This is a general representation of a transducer and its readout device and is not restricted to an oscillator and an oscilloscope. The impedances are mostly resistive for audio frequencies.

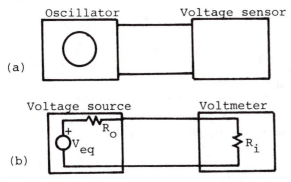

FIGURE E1-6 (a) Voltage measurement. (b) Models of components.

B. Procedure

1. Set the oscillator at a 1-kHz 8-V_{p-p} sine wave.
2. The output impedance R_o is measured as follows:
   a. Measure the open-circuit voltage V_{oc}, shown in Fig. E1-7a, by means of an oscilloscope or a high-impedance voltmeter. Since $R_i \gg R_o$, $V_{oc} \simeq V_{eq}$.
   b. Insert a resistor R ($\simeq 100$ or $600~\Omega$) across terminals a–b, shown in Fig. E1-7b. Since R and R_o form a voltage divider for V_{eq}, we get

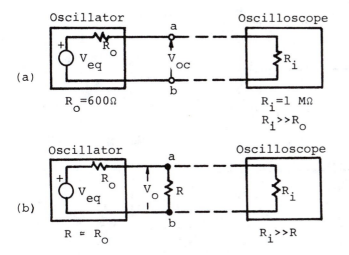

FIGURE E1-7 Measurement of output impedance R_o. (a) Before. (b) After.

$$\frac{V_0}{V_{eq}} = \frac{V_0}{V_{oc}} = \frac{R}{R + R_o} \qquad \text{or} \qquad R_o = R\,\frac{V_{oc} - V_0}{V_0} \qquad \text{(E1-1)}$$

3. The input impedance R_i is measured as follows:

   a. To measure R_i ($\simeq 1\ \text{M}\Omega$) of an oscilloscope, measure V_1 with the oscilloscope, as shown in Fig. E1-8a.

   b. Insert the resistor R ($\simeq 1\ \text{M}\Omega$) shown in Fig. E1-8b across the terminals $a\text{--}b$. The input voltage to the oscilloscope is V_2. Since R_i and R form a voltage divider for V_1, we get

$$\frac{V_2}{V_1} = \frac{R_i}{R + R_i} \qquad \text{or} \qquad R_i = R\,\frac{V_2}{V_1 - V_2} \qquad \text{(E1-2)}$$

4. Repeat step 3 to find R_i for the multimeter. Use the range 0 to 10 V and an external resistor R of 50 kΩ.

5. Repeat step 3 to find R_i for the digital voltmeter. Use $R = 2\ \text{M}\Omega$.

6. Tabulate the results:

   Oscillator: R_o = _____ Oscilloscope: R_i = _____

   Multimeter: R_i = _____ Digital meter: R_i = _____

C. Discussion

1. Compare the measured values with the specifications.

2. With the aid of a sketch, list the steps for finding the input and the output impedance of a hi-fi amplifier.

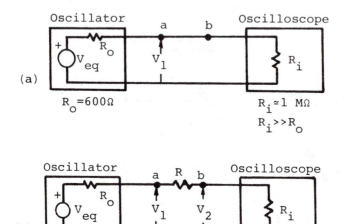

FIGURE E1-8 Measurement of input impedance R_i. (a) Before. (b) After.

E1-6. Loading in Measurements: Experiment 1-3

A. Introduction

The objective of this experiment is to examine loading in voltage measurements. Loading is inherent in all measurements and is caused by the act of measurement itself.

It should be reemphasized that the model in Fig. E1-6b is a general representation of a transducer and its readout device, or any system separated into two parts for the purpose of investigation. Thus the R_o and V_{eq} are due to all components upstream from the point of measurement, and the R_i due to all components downstream. Without loading, the true voltage is the open-circuit voltage, or V_{eq}. With loading, the measured voltage is due to R_i and R_o forming a voltage divider for V_{eq}.

The setups to examine loading in voltage measurements consist of an oscillator, an assortment of resistors, and an oscilloscope, as shown in Fig. E1-9. The oscillator may represent the source of information, the resistors (R_1, R_2), the transducer, and the oscilloscope the readout device. Loading due to other readout devices, such as the mutlimetèr and digital meter, are also examined in the experiment.

B. Procedure

1. Use the suggested values of R_1 and R_2 below. Before connecting the resistors, measure the values of R_1 and R_2 with the digital multimeter. Set the oscillator at a 1-kHz 5-V_{rms} sine wave.
2. Following the steps shown in Fig. E1-9a to d, measure the voltages, and complete the table.
    a. $R_1 = 620\ \Omega$ and $R_2 = 100\ \Omega$

Voltage	Oscilloscope	Multimeter	Digital Meter
V			
V^*			
V_1			
V_2			

b. $R_1 = 1.2\ M\Omega$ and $R_2 = 500\ k\Omega$

Voltage	Oscilloscope	Multimeter	Digital Meter
V			
V^*			
V_1			
V_2			

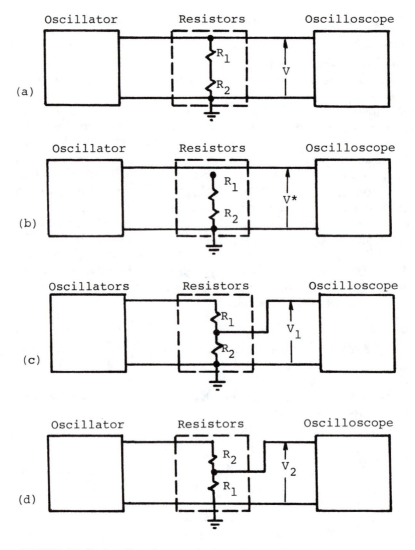

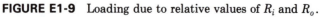

FIGURE E1-9 Loading due to relative values of R_i and R_o.

C. Discussion

1. Is V equal to V^* shown in Fig. E1-9a and b?
2. Is $(V_1 + V_2) = V$ shown in Fig. E1-9c and d?
3. Verify the measured voltages analytically, using the values of the input and output impedance from Expt. 1-2.

E1-7. AC/DC Input Coupling: Experiment 1-4

A. Introduction

The objective of the experiment is to investigate the effects of the ac/dc input coupling modes of instruments. Assume that the signal in Fig. E1-10 from a transducer has a static (dc or average) and a dynamic (ac) component. The dynamic component can be extracted from the signal if the static component is blocked by means of a blocking capacitor (see the ballast circuit in Probs. 1-3 to 1-6). Thus a dynamic component in the signal can be observed independently, as shown in the figure.

For example, assume that the signal v in Fig. E1-11 has a dc and an ac component. If the oscilloscope is in the direct-couple mode, or *dc couple*, the signal v is measured directly across the input resistance R_i as shown in Fig. E-1-11a. The displayed on the screen gives the entire signal. When the control is switched to the *ac couple*, a blocking capacitor C is inserted in the input circuit, as shown in Fig. E1-11b. Thus, only the ac component of the signal is displayed. The input control also has a ground position. This connects the input of the oscilloscope to the ground reference. It is important to note that it does not connect the external input signal to ground.

B. Procedure

1. Set up the circuit as shown in Fig. E1-11. Set the oscillator at a 1-kHz 1-V_p sine wave with a 3-V_{dc} offset.
2. Establish the ground for the oscilloscope (see above) and use the trace of the second channel to mark this datum.
3. Observe the effects of the dc and ac couple.
4. Set the oscillator at a 1-kHz 2-V_p sine wave but without the dc offset. Connect the voltage to channel 1 with a dc couple and to channel 2 with an ac couple. Observe the effects of the dc and ac couple.
5. Repeat step 4 for a 1-kHz 2-V_p square wave.
6. Repeat steps 4 and 5 but with the oscillator at 10 Hz.

C. Discussion

The observations in step 3 are explained in Figs. E1-10 and E1-11. Explain the observations in steps 4, 5 and 6.

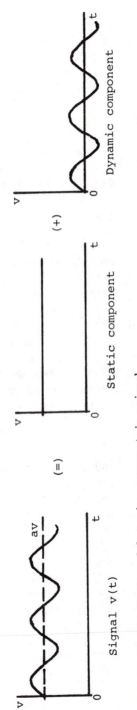

FIGURE E1-10 Static and dynamic components in a signal.

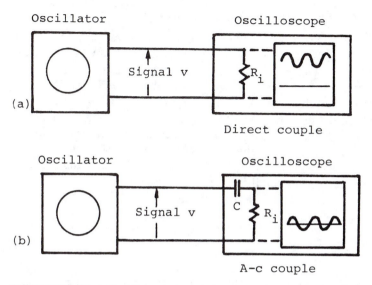

FIGURE E1-11   Coupling modes of an oscilloscope. (a) Direct couple. (b) Ac couple.

E1-8.   Measurement of *C* in the AC Couple: Experiment 1-5

A. Introduction

The objective of the experiment is to measure the value of the coupling capacitor C for the ac couple mode of an oscilloscope. Four methods are used for the measurement. The value of C is of the order of 0.02 μF, and C is in series with R_i, as shown in Fig. E1-11b. Note that this is not the 20-pF capacitance in parallel with R_i in the specification. The 20 pF has no effect for signals at low frequencies, including dc, and is negligible for signals at frequencies below 1 MHz.

B.   Procedure: Direct Comparison—Method 1

1.   Set the oscillator at a 10-Hz 2-V_p square wave without dc offset. Connect the oscillator to both channels of the oscilloscope with an ac couple to show that the traces are identical.
2.   Connect the oscillator to channel 1 of the oscilloscope with an ac couple and to channel 2 with a dc couple but with a capacitor decade box C inserted between the instruments to simulate the coupling capacitor inside the oscilloscope.
3.   Adjust C until the traces of both channels are identical.
4.   Since the channels have identical characteristics, C is the value of the blocking capacitor for the ac couple.

C. Procedure: Time Response—Method 2

1. Set the oscillator at a 10-Hz 2-V_p square wave. Connect this to channel 1 of the oscilloscope with a dc couple, and to channel 2 with an ac couple. The waveforms are as shown in Fig. E1-12a.
2. Obtain values of voltage v versus time t as shown in Fig. E1-12b. Note that the value of v_0 ($t = 0$) is 2 V_p. It may be necessary to adjust the frequency of the oscillator and/or to use the "normal" instead of "auto" trigger for the oscilloscope in order to minimize the flickering on the screen. Corresponding values of (v, t) can be obtained easily by centering v at $t = 0$ on a vertical grid line, and then moving the trace across the screen in small increments of time t.
3. Defining $\tau = R_i C$, it can be shown that

$$v = V_0 e^{-t/\tau} \tag{E1-3}$$

$$\ln \frac{v}{v_0} = -\frac{1}{\tau} t \tag{E1-4}$$

4. Plot Eq. E1-4 and deduce the value of the time constant τ. Find C from the values of τ and R_i measured previously.

D. Procedure: Phase Angle Measurement—Method 3

1. Connect the oscillator to channel 1 of the oscilloscope with a dc couple and to channel 2 with an ac couple.

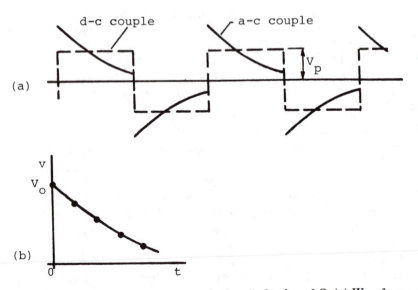

FIGURE E1-12 Time response method to find value of C. (a) Waveforms. (b) Time response.

2. Set the oscillator at a 20-Hz 2-V$_p$ sine wave.
3. Note the phase angle between the sinusoidal traces.

E. Procedure: X-Y Plot—Method 4

1. Set the oscillator at a 10-Hz 2-V$_p$ sine wave. Set the oscilloscope in the x-y mode.
2. Connect the oscillator to channel 1 of the oscilloscope with a dc couple and to channel 2 with an ac couple. Obtain the plot shown in Fig. E1-13.
3. The phsae angle θ between two sine waves of the same frequency can be evaluated from an x-y plot.

$$\sin \theta = \frac{a'}{a} = \frac{b'}{b} \qquad (E1\text{-}5)$$

4. Obtain the phase angle from the x-y plot, using Eq. E1-5.
5. Observe the direction of rotation of the ellipse.

F. Discussion

1. Method 1 is by direct comparison. Is this a variation of the null-balance method in measurement?
2. Derive Eq. E1-3 in method 2.
3. Derive the equation(s) for finding the blocking capacitor C from the phase angle measurement in method 3.
4. Derive Eq. E1-5 in method 4 graphically by combining two sine waves in an x-y plot, and indicate the direction of rotation of the ellipse. Determine which signal is leading the other, and whether Eq. E1-5

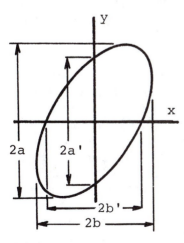

FIGURE E1-13 Phase angle measurement from an x-y plot.

gives the actual phase angle. Calculate the value of the blocking capacitor C.

5. Compare the values of C by the four methods.

E1-9. Trigger System of Oscilloscopes: Experiment 1-6

A. Introduction

The objective of the experiment is to familiarize readers with the trigger system in a general-purpose oscilloscope. An oscilloscope displays the time history of an input voltage. The trigger determines when to switch on the display. This is important for time-related events, such as the pressure–volume signals from an internal combustion engine. One should follow the procedure in the user's manual to familiarize with the controls, and read the specifications for details.

B. Trigger, Slope, and Level

A trigger can be initiated by means of (1) the input voltage itself to either vertical channels (internal), (2) a separate trigger input (external), or (3) the power line (see Figure 6-41). For internal triggering, the beginning of a display is determined by a selected trigger point, which sets the initial slope and the initial level of the trace shown on the screen. For example, the trigger point, shown in Fig. E1-2, is set at a positive slope and a level above the zero datum. Both positive and negative values can be selected for the slope and level control.

The external trigger can be initiated by means of voltage pulse. It synchronizes the display of an input voltage with the pulse from an external event. The line trigger synchronizes the display of an input voltage with the power-line frequency. It is useful when the input is related to the line frequency. These two triggers are for the purpose of synchronizing only and are not displayed.

C. Automatic and Normal Modes

The ramp voltage v_t, shown in Fig. E1-14a, is used to generate the linear time sweep in an oscilloscope. It causes the electron beam in the cathode-ray tube to deflect from left to right to produce a uniform time sweep on the screen. The beam must return to the extreme left of the screen for another sweep. The interval between sweeps is the hold-off, shown in Fig. E1-14b. This consists of a retrace and an inactive zone. The retrace returns the beam to the extreme left of the screen, and the beam is suppressed during the retrace. The oscilloscope waits for another trigger signal at the end of the hold-off.

The *automatic trigger mode* is "signal seeking." Consider a zero input signal. When the sweep ends, the hold-off expires, and a trigger signal is

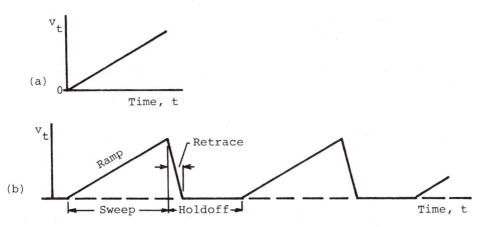

FIGURE E1-14 Ramp voltage for sweep control. (a) Ramp voltage. (b) Trigger waveform.

not found after a time, a trigger is generated anyway, causing the screen to display a baseline for the zero input. This mode allows the oscilloscope to trigger, without making adjustments for the level control.

The *normal trigger mode* does not allow a trace to be displayed when there is no trigger signal. Hence it is not possible to show a zero baseline. This mode of triggering can handle a wider range of trigger signals and is particularly useful for low frequencies.

Again, it is instructive to consult the specifications of the particular oscilloscope and to compare the observations for different settings. It is interesting to try (1) the internal trigger modes for signals below 20 Hz, and (2) the external trigger for signals below 100 mV and 20 Hz.

D. Single Sweep

The single-sweep feature enables an oscilloscope to capture a transient signal. The signal is displayed only once, but it can be photographed as it sweeps across the screen. Since the price of digital oscilloscopes has become competitive, it is more convenient to capture a transient by digital means. The transient can then be examined or reproduced readily.

E1-10. Probe Compensation: Experiment 1-7

A. Introduction

The objective of the experiment is to familiarize readers with probes for voltage measurements. A probe has two advantages compared with simple test leads. First, the probe increases the input impedance of the oscilloscope to minimize loading. Second, it provides the proper shielding and earth gound for test leads. Furthermore, a probe is commonly supplied with different test tips for convenience.

To minimize loading, the input impedance R_i of a sensor should be two orders of magnitude higher than the output impedance R_o of the voltage source (i.e., $R_i > 100R_o$). A 10× probe increases the R_i of the oscilloscope to $10R_i$, but it also reduces the sensitivity by a factor of 10. Note that circuit loading is mostly resistive for frequencies under 5 kHz. The impedance of a probe is both resistive and reactive at high frequencies and is therefore frequency dependent. The user should refer to the specifications for details.

A probe is also an intermediate component between the voltage source and the oscilloscope. A probe for a 1-MΩ oscilloscope is designed to terminate at 1 MΩ to avoid unwanted reflections of an input signal. A probe is trimmed for optimum transient response by a capacitance adjustment to match a particular input channel of the oscilloscope. The adjustment must be repeated when a probe is used for another channel.

B.   Procedure: Probe Adjustment

1.   Set channel 1 of the oscilloscope at an ac couple and sensitivity at the 0.1 V/division. Set the sweep rate at 0.2 ms/division.
2.   Connect the probe to channel 1 and its ground to the BNC collar of channel 2. Insert the tip of the probe into the "probe adjust" of the oscilloscope. A square wave of about 0.5 V at 1 kHz is displayed on the screen.
3.   Use a small screwdriver to trim the probe. Do not turn more than one eighth of a turn at a time. The waveform of a correctly compensated probe is as shown in Fig. E1-15.

E1-11.   Miscellaneous Controls: Experiment 1-8

A.   Introduction

The objective of this experiment is to examine the controls common in a general-purpose oscilloscope. Most oscilloscopes have some convenient features. Since the additional controls may not be standardized, students should refer to the user's manual for details.

B.   Alternate/Chopped, Add/Inverse

The signals displayed for both vertical channels are generally at the same time sweep rate. The displays can be in the alternate or chopped mode. The *alternate mode* displays the traces of the signals alternately, one at a time. This mode is suitable for high sweep rates, from 0.2 ms/division to 0.05 μs/division. To observe the alternate mode at slow motion, connect a

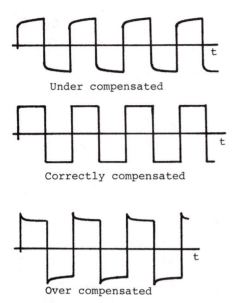

Under compensated

Correctly compensated

Over compensated

FIGURE E1-15 Probe compensation.

2-Hz sine wave from the oscillator to channel 1 of the oscilloscope and the "syn-out" (TTL) from the oscillator to channel 2. Set the oscilloscope at 1 s/division and normal trigger. A sine wave and a square wave are alternately displayed.

The *chopped mode* displays both traces of the input signals simultaneously, but the traces are switched alternately at 500 kHz. In other words, a small piece of trace 1 is shown and then a small piece of trace 2 alternately at 500 kHz during a time sweep. This mode is used for slower signals, from 5 ms/division to 0.5 s/division. The chopped mode can be observed by connecting a 1-kHz sine wave to both vertical channels and setting the oscilloscope at a high sweep rate. If trace 1 is positioned close to trace 2, pieces from the traces would form a continuous display.

The *add control* sums the input signals to the two channels. The channel 2 inverse changes the sign of the channel 2 display. Hence the sum or difference of two signals can be obtained. This can be verified by connecting the same signal to both channels.

C. Delay-Time Control

The delay control changes the time interval between the trigger point and the beginning of a display. The signal for this interval is not shown. In other words, it discards this portion of the signal and displays only the portion of interest, starting from the beginning of a delayed sweep. This mode is a convenience, particularly when the only available trigger point is far from the desired portion of the signal.

The range of the delay is set by switching, and the delay between ranges is adjustable by means of a multiplier control. The amount of delay is selected in three steps.

1.   Trigger the oscilloscope in the usual manner.
2.   Use the intensified mode to brighten the desired portion of the trace to be displayed, and adjust the delay accordingly. This identifies the desired delayed portion. The interval between the beginning of the trace (trigger point) and the brightened portion is the delay time.
3.   Then the oscilloscope is switched to the delay mode to display only the desired portion. The sweep rate can be adjusted in the delay mode.

D.   Variable Trigger Hold-off

A hold-off, as shown in Fig. E1-14b, is the duration between the end of a sweep and the beginning of a trigger for the next sweep. The variable hold-off controls this duration.

A repetitive signal may have many possible trigger points. A variable hold-off prevents a premature trigger, in order to obtain a repetitive trace on the screen. For example, a pulse train is shown in Fig. E1-16a. Every leading (rising) edge of a pulse is a possible trigger point. The traces for a shorter hold-off, as shown in Fig. E1-16b, are not repetitive; therefore, the trace shown is not stable. A repetitive trace is obtained by increasing the duration of the hold-off, as shown in Fig. E1-16c.

10-3.   EXPERIMENT 2. TRANSDUCER AND PHYSICAL LAWS

E2-1.   Introduction

The objective of the experiment is to describe principles of transducers and to demonstrate the hardware and applications of the principles. Physical laws and effects are discussed in Chapter 2, and transducers are described in more detail in the text. The experiment serves to complement the prior chapters.

This study is an overview, but is a necessary step for forming a framework to classify transducers, that is, to gain a perspective of transducers available for measurement. The demonstrations themselves are inadequate to give the student a feel for the subject, but this is

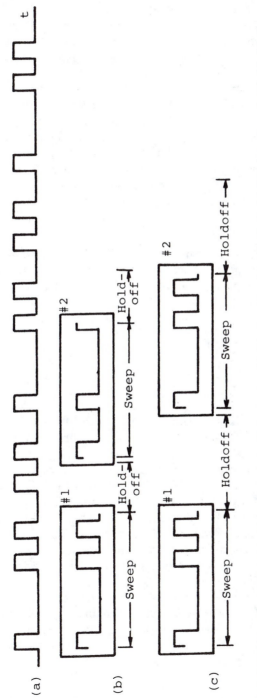

FIGURE E1-16 Variable holdoff control. (a) Pulse train. (b) Displays for short holdoff. (c) Displays for extended holdoff.

followed by an in-depth study of application of strain gages in a later experiment. Hopefully, the combination will introduce readers to the laboratory. This approach is cost-effective, because the cost of a large number of transducers is prohibitive. Neither is there time to go into details in a short course.

Schemes to classify transducers are open ended and at best arbitrary. Transducers can be classified by (1) the function performed, such as a pressure transducer; (2) types of energy involved in the transduction; (3) the principles employed, such as a turbine flowmeter; and (4) others methods. In view of the ramifications, it is expedient to classify transducers on the basis of physical laws and effects. Surprisingly few effects are used in transducers for common applications.

Physical laws are described in Chapter 2 by the relation

$$Y = f(D, U) \tag{E2-1}$$

where Y is the output, D the desired input, U all the undesired inputs, and D and U are independent. There is one and only one desired input in the measurement of a particular variable. $Y_0 = f(D_0, U_0)$ at the quiescent point (0). Expanding Eq. E2-1 in Taylor series about (0), and retaining only the linear terms as a first approximation, we get

$$\Delta Y = \underbrace{\frac{\partial Y}{\partial D}\bigg|_0 \Delta D}_{\text{signal}} + \underbrace{\frac{\partial Y}{\partial U}\bigg|_0 \Delta U}_{\text{noise}} \tag{E2-2}$$

where ΔY is a detectable output due to both the desired ΔD and the undesired inputs ΔU. The desired input is the measurand, or signal. The undesired ones are sources of noise in a measurement.

The equation gives the noise/signal ratio. Another view of the equation is that the output ΔY can be used to measure the variable ΔD or ΔU. Depending on the application, an undesired input for one problem may be the desired one for another. For example, a resistance wire can be used as a temperature sensor as well as a strain gage. A resistance change due to temperature is a noise input in a strain measurement; a resistance change due to strain is a noise input in temperature sensing. This view is used in the demonstrations.

Furthermore, Eq. E2-1 shows that Y is an implicit function, consisting of as many variables as one wishes to place into the equation (see Table 1-2). If Y is for a temperature measurement, temperature can be measured by means of many variables, or in many ways, such as a resistance wire or an optical pyrometer.

A transducer may consist of a primary and a secondary sensor. Consider a tool dynamometer for measuring the cutting force in a lathe (see Fig. 1-3). The force causes an elastic deformation of the tool holder; therefore, the primary sensor is the tool holder. The secondary sensor is the strain gage for detecting its elastic deformation. No attempt is made to identify the primary and the secondary sensors in the demonstrations to follow.

E2-2. Equipment

Equipment for the demonstrations consists of a collection of miscellaneous items. For example, a thermocouple from a home furnace, a thermostat, and a phono pickup could serve the purpose. Transducers for the demonstration should not be restricted to those in this write-up. It may be of interest to identify the transducers around us, or in an automobile, as necessities for daily living.

It is understood that the demonstrations cannot and need not be inclusive. The objective is to show sufficient devices and applications to broaden the perspective for transducers. There is no time to study constructional details or their static and dynamic characteristics.

E2-3. Resistive Transducers

The resistance change of an element is the basis for many transducers. Following Eq. E2-1, the resistance R of a metallic wire is expressible as an implicit function of many variables:

$$R = f(\rho, L, D, P, T, x, \ldots) \tag{E2-3}$$

where ρ is the resistivity of the metal, L the length of the wire, D a characteristic dimension of the cross section, P an external applied pressure, T the wire temperature, and so on.

The equation states that a resistive device can be used to measure many variables, such as a pressure P or temperature T. It does not imply, however, that all resistive transducers have the same physical form. For example, a resistive element, such as a strain gage, can be a wire or a foil type. It can even be a liquid, such as an electrolyte. Obviously, a liquid resistive element is not restricted to simply geometry.

A. Dimensional Effects

If the dimensions L (length) and x (displacement) in Eq. E2-3 are the input variables and the effects of others are negligible, the equation for a resistive transducer becomes

$$R = f(L, x) \tag{E2-4}$$

Several transducers described by the equation are shown below.

1. The *potentiometers* in Fig. E2-1a represent the unbalance method of measurement (see Sec. 3-1B). The output v_o is proportional to the rectilinear displacement input x, or an angular displacement θ. Note that x may be from a liquid level to measure a flow, from a bimetallic strip to sense a temperature, or from a cantilever to indicate a force. The displacement x may be from a position feedback in a control system (see Sec. 3-6).

2. The *differential potentiometer,* or bridge circuit, shown in Fig. E2-1b represents a differential measurement (see Sec. 3-1C). The output v_o is proportional to the difference of the inputs x_1 and x_2. Applications of the circuit are left as an exercise.

3. The *thermocouple potentiometer* in Fig. E2-1c is an example of the null-balance system (see Sec. 3-1A). The system is nulled or balanced before a reading. First, the galvanometer is switched to position a, and R_{adj} is adjusted until the galvanometer is nulled. Thus the voltage across R_s is equal to that of the precision standard cell. At the same time, the slide wire R is calibrated for mV/cm. Now the galvanometer is switched to position-b, and the contact d is moved along the slide wire until the galvnometer is again nulled, that is, the voltage across section c–d of the wire is equal to v_o, or the voltage of the thermocouple (TC). To balance the potentiometer manually for a large number of reading is tedious. It can be mechanized, however, by means of a servomechanism, as in a temperature recorder.

4. The *liquid-level indicator* in Fig. E2-1d consists of two metal rods separated by a fixed distance. The resistance between the rods decreases with height H of the electrolyte in the container. In this case the electrolyte is city water, which is mildly conducting. The output v_o is proportional to H. Alternatively, if H is constant and table salt is added to the water, the device becomes a salt concentration gage (see Fig. 8-8 for the principle of a magnetic flowmeter).

5. The *phone mouthpiece* in Fig. E2-1e is a variable-resistance device. The wave pattern of a voice can be seen on the screen of an oscilloscope.

6. A potentiometer as a *function generator* is shown in Fig. E1-2f. It consists of a resistance wire wound on an insulating card of irregular shape. The output v_o is a function of the displacement input x. The *resolver* in Fig. E2-1g works on the same principle. It consists of a uniform wire-wound potentiometer on a rectangular insulating card. The arm carries two brush contacts, and v_o is measured across the brushes. This is a function generator because the output v_o is proportional to the sine or cosine of the shaft rotation, which determines the number of turns of the wire bracketed by the brushes.

7. A *depth gage* for measuring the height of grain in a silo is shown in Fig. E2-1h. It consists of a potentiometer encased in a plastic envelope like a sheath. The inner surface of the envelope is lined with aluminum foil. The grain pushes the envelope against the potentiometer, thereby causes a change in resistance.

B. Strain Gages

Assume that the variables in a resistive strain gage are its dimensions and the resistivity ρ of the material. In this case, the variables are dependent, and only one type of transducer is possible. Following Eq. E2-3, we get

$$R = f(\rho, L, D) \tag{E2-5}$$

where L is the length and D is a characteristic dimension. For a wire- or a foil-type strain gage, the resistance R is

$$R = \frac{\rho L}{area} = \frac{\rho L}{cD^2} \tag{E2-6}$$

where c is a constant. The sensitivity, or the ratio of $\Delta R/R$ to the applied strain ε is the gage factor G_f (see Eq. 1-6).

$$G_f = \frac{\Delta R/R}{\varepsilon} \tag{E2-7}$$

A foil-type strain gage is shown in Fig. E2-2a. The gages, shown in Fig. E2-2b, are often used in a quarter, half, or a full bridge (see Sec. 5-7C). The application of a strain gage bridge in a force transducer is shown in Fig. E2-2c. It consists of a cantilever as the primary sensor and the gage bridge as the secondary sensor. A strain indicator is used with the bridge to measure the a static force F. A dynamic force can be measured by connecting an oscilloscope to the strain indicator. The pressure transducer, shown in Fig. E2-2d, uses a strain gage to detect the deflection of a diaphragm due to pressure.

C. Temperature Effects

If temperature is the desired input and the effects of the other variables in Eq. E2-3 are negligible, we have

$$R = f(T) \tag{E2-8}$$

An application is the resistance temperature detector (RTD), which is used extensively in industry (see Secs. 9-5 and 9-6). Another example is a *themistor*. It is a semiconductor and has a large resistance change with temperature. They are used for temperature sensing as well as for temperature compensation in instruments. Thermistors are made in numerous forms, and a bead type is shown in Fig. E2-3a.

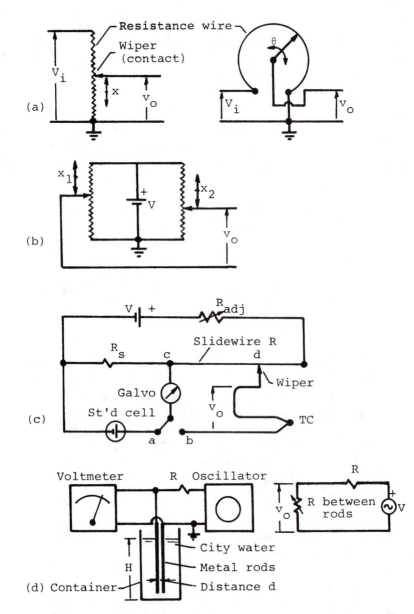

FIGURE E2-1 Examples of resistive transducers. (a) Potentiometers. (b) Differential potentiometer. (c) Thermocouple potentiometer. (d) Liquid-level indicator. (e) Phone mouthpiece as a transducer. (f) Potentiometric function generator. (g) Sine–cosine resolver. (h) Height gage in silo.

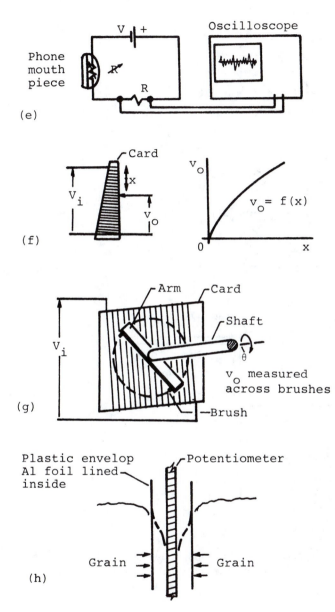

(e)

(f)

(g)

(h)

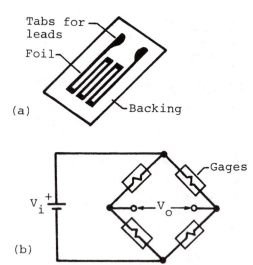

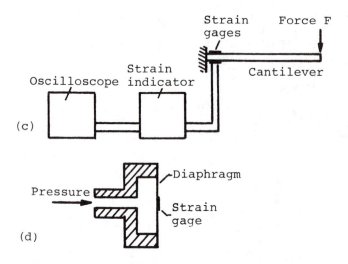

FIGURE E2-2   Strain gages and sample applications. (a) Foil-type strain gage. (b) Strain gage bridge. (c) Force measurements. (d) Pressure transducer.

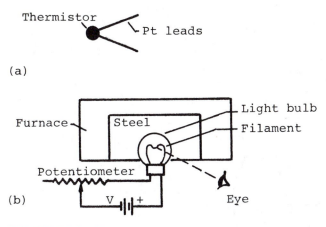

FIGURE E2-3 Examples of resistive temperature sensors. (a) Thermistor. (b) Optical pyrometer.

The *optical pyrometer,* shown in Fig. E2-3b, compares the temperature of a piece of steel in a heat-treating furnace with that of the filament of a light bulb under monochromatic conditions at an effective wavelength of 0.65 μm (see Sec. 9-8A). The temperature of the steel is the unknown and that of the filament is the reference standard. A potentiometer is used to adjust the current I through the filament and therefore its I^2R heating and temperature. The steel and the filament are viewed simultaneously. The filament appears darker than the steel when it is at a lower temperature, it appears brighter when at a higher temperature, and it seems to disappear when the filament and the steel are at the same temperature.

D. Discussion

1. With the aid of a sketch, briefly describe each of the following: (a) a hot-wire anemometer, (b) a vacuum gage using a resistive element, and (c) a Bridgman-type pressure transducer.
2. Briefly describe additional examples of resistive transducers for measurements of flow, pressure, temperature, force, torque, and acceleration.
3. Give two examples of applications of the differential potentiometer, shown in Fig. E2-1b.
4. Sketch the pattern of a special strain gage for pressure transducers.

E2-4. Capacitive Transducers

A. Introduction

A capacitor C consists of two parallel conducting plates of area A separated by a distance d, as shown in Fig. E2-4a.

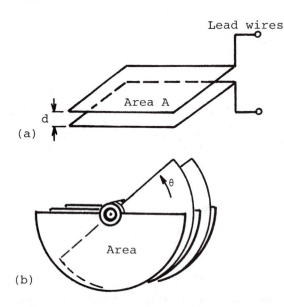

FIGURE E2-4 Schematic of capacitive transducer. (a) Capacitor. (b) Angular measurement.

$$C = f(k,A,d) = \frac{kA}{d} \tag{E2-9}$$

where k is the dielectric constant of the medium between the plates. Most capacitive transducers employ A or d as the input variables and are essentially displacement transducers. Since a displacement is a useful intermediate step for measurements, capacitive transducers have numerous applications. They are generally more expensive because of the associated electronics.

The capacitance C in Eq. E2-9 is a function of three independent variables; therefore, three types of capacitive transducer are obtainable by expanding the equation in a Taylor series.

$$\Delta C = \underbrace{\frac{\partial C}{\partial k} \Delta k}_{\text{type 1}} + \underbrace{\frac{\partial C}{\partial A} \Delta A}_{\text{type 2}} + \underbrace{\frac{\partial C}{\partial d} \Delta d}_{\text{type 3}} \tag{E2-10}$$

Each term on the right side of the equation is a basis for a type of transducer. If k and A are constant and d is the variable, we have a capacitance micrometer. If k and d are constant and A is the variable, we obtain a transducer for angular measurements, as shown in Fig. E2-4b. If A and d are constant and k is the variable, we obtain a pressure transducer for high-pressure measurement, since the dielectric of air is a function of pressure. Note that a parameter, or a constant under normal conditions, is not invariable for other applications.

B. Demonstration

The capacitance micrometer for the demonstration is shown in Fig. E2-5a. It consists of two metal plates separated by sheets of paper. The oscillator for the power supply is at about 4 kHz. The capacitance is varied by inserting sheets of paper between the plates. The circuit is shown in Fig. E2-5b. The output V_o is measured across the drop resistor R with a sensitive voltmeter. Light pressure is applied to the top plate in order to obtain a consistant V_o readout. The static characteristics of the transducer is obtained by plotting V_o versus number of sheets of paper. The thickness of paper is about 0.003 in., and a resolution of 0.001 in. can readily be obtained.

C. Discussion

Briefly describe a capacitive device for each of the following applications: (1) a microphone, (2) a pressure transducer for an internal-combustion engine, (3) an extensometer for tensile testing, and (4) a liquid-level indicator.

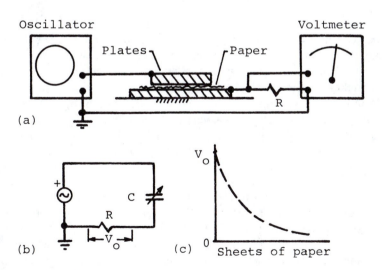

FIGURE E2-5 Simple capacitance micrometer. (a) Schematic of capacitance micrometer. (b) Circuit. (c) Characteristic.

E2-5.   Inductive Transducers

A.   Introduction

Physical laws relating the self-inductance L of a coil and the mutual-inductance M between two coils are

$$L = \mu N^2 K \qquad\qquad\qquad\qquad\qquad\qquad\qquad \text{(E2-11)}$$

$$M = \mu N_1 N_2 K \qquad\qquad\qquad\qquad\qquad\qquad \text{(E2-12)}$$

where μ is the effective permeability of the medium around the coil, N the number of turns of each coil, and K a geometric factor. Inductive transducers are also called variable reluctance devices, because their operations depend on the reluctance of the magnetic path. All the parameters in the equations above can be varied for measurements. Many transducers can be constructed on the basis of the same principle, but with different physical configurations (see Secs. 5-3 and 5-6).

Inductive transducers also include a large class of devices based on the "motor-generator" effect. *Generator effect* describes the voltage v induced in a moving conductor of length L with a velocity $\dot{x}$ in a magnetic field of flux intensity B.

$$v = BL\dot{x} \qquad\qquad\qquad\qquad\qquad\qquad\qquad\qquad \text{(E2-13)}$$

Conversely, *motor effect* describes the force f acting on a conductor of length L carrying a current i in a magnetic field.

$$f = BLi \qquad\qquad\qquad\qquad\qquad\qquad\qquad\qquad \text{(E2-14)}$$

In fact, a dc generator can be converted into a dc motor, and vice versa. For example, the dc generator in an engine dynanometer can also be used as a motor for finding the friction of the engine. The small electro-mechanical shaker for the next experiment is an inverted velocity pick-up. It is designed as a voltage generator for a velocity pickup, and its voltage output is proportional to the velocity of the mechanical input. It is a motor when used as a shaker. The mechanical force output is proportional to the voltage applied to the electrical terminals.

B.   Demonstrations

1. The *comparator* for dimensional inspection shown in Fig. E2-6a consists of two inductance coils, a heavy steel lever with a fulcrum, and a push rod. First, a precision gage block is used to set the datum, or a zero reference for the comparator. The output of the comparator is zero when the gage block is replaced with a test piece of the same dimension. Otherwise, the test piece causes the lever to pivot about the fulcrum, thereby changing the geometric factor at the coils to give the deviation from the

reference. The *differential pressure transducer* shown in Fig. E2-6b works on the same principle. A differential pressure $(P_2 - P_1)$ deflects the steel diaphragm, thereby changing the geometric factor of the two inductance coils.

2. The *linear variable differential transformer* (LVDT) shown in Fig. E2-6c converts a mechanical displacement input x into an electrical voltage output (see Sec. 5-2). The transformer consists of a primary coil P, two secondary coils S_1 and S_2, and a core of magnetic mateiral. The position x of the core determines the magnetic flux coupling, or the mutual inductance, between P and S_1 and S_2. When the core is above its electrical neutral position, the flux coupling between P and S_1 is larger than that between P and S_2, and the voltage V_1 is larger than V_2. V_1 equals to V_2 when the core is at neutral. V_1 is less than V_2 when the core is below neutral. The circuit to obtain a differential output voltage $(V_1 - V_2)$ is shown in Fig. E2-6d. The LVDT as a carrier system (see Sec. 5-4) will be demonstrated in Expt. 7.

3. Generator effect is the basis for the *magnetic counter* shown in Fig. E2-6e. It consists of a permanent magnet and a stationary coil. Normally, the coils in a generator rotates in a stationary magnetic field. Here the magnetic flux changes about a stationary coil. When a gear tooth of magnetic material crosses the face of the permanent magnet, as shown in Fig. E2-6f, it interrupts the flux pattern momentarily, and a voltage pulse is induced in the coil. The pulses can be counted with an electronic counter. This is a common method for finding the rotational speed of a shaft.

4. Motor effect is the basis for the electromechanical *loudspeaker,* shown in Fig. E2-6g. It consists of a movable coil in the field of a permanent magnet. The motion of the cone is governed by the voltage applied to the coil. Another example is the *phone earpiece* shown in Fig. E2-6h. The motion of the steel diaphragm is governed by the voltage applied to the coil. Similarly, a hi-fi speaker makes a good electromechanical shaker. It is necessary to cut out parts of the cone, however, to reduce the noise output.

5. Other inductive transducers in the demonstration are tachometers, galvanometers, and synchros.

C. Discussion

Describe the operations of the following transducers: (1) an eddy-current dynanometer, (2) an ac tachometer, (3) an inductive transducer for measuring the thickness of paint, and (4) two synchros acting as input and output transducers.

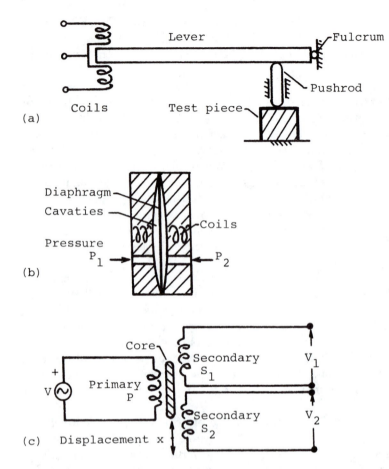

FIGURE E2-6 Examples of inductive transducers. (a) Comparator. (b) Differential pressure transducer. (c) Differential transformer. (d) Differential output $(V_1 - V_2)$. (e) Magnetic pickup. (f) Setup for counting. (g) Speaker. (h) Phone earpiece.

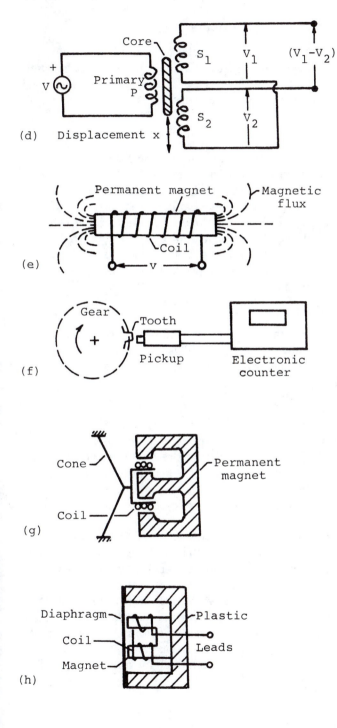

(d) Displacement x

(e)

(f)

(g)

(h)

E2-6. Other Common Transducers

Virtually unlimited physical laws and effects can be used for transducers. If an input produces a detectable effect on an object, the object is potentially a transducer for that input. Several useful transducers are described below, but they are not as versatile as the resistive, capacitive, and inductive types.

A. Flexures

A flexure is a spring, or an elastic member. It is often used as the primary sensor in a transducer for mechanical measurements (see Fig. 1-3). Common secondary sensors to provide an electrical output are the resistive, inductive, and capacitive types. The steel diaphragm for pressure sensing shown in Fig. E2-6b is an example of a flexure. The inductance coils are the secondary sensors to give an electrical output. A torque transducer consisting of a torsional shaft with strain gages is another example. A flexure can be made into very intricate shapes and is designed to respond mainly in certain desired directions. For example, the force applied to an aircraft in a wind tunnel can be resolved in the (x, y, z) directions by means of flexures.

The performance of a transducer for mechanical measurements depends largely on the flexure employed. This includes the static and dynamic characteristics, the cross sensitivity, as well as its long-term stability of the transducer.

B. Resonances

The speed indicator, shown in Fig. E2-7a and b, is for large stationary engines, such as a steam turbine. The indicator consists of a bank of reeds or cantilevers. Each reed is tuned to a different resonant frequency, and a range of frequency is covered by the bank of reeds. Thus the engine speed is indicated by the reed in resonance. This may be regarded as an elementary spectrum analyzer for the frequency of the engine (see Sec. 6-9B).

The resonant frequency of a grinding wheel is determined by a direct comparison method, shown in Fig. E2-7c. There is a correlation between the quality of a grinding wheel, its modulus of elasticity, and the natural frequency. The suspended wheel is struck with a small mallet. Since the higher harmonics will soon die out, the wheel rings at its fundamental frequency. A microphone picks up the ringing and compares this frequency with that from an oscillator, using an oscilloscope in the x-y mode for comparison. The natural frequencies of a system can also be determined by means of feedback. This is a useful technique for finding natural frequencies and is demonstrated in Expt. 3-3.

C. Stroboscope and Photocell

The rotational speed of a small fan is measured by means of an electronic stroboscopic lamp, as shown in Fig. E2-7d. The lamp flashes with a strong light of very short duration. Its rate of flashing can be adjusted easily. When the stroboscope and the fan are synchronized (i.e., when the flashing rate of the lamp and the rotational speed of the fan are identical), the fan appears stationary. The rotational speed is read directly from the stroboscope (see Sec. 6-8E).

A stationary image is also observed when the flashing rate of the stroboscope is a multiple or submultiple of the rotational speed of the fan. The procedure for a speed measurement is to flash the stroboscope from a high to a low rate. The proof of this statement is left as an exercise. The correct measurement is the highest setting of the stroboscope at which a single stationary image is achieved.

The range of a stroboscope is generally from 110 to 25,000 flashes per minute. Speeds beyond the operating range are found by means of submultiple measurements. The procedure to find the unknown speed N is as follows:

1. Synchronize the rotating object with the highest flashing rate X. Assume that this is the nth submultiple, but the value of the submultiple s_n is unknown.
2. The flashing rate is decreased slowly, until it synchronizes at m additional submultiples. This flashing rate is Y and the submultiple is s_{n+m}.
3. The flashing rates X and Y, and the value of $m = s_n - s_{n+m}$ are known. It can be shown that the speed N is determined from

$$N = \frac{mXY}{X - Y} \tag{E2-15}$$

The flashing rate can also be synchronized by means of a trigger. For example, a reflector shown in Fig. E2-7e is attached to a rotating object. A phototube picks up the reflection from a steady light source to trigger the stroboscope. Since the rate of flashing is synchronized, any part of the system can be viewed with respect to a reference. Furthermore, the flashing can also be delayed from the point of trigger in order to examine parts of the rotating cycle.

D. Piezoelectric Effect

A piezoelectric element (see Sec. 7-5B) is a slab or a cut from a solid crystalline dielectric. The crystal is sandwiched between two plate electrodes, as shown in Fig. E2-7f. Piezoelectric effect describes the charge Q

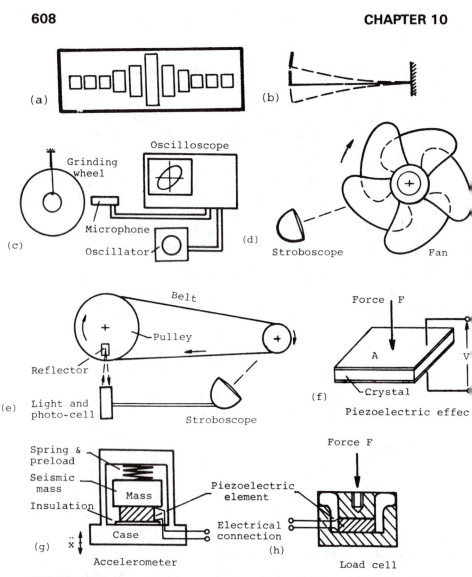

FIGURE E2-7 Examples of common transducers. (a) Speed indicator. (b) Vibrating reed. (c) Setup for finding resonance. (d) Rotational speed measurement. (e) Rotational measurements with trigger. (f) Piezoelectric effect. (g) Accelerometer. (h) Load cell.

generated within a crystal due to its deformation, and Q is proportional to the applied force F. Since the electrodes also form a capacitor, the voltage V across the electrodes is proportional to Q and therefore the force F.

A piezoelectric accelerometer with an input $\ddot{x}$ is shown in Fig. E2-7g. The input to the piezoelectric crystal is the inertial force of the seismic mass due to the acceleration $\ddot{x}$. The system is preloaded by means of a spring above the seismic mass so that crystal is always in compression. The sensitivity is expressed in coulombs/g or volts/g of acceleration.

Piezoelectric effect is the basis for both the accelerometer and the load cell, shown in Fig. E2-7h. For the load cell, an external force F is applied to the crystal instead of the inertial force of the seismic mass. In fact, the two devices can be combined into one unit called the impedance head (see Fig. 7-19). The voltage output of a load cell is proportional to the applied force F. Load cells have high stiffness and high natural frequencies and are generally used for dynamic force measurements.

It should be mentioned in passing that the inverse of a load cell is an electromechanical shaker (not shown). The input is an electrical voltage and the output is a mechanical force. The device is suitable for high-frequency applications.

E. Miscellaneous

The demonstrations include several transducers not in the write-up. Some examples are:

1. Air velocity can be measured by means of a turbine (fan) and a mechanical counter. A turbine flow meter works on the same principle (see Sec. 8-6).
2. A mercury-in-glass thermometer uses the differential coefficient of expansion between mercury and glass for temperature measurements (see Sec. 9-3). A bimetallic thermometer employs the same principle but utilizes the differential expansion of two dissimilar metals (see Fig. 9-1).
3. Temperature can be measured with a thermocouple, the output of which can be read directly with a digital voltmeter (see Sec. 9-4).

F. Discussion

1. Does loading exist when using a mechanical revolution counter for measuring the speed of a small electric motor?
2. Loading is negligible with a non-contact-type instrument such as a stroboscope to measure the speed of a small fan for the cooling of electronic instruments. Show the possibility of loading in this measuring process.
3. A four-bladed fan is at 3600 rpm. One of the blades is marked with a

white dot. Sketch the possible stationary images of the fan if the flashing rates of a stroboscope is from 700 to 25,000 flashes per minute.

4. Assume that the fan shown in Fig. E2-7d is at 1800 rpm. Obtain the speeds for three harmonics and three subharmonics. Then verify the fan speed from the three subharmonics.
5. Derive Eq. E2-15.
6. Give two additional examples of flexures in transducers.
7. Give two examples of applications of thermocouples in which the transducers are for measuring variables other than temperature.

10-4. EXPERIMENT 3. MEASURING SYSTEMS

E3-1. Introduction

The objective of the experiment is to illustrate the functional stages of a measuring system (see Fig. 1-2). The description by functional stage gives an overview but only loosely defines the components. For a given system, a practical description must be more specific, including the identification of components, how the functions are being performed, and the characteristics and compatibility of components.

The transducer-detector constitutes the *first stage* of a measuring system, shown in Fig. E3-1. Its function is to sense the output from an information soruce, to transduce this energy input into a suitable form, and to condition the signal for transmission to the next stage. A transducer for mechanical measurements often consists of a primary and a secondary sensor (see Sec. E2-6A). The transducer output is generally electrical, for the convenience of signal transmission and manipulation. Aside from the input sensor and with few exceptions, all components of a measuring system are electrical.

The function of the *intermediate stage* is to receive the signal from the first stage, perform the necessary operations, and then prepare the signal for transmission to the next stage. This includes a host of operations, such as signal conditioning, amplifying, filtering, manipulation, analog-to-digital conversion, and signal transmission. These are described in the text. It is necessary to refer to a specific appliation, however, in order to describe the operations performed in a functional stage. The *terminal stage* prepares the signal for utilization, such as monitoring, recording, and storage.

1. Note that the demarcation between stages is arbitrary. Components are grouped into functional stages for the convenience of description

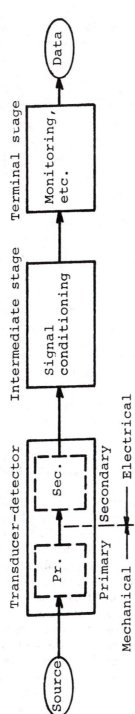

FIGURE E3-1 Functional stages of measuring systems.

only, and an operation may be performed in one stage or another. For example, the amplification of a signal can be performed in any of the three stages.

2. It is important that the stages are compatible. For analog signals, compatibility is impedance matching (see Secs. E1-5 and E1-6). For digital signals, the stages must be able to communiate with each other (see Sec. 6-9E).

Systems selected for the demonstrations depend on the equipment on hand. Even a simple pressure gage can be divided into three functional stages. The measuring systems for the demonstration are described below.

E3-2. Acoustic Pressure Detection

The objective of the experiment is to demonstrate the functional stages of a specific measuring system. It shows that the acoustic pressure from a human voice is sufficient to produce a measurable motion in a glass panel.

The components are as shown in Fig. E3-2. The glass panel of a door and the accelerometer form the first stage of the system. The accelerometer is attached firmly to the glass with a thin film of beeswax, using a wringing motion. The input to the system is the acoustic pressure from a human voice. The glass panel transduces the energy from the acoustic pressure into mechanical energy in the form of a motion. This is further transduced into electrical energy by means of the accelerometer. Hence the glass panel is the primary transducer and the accelerometer the secondary. Note that the acoustic pressure is the excitation input to the system, and it causes the panel to vibrate at the voice frequency. This is not a resonance phenomenon, because resonances occur only at the natural frequencies of the glass panel.

The intermediate stage consists of the preamplifier, amplifier, and filter. The preamplifier has a high input impedance and low output impedance for impedance matching (see Sec. E1-6). It is adjacent to the accelerometer to avoid its loading by the cabling and the components downstream. In other words, the preamplifier prepares the signal for transmission. It has sufficient power to deliver the signal, and longer cables can be used for the components down stream. The preamplifier can be grouped with the first or the second stage. In fact, with the advent of microelectronics, the preamplifier is often an integral part of the transducer. The function of the amplifier is self-evident. The filter is for additional signal conditioning. The characteristics of a laboratory variable filter are shown in Fig. E3-3. It may be of interest to adjust the bandwidth

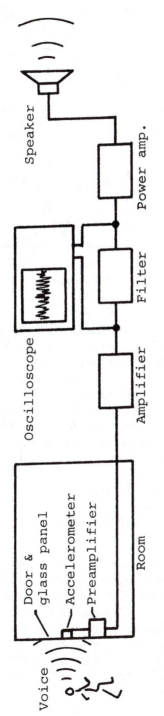

FIGURE E3-2 Vibrations due to acoustic pressure.

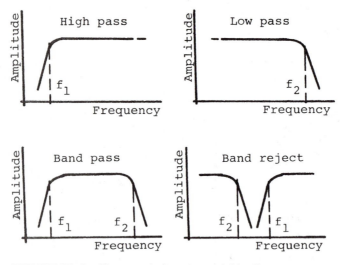

FIGURE E3-3   Characteristics of a variable filter.

of the band pass filter to find the minimum frequency range for which the voice output from the loudspeaker remains intelligible, or to use the high-pass and low-pass sections to alter the pitch of the voice from the speaker. The waveforms of the signals before and after filtering are compared by means of the dual-channel oscilloscope.

The power amplifier and loudspeaker constitute the terminal stage. They are salvaged from the output stage of a portable AM radio. The oscilloscope can also be used to display the waveforms of the signal to the speaker. The glass panel and the speaker must be in separate rooms to avoid resonances due to positive feedback. Resonances due to feedback can be demonstrated in the experiment.

E3-3. Vibration of a Simple Beam

A. Description

The objectives of the experiment are (1) to demonstrate the functional stages of a measuring system, using a simple force and a motion measuring system; and (2) to find the sinusoidal steady-state frequency response and the natural frequencies of a simply supported beam (see Sec. 4-5). Since the demarcation between functional stages is arbitrary, the discussion of functional stages is not repeated here.

The force measuring system, shown in Fig. E3-4, consists of the oscillator as a power source for the shaker, the piezoelectric load cell for the

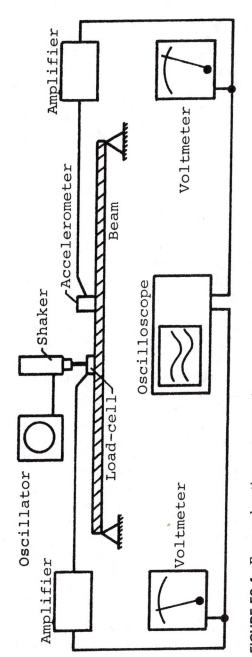

FIGURE E3-4 Force and motion measurements.

force detection, the amplifier for signal conditioning, and the voltmeter for the readout. The electromechanical shaker for the force input to the system is an inverted velocity pickup (see Sec. E2-5). The waveform of the force signal is monitored with an oscilloscope. It is desired to keep the force signal reasonably sinusoidal for the frequency response test.

The motion measuring system consists of the piezoelectric accelerometer for the motion detection, the amplifier for signal conditioning, and the voltmeter for the readout. Both the force and the acceleration signals are monitored and compared by means of the dual-channel oscilloscope.

B. Procedure

1. Measure the dimensions of the beam and calculate its first four natural frequencies.
2. Set up the system shown in Fig. E3-4. Preload the load cell with the shaker. Position the load cell and the accelerometer along the beam to avoid nodal points for the resonances anticipated in the test. Attach the accelerometer firmly to the beam with beeswax, using a wringing motion.
3. Apply a sinusoidal voltage to the shaker, proceeding from low to high frequencies. It may take a short time for the beam to response to the excitation, particularly at the lower frequencies. It takes time to transfer energy to the beam, even at a resonant frequency. Adjust the voltage level near resonances, so that both the force and the motion signals are reasonably sinusoidal.
4. Obtain frequency response data through the first four resonances: the values of force, acceleration, and phase angle at the excitation frequencies. The phase angle can be obtained by means of the oscilloscope (see Fig. E1-13) or a phase meter. The load cell and the accelerometer are not calibrated for the experiment because the amplitude ratio of acceleration to force are used to plot the data.
5. Present the data in Bode plots: (a) the magnitude ratio of (acceleration/force) versus frequency in a log-log plot, and (b) the phase angle versus log frequency in a semilog plot (see Sec. 4-6).

C. Discussion

1. False resonances may occur if the setup is not ideal in a vibration test. Note the false resonances, if any.
2. The input to the shaker from the oscillator is sinusoidal. From the observations, explain whether it is possible for the force signal from the load cell to be nonsinusoidal.

3. Compare the resonant frequencies of the beam with predicted values. Explain the discrepancy, if any.
4. Why is a free-free beam commonly used in vibration testing?

10-5. EXPERIMENT 4. STRAIN GAGE APPLICATIONS

E4-1. Objectives

1. To provide the student with sufficient hands-on experience in one field of study, to get a feeling for the subject, and to acquire some proficiency in experimentation
2. To demonstrate applications of the strain gage (a) in strain measurements, and (b) in transducers
3. To illustrate the problem of coupled modes of elastic deformation in applications

E4-2. Introduction

This experiment is a detailed study of the application of strain gages. The objectives are as stated above. In essence, the purpose is to give students a laboratory experience with which they can relate to real problems.

The electrical resistance strain gage is selected for the study because (1) stress is an important consideration in machines, (2) the strain gage is a common secondary sensor in transducers (see Sec. 5-7C), and (3) the strain gage and its associated basic instrumentation are relatively inexpensive compared with transducers and instruments of comparable quality for other measurements. A modest laboratory budget could afford multiple setups and the experience of applying gages to a test specimen. Due to the versatility of strain gages, the instrumentation can be shared with other experiments or another laboratory.

This laboratory consists of nine simple setups. The rationale for their selection are discussed in Sec. E4-7 before the detailed write-up of experiments. Only static strains are measured, but dynamic strain measurements are demonstrated. Similar instruments are used for most setups. The uniformity of equipment alleviates the instructional time on laboratory instruments. Two or three students are assigned to a setup, and students rotate from one setup to another.

Topics common to the experiments, such as stress–strain relations and circuits, are reviewed in the sections below.

E4-3. Stress–Strain Relations: A Review

Two-dimensional stress–strain relationships are reviewed briefly in this section. When a strain gage is bonded to the free surface of an object, the gage becomes essentially a part of the surface. Hence this is a two-dimensional study. The state of stress of an object is usually of interest, but stresses are not measured directly and are deduced from strain measurements.

With few exceptions, such as a thin plastic object, the effect of the bonded gage on an object is negligible. In other words, if the bonded gage stiffens an object, loading in the measurement is due to a structural change at the source of information. Loading due to the bonded gage is ignored in the experiments.

A. Simple Stress–Strain Relations

Stress–strain relations are introduced by considering normal stresses applied to an object in orthogonal directions. The general case is examined in Sec. E4-3B. Let an uniaxial force F in the x direction be applied to a bar of uniform cross-sectional area A, shown in Fig. E4-1a. Assume that no

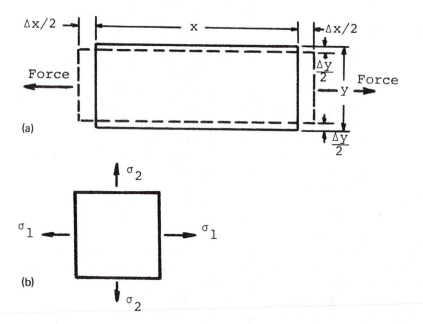

FIGURE E4-1 Simple stress–strain relations. (a) Unixial stress. (b) Biaxial stresses.

force is applied in the y direction. The *uniaxial* stress–strain relation is described by *Hooke's law:*

$$\text{stress} = E(\text{strain}) \tag{E4-1}$$

$$\sigma_x = E\varepsilon_x \quad \text{or} \quad \varepsilon_x = \frac{1}{E}\sigma_x \tag{E4-2}$$

where E is Young's modulus of elasticity of the material, $\sigma_x = F/A$ is the axial stress, and $\varepsilon_x = \Delta x/x$ is the unit axial strain, or the deformation per unit length in the x direction. If the volume of the bar remains constant, this produces a unit strain $\Delta y/y$, or a change in the lateral dimension as shown in the figure. In other words, a two-dimensional strain is caused by a one-dimensional stress. The ratio of the strains is *Poisson's ratio:*

$$\text{Poisson's ratio} = -\frac{\text{lateral strain}}{\text{axial strain}} = -\frac{\Delta y/y}{\Delta x/x}$$

$$\nu = -\frac{\varepsilon_l}{\varepsilon_a} \tag{E4-3}$$

Let two orthogonal normal forces be applied to an element of unit thickness in an object, shown in Fig. E4-1b, and let the *biaxial* stresses be σ_1 and σ_2. An axial and a lateral strain are produced by each of the stresses. Introducing the Poisson effect from Eq. E4-3, the corresponding strains ε_1 and ε_2 are

$$\varepsilon_1 = \frac{1}{E}(\sigma_1 - \nu\sigma_2) \quad \text{and} \quad \varepsilon_2 = \frac{1}{E}(\sigma_2 - \nu\sigma_1) \tag{E4-4}$$

Conversely, the equations can be solved simultaneously to find the normal stresses σ_1 and σ_2 from the normal strains.

$$\sigma_1 = \frac{E}{1 - \nu^2}(\varepsilon_1 + \nu\varepsilon_2) \quad \text{and} \quad \sigma_2 = \frac{E}{1 - \nu^2}(\varepsilon_2 + \nu\varepsilon_1) \tag{E4-5}$$

B. General Plane Stress–Strain Relations

The general two-dimensional deformation of an element $(\Delta x, \Delta y)$ due to arbitrary external forces is shown in Fig. E4-2a, where (u, v) are translations in the (x, y) directions, $(\varepsilon_x, \varepsilon_y)$ the normal strains, and $\gamma_{xy} = \gamma_{yx} = \gamma$ the shear strains. The strains are defined in the equations

$$\varepsilon_x = \frac{\partial u}{\partial x} \quad \varepsilon_y = \frac{\partial v}{\partial y} \quad \gamma_{xy} = \frac{\partial v}{\partial x} + \frac{\partial u}{\partial y} \tag{E4-6}$$

Let a strain gage of length L, shown in Fig. E4-2b, be placed along an arbitrary θ direction to measure the strains defined in Eq. E4-6. The con-

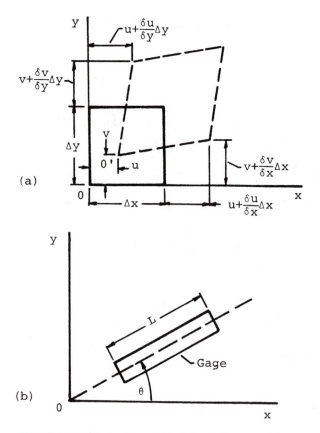

FIGURE E4-2 (a) Two dimensional deformation of an element. (b) Gage of length L subjected to strains ε_x, ε_y, and γ_{xy}.

tribution of each of the terms to ε_θ is shown in Fig. E4-3. Note that the Poisson strain is not introduced in the derivations, because only the geometry of the problem is being considered. The strain ε_θ along the strain gage is the sum of the strains shown in the figure.

$$\varepsilon_\theta = \varepsilon_x \cos^2\theta + \varepsilon_y \sin^2\theta + \gamma_{xy} \sin\theta \cos\theta \qquad (E4\text{-}7)$$

Substituting the identities

$$\cos^2\theta + \sin^2\theta = 1 \qquad \cos 2\theta = 1 - 2\sin^2\theta \qquad \sin 2\theta = 2\sin\theta\cos\theta$$

into Eq. E4-7 and simplifying, we obtain

$$\varepsilon_\theta = \tfrac{1}{2}[(\varepsilon_x + \varepsilon_y) + (\varepsilon_x - \varepsilon_y)\cos 2\theta + \gamma \sin 2\theta] \qquad (E4\text{-}8)$$

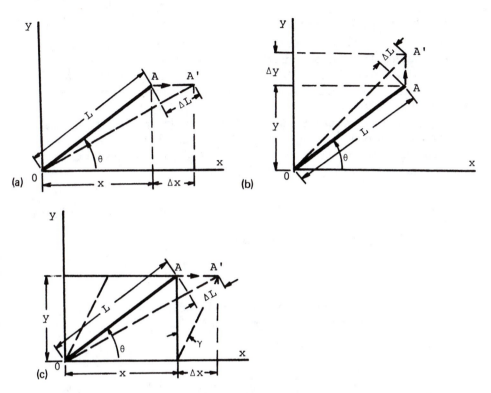

FIGURE E4-3 Derivation of strain for gage in θ direction. (a) ε_θ due to Δx. (b) ε_θ due to Δy. (c) ε_θ due to shear γ.

Equating $d\varepsilon_\theta/d\theta$ to zero to find the maximum and minimum values of ε_θ, we get

$$\frac{d\varepsilon_\theta}{d\theta} = -(\varepsilon_x - \varepsilon_y)\sin 2\theta + \gamma \cos 2\theta = 0$$

$$\tan 2\theta_{1,2} = \frac{\pm \gamma}{\pm(\varepsilon_x - \varepsilon_y)} \tag{E4-9}$$

Two angles $2\theta_{1,2}$ are determined by Eq. E4-9 and they are 180° apart. Hence $\theta_{1,2}$ are 90° apart, and $\varepsilon_{1,2} = \varepsilon_\theta(\text{max,min})$ and are orthogonal. The maximum and minimum normal strains $\varepsilon_{1,2}$ are called the *principal strains*. The orthogonal planes over which they act are called the *principal planes*. Substituting $2\theta_{1,2}$ into Eq. E4-8, the value of the principal strains are

$$\varepsilon_\theta(\text{max,min}) = \varepsilon_{1,2} = \tfrac{1}{2}[(\varepsilon_x + \varepsilon_y) \pm \sqrt{(\varepsilon_x - \varepsilon_y)^2 + \gamma^2}] \tag{E4-10}$$

C. Rectangular Rosettes

A strain gage rosette generally consists of three gages packaged as a unit (see Table 5-1). It can be bonded to a surface in an arbitrary orientation to find the direction and magnitude of the principal strains and the maximum shear strain. The stresses are deduced from the strain measurements.

A rectangular rosette is shown in Fig. E4-4. The gages a, b, and c are arranged with $\theta_a = 0°$, $\theta_b = 45°$, and $\theta_c = 90°$. Let the values of the measured strain be ε_a, ε_b, and ε_c. Let ε_a be aligned with the x axis. Substituting the values in Eq. E4-8 gives

$$
\begin{aligned}
\theta_a &= 0° & \varepsilon_a &= \varepsilon_x \\
\theta_c &= 90° & \varepsilon_c &= \varepsilon_y \\
\theta_b &= 45° & \varepsilon_b &= \tfrac{1}{2}[(\varepsilon_x + \varepsilon_y) + \gamma]
\end{aligned} \qquad \text{(E4-11)}
$$

or

$$\gamma = 2\varepsilon_b - (\varepsilon_a + \varepsilon_c)$$

The equations above are substituted into Eqs. E4-9 and E4-10 to find the direction and magnitude of the principal strains.

$$2\theta_{1,2} = \tan^{-1} \frac{2\varepsilon_b - \varepsilon_a - \varepsilon_c}{\varepsilon_a - \varepsilon_c} \qquad \text{(E4-12)}$$

$$\varepsilon_{1,2} = \tfrac{1}{2}[(\varepsilon_a + \varepsilon_c) \pm \sqrt{2(\varepsilon_a - \varepsilon_b)^2 + 2(\varepsilon_b - \varepsilon_c)^2}] \qquad \text{(E4-13)}$$

The principal stresses $\sigma_{1,2}$, in terms of the measured strains from a rec-

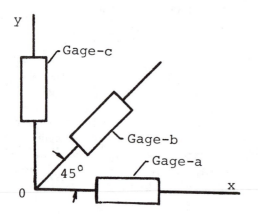

FIGURE E4-4   Rectangular rosette strain gage.

tangular rosette gage, are obtained by substituting Eq. E4-13 into Eq. E4-5.

$$\sigma_{1,2} = \frac{E}{2}\left[\frac{\varepsilon_a + \varepsilon_c}{1 - v} \pm \frac{1}{1 + v}\sqrt{2(\varepsilon_a - \varepsilon_b)^2 + 2(\varepsilon_b - \varepsilon_c)^2}\right] \qquad \text{(E4-14)}$$

By Eqs. E4-9 and E4-12, the direction of principal stresses coincide with those of the principal strains. The planes over which they act are called the principal planes, as before. It is intuitive that if only normal stresses are applied to principal planes, shear stresses do not exist on the principal planes.

The maximum shear stress is also of interest in machines. Let an element of thickness t be subjected to the principal stresses $\sigma_{1,2}$, as shown in Fig. E4-5a, and τ_θ be the shear stress on the plane at an angle θ, as shown in Fig. E4-5b. By resolution of forces, we have

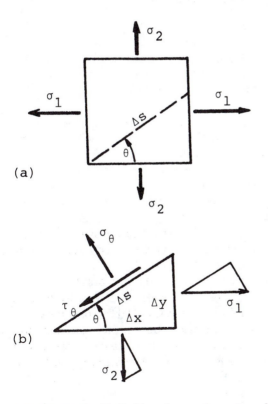

(a)

(b)

FIGURE E4-5 To find maximum shear stress by resolution. (a) Element subjected to principal stresses α_1 and α_2. (b) Resolution of forces. Normal stress = σ_θ, shear stress = τ_θ.

$$\tau_\theta t \, \Delta s = \sigma_1 t \, \Delta y \cos \theta - \sigma_2 t \, \Delta x \sin \theta$$

$$\tau_\theta = (\sigma_1 - \sigma_2) \sin \theta \cos \theta$$

$$= \tfrac{1}{2}(\sigma_1 - \sigma_2) \sin 2\theta \tag{E4-15}$$

Maximum τ_θ occurs at $\sin 2\theta = 1$, or at $\theta = 45°$ relative to the principal axes. Hence

$$\tau_{\max} = \tfrac{1}{2}(\sigma_1 - \sigma_2) \tag{E4-16}$$

The maximum shear stress $\tau_{\max}$ can be expressed in terms of the strains from a rectangular rosette by substituting Eq. E4-14 into Eq. E4-16.

$$\tau_{\max} = \frac{E}{2(1 + v)} \sqrt{2(\varepsilon_a - \varepsilon_b)^2 + 2(\varepsilon_b - \varepsilon_c)^2} \tag{E4-17}$$

E4-4. Strain Gages and Gage Circuits

A. Strain Gages

General-purpose foil gages for the experiment are shown schematically in Fig. E4-6. The resistance strain gage is almost an ideal transducer in terms of stability of calibration, accuracy, linearity in response, size, small mass, remote sensing, versatility, simplicity in installation, and low cost.

Temperature-compensated strain gages are used for the experiment. If the material and the bonded gage have the same thermal coefficient of expansion, there is no apparent strain due changes in temperature, and a "dummy" gage for temperature compensation is not necessary. Temperature-compensated gages are available for different materials. The compensation for steel is 06 ppm/°F. If the test piece is of another material, however, a gage with a different compensating number must be used. It should be cautioned that neither the temperature-compensated gage nor the dummy gage are designed to compensate for the effect of temperature on the leads and other components in the circuit.

Specifications for the gages, such as fatigue life and strain limits, are more than adequate for the experiment. For example, the strain limit of constantan gages is 3%, corresponding to a stress of 900,000 psi for steel. A strain gage measures strain in one direction, but surface strain is two-dimensional, even for a uniaxial stress. The effect of the two-dimensional strain on a gage is included in the gage factor. Other considerations for the selection of gages are discussed in Chapter 6.

B. Gage Circuits

Basic circuits for strain measurements, the effect of lead wires, and the switch unit for connecting several circuits to a common strain indicator are described in this section.

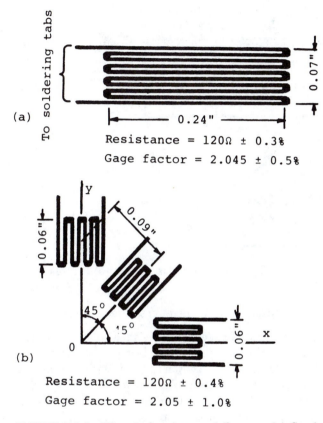

(a)

Resistance = 120Ω ± 0.3%
Gage factor = 2.045 ± 0.5%

(b)

Resistance = 120Ω ± 0.4%
Gage factor = 2.05 ± 1.0%

FIGURE E4-6 Electrical resistance foil gages. (a) Strain gage. (b) Rectangular rosette.

The *ballast circuit*, shown in Fig. E4-7a, is for dynamic measurements. It can be shown that if R is the gage resistance, $R = R_b$, and $\varepsilon =$ applied strain, the output voltage v_o is

$$\frac{v_o}{V_i} = \frac{1}{4} G_f \varepsilon \tag{E4-18}$$

where G_f is the gage factor. The bridge circuit shown in Fig. 4-7b is suitable for both static and dynamic measurements. A bridge for strain measurements is generally used in the unbalanced mode. If all the gages are of equal resistance R and the strain applied to the gages are from ε_1 to ε_4, the output voltage v_o is

$$\frac{v_o}{V_i} = \frac{1}{4} G_f (\varepsilon_1 - \varepsilon_2 + \varepsilon_3 - \varepsilon_4) \tag{E4-19}$$

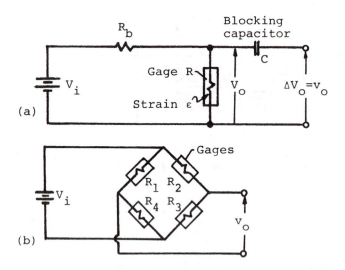

FIGURE E4-7   Basic strain gage circuits. (a) Ballast circuit. (b) Bridge circuit.

The examples below are used to show the effect of lead wires.

1. The temperature-compensated gage is not designed to eliminate the effect of temperature on the leads. A quarter bridge is shown in Fig. E4-8a. Any ΔR in the leads, or in a switch in series with the gage, is an apparent strain in the measurement. Assume a 120-Ω gage with a gage factor $G_f = 2$, and a $\Delta R = 0.012\ \Omega$ in the leads. This will introduce an apparent strain of 50 μstrain in the measurement.

2. The lead wires may reduce the sensitivity of a gage. The resistance of the leads adds to the resistance of the gage. This is equivalent to a gage of higher resistance R, and a smaller gage factor G_f, because $G_f = (\Delta R/R)/\varepsilon$. For example, the resistance of a pair of 75-ft leads of No. 20 AWG copper wire is about 1.5 Ω ($\simeq$ 10 Ω/1000 ft). This resistance in series with a 120-Ω gage will reduce the gage factor by 1.25%. The change in G_f can be corrected by changing the G_f setting at the strain indicator.

3. The lead wires may contribute to a calibration problem. The calibrating resistor R_c is generally at the strain indicator rather than at the bridge, as shown in Fig. E4-8b. Similar to the example above, the resistance of the leads adds to the gage resistance to cause an error.

The schematic of a *switch-and-balancing unit* for several bridges using a common strain indicator is shown in Fig. E4-9. It is recommended that

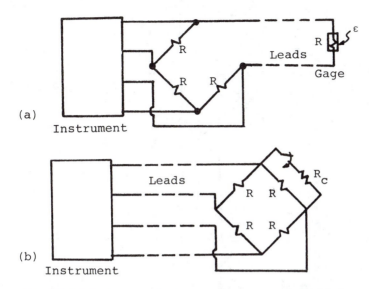

(a) Instrument

(b) Instrument

FIGURE E4-8 Effect of lead wires in circuit analysis. (a) Quarter bridge. (b) Calibration resistor.

all switching contacts be placed completely outside the bridge ring, instead of the switching individual gages (i.e., complete bridges are switched into and out of a measuring system). Due to the small ΔR in strain measurements and the uncertainty of the contact resistance in switching, the effect of a switch in series with a strain gage is unpredictable. A potentiometer is used in the switch unit for balancing the bridge, as shown in the figure. A bridge is balanced when $R_1 R_3 = R_2 R_4$. This condition is not precisely correct for the balancing circuit shown, and the equivalent R_1 and R_4 should be used in the equation. The circuit tends to desensitize R_1 and R_4, but the effect is negligible for a well-designed circuit.

C. Dynamic Strain Measurements

The ballast circuit or the strain gage bridge can be used to demonstrate dynamic strain measurements. The setup consists of a cantilever beam of fairly low natural frequency shown in Fig. E4-10a. The bonded strain gages can be wired for either circuit.

The sinusoidal excitation input is from an electromechanical shaker (see Fig. E3-4). For the ballast circuit, an oscilloscope with ac couple is used for the readout. It is necessary to take steps to suppress excessive noise pickups. For the bridge circuit, the output of the strain indicator is connected to an oscilloscope for the readout.

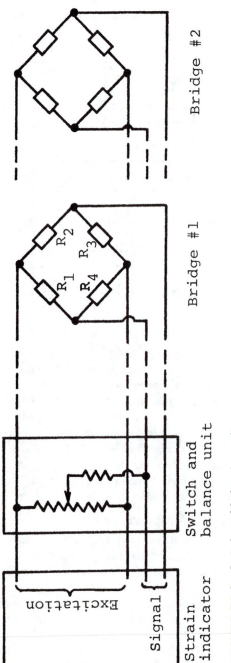

FIGURE E4-9 Switch and balancing unit.

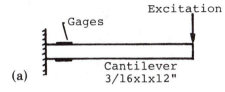

(a)

Test setup

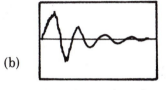

(b)

A-c signal

FIGURE E4-10 Dynamic strain measurement. (a) Test setup. (b) Ac signal.

A pulse input can also be applied to the cantilever by tapping near its free end with a small hammer. If the oscilloscope is set at the single-sweep mode, a typial trace for the transient response is as shown in Fig. E4-10b. The response consists of harmonics superposed on the fundamental frequency, but the harmonics will soon die out. Either an internal or an external trigger can be used for the oscilloscope. The level of the internal trigger is set by trial and error. The contacting between the hammer and the cantilever can be used as a signal for the external trigger. The transient response can be observed more conveniently by means of a digital oscilloscope.

E4-5. Flexures and Cross-Sensitivity

The purpose of this section is to discuss (1) applications of the strain gage in transducers, and (2) coupled modes of elastic deformation.

The strain gage is often employed as the secondary sensor in transducers for detecting the deformation in a flexure (see Chapters 5, 7, and 8 and Expts. 2 and 3). The soft cantilever in the displacement gage in Fig. E4-11a is a flexure and the primary sensor, and the strain gages are the secondary sensor. The accelerometer in Fig. E4-11b uses a stiff cantilever as a force transducer to measure the inertia force of a seismic mass. The torque table in Fig. E4-11c is a dynamometer for measuring the reactive

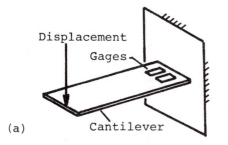

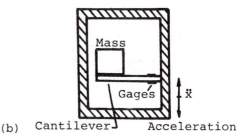

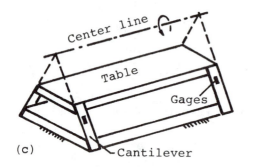

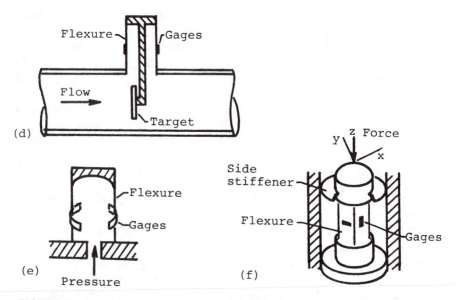

FIGURE E4-11   Example of strain gage transducers. (a) Displacement gage. (b) Accelerometer. (c) Torque table. (d) Flowmeter. (e) Pressure transducer. (f) Force transducer.

torque about the centerline of a rotating machine. The target flowmeter in Fig. E4-11d uses a cantilever as the primary sensor. The common denominator in these transducers is that a flexure in the form of a beam is the primary sensor.

The flexure is not restricted to a beam or an object of particular shape, since it is simply an elastic member. The pressure transducer in Fig. E4-11e measures pressure by means of the circumferential strain of a cylinder. The force transducer in Fig. E4-11f measures force by means of the axial strain of a short column. Only the force in the z direction is of interest, and side stiffeners are used to constrain the x and y motions.

The *cross-sensitivity* in a transducer is caused by the coupled modes of deformation in a flexure, or by extraneous conditions, such as asymmetry loading, resonances in the system, and thermal effect on the components. An input may have forces in the (x, y, z) directions and/or torques about the three axes. The flexure in a transducer is designed to respond mainly in a desired direction, or to resolve the input into the desired directions. For example, if a flexure is designed to measure torque, the output due to bending in the flexure is noise in a torque measurement. Normally, an input F_x in the x direction causes an output V_x. Cross-sensitivity is evident when F_x is zero and an output V_x is caused by an extraneous input F_y in the y direction. Cross-sensitivity occurs under both static and dynamic conditions. It is a source of noise in measurements, and often a part of the specification of transducers.

Many methods are used to minimize cross-sensitivity in transducers. The technique described here is by the orientation of strain gages such that only the desired input is detected. Note that this is a method of noise reduction by insensitivity (see Sec. 3-7B). Consider the alternatives for using the gages R_1 to R_4 in the bridge shown in Fig. E4-12a for a force measurement. If R_1 is used for sensing the horizontal force H, as shown in Fig. E4-12b, the transducer is sensitive to H, the bending due to asymmetry in loading, and the thermal effect on the leads. The effect of bending is minimized by using R_1 and R_3, as shown in Fig. E4-12c. Effects of both bending and temperature are minimized by using all four gages, as shown in Fig. E4-12d. Similarly, the setup in Fig. E4-12e is sensitive only to the vertical force V.

An input can also be resolved into the desired directions by using more than one bridge circuits. For example, let a force F consisting of a horizontal component H and a vertical component V be applied to a cantilever. The H and V components can be resolved by using gages from two bridges, as shown in Figs. E4-12d and e, in the same transducer. Similarly, the force F applied to a circular bar, shown in Fig. E4-12f, can be resolved into its (x, y) components by using two bridge circuits. It is obvious that a force

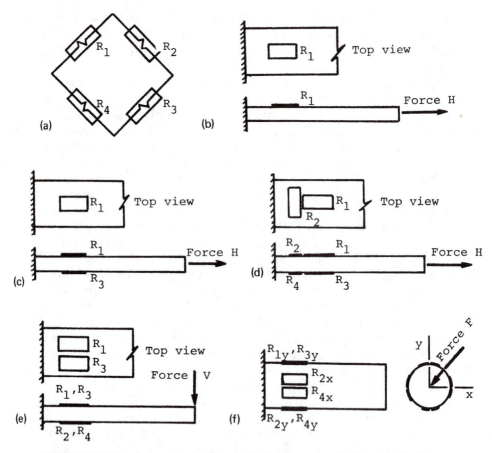

FIGURE E4-12   Gages arranged to minimize cross-sensitivity. (a) Strain gage bridge. (b) Circuit sensitive to H, asymmetry, and resistance change in leads. (c) Circuit sensitive to H and resistance change in leads but not asymmetry. (d) Circuit sensitive to H only. (e) Circuit sensitive to V only. (f) Resolution of force F in x and y directions.

in an arbitrary direction can be resolved into its (x, y, z) components by means of three bridge circuits.

E4-6. Equipment

Equipment for the experiment are strain indicators, switch units, and simple setups. The basic strain indicator is a single-channel dc instrument for static strain measurements. It consists of a dc amplifier, a balancing circuit, a shunt calibrator, an analog or a digital display, and a connector for dynamic readout. The switch unit allows several bridge circuits to use a common strain indicator. The setups are described in the experiments to follow.

The operation of a strain indicator has essentially two main steps: (1) balancing the unit, and (2) setting the desired sensitivity for the measurement. For some indicators, it may be necessary to balance the amplifier without the bridge circuit, and then balance the setup with the bridge circuit. The sensitivity is set by adjusting the gain of the amplifier to obtain the correct microstrain reading for an input. For example, the gage factor setting at the indicator is an amplifier gain control.

Simplified procedures for operating the strain indicator and the switch unit are usually given on the cover of the instrument. The first step is to make the connections for the selected circuit, which is a quarter, half, or full bridge. The step-by-step simplified instructions are easy to follow, but the reader should refer to the instruction manual for the particular instrument for details and specifications.

It may be remarked in passing that it is preferable to use mechanical instead of hydraulic devices for loading the specimen. An inexpensive commercial hydraulic loading device does not maintain the load, and lacks both sensitivity and accuracy. Some old equipment, such as a Brinell hardness tester, can be modified for this purpose.

E4-7. Laboratory Experiments

The laboratory consist of nine simple experiments for static strain measurements. Experiment 4-1 shows an application of strain gages in a commercial strain gage pressure transducer. Its static characteristics are obtained from a calibration with a dead-weight tester. The remaining experiments can be divided into three overlapping groups.

The first group includes four experiments in which *symmetrical loading* is intended. (1) Experiment 4-2 calibrates a cantilever as a force transducer by means of a direct and an indirect calibration. (2) The simply supported beam in Expt. 4-3 verifies the beam equation: stress = Mc/I. It is a textbook problem and students like the test results. The two-dimensional surface strain due to a single bending load is shown by means of a rosette gage. Data from two gages along the principal axes are used to deduce the Poisson ratio. (3) Experiment 4-4 is a tensile test, and (4) Expt. 4-5 a compression test.

The second group demonstrates *asymmetrical loading* in applications. It includes Expts. 4-4, 4-5, and 4-6. Asymmetry may be detected in the tensile test in Expt. 4-4. It is difficult to obtain symmetrical loading in a compression test in Expt. 4-5. Experiment 4-6 uses a unsymmetrical flat specimen in a tensile test. Bending due to the asymmetry is evident in the tensile test. The gage on one side of the specimen first shows a compression and reverses to a tension with increasing tensile force.

The third group illustrates *biaxial stresses*. (1) A combination of torsion and bending is applied to a torsional shaft in Expt. 4-7. It shows that the effect of bending can be minimized by the proper orientation of strain gages. (2) Experiment 4-8 uses a can of soft drink to show biaxial stresses and the ratio of axial-circumferential stresses in a long cylinder. Students install the gages on the can, and then estimate the pressure in the can by releasing its pressure. (3) This is followed by Expt. 4-9, in which pressure is applied to a gas tank by means of a dead-weight tester. It demonstrates biaxial stress measurements, and is more realistic and complex than the soft-drink-can problem.

E4-8. Pressure Transducer Calibration: Experiment 4-1

The objective of this experiment is to show an application of strain gages and to obtain the static characteristics of a commercial strain gage pressure transducer (see Sec. 2-2A). The setup, shown in Fig. E4-13, consists of a dead-weight tester, a pressure gage for monitoring, the pressure transducer, its power supply, and a digital voltmeter for the readout. The reference for the calibration is the dead-weight tester, which is basically a hydraulic balance. The piston is actuated to compress the oil in the cylinder. When the plunger with the dead-weight floats freely, the pressure exerted by the plunger on the oil is equal to that transmitted to the pressure transducer.

A. Procedure

1. Set up the equipment. Remove the plunger from the dead-weight tester so that zero pressure is applied to the transducer. Note the tare reading from the voltmeter.
2. Apply pressure to the transducer to cover its working range. Ensure that the plunger is floating freely and has a slight rotational motion before taking readings. Take data from the dead-weight tester, the

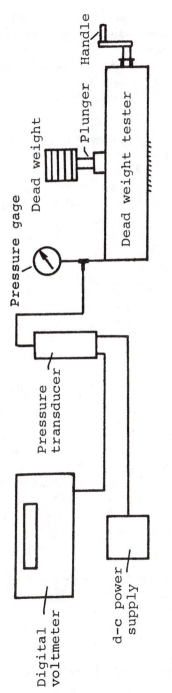

FIGURE E4-13 Calibration of pressure transducer.

pressure gage, and the voltmeter for 20-psi increments, as indicated by the dead-weight tester (about 10 readings).
3. Repeat step 2 for 20-psi decrements.
4. As a preliminary check, plot the running data for voltage versus pressure before leaving the workstation.
5. Include all original data in the appendix of the report.

B. Discussion

1. Use the pressure from the dead-weight tester to obtain a voltage versus pressure plot. This is the static characteristic of the pressure transducer.
2. Find the sensitivity of the transducer.
3. With the aid of a sketch, give three examples of applications of strain gages other than those described in this write-up.

E4-9. Direct and Indirect Calibrations: Experiment 4-2

The objective of this experiment is to show the direct and indirect calibration of a cantilever, or a force transducer. Strain gage load cells, or force transducers, are often calibrated directly (see Sec. 2-4). Both static and dynamic calibrations may be necessary, depending on the application. The principle of a direct calibration is that a known input gives a known output. The corollary is that a zero input yields a zero output. Any output with a zero input is due to the inherent noise in the measuring system. Strain gages for strain measurements, however, are generally calibrated indirectly by means of a shunt resistor (see Sec. 3-3B and Fig. E4-8b), because a direction calibration is generally not possible.

The setup consists of a cantilever with a strain gage bridge, the dead weight, and a hanger, as shown in Fig. E4-14. The cantilever is a force transducer, and the dead weight gives the known input for the direct calibration. The strain at the gages is calculated by the beam equation. The Ellis BAM-1 bridge amplifier (not shown) is for the bridge excitation, indirect calibration, and the readout. It is more convenient to use the

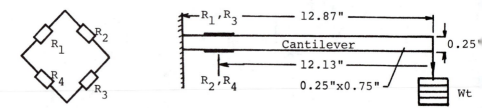

FIGURE E4-14 Direct and indirect calibration of force transducer.

shunt calibrator in the BAM-1 bridge than to devise a separate circuit for the indirect calibration.

A. Procedure

1. Check the battery in the BAM-1 bridge amplifier. Set up the equipment and use the four-active-arm setting at the amplifier. Balance the amplifier without the strain gage bridge and then balance the setup with the bridge circuit.

2. Steps for the direct calibration are: Calculate the strain at the gages due to the applied dead weight. (Do not exceed 800 μstrain at the gages). Apply the weight. Set the gain of the amplifier to obtain an output corresponding to the calculated strain. The system is now calibrated. Reduce the weight and repeat to check the calibration.

3. Steps for the indirect calibration are: Balance the system as above. Select a calibrator setting (i.e., the shunt calibrating resistors). Find the calibrating factor from an equation in the instruction manual. Actuate the calibrator. This places the shunt calibrating resistor in parallel with a gage in the strain gage bridge. Set the gain of the amplifier to give the calculated strain. Verify the indirect calibration by applying a known weight to the cantilever.

E4-10. Simply Supported Beam: Experiment 4-3

The objectives of this experiment are (1) to verify the beam equation, and (2) to show the two-dimensional strain and Poisson ratio. The simply supported beam with a concentrated load is shown in Fig. E4-15a. The force F is applied at the center of the beam by means of a mechanical fixture. The bending moment at F is $(F/2)(12)$ in.-lb$_f$. The strain gages located on the bottom side of the beam are as shown in Fig. E4-15b. A Measurements Group P3500 indicator and a SB10 switch unit are used for the experiment.

A. Procedure

1. Connect the gages to the switch unit and then to the strain indicator. Follow the instructions to initialize the instrument to zero for $F = 0$.

2. Take data from gages 1, 2 and a, b, c of the rosette gage for 100-lb$_f$ increments in F until $F = 1000$ lb$_f$. *Caution:* Do not exceed the 1000 lb$_f$ to avoid damage to the beam.

3. Repeat step 2 for 100-lb$_f$ decrements in F.

4. Check the preliminary data before leaving the workstation. (a) Use the running data from step 2 for gage 1 to plot strain versus F. (b)

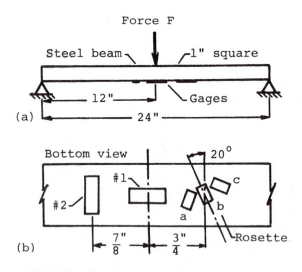

FIGURE E4-15 Setup to verify the beam equation. (a) Simply supported beam. (b) Location of gages.

Calculate the strain at $F = 500\ \text{lb}_f$ from the beam equation, and compare with the measured strain by gage 1.

5. Include all original data in the appendix of the report.

B. Discussion

1. Use the average strain from steps 2 and 3 for gage 1 and the same for gage 2 to plot Poisson ratio versus F.
2. In the same graph, use the average strain from steps 2 and 3 for gage 1 to calculate the stresses, and plot stress versus F. Plot stress versus F from the beam equation in the same graph for comparison.
3. Use average values from steps 2 and 3 to plot the strains for the individual gages in the rosette to ensure that the data are not erratic. For $F = 500\ \text{lb}_f$, select values from the plot to calculate the direction and magnitude of the principal strains from Eq. E4-13. Compare with the values from gages 1 and 2.

E4-11. Tensile Test: Experiment 4-4

The objective of this experiment is to compare the strain measured mechanically with that by strain gages. The steel specimen in Fig. E4-16 is used for a tensile test in a commercial hydraulic test stand. An old Baldwin SR-4 strain indicator is employed. Strain can also be deduced from $\Delta L/L$, or from simple stress–strain relations. Unfortunately, the

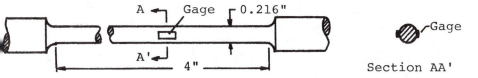

FIGURE E4-16 Specimen for comparison of methods of strain measurement.

hydraulic stand does not work satisfactorily, and it does not maintain the load. Hence the results from the mechanical measurements are only qualitative. Steps are taken to replace the hydraulic with a mechanical loading device.

The axial deformation is measured by means of a micrometer with a 0.0002-in. resolution, but there is a problem of asymmetry in the setup. The equations for the stress–strain relations are: force = (pressure)(area of hydraulic piston), stress = (force)/(area of specimen), and strain = (1/E)(stress). The area of the hydraulic piston is 0.223 in². The pressure reading is far from precise, and the friction in the system is unknown.

A. Procedure

1. Set up the equipment. *Caution:* Back-up the micrometer before a load change to avoid damage to the micrometer. The SR-4 strain indicator has provision for two inputs. The operating instruction is on the cover. Set the gage factor on the SR-4 indicator. Initialize the indicator by adjusting the dial to read zero, and obtain an initial reading ε_0 with zero load, where ε_0 = (switch reading) + (dial reading). For example, ε_0 = (10 × 1000) + 820 = 10,820.
2. Apply tension to the specimen by means of the hydraulic pump. Balance the indicator for each load increment before taking strain readings. The strain is [(indicated strain) − ε_0]. Take data from (a) both strain gages with the RS-4 strain indicator for 150-μstrain increments until 750 μstrain, (b) the pressure gage, and (c) the micrometer.
3. Repeat step 2 for 150-μstrain decrements.
4. As a preliminary check, plot the data for gage 1 from step 2 versus micrometer readings before leaving the workstation.
5. Include all original data in the appendix of the report.

B. Discussion

1. Using the average values from steps 2 and 3, calculate the strain (a) $\Delta L/L$ from the micrometer readings for L = 4 in., and (b) from the stress–strain equations.

2. Plot the strain calculated from the micrometer versus the average strain from the gages. Use the method of least squares for data fitting (see Sec. 2-2).
3. In the same graph, plot the strain calculated from the pressure gage versus the average strain from the strain gages.
4. Explain whether the method of least squares should be used for the data. Do the readings from the strain gages indicate asymmetry? Discuss the sources of error in the experiment.
5. When two gages are in series in one arm of a bridge, is the strain indicated twice that of a single gage?
6. Estimate the root-sum-square error in the three methods for the load at 500 μstrain.

E4-12. Compression Test: Experiment 4-5

The objective of this experiment is to demonstrate asymmetry in an application. The specimen, shown in Fig. E4-17, is a section of steel tubing with three equally spaced strain gages. Strain from the individual gages are measured to investigate the extend of asymmetry due to loading, which is almost unavoidable in a simple compressive application.

A commercial hydraulic test stand is used for the test. The area of the hydraulic piston is 0.375 in^2 for compressive loading. Similar to Expt. E4-4, strain can be measured mechanically or by means of strain gages. Since the height of the specimen is 1.0 in., the resolution of the micrometer is

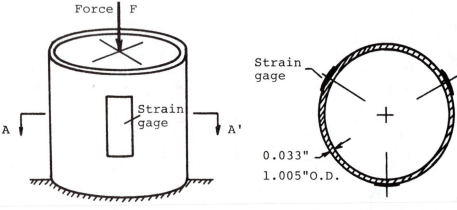

Test specimen   Cross-section A-A'

FIGURE E4-17 Compression test to show asymmetry in loading.

0.0002 in., and the suggested strain limit is 800 µin. (stress = 24,000 psi), only four data points can be expected for the test, and the results from the mechanical measurements are only qualitative. The hydraulic test stand is not satisfactory, as discussed above. Steps are taken to replace it with a mechanical setup.

A. Procedure

1. Set up the circuit. *Caution:* Back-up the micrometer before each load change to avoid damage. Initialize the strain indicator (Measurements Group P3500) and the switch unit (SB-10) to zero for a zero load. Apply load by means of the hydraulic pump.
2. Using the strain gage with the highest output as reference, take data from the micrometer, the pressure gage, and the individual strain gages for 200-µstrain increments until 800 µstrain.
3. Repeat step 2 for 200-µstrain decrements.
4. As a preliminary check, plot the running data from the reference strain gage versus applied pressure before leaving the workstation.
5. Include all original data in the appendix of the report.

B. Discussion

1. Using the average values from steps 2 and 3, find the average strain for the three strain gages at each loading. Plot the strain from the individual gages versus the average strain to show the extent of asymmetry in loading.
2. Plot the strain $\Delta L/L$ from the micrometer readings versus the average strain above. Comment on the plot.
3. In the same graph, plot the strain from the pressure gage readings versus the average strain. Comment on the plot.
4. Estimate the root-sum error in the three methods for the load at 600 µstrain.

E4-13. Cross-Sensitivity: Experiment 4-6

The preceding experiment illustrates asymmetry in compressive loading. The objective of this experiment is to demonstrate a severe case of asymmetry in a flexure. A hydraulic test stand is used for the tensile test of the unsymmetrical specimen shown in Fig. E4-18. The holes for applying the tensile force F are slanted about 10° to accentuate the asymmetry. As F is applied to the specimen, gage 1 initially shows a compression due to bending. The output of gage 1 then changes sign to show a tension with a further increase in F.

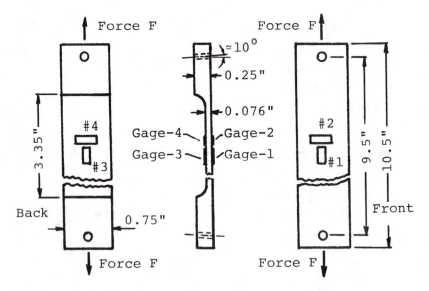

FIGURE E4-18 Tensile test of unsymmetrical specimen.

A. Procedure

1. Connect the strain gages to the switch unit (Measurements Group SB-10) and then to the strain indicator (P3500). Initialize the strain indicator and the switch unit to zero for zero load.
2. Touch the specimen at gage 1 to feel the order of magnitude in a strain measurement.
3. .Apply the load F by means of the hydraulic pump.
4. Take data from the pressure gage and strain gages 1 to 4 for 150-μstrain increments, until gage 3 shows 900 μstrain. *Caution:* Do not exceed 900 μstrain for gage 3.
5. Repeat step 4 for 150-μstrain decrements.
6. As a preliminary check, plot the data from gage 3 versus the applied pressure before leaving the workstation.
7. Include all original data in the appendix of the report.

B. Discussion

1. Plot the strain calculated from the pressure gage readings versus the average strain by the strain gages. Note that the bridge constant is 2.6 for the setup (see Eq. 3-61).
2. Plot the individual strain from gages 1 through 4 versus the average strain from above to show the effect of asymmetry in the flexure.

E4-14. Torsion and Bending: Experiment 4-7

The purpose of the experiment is to show (1) biaxial stresses due to coupled modes of deformation, and (2) the resolution of the modes by means of two bridge circuits (see Fig. E4-12).

The shaft, shown in Fig. E4-19a, is subjected to both bending and torsion when the weights are applied at position A. The specimen, shown in Fig. E4-19b, is a steel tubing of 1.05 in. OD and 1/32 wall thickness. The torsional shaft from thin-walled tubing has a fairly low torsional stiffness, and reasonable weights can be used for the load. The tubing is plugged at both ends to facilitate the clamping and loading.

The experiment shows that the bending and torsional modes can be resolved by positioning the strain gages as shown in Fig. E4-19b. The *bending mode* is sensed by gages a and b in a half bridge. This mode can be identified by placing the weights at B. It is evident that the effect of torsion is negligible for these gages. The *torsional mode* is sensed by gages 1 to 4 in a full bridge. For a shaft in torsion, maximum compressive and tensile strains lie along 45° helices on the shaft surface. Hence the orientation of the gages are: (1) gages 1 and 3 are along one helix, and gages 2 and 4 along another helix; (2) the helices are at 90° with one another; and (3) gages 1 and 2 are on a centerline of the shaft, and gages 2 and 4 on the opposite centerline. The orientation of gages 1 to 4 is chosen to minimize the effect of bending as well as the bending gradient along the shaft. The rosette gage on the shaft can also be used to verify the direction and magnitude of the principal strains.

A. Procedure

1. Set up the equipment. The rosette gage is not used in order to save time. Initialize the strain indicator (Measurements Group P3500) and the switch unit (SB-10) to zero for zero load.
2. Apply the weights at position A in five steps. Total weight $\simeq$ 13 lb. The output from the full bridge should not exceed 3000 µstrain. Record the output from the bridges. *Note:* The indicator must be switched to half bridge to read gages a and b for bending, and switched back to full bridge to read gages 1 to 4 for torsion.
3. Decrement the load.
4. Repeat steps 2 and 3 with the load applied at B.
5. As a preliminary check, (a) use the data from step 2 to plot the output from the full bridge versus increasing load, and (b) compare the calculated torque with the applied torque at the maximum load.
6. Include all data in the appendix of the report.

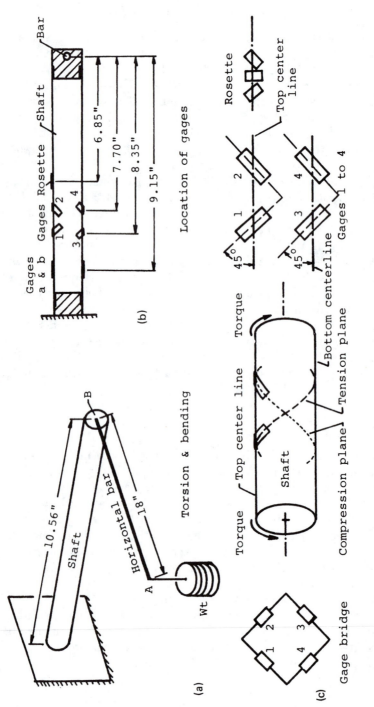

FIGURE E4-19 Gages arranged to uncouple bending and torsional modes. (a) Torsion and bending. (b) Location of gages. (c) Detail location of gages.

B. Discussion

1. For the bending mode with weights at A, (a) calculate the strain due to the weights, and (b) plot the calculated strain versus the average strain in gages a and b in steps 2 and 3 for each loading. Note that the bridge constant is 2 for bending.
2. Repeat (1) for the weights at B, but use the data in step 4 of the procedure. Use the same graph for the plots to compare the bending with weights at A and B.
3. For the torsional mode with the weights at A, (a) use the average values from steps 2 and 3 of the procedure for each loading to calculate the torque from the bridge output, and (b) plot the calculated torque versus the applied torque.
4. From step 4 of the procedure with the weights at B, calculate the bending measured by the full bridge for each loading, and the ratio of bending to applied torque in step 3. Plot the results in the same graph to show the cross-sensitivity.
5. A configuration of the full bridge for a torque measurement is shown in Fig. E4-19c. There are six possible configurations. Sketch all six and identify the gages in the bridge. Select one of the six and investigate whether it is sensitive to bending as well as to the strain gradient along the shaft due to bending.
6. Torsion can also be sensed by means of a half bridge, with one strain gage along the compression and the other along the tension plane. Select a possible configuration and investigate whether it is sensitive to bending and to the strain gradient due to bending.

E4-15. Long Thin-Walled Cylinder: Experiment 4-8

The objectives of this experiment are (1) to give students an experience in applying strain gages to a specimen, and (2) to investigate the ratio of axial/circumferential stresses in a long thin-walled cylinder. The setup consists of a can of soft drink, a strain indicator, and a switch unit. Students apply the gages on the can. Either two gages along the principal axes or a rosette gage can be used, as shown in Fig. E4-20a [see *Experimental Stress Analysis Notebook* (Rayleigh, N.C.: Measurements Group, Inc., Issue 1, Oct. 1985), pp. 3–6].

A. Procedure

1. Let the can of soft drink be at room temperature. Prepare an adequate surface area for the gages, but do not thin the can unnecessarily. Get instructions for applying strain gages.
2. Set up the equipment but do not release the pressure in the can. In-

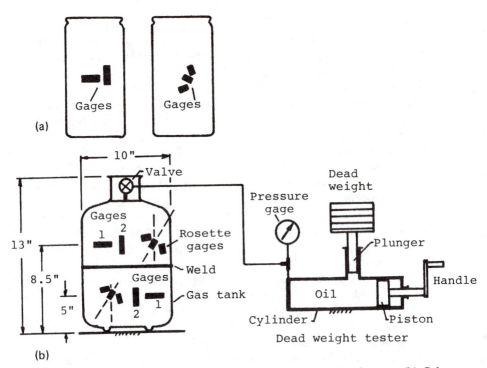

FIGURE E4-20 Biaxial strain measurements. (a) Soft-drink cans. (b) Schematic of test setup.

itialize the strain indicator (Measurements Group P3500) and the switch unit (SB-10) to zero.

3. Release the pressure in the can. Take strain measurements. Avoid spraying soft drink on the equipment.

4. Measure the overall dimensions of the can. Rip open the can and measure its wall thickness with a micrometer.

B. Discussion

1. Find the ratio of axial/circumferential strains in the can before the pressure is released.

2. If the strain gages are along the principal axes, use Eq. E4-5 to calculate the principal stresses. Compare the ratio of principal stresses with that for a long thin-walled cylinder. Comment on the discrepancy, if any. Note the difference between the ratio of principal stresses and that for principal strains. Estimate the pressure in the can before the pressure is released.

3. If a rosette gage is used, calculate the principal stresses by Eq. E-14. Find the ratio of the principal stresses and compare with that of a long thin-walled cylinder. Estimate the pressure in the can before the pressure is released.
4. If a rosette gage is used, calculate the principal strains by Eq. E4-13. Compare with the values those from two gages along the principal axes.

E4-16. Biaxial Strain Measurements: Experiment 4-9

The objective of this experiment is to present a realistic study in biaxial stresses. The setup, shown in Fig. E4-20b, consists of a gas tank filled with water, a dead-weight tester for applying hydrostatic pressure to the tank, a strain indicator, and a switch unit. Although the wall thickness of the tank can be estimated by its empty weight, the geometry of the tank is more complex than that of a soft-drink can in the last experiment, and a long thin-walled cylinder cannot be assumed. The dead-weight tester was described in Sec. E4-8. The pressure gage is for monitoring only. The strain gages are attached to the tank as shown in the figure.

A   Procedure

1. Set up the equipment. Use the rosette gage and the single strain gages along the principal axes in the same general area of the tank. Initialize the strain indicator (Measurements Group P3500) and the switch unit (SB-10) to zero for zero load.
2. Using the dead-weight tester as reference, take data from the pressure gage, strain gages 1 and 2, and the gages in the rosette for 20-psi increments, until 200 psi (approximately 10 readings). Ensure that the plunger of the dead-weight tester is floating freely and rotating slowly before each reading.
3. Repeat step 2 for 20-psi decrements.
4. As a preliminary check, plot the values for gage 1 from step 2 versus the pressure from the dead-weight tester before leaving the workstation.
5. Include all original data in the appendix of the report.

B.   Discussion

1. Plot the average values in steps 2 and 3 for gage 1 versus pressure from the dead-weight tester. Repeat for gage 2. These are the principal strains.
2. Select values at 120 psi from the plot above to calculate the principal stresses from Eq. E4-5. Find the ratio of the principal stresses.
3. From the average values in steps 2 and 3, plot the strain from the

gages in the rosette versus pressure from the dead-weight tester to ensure that the data are not erratic. Select values at 120 psi from the plot to calaculate the principal strains from Eq. E-13 and the principal stresses from Eq. E-14. Find the ratio of the principal stresses.
4. Compare the results from steps 2 and 3.
5. For a long thin-walled cylinder, the ratio of the axial and hoop stress is 1:2. Compare this with the calculations above and discuss the deviations, if any.

10-6. EXPERIMENT 5. TIME RESPONSE OF INSTRUMENTS

E5-1. Introduction

The objective of this experiment is (a) to study the dynamic characteristics of instruments in the time domain by means of a model study, and (b) to evaluate the system parameters from time response data.

The model study, as a first step for dynamic analysis, has many advantages.

1. A real system, such as a hydraulic machine, must be fully instrumented with transducers to be useful for this study. This would introduce ramifications to mask the objective. The time and expense for multiple setup of real systems for a beginning laboratory would be prohibitive. This should be postponed for a follow-up laboratory.
2. A model is stripped of non-essentials, and easy to comprehend. The student can determine the basic system parameters readily. Moreover, the model is representative of many physical systems.
3. The model study with simple electrical components allows the student additional opportunity to use basic laboratory instruments and to recognize the limitation of a model. This cannot be accomplished with a computer simulation.

The required equipment are general laboratory instruments from previous experiments, such as R, L, and C decade boxes and the oscilloscope. Dynamic characteristics of instruments are also demonstrated with simple hardware, such as a thermocouple.

E5-2. The Model

The mathematical model for the study, shown in Fig. E5-1, is a linear single-input, single-output system (see Sec. 4-2). The system equation is

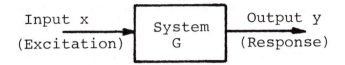

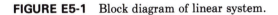

FIGURE E5-1 Block diagram of linear system.

response function = (system function)(input function) (E5-1)

or

$$y = Gx \qquad (E5\text{-}2)$$

where G is the system function and x and y are as defined in the equations. G is called the transfer function in the frequency response method.

A study in the time domain is by means of differential equations. The equations above can be expressed as

$$\frac{1}{G} y = x \qquad (E5\text{-}3)$$

$$a_n \frac{d^n y}{dt^n} + \cdots + a_2 \frac{d^2 y}{dt^2} + a_1 \frac{dy}{dt} + a_0 y = b_0 x(t) \qquad (E5\text{-}4)$$

Three basic models are deduced from the general nth-order differential equation. The models are justified if they describe the behavior of real systems for the intended applications.

For $n = 0$:	$a_0 y = b_0 x(t)$	(E5-5)
For $n = 1$:	$a_1 \dot{y} + a_0 y = b_0 x(t)$	(E5-6)
For $n = 2$:	$a_2 \ddot{y} + a_1 \dot{y} + a_0 y = b_0 x(t)$	(E5-7)

where $\dot{y}$ denotes dy/dt. The equations represent a zero-order instrument for $n = 0$, a first-order instrument for $n = 1$, and a second-order instrument for $n = 2$. System parameters, for describing the dynamic behavior of instruments, can be obtained from these equations. A step function is used in this experiment for the input $x(t)$ to deduce the system parameters.

The advantage of the time-domain study is that time is intuitive, since events occur in real time. The limitation is that more complex systems must be represented by higher-order differential equations (see Sec. 4-5), from which the effect of basic system parameters cannot be examined readily. Complex systems, however, are generally examined in the frequency domain. It can be shown that the basic models are also components of complex systems and the same system parameters are involved in the study.

E5-3. Demonstration of Zero-Order Instruments

The basic parameter of a zero-order instruments is the sensitivity K. It is deduced by dividing Eq. E5-5 by a_0 to yield

$$y = Kx(t) \tag{E5-8}$$

where $K = b_0/a_0$ is the sensitivity. The equation shows that the output y is equal to the input $x(t)$ except for a scale factor K. The response is instantaneous without distortion. This describes an almost "ideal" instrument.

Two examples are used for the demonstration. An instrument is zero-order when its response is much faster than the input. For example, a 1-kHz square wave is applied to a 10-MHz oscilloscope with dc couple, as shown in Fig. E5-2a. Since the 10-MHz response is much faster than the 1-kHz input voltage, the input is displayed instantaneously without distortions. The scale factor K is the volts/division setting at the oscilloscope. In other words, the oscilloscope is zero-order for this application. However, the oscilloscope ceases to be zero-order for a 20-MHz signal.

The second demonstration in Fig. E5-2b shows the step response of a thermocouple (TC) when transferred from an ice bath to a hot-water bath. The strip-chart recorder is set at a slow speed. Since the chart speed is slow, the TC has a fast response relative to the time frame of the recorder. Hence a TC is zero-order for measuring slowly varying temperatures. By the same token, a barometer is zero-order for measuring atmospheric pressure. It will be shown presently that a TC is a first-order instrument.

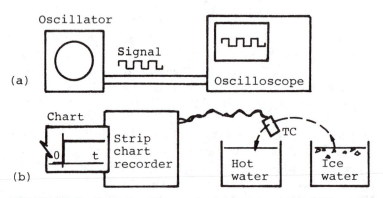

FIGURE E5-2 Examples of zero-order instrument. (a) 20-MHz oscilloscope for 1 kHz signal. (b) Strip chart recorder at slow speed.

E5-4. Demonstration of First-Order Instruments

The basic parameter of first-order instruments is the time constant τ. The model is discussed in Sec. E5-5. The thermocouple TC was shown in Fig. E5-2b as a zero-order instrument. Let us set the recorder at a high speed and repeat the demonstration to show (1) that a TC is not zero-order for this application, (2) the effect of τ, and (3) that τ is not an absolute constant.

Two TCs of the same wire size are used for the demonstration. They are embedded in small brass cylinders of different sizes to control the response time. Evidently, the response of TC-1 with a larger brass cylinder is slower than TC-2 with a smaller brass cylinder. *Note:* Save the plots from the recorder for Expt. 5-1.

Steps for the demonstration are as follows:

1. Place TC-1 with the larger brass cylinder in the ice bath until a steady state is reached. Plunge TC-1 in the hot bath and agitate the bath vigorously. This gives curve 1 in Fig. E5-3a.

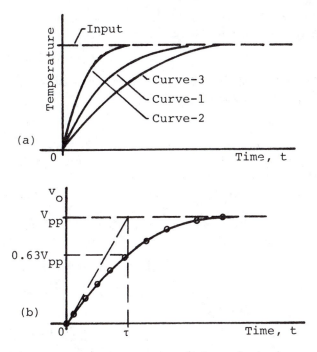

FIGURE E5-3 Step response of first-order instruments. (a) Step response of thermocouple. (b) Methods for finding time constant τ.

2. Repeat step 1 for TC-2 with the smaller brass cylinder. This gives curve 2 in the figure.
3. Now repeat step 1 for TC-1, but without the agitation. This gives curve 3 in the figure.

The observations from Fig. E5-3a are: (1) The responses are not instantaneous, and the TCs cannot be zero-order. (2) TC-2 with a smaller mass has a faster response than TC-1 with a larger mass. (3) Curves 1 and 3 are responses for the same TC under identical conditions except for the agitation of the hot bath. The agitation increases the rate of heat transfer and decreases the time constant τ. Hence τ is not an absolute constant and is dependent on operating conditions.

E5-5. First-Order Instruments

A. Introduction

The purpose of this section is (1) to introduce the model of first-order instruments, (2) to evaluate the time constant, and (3) to show the integrating and differentiating characteristics of a first-order instrument.

The model of a first-order instruments is obtained by dividing the first-order differential equation in Eq. E5-6 by a_0:

$$\tau \dot{y} + y = Kx(t) \tag{E5-9}$$

where $\tau = a_1/a_0$ is the time constant and $K = b_0/a_0$ is the sensitivity, defined in Eq. E5-8. The characteristic equation is

$$\tau D + 1 = 0 \tag{E5-10}$$

where $D = d/dt$ is a differential operator. The time lag in the response, shown in Fig. E5-3, is governed by the time constant τ. If τ approaches zero, a first-order instrument degenerates to zero-order.

The time constant τ can be determined by three methods. Let $K = 1$ in Eq. E5-9, the input be a step function of magnitude V_{p-p}, and the response v_o be as shown in Fig. E5-3b. The equation of a first-order instrument and the solution are

$$\tau \dot{v}_o + v_o = V_{p-p} \tag{E5-11}$$

$$v_o = V_{p-p}(1 - e^{-t/\tau}) \tag{E5-12}$$

Method 1: When the time $t = \tau$, we get

$$v_o = 0.632 \, V_{p-p} \qquad \text{for} \quad t = \tau \tag{E5-13}$$

The corresponding value of τ is shown in Fig. E5-3b.
Method 2: The initial slope of the response from Eq. E5-12 is

$$\frac{dv_o}{dt}\bigg|_{t=0} = \frac{1}{\tau}V_{\text{p-p}} \qquad\qquad\qquad\text{(E5-14)}$$

The value of τ from the initial slope is shown in Fig. E5-3b.

Method 3: Equation E5-12 can rearranged to give

$$\ln\left(1 - \frac{v_o}{V_{\text{p-p}}}\right) = -\frac{1}{\tau}t \qquad\qquad\qquad\text{(E5-15)}$$

This method is more accurate, because Eq. E5-15 describes a straight line in a semilog plot (not shown), and it is easier to estimate τ from a straight line than from a curve.

B. Time-Constant Evaluation: Experiment 5-1

The procedure for obtaining data to evaluate the time constants is similar to that described in Sec. E1-8C.

1. Setup the circuit shown in Fig. E5-4a. The $v_i(t)$ is from an oscillator, R and C are decade boxes, and v_o is measured by means of an oscilloscope with dc couple.

2. Set $v_i(t)$ at a 1-kHz square wave. Adjust R and C to get v_o as shown (say, $R = 20\ \text{k}\Omega$ and $C = 0.005\ \mu\text{F}$). This gives a repeated step input to the first-order RC network.

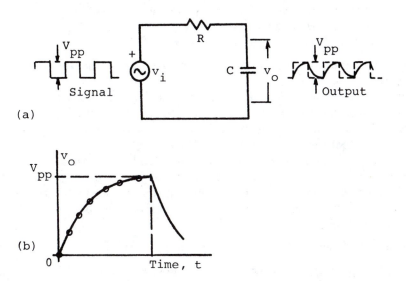

FIGURE E5-4 Evaluation of time constant. (a) RC network. (b) Time response v_0 across C.

3. Obtain the values of (v_o, t) for v_o measured across C, as shown in Fig. E5-4b.
4. Using the same values for R and C, repeat the procedure but with v_o measured across R (not shown). *Caution:* Check the circuit for proper ground connections. Note the waveform of v_o, and obtain values of (v_o, t) as before.

Include the following items in the laboratory report:

1. Using the data from step 3 above, find the value of τ by the three methods enumerated. Compare τ with the product RC.
2. Using the data from step 4 in the procedure, find τ by the three methods enumerated. Why should the value of τ be the same when v_o is measured across C or across R?
3. Explain the waveform of v_o versus time t for v_o measured across C.
4. Repeat (3) for v_o measured across R.
5. Estimate the values of τ for the thermocouples TCs from the step responses shown in Fig. E5-3. Why are the time constants different for TCs with identical wire size?
6. Discuss whether the oscilloscope should be at an ac or dc couple to measure v_o.

C. Integration and Differentiation: Experiment 5-2

The purpose of this experiment is to show the integrating and differentiating characteristics of a first-order instrument. This will be discussed further in the frequency-domain study.

The steps for the experiment are as follows:

1. Use the same values of R and C as in Expt. 5-1, and set up the circuit shown in Fig. E5-4a.
2. Monitor v_o across C. Set the oscillator at 20 kHz and the oscilloscope at a dc couple. Display the waveforms of v_i and v_o simultaneously. Sketch the waveforms of v_o in Fig. E5-5b when v_i is (a) a square-wave, (b) a sawtooth wave, and (c) a sine wave.
3. Change the ground connection if necessary, and monitor v_o across R. Set the oscillator at 200 Hz and the oscilloscope at a dc couple. Display the waveforms of v_i and v_o simultaneously. Sketch the waveforms of v_o in Fig. E5-5c when v_i is (a) a square wave, (b) a sawtooth wave, and (c) a sine wave.

Include the following items in the laboratory report:

1. Explain the waveforms shown in Fig. E5-5b and c.
2. Derive the differential equation for v_o across C. Justify that v_o is the integral of v_i when v_i is at high frequencies.

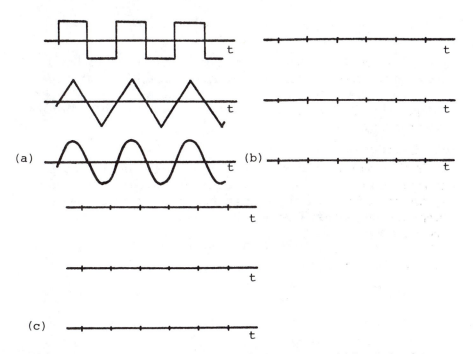

FIGURE E5-5   Characteristics of first-order instruments. (a) Input signal waveforms. (b) Output v_o across C. (c) Output v_o across R.

3.   Derive the differential equation for v_o across R. Justify that v_o is the derivative of v_i when v_i is at low frequencies.

E5-6.   Demonstration of Second-Order Instruments

The purpose of this section is to demonstrate the effects of the damping and natural frequency on the behavior of second-order instruments. The system equation is obtained by dividing Eq. E5-7 by a_0:

$$\frac{\ddot{y}}{\omega_n^2} + \frac{2\zeta}{\omega_n}\dot{y} + y = Kx(t) \tag{E5-16}$$

where $\omega_n^2 = a_0/a_2$, $2\zeta/\omega_n = a_1/a_0$, and $K = b_0/a_0$. The sensitivity K is already defined in Eq. E5-8. The remaining parameters in the equation are the damping ratio ζ, and the undamped natural frequency is ω_n in rad/s.

The characteristic equation from Eq. E5-16 is

$$\frac{1}{\omega_n^2} D^2 + \frac{2\zeta}{\omega_n} D + 1 = 0 \tag{E5-17}$$

where $D = d/dt$ is a differential operator. Hence the parameters that characterize a second-order instrument are ζ and ω_n. The instrument is overdamped when $\zeta > 1.0$, critically damped when $\zeta = 1.0$, and underdamped when $\zeta < 1.0$. Instruments are generally underdamped or have little damping.

A. Effect of Damping

An RLC circuit is used to simulate a second-order instrument. The effect of damping is demonstrated by comparing the step response (indicial response) of two RLC circuits of same natural frequency. It is convenient to keep one circuit as the reference and to vary the damping in the other for the comparison.

The equations for the RLC circuits in Fig. E5-6, and that for ζ and ω_n in terms of R, L, and C, are

$$LC\ddot{v}_o + RC\dot{v}_o + v_o = V_i \tag{E5-18}$$

or

$$\frac{1}{\omega_n^2} \ddot{v}_o + \frac{2\zeta}{\omega_n} \dot{v}_o + v_o = V_i \tag{E5-19}$$

where

$$\frac{1}{\omega_n^2} = LC \qquad \frac{2\zeta}{\omega_n} = RC \tag{E5-20}$$

$$\zeta = \frac{R}{2} \sqrt{\frac{C}{L}} \qquad R = 2\zeta \sqrt{\frac{L}{C}} \tag{E5-21}$$

The damping ratio ζ is proportional to R when L and C are constant, as shown in Eq. E5-21. Suggested values of R, L, and C for the demonstration are shown in Table E5-1. For example, if $L = 1.0$ H, $C = 0.0063$ μF, and the natural frequency is $f_n = 2$ kHz, the resistance is $R = 1256\Omega$ for $\zeta = 0.05$, and $R = 5027\Omega$ for $\zeta = 0.20$. *Caution:* The resistance in the inductor L must be included in the calculation.

The step input of magnitude V_{p-p} is simulated by a low-frequency square wave from the oscillator, shown in Fig. E5-6. The display of the responses are shown in Fig. E5-7a. The frequency of the oscillator should be sufficiently low such that the oscillations in the response die out before a reverse step is applied. Evidently, the responses have the same frequency, and the effect of damping is to decrease the amplitude of the oscillations and to cause the oscillations to die out at a faster rate.

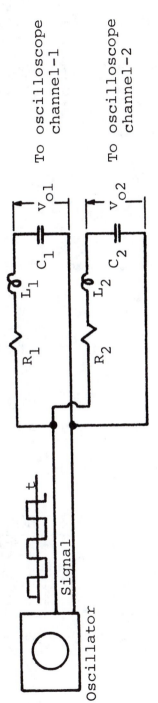

FIGURE E5-6 Demonstration of effects of ζ and ω_n.

TABLE E5-1 Suggested RLC Values for Simulation of Second-Order Instruments

(a) Effect of Damping Ratio ζ

$L = 1.0$ H, $C = 0.0063$ µF, natural frequency $f_n = 2.0$ kHz

Damping ratio, ζ	Total resistance, R (Ω)	Maximum overshoot (%)
0.05	1,256	85
0.10	2,513	73
0.20	5,027	53
0.30	7,540	37
0.50	12,566	16
1.00	25,133	0

(b) Effect of Natural Frequency ($L = 1.0$ H)

f_n (kHz)	1.0	2.0	3.0	4.0	5.0	6.0
C (µF)	0.0253	0.0063	0.0028	0.0016	0.0010	0.0007
For $\zeta = 0.05$:						
R (Ω)	628	1,256	1,885	2,513	3,142	3,770
For $\zeta = 0.10$:						
R (Ω)	1,257	2,513	3,770	5,027	6,283	7,540
For $\zeta = 0.20$:						
R (Ω)	2,513	5,027	7,540	10,053	12,566	15,080
For $\zeta = 0.30$:						
R (Ω)	3,770	7,540	11,310	15,080	18,850	22,619
For $\zeta = 0.40$:						
R (Ω)	5,027	10,053	15,080	20,160	21,133	30,159

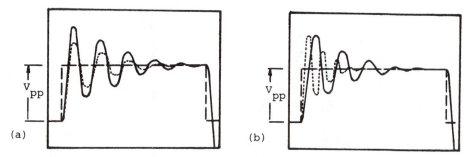

(a)   (b)

FIGURE E5-7 Effects of ζ and ω_n. (a) Effect of ζ, same ω_n. (b) Effect of ω_n, same ζ.

B.   Effect of Natural Frequency

The effect of natural frequency is demonstrated by comparing the step response of two RLC circuits in Fig. E5-6 of the same damping ratio. It is convenient to keep one circuit as the reference and to vary the natural frequency of the other for the comparison.

Suggested values for R, L, and C are shown in Table E5-1. The natural frequency can be changed conveniently by varying the value of C rather than that of L. *Caution:* The resistance in the inductor L must be included in the calculation.

The change in frequency is evident in the responses shown in Fig. E5-7b. It will be shown in Sec. E5-7 that the amplitude of the oscillations, or overshoot, about V_{p-p} is a function of the damping ratio ζ only. Hence it is anticipated that the response with the same ζ will have the same overshoot.

C.   Coupled Systems

The phenomenon of beats in a coupled system is demonstrated in Fig. E-5-18. The mass m in Fig. E5-8a is attached to a coil spring of stiffness k. When m is given an initial displacement y_0 and released from rest, the system vibrates at its natural frequency, as shown in Fig. E5-8b. Now the system shown in Fig. E5-8c has a vertical motion and a rotational motion about its longitudinal axis. Each mode of motion is described by a second-order differential equation and has its own natural frequency. The system is tuned such that the natural frequencies for the two motions are almost equal.

The two modes of motion are coupled, because the spring gives a slight torque when elongated, and it gives a slight axial force when twisted. When the mass in Fig. E5-8c is given an initial displacement and released from rest, the vertical motion beats as shown in Fig. E5-8d (i.e., the oscillations gradually diminish to almost zero and then start all over again). The rotational motion also displays the beat phenomenon. Maximum angular displacement occurs when the vertical motion is minimum, and vice versa.

D.   Discussion

1.   Solve Eq. E5-19 for a unit step input, assuming zero initial conditions (i.e., $v_o = 0$ and $\dot{v}_o = 0$ for $t = 0^-$). Program the solution in a calculator. If the natural frequency is $f_n = 2$ kHz, plot the time response for $\zeta = 0.10$ and $\zeta = 0.40$. Obtain the overshoot from the plots and compare with that in Table E5-1.

2.   In the same graph, plot the response for $f_n = 3$ kHz and $\zeta = 0.10$. Compare the overshoot with that in Table E5-1.

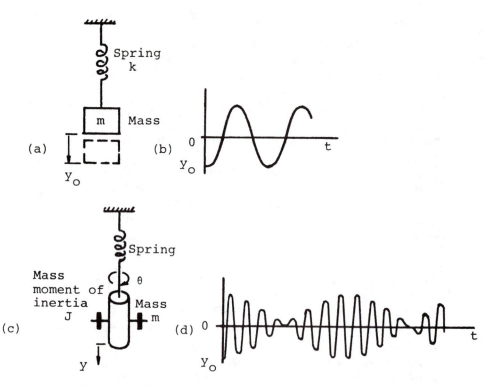

FIGURE E5-8   Vibrations of mass–spring systems. (a) Mass–spring system. (b) Free vibrations. (c) Coupled system. (d) Oscillations of the mass.

3. Derive the equation of motion for the system of mass m and spring constant k, shown in Fig. E5-8a. Solve the equation and express ζ and ω_n in terms of m and k.

4. Derive the equations of motion for the coupled system, shown in Fig. E5-8c, of mass m, spring constant k, mass moment of inertia J, and torsional spring constant k_t. Assume a small difference in the natural frequencies between the two modes of motion. Define the beat frequency, and relate the beat frequency to the natural frequencies of the system.

E5-7. Second-Order Systems: Experiment 5-3

A. Introduction

The effects of the damping ratio ζ and natural frequency ω_n on the behavior of a second-order instrument were demonstrated in the preceding section. The objective of this experiment is to measure ζ and ω_n, and to compare them with those predicted from the values of R, L, and C in the circuit. A step function is used for the input. The step response (indicial response) is also the basis for time-domain specifications.

For a step input of magnitude V_{p-p}, the solution of Eq. E5-19 for a second-order instrument is

$$v_o = V_{p-p}\left[1 - \frac{1}{\sqrt{1 - \zeta^2}} e^{-\zeta\omega_n t} \sin(\omega_d t + \phi)\right] \qquad \text{(E5-22)}$$

where

$$\phi = \sin^{-1}\sqrt{1 - \zeta^2} \qquad \omega_d = \omega_n\sqrt{1 - \zeta^2} \qquad \text{(E5-23)}$$

The step response v_o is plotted in Fig. E5-9. It can be shown that the damping ratio ζ is related to the maximum percentage overshoot in the equation

$$\text{maximum overshoot (\%)} = 100 \times \frac{\text{maximum overshoot}}{\text{input magnitude}}$$

$$= 100e^{-\zeta\pi/\sqrt{1-\zeta^2}} \qquad \text{(E5-24)}$$

Thus ζ can be determined from the maximum overshoot in the response, and the undamped natural frequency ω_n deduced from the period T of the oscillations by the equation

$$\omega_n = \frac{2\pi}{T}(1 - \zeta^2)^{-1/2} \qquad \text{(E5-25)}$$

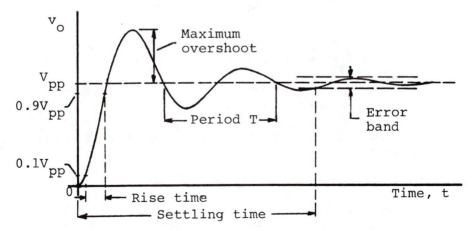

FIGURE E5-9 Time-domain specifications.

In addition to ζ and ω_n, common time-domain specifications are the rise time, percentage overshoot, and settling time, as shown in Fig. E5-9. Settling time is the time for the output to settle within a given error band about the steady-state value. This is a measure of the time for an instrument to meet the error limit.

B. Procedure

1. Set up a series RLC circuit (see Fig. E5-6). Adjust the components such the output is similar to that shown in Fig. E5-9, say, $L = 1.0$ H, $C = 0.0028$ µF, and the total circuit resistance $R = 1885$ Ω.
2. Measure the maximum percentage overshoot, the period, the rise time, and the settling time for an error band of 10%.
3. Set the oscillator at 10 kHz. Display v_o and v_i simultaneously. Sketch the waveforms of the output v_o when v_i is a square wave, sawtooth wave, and a sine wave. For the sine-wave input, note the amplitude ratio v_o/v_i and the phase angle of v_o relative to v_i.

C. Discussion

1. Derive Eq. E5-24. Plot the maximum percentage overshoot versus damping ratio ζ.
2. Calculate ζ from the observed data. Compare the observed value with that predicted in Eq. E5-21.
3. Calculate the undamped natural frequency ω_n in rad/s and compare with that predicted from the values of L and C.
4. For the 10-kHz input, justify the observed waveforms for v_o when v_i is a square wave, a sawtooth wave, and a sine wave. For the 10-kHz sine-wave input, calculate the amplitude ratio v_o/v_i and the phase angle of v_o relative to v_i, and compare with the observed values.

10-7. EXPERIMENT 6. FREQUENCY RESPONSE OF INSTRUMENTS

The objectives of this experiment are (1) to examine the dynamic characteristics of instruments in the frequency domain by means of a model study, and (2) to evaluate the system parameters from frequency response data. The rationale for the model study was discussed in the last experiment.

E6-1. Frequency Response Method

The frequency response method is a sinusoidal steady-state study. The block diagram of a single-input, single-output linear system is shown in Fig. E6-1. The transfer function $G(j\omega)$ relates the output/input quantities

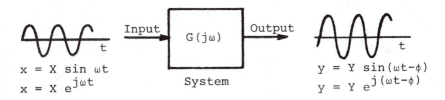

x = X sin ωt
x = X e^{jωt}

Input G(jω) Output

System

y = Y sin(ωt−φ)
y = Y e^{j(ωt−φ)}

FIGURE E6-1 Block diagram of a linear system.

under steady-state conditions. It will be shown that $G(j\omega)$ can also be obtained by means of transient methods, such as a pulse testing. The test data, however, are ultimately expressed in terms of frequency response.

The basic method applies an input $x = X \sin \omega t$ to the system shown in Fig. E6-1 to obtain an output $y = Y \sin (\omega t - \phi)$, where X is the amplitude of the input, ω the excitation frequency in rad/s, Y the amplitude of the output, and ϕ the phase angle of y relative to x. For a linear system, both x and y must be of the same frequency. Expressing x and y in vectorial form for convenience (see Sec. 4-6), we get

$$x = Xe^{j\omega t} \quad \text{and} \quad y = Ye^{j(\omega t - \phi)} \tag{E6-1}$$

$$\frac{y}{x}(j\omega) = G(j\omega) \tag{E6-2}$$

$$\frac{Y}{X}(j\omega) = |G(j\omega)| \tag{E6-3}$$

$$\phi(j\omega) = \underline{/G(j\omega)} \tag{E6-4}$$

The quantity $j\omega$ in the equations denotes that the transfer function $G(j\omega)$, the amplitude ratio $Y/X(j\omega)$, and the phase angle $\phi(j\omega)$ are frequency dependent. It does not imply a product of these quantities with $j\omega$.

The general transfer function in Eq. E6-2 can be expressed as a ratio of two polynomials.

$$\frac{y}{x}(j\omega) = \frac{b_m(j\omega)^m + \cdots + b_2(j\omega)^2 + b_1(j\omega) + b_0}{a_n(j\omega)^n + \cdots + a_2(j\omega)^2 + a_1(j\omega) + a_0} = G(j\omega) \tag{E6-5}$$

where a's and b's are real coefficients and m and n integers. It can be shown that $m \leqslant n$ for the system to be realizable (i.e., to be constructed physically).

The polynomials in Eq. E6-5 can be factorized into elementary transfer functions (see Sec. 4-6A). The type of factors with real coefficients are

$$G(j\omega) = K \tag{E6-6}$$

$$G(j\omega) = j\omega^{\pm 1} \tag{E6-7}$$

$$G(j\omega) = (1 + j\omega\tau)^{\pm 1} \tag{E6-8}$$

$$G(j\omega) = \left[\left(\frac{j\omega}{\omega_n} \right)^2 + 2\zeta \frac{j\omega}{\omega_n} + 1 \right]^{\pm 1} \tag{E6-9}$$

The $\pm$ signs above denote that the factors can be in the numerator or the demoninator of the general transfer function $G(j\omega)$. In other words, a complex system is a combination of the elementary transfer functions. For example, Eq. E6-6 represents a zero-order instrument, Eq. E6-7 an integrator or differentiator (see Sec. E5-5C), Eq. E6-8 a first-order instrument, and Eq. E6-9 a second-order instrument.

The behavior of a system must be independent of the method for its investigation. The time- and frequency-domain methods are merely two views of the same physical system, because the same elementary functions in Eqs. E6-6 to E6-9 are used for both. The advantage of the frequency method is that the elementary functions can be examined individually and then recombined to obtain the behavior of a complex system shown in Eq. E6-5 (see Sec. 4-6B). There is insufficient time in this experiment for this detailed study, but the behavior of a simple combination of the elementary function in a "black box" can be demonstrated.

The setup for a frequency response test is shown in Fig. E6-2. The amplitude ratio Y/X in Eq. E6-3 and the phase angle ϕ in Eq. E6-4 for an excitation frequency ω can be measured by means of a dual-channel oscilloscope (see Expt. 1). This is performed for each frequency over the frequency range of interest. The method is basic, but is also tedious. It will be demonstrated that the procedure can be mechanized.

The test data are often presented in the Bode plots shown in Fig. E6-3. At a frequency ω, the transfer function $G(j\omega)$ in Eq. E6-5 is a complex number, involving three quantities: the frequency ω, the amplitude ratio Y/X, and the phase angle ϕ. The Y/X versus ω is in a log-log plot (as in a log-log graph paper and not the log of a number). The phase angle ϕ ver-

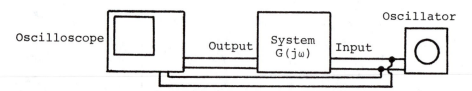

FIGURE E6-2   Schematic of test setup.

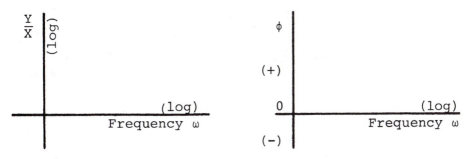

FIGURE E6-3 Bode plots.

sus ω is in a rectangular versus log plot. The ordinate Y/X is often in decibels (dB), where dB is defined as

$$\text{decibel (dB)} = 20 \log_{10}(\text{ratio}) \qquad \text{(E6-10)}$$

E6-2. Demonstration of Zero-Order Instruments

The transfer function of a zero-order instrument from Eq. E6-6 is $G(j\omega) = K$, which is a real number. The amplitude ratio Y/X is independent of frequency ω, and the phase angle φ is zero.

Procedure for the demonstration is as follows:

1. Set up the voltage-divider circuit shown in Fig. E6-4. Let $R = 2\ k\Omega$, and $v_i = 6 \sin \omega t$. In fact, a sawtooth wave, a square wave, or any periodic waveform can be used, because the waveform can be expressed in harmonic components in a Fourier series.
2. Connect v_i and v_o to an oscilloscope with dc couple. Observe the waveforms of v_i and v_o for the frequencies of 20 Hz, 200 Hz, 2 kHz, and 20 kHz.
3. Set the oscilloscope in the X–Y mode, and record the amplitude ratio Y/X and the phase angle for the frequencies in step 2.
4. Plot the data in the Bode plots shown in Fig. E6-3.

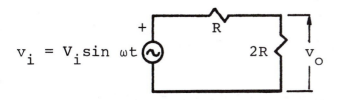

FIGURE E6-4 Simulation of zero-order instruments.

E6-3. First-Order Instruments

A. *Low-Pass Characteristics: Experiment 6-1*

The model of a first-order instrument is shown in Fig. E6-5a. The time constant is $\tau = RC$. The transfer function $G(j\omega)$ for v_o measured across the capacitor C is

$$\frac{v_o}{v_i}(j\omega) = \frac{1}{1 + j\omega\tau} = G(j\omega) \tag{E6-11}$$

$$\frac{V_o}{V_i}(j\omega) = |G(j\omega)| = \frac{1}{\sqrt{1 + (\omega\tau)^2}} \tag{E6-12}$$

$$\phi(j\omega) = \underline{/G(j\omega)} = -\tan^{-1}\omega\tau \tag{E6-13}$$

The following characteristics are observed from the Bode plots shown in Fig. E6-5b and c:

1. In the V_o/V_i versus ω plot, the asymptotes intersect at a breakpoint at the break frequency $\omega = 1/\tau$. For $\omega \ll 1/\tau$, V_o/V_i is a horizontal line, indicating that the system is zero-order. For $\omega \gg 1/\tau$, V_o/V_i attenuates with increasing frequency and a slope of -1. It can be deduced from Eq. E6-7 that a -1 slope corresponds to one integration. The overall transfer characteristic shown in the figure is that of low-pass filter.

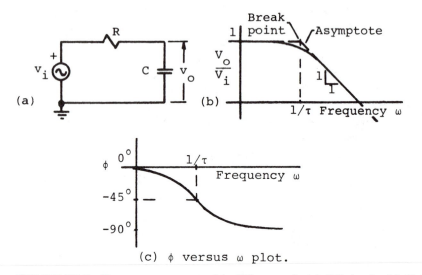

FIGURE E6-5 Frequency response of an RC network. (a) RC circuit. (b) V_o/V_i versus ω plot. (c) ϕ versus ω plot.

2. In the phase angle ϕ versus ω plot, $\phi = -45°$ at the break frequency $\omega = 1/\tau$. For $\omega \ll 1/\tau$, $\phi = 0$, and the system is zero-order. For $\omega \gg 1/\tau$, $\phi = -90°$. In other words, a phase shift of $-90°$ corresponds to an integration and a slope of -1 in the V_o/V_i plot.

Procedure for the experiment is as follows:

1. Set up the circuit shown in Fig. E6-5a. Use the R and C values from Expt. 5-1, say, $R = 20$ kΩ and $C = 0.005$ μF.
2. With the oscilloscope in the X–Y mode, obtain values of V_o/V_i and ϕ versus ω for the Bode plots, shown in Fig. E6-5b and c. Generally, V_i is constant. The selected frequencies should be equally space in a logarithmic scale, and the range be sufficient to plot the symptotes.
3. Set the oscillator at 20 kHz. Display v_o and v_i simultaneously. Find V_o/V_i and the phase angle ϕ.

Include the following discussion topics in the report:

1. Obtain the Bode plots. Compare τ from the plots with the R and C product.
2. Derive Eq. E6-11 for the circuit shown in Fig. E6-5a.
3. From Eq. E6-11, show that (a) $\phi = -45°$ at $\omega = 1/\tau$, (b) the low-frequency asymptote has a zero slope, and (c) the high-frequency asymptote has a -1 slope.
4. Does a -1 slope denote an integration? Using $\tau = RC$, calculate V_o/V_i and ϕ from Eq. E6-11, and compare with those from the data in step 3 of the procedure.

B. High-Pass Characteristics: Experiment 6-2

The model of a first-order instrument with v_o measured across R is shown in Fig. E6-6a. The time constant is $\tau = RC$. The transfer function $G(j\omega)$ is the product of two elementary functions, namely, $j\omega$ from Eq. E6-7 and $1/(1 + j\omega\tau)$ from Eq. E6-8.

$$\frac{v_o}{v_i}(j\omega) = \frac{j\omega\tau}{1 + j\omega\tau} = G(j\omega) \tag{E6-14}$$

$$\frac{V_o}{V_i}(j\omega) = |G(j\omega)| = \frac{\omega\tau}{\sqrt{1 + (\omega\tau)^2}} \tag{E6-15}$$

$$\phi(j\omega) = \underline{/G(j\omega)} = 90 - \tan^{-1}\omega\tau \tag{E6-16}$$

The following characteristics are observed from the Bode plots in Fig. E6-6b and c.

1. In the V_o/V_i versus ω plot, the break frequency occurs at $\omega = 1/\tau$, as before. For $\omega \ll 1/\tau$, the V_o/V_i asymptote has a $+1$ slope, which corre-

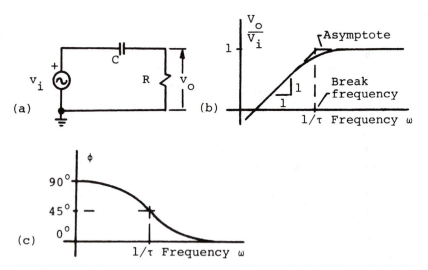

FIGURE E6-6   Frequency response of an RC network. (a) RC circuit. (b) V_o/V_i versus ω plot. (c) ϕ versus ω plot.

sponds to a differentiation. For $\omega \gg 1/\tau$, the V_o/V_i asymptote is horizontal, indicating zero-order.

2. For the phase angle ϕ versus ω plot, $\phi = +45°$ at $\omega = 1/\tau$. For $\omega \ll 1/\tau$, $\phi = +90°$ corresponding to a $+1$ slope. In other words, a $+90°$ denotes a differentiation. For $\omega \gg 1/\tau$, $\phi = 0$. Again, this is a zero-order characteristic.

The procedure for the experiment is as follows:

1. Use the same values for R and C as in Expt. 6-1 to set up the circuit shown in Fig. E6-6a.
2. With the oscilloscope in the X-Y mode, obtain values of V_o/V_i and ϕ versus ω for the Bode plots shown in Fig. E6-6b and c. The selected frequencies should be equally spaced in a logarithmic scale, and the range be sufficient to plot the asymptotes.
3. Set the oscillator to 200 Hz. Observe v_o and v_i simultaneously. Find V_o/V_i and the phase angle ϕ.

Include the following discussion topics in the report:

1. Obtain the Bode plots. Compare τ from the plots with the value of the RC product.
2. Derive Eq. E6-14 from the circuit shown in Fig. E6-6a.
3. From Eq. E6-14, show that (a) the low-frequency asymptote has a $+1$

slope, (b) the high-frequency asymptote has a zero slope, and (c) the phase angle at $\omega = 1/\tau$ is $+45°$.

4.   Using the values of R and C from the circuit, plot V_o/V_i and ϕ versus frequency ω. Compare the plots with that from the test data. Why does a +1 slope denote a differentiation?

E6-4. Second-Order Instruments

A. *Introduction*

The *RLC* circuits, shown in Fig. E6-7, can be used to simulate second-order instruments. They have distinct characteristics and are models for different applications. The inductor L, however, is not an ideal circuit component because of its inherent resistance. The operational amplifier described in Expt. 7 can be used to overcome this difficulty. Hence the

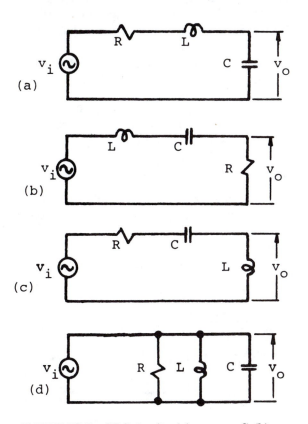

FIGURE E6-7   *RLC* circuits. (a) v_o across C. (b) v_o across R. (c) v_o across L. (d) v_o across L and C.

circuit in Fig. E6-7a is employed for this experiment. It represents a simple mass–spring system (see Fig. E5-8), in which the displacement of the mass is the output. The transfer function $G(j\omega)$ is

$$\frac{v_o}{v_i}(j\omega) = \frac{1}{(j\omega/\omega_n)^2 + 2\zeta(j\omega/\omega_n) + 1} = G(j\omega) \tag{E6-17}$$

$$\frac{V_o}{V_i}(j\omega) = |G(j\omega)| = \frac{1}{\{[1 - (\omega/\omega_n)^2]^2 + (2\zeta\omega/\omega_n)^2\}^{1/2}} \tag{E6-18}$$

$$\phi(j\omega) = \underline{/G(j\omega)} = -\tan^{-1}\frac{2\zeta(\omega/\omega_n)}{1 - (\omega/\omega_n)^2} \tag{E6-19}$$

where $\omega_n^2 = 1/LC$ and $2\zeta/\omega_n = RC$ (see Eqs. E5-18 to E5-21).

B.　Demonstrations

The objectives of the demonstrations are (1) to examine the frequency response of a strip-chart recorder, (2) to illustrate an application of the spectrum analyzer, and (3) to find the transfer function of a "black box," consisting of two elementary transfer functions.

Strip-chart recorder.　The mechanical section of a strip-chart recorder can be approximated as a damped second-order system. Since the response of the electronics is fast compared with the input signal, the dynamics of the recorder is determined by the servomechanism of the writing pen (see Fig. 6-46). The Bode plot of the amplitude ratio Y/X versus excitation frequency ω can be obtained from the setup shown in Fig. E6-8.

The procedure for the demonstration is as follows:

1.　An input voltage $x = X \sin \omega t$ from an oscillator is applied to the recorder. Set the input at a low frequency to verify the sinusoidal waveform at the output.
2.　Set the recorder at a low speed to conserve paper. Increment the input frequency and record Y_{p-p} as shown in the figure. Only V_{p-p} is

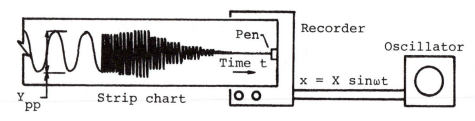

FIGURE E6-8　Frequency response of a strip-chart recorder.

recorded, because the amplitude X is constant for all frequencies and only the amplitude ratio Y/X is of interest.
3. When Y_{p-p} is fully attenuated, switch the input freqeuncy to a high value, say, 1 kHz.

The following discussion topics should be included in the report.

1. Plot Y/X versus frequency in hertz from the data. Compare the plot with the specification of the recorder.
2. It was stated above that the response of the recorder is that of a damped second-order system. Estimate the damping from the plot above, and explain why damping is necessary.
3. Why does the pen not respond to a high-frequency input?

Spectrum analysis. A pulse of very short duration (see Sec. 4-8 and App. C) is substituted for the sinusoidal input in this demonstration. A train of pulses is applied to the series RLC circuit shown in Fig. E6-9. Thus each pulse in the train is an input to the system. The pulse train is obtained by using an oscillator to trigger a second oscillator. The procedure for triggering an oscillator is given in its description. Alternatively, a differentiating circuit (see Fig. E6-6) and a dioide can be used to generate the pulse train, but this is more awkward. The response v_o across C is a sequence of transients. If v_o is applied to a spectrum analyzer (see Sec. 6-9B), the output is a discrete frequency spectrum as shown in the figure. The envelope of the spectrum is the frequency response of the circuit. The phase information is also available for some spectrum analyzers.

The pulse described above is an approximation of an impulse. If the duration of the pulse is extremely short, the pulse becomes an impulse, which has equal magnitude for all frequencies. The response v_o is the superposition of the responses due to all frequencies. The frequency spectrum of v_o is continuous, and it gives the magnitude of the responses at the individual frequencies. This is precisely the definition of a frequency response test. Thus by means of a single pulse input, the output v_o has essentially the information for all the data points for a frequency response test.

Frequency response of a "black box." The test of a complex system is more realistic than that of an elementary transfer function. The black box consists of two or more stages of the elementary transfer functions described in Eqs. E6-6 to E6-9. Either the basic test setup in Fig. E6-2 or a spectrum analyzer can be used to obtain the frequency response plots.

The circuit can be set up conveniently in a breadboard, and the hardware for modeling depends on components on hand. For ease of

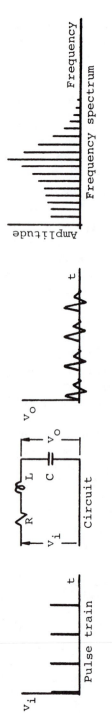

FIGURE E6-9   Impulse response and frequency spectrum.

analysis, the break frequencies of the stages should be sufficiently separated, and light damping be used in a second-order elementary transfer function. The impedance level of the stages should go up about 10 to 1 for each successive stage to avoid loading. The circuit in the black box can be changed as often as desired. From past experience, the black box can also be used as a hands-on experiment.

C. Frequency Response: Experiment 6-3

The circuit in Fig. E6-7a is used in this experiment to simulate a second-order instrument. The Bode plots of typical frequency data are shown in Fig. E6-10.

The procedure for the test with the output v_o measured across C is as follows:

1. Use the basic test setup in Fig. E6-2 for the experiment. Set up the circuit for a damping ratio $\zeta = 0.05$ and an undamped natural frequency $f_n = 3$ kHz. From Table E5-1, suggested values for the components are $L = 1.0$ H, $C = 0.0028$ μF, and $R = 1885$ Ω.
2. Obtain data for the amplitude ratio V_o/V_i and the phase angle ϕ versus frequency ω for the Bode plots. The selected frequencies should be equally spaced on a logarithmic scale. More data points should be taken near resonance.
3. Repeat step 2 for $\zeta = 0.15$ and $f_n = 3$ kHz. Use the same values for L and C, but change R to 5655 Ω.
4. Repeat step 2 for $\zeta = 0.15$ and $f_n = 1$ kHz. Suggested values are $L = 1.0$ H, $C = 0.0253$ μF, and $R = 1884$ Ω.

The following discussion topics should be included in the report:

1. Obtain the Bode plots, as shown in Fig. E6-10, for the data from step

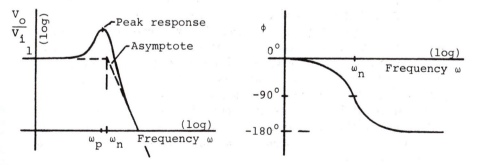

FIGURE E6-10 Bode plots from Eqs. E6-18 and E6-19.

3 above, Draw the high- and low-frequency asymptotes. Find ω_n from the intersection of the asymptotes, and compare with that predicted from the values of L and C in the circuit. Verify from the plots that $\phi = -90°$ at $\omega = \omega_n$, $\phi = 0°$ for $\omega \ll \omega_n$, and $\phi = -180°$ for $\omega \gg \omega_n$.

2. From the model shown in Fig. E6-7a, $(V_0/V_i)_{max}$ occurs at $\omega = \omega_p$, where

$$\omega_p = \omega_n \sqrt{1 - 2\zeta^2} \tag{E6-20}$$

$$\frac{V_0}{V_i}\bigg|_{max} = \frac{1}{2\zeta(1 - \zeta^2)^{1/2}} \tag{E6-21}$$

Derive the equations. Measure $(V_0/V_i)_{max}$ and ω_p from the Bode plots above, and compare with the predicted values from the equations.

3. Use the data from steps 2 to 4 of the test procedure above to plot V_0/V_i versus ω/ω_n, and ϕ versus ω/ω_n. What is the advantage of these normalized plots?

4. Assuming ideal circuit elements, derive the transfer function for the systems shown in Fig. E6-7b and c. Do the systems have the same characteristic equations?

10-8. EXPERIMENT 7. SIGNAL CONDITIONING

E7-1. Introduction

The objective of this experiment is to examine signal conditioning, or the functions performed by the intermediate stage of a measuring system (see Fig. E3-3). Signal conditioning prepares the output from a transducer for subsequent applications. To limit the scope of this broad study, (1) only analog signals are treated, and (2) the selected topics are based mainly on equipment on hand and devices that can be constructed readily.

General-purpose equipment are used whenever possible. For example, amplifiers and filters for the experiments are typical components for signal conditioning in the laboratory. Many meaningful and inexpensive experiments in instrumentation can be devised by using operational amplifiers and integrated circuits with breadboards. The verification of specifications of instruments and some demonstrations can also be utilized as hands-on experiments. To this end, this is a write-up of suggested experiments rather than a comprehensive set of experiments for signal conditioning.

E7-2. Demonstration of Thermocouple Compensation

A. Introduction

Series compensation is a simple technique to improve the dynamic performance of a system (see Sec. 4-6C). Let G_1 be a given system and G_2 the series compensator, as shown in Fig. E7-1a. If loading is negligible, the compensated transfer function G_c with the desired characteristics is

$$\frac{y}{x} = G_1 \qquad \frac{v_o}{y} = G_2 \qquad \frac{v_o}{x} = G_c = G_1 G_2 \qquad (E7\text{-}1)$$

The objective of the thermocouple compensation is to increase its time rate of response. The series compensation of a thermocouple is shown in Fig. E7-1b. Let G_1 be its transfer function (see Sec. E5-5).

$$\frac{y}{x} = G_1 = \frac{K}{1 + \tau_1 D} \, \mathrm{mV/^\circ C} \qquad (E7\text{-}2)$$

where $D = d/dt$ is a differential operator, τ_1 the time constant, and K the sensitivity in mV/°C. If the transfer function G_2^* of the compensator is of the form

$$\frac{v_o}{y} = G_2^* = (1 + \tau_1 D) \qquad (E7\text{-}3)$$

the compensated transfer function G_c^* from Eq. E7-1 becomes

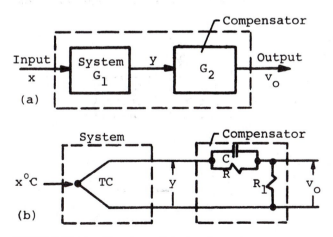

FIGURE E7-1 Signal conditioning by series compensation. (a) Compensated system G_c. (b) Series compensation of a thermocouple.

$$\frac{v_o}{x} = G_1 G_2^* = \frac{K}{1 + \tau_1 D} \frac{1 + \tau_1 D}{1} = K = G_c^* \qquad \text{(E7-4)}$$

This shows the compensated system would be zero-order and a perfect instrument. Unfortunately, G_2^* in Eq. E7-3 is not realizable (see Sec. 4-2); that is, a corresponding physical system cannot be built.

Let the lead network G_2 in Fig. E7-1a be used for the compensation. It can be shown that

$$G_2 = \frac{\alpha(1 + \tau_2 D)}{1 + \alpha \tau_2 D} \qquad \text{(E7-5)}$$

where $\alpha = R_1/(R + R_1) < 1$, and $\tau_2 = RC$. The thermocouple is correctly compensated if $\tau_2 = \tau_1$.

$$\frac{v_o}{x} = G_1 G_2 = \frac{K}{1 + {}_1 D} \frac{\alpha(1 + \tau_2 D)}{1 + \alpha \tau_2 D} \qquad \text{(E7-6)}$$

$$\frac{v_o}{x} = G_c = \frac{\alpha K}{1 + \alpha \tau_1 D} \qquad \text{for} \quad \tau_1 = \tau_2 \qquad \text{(E7-7)}$$

The compensated transfer function G_c is first-order with a time constant $\alpha \tau_1$. Since $\alpha \tau_1$ is less than τ_1, the compensated system has a faster response than the original thermocouple (see Fig. E5-3). The output v_o is attenuated by a factor of α. This is not critical, because v_o can be amplified.

The setup for the demonstration in Fig. E7-2a consists of an ice bath, a hot-water bath, a lead compensating network, a thermocouple (TC), and

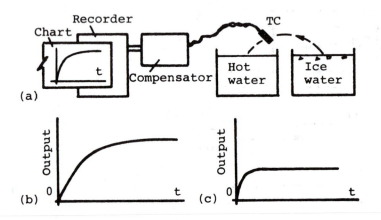

FIGURE E7-2 Step response of TC with and without compensation. (a) Step input to thermocouple. (b) Uncompensated. (c) Compensated.

a strip-chart recorder. The TC is embedded in a small brass cylinder to control its time constant.

B. Procedure

A step response is used to demonstrate the effect of the compensation. First, predetermine τ_1 for the TC, and prepare two compensating networks shown in Fig. E-7-1b, one with $\tau_2 = \tau_1$ and another with $\tau_2 \neq \tau_1$. The procedure is as follows:

1. Set up the equipment as shown in Fig. E7-2a.
2. Apply a step input to the uncompensated TC by transferring from the ice bath to the hot-water bath and agitate vigorously. The uncompensated response is as shown in Fig. E7-2b.
3. Connect the TC and the series compensating network with $\tau_2 = \tau_1$. Repeat step 2. The step response is shown in Fig. E7-2b.
4. Connect the TC and the series compensating network with $\tau_2 \neq \tau_1$. Repeat step 2. The step response is similar to that shown in Fig. E7-2c.

C. Discussion

1. Derive Eq. E7-5 for the lead network shown in Fig. E7-1b.
2. From the step responses, find the time constant of (a) the uncompensated TC and (b) the compensated TC with $\tau_2 = \tau_1$.
3. For $\tau_2 = \tau_1$, (a) compare the measured value $\alpha\tau_1$ of the compensated TC with the predicted value in Eq. E7-7, and (b) divide the data from the compensated response by α, and plot this normalized response and the uncompensated response in the same graph, as shown in Fig. E7-3a.
4. If $\tau_2 > \tau_1$, the step response of the compensated system resembles an "overshoot," as shown in Fig. 7-3b. An overshoot is characteristic of an underdamped second-order system (see Fig. E5-9). Can the response of the compensated TC be oscillatory?
5. From Eq. E7-6, assume that $\alpha = 0.5$ and calculate the normalized step response curves for (a) $\tau_2 = 1.2\,\tau_1$, (b) $\tau_2 = \tau_1$, and (c) $\tau_2 = 0.8\,\tau_1$. Does the step response from the demonstration for $\tau_2 \neq \tau_1$ resemble one of the curves shown in Fig. E7-3b?
6. From Eq. E7-6, assume that $\alpha = 0.5$ and obtain the normalized Bode plots for (a) $\tau_2 = 1.2\,\tau_1$, (b) $\tau_2 = \tau_1$, and (c) $\tau_2 = 0.8\,\tau_1$. Comment on the correlation, if any, between the time and frequency response curves.

E7-3. Demonstration of Carrier Systems

A carrier system is non-self-generating, and requires an auxiliary input for its operation (see Sec. 2-3A). The auxiliary input is a carrier, which

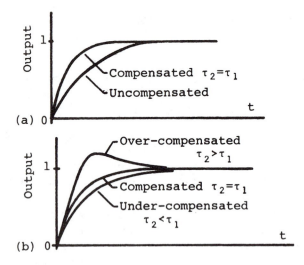

FIGURE E7-3 Normalized step response of TC.

conditions or transforms the signal into a suitable form for transmission. For example, an audio signal is transformed into radio frequency for broadcasting. Frequency modulation (FM) and a carrier amplifier are demonstrated briefly. A suppressed carrier system commonly used in instrumentation is examined in greater detail in the next section.

A. Frequency Modulation

A simple setup to demonstrate frequency modulation is shown in Fig. E7-4a. The desired signal is v_{o1}. Let v_{o1} be a 2-V_{pp} sawtooth wave at 200 Hz from oscillator 1. The modulated signal is v_{o2}. Let v_{o2} be a sine wave at a center frequency of 10 kHz from oscillator 2. The signal v_{o1} is connected to the voltage-controlled generator VCG input of oscillator 2. Thus v_{o1} controls the frequency of v_{o2} to give a FM output. The voltages v_{o1} and v_{o2} are shown in Fig. E7-4b.

The voltage v_{o2} is at a high frequency and is more suitable for transmission. The characteristics of FM signal are (1) an increasing amplitude of v_{o1} increases the frequency of v_{o2} about a center frequency, and (2) a decreasing amplitude of v_{o1} decreases the frequency of v_{o2}. The desired signal v_{o1} can be recovered from the modulated signal v_{o2} by means of a demodulator (see Fig. 5-15).

B. Carrier Amplifiers

The purpose of the demonstration is to show the steps for calibrating a carrier amplifier. The calibration of a dc amplifier, as described in Expt.

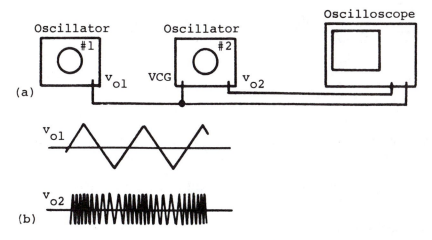

FIGURE E7-4 Demonstration of frequency modulation. (a) Schematic of test setup. (b) Output waveforms.

4-2, consists of (1) balancing the setup, and (2) setting the sensitivity for magnitude scaling. The steps are essentially the same for a carrier amplifier, except that for an ac system, both the resistive and the reactive components in the circuit must be balanced. The procedure for the calibration of a particular carrier amplifier is given in the user's manual.

A general-purpose carrier amplifier from a recorder is used for the demonstation. It is designed to drive non-self-generating transducers, such as the strain gage. The setup consists of a 2.5-kHz carrier amplifier, a strain gage bridge on a cantilever, and an oscilloscope for the readout, as shown in Fig. E7-5.

A low-frequency excitation is applied to the cantilever by means of the shaker (see Fig. E3-4). Due to the residual unbalance of the carrier amplifier, the oscilloscope display has a small ripple at the carrier frequency superposed on the signal, as shown in Fig. E7-6a. This can be eliminated

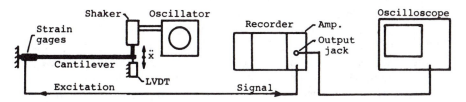

FIGURE E7-5 Demonstration of a carrier system.

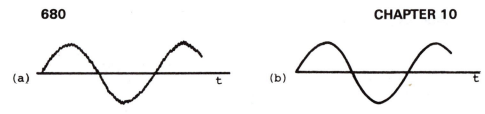

FIGURE E7-6 Residual unbalance in carrier system.

by alternately readjusting the R and C unbalance of the amplifier, or by means of a low-pass filter. The filtered signal is shown in Fig. E7-6b.

E7-4. Demonstration of Amplitude Modulation and Demodulation

A. Introduction

The suppressed carrier modulation–demodulation process for a LVDT (see Sec. 5-2) is a special amplitude modulation AM system for instrumentation. The purpose of the demonstration is to show additional applications of the carrier amplifier and to examine the process in greater detail.

The general block diagram of the process is shown in Fig. E7-7 (see Fig. 5-12). It consists of a modulator, a demodulator, and a low-pass filter. The modulator is a non-self-generating transducer (see Sec. 2-3). The low-frequency signal $x(t)$ is the desired input. The high-frequency carrier v_c is sinusoidal with constant magnitude and frequency. Both $x(t)$ and v_c are applied to the modulator. The signal $x(t)$ modulates the carrier to yield a modulated signal $v_1 = [x(t)][v_c]$ for signal transmission.

The demodulator is basically a rectifier. It receives the signal v_1, and converts it to v_2 as shown in Fig. E7-7. The carrier v_c is also the synchronizing signal for the demodulation, which must be phase-sensitive in order to retrieve the $(\pm)$ sign of $x(t)$ from v_1. Since $v_2 = (v_1)(v_c) = [x(t)][v_c^2]$, v_2 and $x(t)$ have the same sign. The original signal $x(t)$ is then recovered from v_2 by means of a low-pass filter.

B. Suppressed Carrier Systems

The linear variable differential transformer (LVDT) for the demonstration consists of a primary winding, a movable core of magnetic material, and two secondary windings S_1 and S_2, as shown in Fig. E7-8a. The secondary voltages V_1, V_2, and the differential output $V_{net} = V_1 - V_2$ are shown in Fig. E7-8b. The induced voltage depends on the magnetic flux coupling between the primary and the secondary. If the core is above its electrical neutral position, there is more flux coupling between the

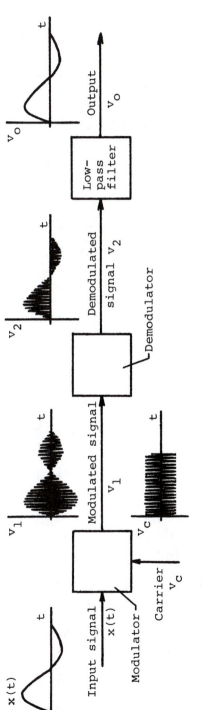

FIGURE E7-7 Amplitude modulation–demodulation.

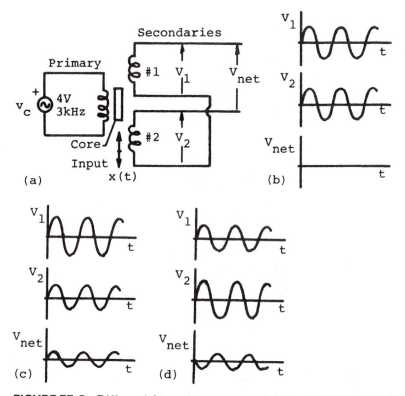

FIGURE E7-8 Differential transformer as a modulator. (a) LVDT with differential output. (b) Core at neutral. (c) Core above neutral. (d) Core below neutral.

primary and S_1 than S_2, and $V_1 > V_2$. Hence V_{net} is positive. If the core is at neutral, $V_{net} = V_1 - V_2 = 0$. The voltages for the core at neutral, above neutral, and below neutral are shown in the figure.

The LVDT is a modulator. The carrier voltage v_c at 3 kHz is applied voltage to the primary. The motion $x(t)$ of the core modulates the carrier to give an output V_{net}. If $x(t)$ is oscillatory, the waveform of V_{net} is similar to that for v_1 in Fig. E7-7. It is not possible, however, to detect the ($\pm$) sign of $x(t)$ from V_{net} without a phase-sensitive demodulator.

The phase-sensitive demodulator is implemented with two diode bridge rectifiers, shown in Fig. E7-9a. The diodes are assembled in a small chassis with provisions to plug-in the resistors and other connections. One of the bridges is shown in Fig. E7-9b to explain the process. A voltage V_1 from S_1 is applied to the bridge. Since the current through the resistor R is always in one direction, V_1 is rectified to yield V_1^* across R, as shown in the

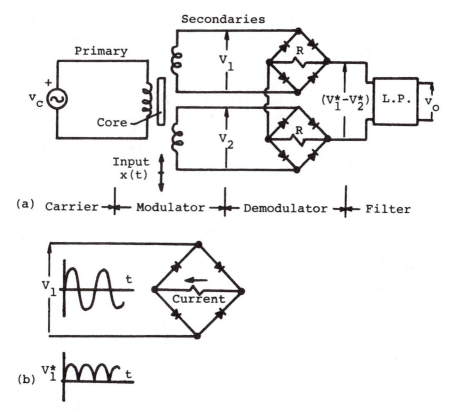

FIGURE E7-9 Suppressed carrier amplitude modulation–demodulation. (a) Amplitude modulation–demodulation. (b) Diode bridge rectifier.

figure. Similarly, V_2 is rectified to yield V_2^*. The differential $(V_1^* - V_2^*)$ is the phase-sensitive demodulated signal. This can be filtered to yield an output v_o proportional to the core displacement $x(t)$.

The modulation–demodulation process in Fig. E7-7 can be identified with its hardware implementation shown in Fig. E7-9a. The procedure for the demonstration can be referred to both figures.

1. Set up the LVDT and the cantilever shown in Fig. E7-5. The core of the LVDT is attached to the free end of the cantilever, and the body of the LVDT with the windings is stationary. Preload the cantilever with the shaker. Set up the circuit in Fig. 7-8a, and adjust the body of the LVDT so that V_{net} is zero with the preload. The voltages are measured by an oscilloscope with dc couple.

2. Move the cantilever vertically by hand to demonstrate each of the waveforms shown in Figs. E7-8b to d.

3. Excite the cantilever with the shaker to demonstrate that the LVDT is a modulator, and its output waveform is similar to v_1 shown in Fig. E7-7.

4. Connect the diode bridges shown in Fig. E7-9a to the LVDT. Move the cantilever by hand to demonstrate the voltages V_1, V_1^*, V_2, and V_2^* in Fig. E7-9.

5. Demonstrate the differential output $[V_1^* - V_2^*]$ by moving the cantilever by hand to show the waveform.

6. Excite the cantilever with the shaker, and compare the waveform of $[V_1^* - V_2^*]$ with that of v_2 in Fig. E7-7.

7. Insert a 500-Hz low-pass RC filter between $[V_1^* - V_2^*]$ and the oscilloscope to recover the original signal $x(t)$.

E7-5. Characteristics of Filters

A. Introduction

The objectives of this experiment are to examine (1) the effect of filtering on the waveform of a signal, and (2) the frequency response of a commercial variable filter. In fact, the first objective can be explained by means of the frequency response data.

Filters for electrical signals are frequency selective. They are classified as low-pass (LP), high-pass (HP), bandpass (BP), and band-reject (BR), shown in Fig. E7-10a. For example, a low-pass filter passes the low-frequency component in a signal, but rejects the high-frequency components. A commercial variable filter, shown in Fig. E7-10b, consists of a HP and a LP section. The type of filters enumerated can be obtained from combinations of the two sections.

The cutoffs, or break frequencies f_1 and f_2, of an active filter (with built-in amplification) can be varied conveniently, but a passive filter (with passive components only) is generally designed for a fixed frequency range. The cutoffs are rounded, and the bandwidth is measured at the -3-dB points, as shown in the figure. The rate of attenuation of an active variable filter is either 24 dB or 48 dB per octave, corresponding to a 4:1 or 8:1 slope in a Bode plot.

B. Effect of Filtering on Waveforms: Experiment 7-1

This experiment is a qualitative study of the effect of filtering on the waveform of a signal. It is followed by a quantitative frequency response study in the next experiment. It was concluded in Expt. 6 that a zero-order instrument has no distortion and a flat response (i.e., the amplitude

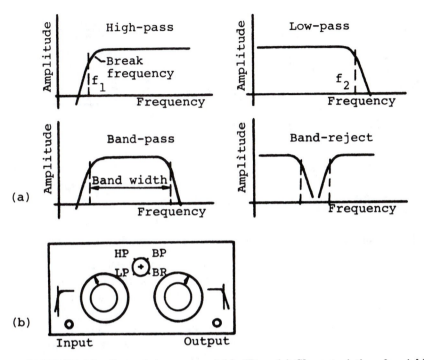

FIGURE E7-10 General-purpose variable filter. (a) Characteristics of variable filter. (b) Schematic of variable filter.

ratio versus frequency plot is horizontal). Hence signal distortions are due to the slopes at the cutoff, as shown in Fig. E7-10a. Due to the phase shift at the cutoff, however, signal distortion is also extended to the flat portion of the response near the cutoff frequencies.

A bandpass filter (see Fig. E7-10a) is used in the experiment in lieu of a separate high-pass and low-pass filter, since the break frequencies can be varied to reveal the effects of both. The procedure is as follows:

1. Set up the equipment as shown in Fig. E7-11. Set (a) the filter at bandpass with the low-frequency cutoff f_1 at 200 Hz and the high-frequency cutoff f_2 at the highest setting, (b) $v_i = 10\,\text{V}_{\text{p-p}}$ square wave at 200 Hz, and (c) the oscilloscope to dc couple to compare the waveforms of v_o from the filter with v_i.
2. Sketch the waveforms of v_i and v_o (See Fig. 4-21.)
3. To observe the effect of the low-frequency cutoff f_1, (a) increase f_1 to 400 Hz, then (b) decrease f_1 to 20 Hz, 2 Hz, until reaching the lowest setting. Sketch v_i and v_o for each of the steps.

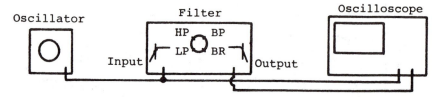

FIGURE E7-11   Test setup for filters.

4. To observe the effect of the high-frequency cutoff f_2, keep f_1 at the lowest setting as before, but decrease f_2 by an order of magnitude in steps until $f_2 = 20$ Hz. Sketch the waveforms of v_i and v_o for each of the steps.

The following discussion topics should be included in the report.

1. What is the percentage change in magnitude at -3 dB?
2. With the aid of a block diagram, show the combination of a low-pass and a high-pass filter to obtain (a) a bandpass filter, and (b) a band-reject filter.
3. A $v_i = 200$ Hz square wave has a fundamental at 200 Hz and all odd harmonics. If the bandwidth of a filter is from 200 Hz to 200 kHz, v_o should be a reproduction of v_i, because the filter passes the fundamental at 200 Hz and the 1000th harmonic. Explain the waveform of v_o observed in step 2 of the procedure.
4. From step 3 of the procedure, explain why the waveform of v_o is improved by lowering the cutoff frequency f_1.
5. Explain the waveform of v_o from step 4 of the procedure, when v_i is a 200-Hz square wave and the high-frequency cutoff f_2 is at 20 Hz.

The procedure to examine a band-reject filter is as follows:

1. Set up the equipment in Fig. E7-11 as before. Set (a) v_i from the oscillator at a 200-Hz sine wave, and (b) the filter at band-reject with the lower cutoff frequency f_2 at 100 Hz and the higher f_1 at 400 Hz.
2. Compare the magnitude of v_i and v_o.

C. Frequency Response of Bandpass Filters: Experiment 7-2

Both amplitude and phase distortion must be considered for a true waveform reproduction in measurements (see Sec. 4-4A). The objectives of this experiment are (1) to obtain the Bode plots of a bandpass filter, and (2) to show that the effect of phase shift at a break frequency can extend far into the flat portion of the amplitude ratio versus frequency response curve.

The phase angle introduced in filtering is examined by means of a frequency response test. Let x be the input to the bandpass filter, shown in Fig. E7-11, and y the corresponding output. The oscilloscope is in the x–y mode. Let $x = X \sin \omega t$ be the reference, $y = Y \sin (\omega t + \phi)$, and the x–y plot of the sinusoids be an ellipse shown in Fig. E7-12a. Beginning at $t = 0$, x starts from $x = 0$ and increases with time. This fixes the initial point $t = 0$ on the ellipse and dictates the value of y at $t = 0$. The ellipse rotates clockwise; therefore, initially y also increases with time. The phase angle ϕ is shown in the figure.

It is expedient to represent the sinusoids by rotating vectors with an angular velocity ω rad/s, shown in Fig. E7-12b. The value of $x = X \sin \omega t$ is the projection of the X vector on the x axis. The ellipse in the x–y plot is constructed from the projections of the vectors at corresponding values of ωt. The ellipse has a clockwise rotation in this example, and $\theta = \sin^{-1}(a'/a)$. Note that θ is related to the actual phase angle ϕ in Table E7-1.

The procedure for the experiment is as follows:

1. Set up the equipment shown in Fig. E7-11. Set (a) the oscilloscope in the X–Y mode with dc couple, (b) the filter at bandpass with f_1 at 4 Hz and f_2 at 2 kHz, and (c) the oscillator at 8-V_{pp} sine wave.
2. Starting the oscillator at about 2 Hz, measure the amplitude ratio Y/X and the phase angle ϕ. Sketch the x–y plot and note the direction of rotation of the ellipse. Deduce the phase angle ϕ from Fig. E7-12 and Table E7-1.
3. Repeat step 2 by increasing the frequency of x. Using Fig. E7-13 as a guide, the increment in frequency should be limited to about 45° change in ϕ. Obtain sufficient data for the Bode plots.

The following discussion topics should be included in the report:

1. Obtain the Bode plots for the bandpass filter. Deduce from the plots the approximate slope of the cutoffs.

TABLE E7-1 Find the Phase Angle ϕ from the X–Y Plot for $x = X \sin \omega t$ and $y = Y \sin (\omega t + \phi)$[a]

y leading x		y lagging x	
$0° < \phi < 90°$	$\phi = 0$	$-360° < \phi < -270°$	$\phi = -(360° - \theta)$
$90° < \phi < 180°$	$\phi = 180° - \theta$	$-270° < \phi < -180°$	$\phi = -(180° + \theta)$
$180° < \phi < 270°$	$\phi = 180° + \theta$	$-180° < \phi < -90°$	$\phi = -(180° - \theta)$
$270° < \phi < 360°$	$\phi = 360° - \theta$	$-90° < \phi < 0°$	$\phi = -\theta$

[a]Where $\theta = \sin^{-1}(a'/a)$, see Fig. E-7-12.

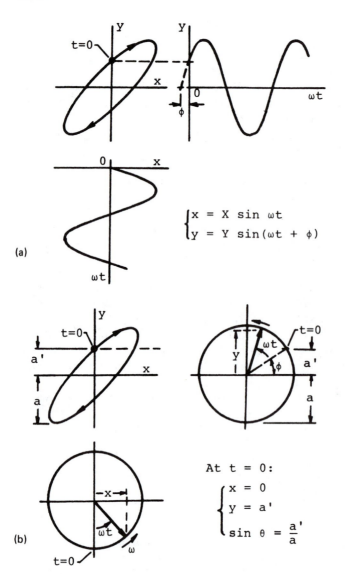

$$\begin{cases} x = X \sin \omega t \\ y = Y \sin(\omega t + \phi) \end{cases}$$

(a)

(b)

At $t = 0$:

$$\begin{cases} x = 0 \\ y = a' \\ \sin \theta = \dfrac{a'}{a} \end{cases}$$

FIGURE E7-12 Deducing phase angle θ from x–y plot. (a) x and y as sine waves. (b) x and y as rotating vectors.

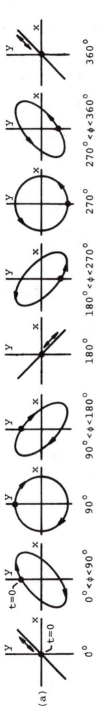

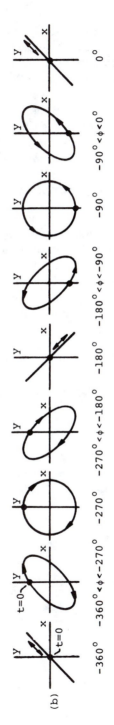

FIGURE E7-13 X–y plots of sinusoidal functions of same frequency. (a) X–y plot: phase angle ϕ for y leading x. (b) X–y plot: phase angle ϕ for y lagging x.

2. If it is specified that $\phi \leqslant 10°$ in a measurement, what is the usable frequency range of this bandpass filter?
3. Specify the conditions for true waveform reproduction when an instrument has a phase shift. When the phase angle ϕ is positive, is it possible to have true waveform reproduction?
4. It was stated in Expt. 6 that a $+1$ slope in the Bode plot corresponds to a differentiation in the time domain and a -1 slope to an integration. Explain the waveforms in Expt. 7-1 from the viewpoint of differentiation and integration.

E7-6. Operational Amplifiers

A. Introduction

Operational amplifiers, or op-amps, are extensively used in instruments (see Secs. 6-2 to 6-4). The objective of this study is to show applications of the op-amp as an inverting voltage amplifier and as a preamplifier.

An integrated-circuit IC op-amp in a dual-in-line package (DIP) is shown in Fig. E7-14. The op-amp is a differential amplifier with inverting input V_- at pin 2 and noninverting input V_+ at pin 3. Pin 6 is the output V_o. Inverting means that a positive input to pin 2 gives a negative v_o at pin 6. The power supply is typically $\pm15\ V_{dc}$ applied at $+V_{cc}$ and $-V_{cc}$. Pin 8 is not connected. Pins 1 and 5 are for trimming and op-amp when necessary.

An op-amp is a high-gain differential amplifier. The output voltage V_o is

$$V_o = A(V_+ - V_-) \qquad\qquad\qquad\qquad (E7\text{-}8)$$

where A is the open-loop gain, typically 10^6. Since the magnitude of V_o is about 10 V, the differential voltage is $(V_+ - V_-) = V_o/A = 10/10^6 \simeq 0$. In other words, the high gain A forces $V_- \simeq V_+$.

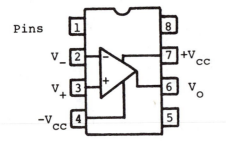

FIGURE E7-14 IC operational amplifier in a DIP package.

B. Inverting Amplifier: Experiment 7-3

The circuit for an op-amp as an inverting amplifier is shown in Fig. E7-15. Summing the currents at node a gives

$$I_1 = I_f + I_2 \quad \text{or} \quad I_1 \simeq I_f \tag{E7-9}$$

The current $I_2 \simeq 0$, because the op-amp has a high input impedance. Since V_+ is at ground potential and $V_+ \simeq V_-$, V_a is a "virtual" ground (i.e., node a is almost at the ground potential but is not connected to ground). From Eq. E7-9 we get the basic op-amp equation

$$\frac{V_i - 0}{R_1} = \frac{0 - V_o}{R_f} \quad \text{or} \quad \frac{V_o}{V_i} = -\frac{R_f}{R_1} \tag{E7-10}$$

The procedure for the experiment is as follows:

1. Breadboard the circuit shown in Fig. E7-15, with $R_1 = 10$ kΩ and $R_f = 20$ kΩ. From Eq. E7-10 we get

$$\frac{V_o}{V_i} = -\frac{2}{1} \tag{E7-11}$$

2. Set the oscillator at a 2-V_{p-p} sine wave and 1 kHz. Observe the input and output voltages with an oscilloscope at dc couple and verify Eq. E7-11.
3. Set the oscillator at a 2-V_{p-p} sine wave and 1 kHz as before, but with a +4 V_{dc} bias. Observe the input and output voltages of the amplifier.
4. Increase the bias voltage until "clipping" occurs at the output. Note the maximum voltage at which clipping occurs.
5. Set the oscillator at a 2-V_{p-p} square wave and 1 kHz, but without the dc bias. Compare the input and output voltages.
6. Set the oscillator at a 2-V_{p-p} sine wave and the oscilloscope in the X–Y mode. Starting from 1 Hz, measure the output/input amplitude

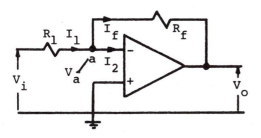

FIGURE E7-15   Inverting amplifier.

ratio and the phase angle for the frequencies of 10 Hz, 100 Hz, and so on, until the phase angle is about 20°. Note the frequency at which the phase shfit occurs.

The following discussion topics should be included in the report:

1. Does the procedure above show that an op-amp is a dc amplifier?
2. What is the significance of the "clipping" in step 4 of the procedure?
3. What is the usable frequency range of this amplifier if the phase shift between V_i and V_o should not exceed 20° (see Sec. 6-4)?
4. What is the input impedance of the inverting amplifier in Fig. E7-15?

C. Preamplifier: Experiment 7-4

Due to its high input impedance and low output impedance, an op-amp can be used as a voltage follower or as a preamplifier for impedance matching. Consider the noninverting circuit in Fig. E7-16a. R_1 and R_f form a voltage divider for V_o, that is,

$$\frac{V_a}{V_o} = \frac{R_1}{R_1 + R_f} \tag{E7-12}$$

Let $V_i = V_b$. Since the op-amp is a high-gain differential amplifier, we deduce from Eq. E7-8 that

$$V_i = V_b \simeq V_a \tag{E7-13}$$

Substituting $V_i = V_a$ into Eq. E7-12 and simplifying, the closed-loop gain of the amplifier is

$$V_o = \left(1 + \frac{R_f}{R_1}\right)V_i \tag{E7-14}$$

If R_1 becomes infinite, or an open circuit, we get

$$V_o = V_i \tag{E7-15}$$

where R_f is any finite arbitrary resistance. The preamplifier circuit in Fig. E7-16b uses $R_f = 0$.

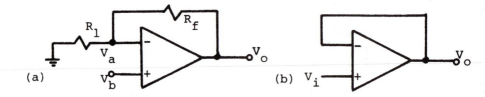

FIGURE E7-16 Voltage follower. (a) Noninverting circuit. (b) Preamplifier.

The procedure for the experiment is as follows:

1. Breadboard the circuit shown in Fig. 7-16b. Set V_i from the oscillator at a 2 V_{p-p} sine wave and 1 kHz. Observe the waveforms of V_o and V_i with an oscilloscope. Repeat V_i for a square wave and then a sawtooth wave.
2. Set the oscillator at a 2-V_{p-p} sine wave and the oscilloscope in the X–Y mode. Starting from 0.2 Hz, measure the output/input magnitude ratio and phase angle for frequencies 2 Hz, 20 Hz, 200 Hz, and so on, until the phase angle is about 20°. Note the frequency at which this phase angle occurs.
3. Set V_i at a 2-V_{p-p} sine wave and 1 kHz. Estimate the input resistance of the preamplifier, using a resistor of about 5 MΩ at the input (see Fig. E1-8).
4. Estimate the output resistance of the preamplifier using a resistance decade box at the output (see Fig. E1-7).

The following discussion topics should be in the report:

1. From the observations in step 1 above, comment on the preamp as a voltage follower, as described in Eq. E7-15.
2. What is the usable frequency range for this follower if the criterion is a 20° phase shift?
3. A preamplifier should have high input impedance and low output impedance. What are the impedance values estimated from steps 3 and 4 in the procedure?

Appendix A
Review of Electrical Networks

A-1. DIRECT-CURRENT CIRCUITS

The basic laws for circuit analysis are Ohm's law and Kirchhoff's voltage and current laws. This appendix first reviews direct-current (dc) circuits with resistors. Then capacitors and inductors are introduced as circuit elements, and the techniques for dc circuit analysis are extended for alternating-current (ac) circuits.

A. Ohm's Law: Resistors in Series and in Parallel

Series and parallel circuits are examined in this section. This leads directly to the useful concepts of voltage and current dividers.

Ohm's law gives the voltage–current relation of a component and it is therefore a component equation. The dc circuit in Fig. A-1a has a battery V, a switch S, and a resistor R. Upon closing the switch, a current I flows through R, as shown in Fig. A-1b. From Ohm's law, we get

$$V = RI \qquad (A\text{-}1)$$

where V is in volts V, I in amperes A, and R in ohms Ω. The current I flows from a positive to a negative potential. The voltage across R is the IR *drop*. A voltage drop is always in the direction of the current flow. A *voltage rise* from b to a is written as V_{ba}, and the arrow denotes a voltage rise.

Power is the time rate of work. The power dissipation P in a resistor R appears as heat.

$$P = VI \qquad (A\text{-}2)$$

The product of volts and amperes has the dimension of joules per second J/s or watts (W). Substituting Eq. A-1 into Eq. A-2 gives

$$P = I^2 R = \frac{V^2}{R} \qquad (A\text{-}3)$$

695

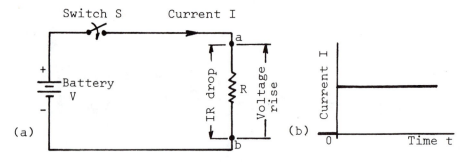

FIGURE A-1 (a) Dc circuit. (b) Current through R.

The resistors R_1 and R_2 in Fig. A-2a are in series, and the same current flows through components in series. The IR drop across R_1 is $V_1 = R_1 I$ and that across R_2 is $V_2 = R_2 I$. Their sum $(V_1 + V_2)$ is the voltage V of the battery.

$$V = V_1 + V_2 \tag{A-4}$$

$$V = R_1 I + R_2 I = (R_1 + R_2)I = R_{eq} I \tag{A-5}$$

$$R_{eq} = \sum_k R_k \quad \text{resistors in series} \tag{A-6}$$

The equivalent resistance R_{eq} is the sum of the individual resistors in series.

The voltage V is divided into V_1 and V_2, and R_1 and R_2 form a *voltage divider*. The voltages in a voltage divider are directly proportional to the values of the resistors.

$$\frac{V_1}{V_2} = \frac{R_1 I}{R_2 I} = \frac{R_1}{R_2} \quad \text{or} \quad \frac{V_1}{V} = \frac{R_1}{R_1 + R_2} \tag{A-7}$$

The series circuit in Fig. A-2 forms a loop. An analogy is a hydraulic loop,

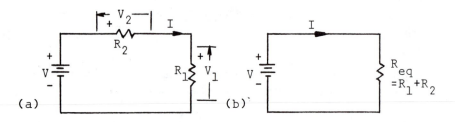

FIGURE A-2 (a) Series circuit. (b) Equivalent circuit.

consisting of a pump and two valves. V is analogous to the pump pressure, and the R's the flow resistance of the valves. The analogy of Ohm's law is that the pressure drop across a valve is proportional to the flow rate through the valve. The valves form a pressure divider, and the pump pressure is the sum of the pressure drops across the valves.

The resistors R_1 and R_2 in Fig. A-3a are in parallel. The same voltage is applied across components in parallel. The sum of the currents I_1 and I_2 is the source current I.

$$I = I_1 + I_2 \tag{A-8}$$

$$I = \frac{V}{R_1} + \frac{V}{R_2} = \left(\frac{1}{R_1} + \frac{1}{R_2}\right)V = \frac{V}{R_{eq}} \tag{A-9}$$

$$\frac{1}{R_{eq}} = \sum_k \frac{1}{R_k} \tag{A-10}$$

The reciprocal of the equivalent resistance R_{eq} is the sum of the reciprocal of the individual resistors in parallel.

The current I in Fig. A-3a is divided into I_1 and I_2. R_1 and R_2 form a *current divider,* and the currents through a current divider are inversely proportional to the values of the resistors.

$$\frac{I_1}{I_2} = \frac{V/R_1}{V/R_2} = \frac{R_2}{R_1} \quad \text{or} \quad \frac{I_1}{I} = \frac{R_2}{R_1 + R_2} \tag{A-11}$$

Example A-1. Find the current and the IR drop for each of the circuit elements in Fig. A-4.

Solution: Resistors in series and in parallel are applied successively to the circuit. In Fig. A-5a, the parallel combination of R_2 and R_3 is 25 Ω and that of R_4 and R_5 is 200 Ω. In Fig. A-5b, R_1 is in parallel with 625 Ω. The total resistance of the circuit in Fig. A-5c is 243.9 Ω. The corresponding current is $I = 100/243.9 = 0.41$ A.

The currents and the IR drops of the resistors are obtained by reversing the procedure above. In Fig. A-5b, the 625-Ω equivalent resistance and R_1

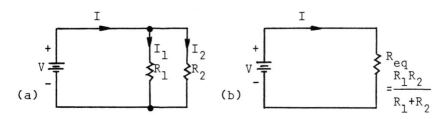

FIGURE A-3 (a) Parallel circuit. (b) Equivalent circuit.

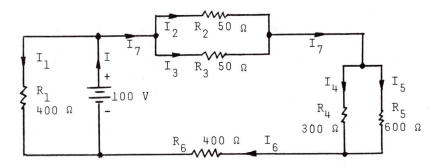

FIGURE A-4 Series–parallel network.

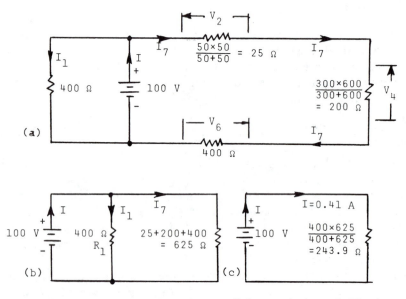

FIGURE A-5 Reduction of the series–parallel network shown in Fig. A-4.

form a current divider. From Eq. A-11, the currents are $I_1 = 0.41 \times 625/(625 + 400) = 0.25$ A and $I_7 = 0.25 \times 400/625 = 0.16$ A. The voltage drop across R_1 is 100 V, which is the source voltage.

In Fig. A-4, $R_2 = R_3$; therefore, $I_2 = I_3 = I_7/2 = 0.08$ A. The IR drop across R_2 and R_3 in parallel is $0.08 \times 50 = 4$ V. R_4 and R_5 form a current divider for I_7, and the currents are $I_4 = 0.16 \times 600/(600 + 300) = 0.107$ A and $I_5 = 0.107 \times 300/600 = 0.053$ A. The drop across R_4 and R_5 in parallel is $0.053 \times 600 = 32$ V. From $I_6 = I_7 = 0.16$ A, the IR drop across R_6 is $0.16 \times 400 = 64$ V.

B. Kirchhoff's Laws

Simple circuits can be analyzed by the successive applications of Ohm's law and the series–parallel technique. More complex circuits are solved systematically by means of Kirchhoff's voltage and current laws.

Kirchhoff's voltage law states that around a closed loop (path, mesh) in an electric network, the algebraic sum of the instantaneous voltage drops (or rises) is zero:

$$\sum_k V_k(t) = 0 \qquad \text{around a closed loop} \qquad \text{(A-12)}$$

where $V_k(t)$ is an instantaneous voltage drop (or rise) along a circuit component or along a path in the loop. For example, the sum of the voltage drops in the loop in Fig. A-2a in the clockwise direction is $V_2 + V_1 - V = 0$.

Kirchhoff's current law states that at a common junction or node in an electric network, the algebraic sum of the instantaneous currents flowing into (or away from) the node is zero:

$$\sum_k I_k(t) = 0 \qquad \text{at a common node} \qquad \text{(A-13)}$$

where $I_k(t)$ is an instantaneous current entering (or leaving) the node. For Fig. A-3a, the current law gives $I - I_1 - I_2 = 0$.

Example A-2. Apply Kirchhoff's voltage law to find the current I_5 in the bridge network in Fig. A-6.
Solution: Select the three loops as shown in the figure. The corresponding loop currents are I_1, I_2, and I_3.

In loop 1, the current through a typical resistor R_1 in the direction of I_1 is $(I_1 - I_2)$. The IR drop across R_1 in loop 1 is $R_1(I_1 - I_2)$. Similarly, the IR drop across R_4 in the direction of I_1 is $R_4(I_1 - I_3)$. Summing the IR drops in loop 1 in the clockwise direction gives

$$R_1(I_1 - I_2) + R_4(I_1 - I_3) - V = 0$$

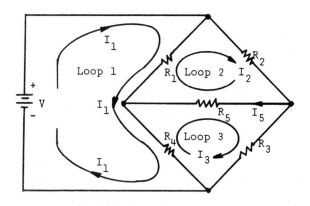

FIGURE A-6   Resistance bridge with a constant-voltage source.

Applying Kirchhoff's voltage law to all three loops shown in the figure, we get

Loop 1: $R_1(I_1 - I_2) + R_4(I_1 - I_3) - V = 0$
Loop 2: $R_1(I_2 - I_1) + R_5(I_2 - I_3) + R_2I_2 = 0$
Loop 3: $R_4(I_3 - I_1) + R_5(I_3 - I_2) + R_3I_3 = 0$

The equations are rearranged to solve for the I's.

$(R_1 + R_4)I_1 - \qquad\qquad R_1I_2 - \qquad\qquad\qquad R_4I_3 = V$
$\qquad -R_1I_1 + (R_1 + R_2 + R_5)I_2 - \qquad\qquad R_5I_3 = 0$
$\qquad -R_4I_1 - \qquad\qquad R_5I_2 + (R_3 + R_4 + R_5)I_3 = 0$

The simultaneous equations can be solved for I_2 and I_3, and the desired current is $I_5 = I_2 - I_3$.

$$I_5 = \frac{V(R_1R_3 - R_2R_4)}{\begin{vmatrix} R_1 + R_4 & -R_1 & -R_4 \\ -R_1 & R_1 + R_2 + R_5 & -R_5 \\ -R_4 & -R_5 & R_3 + R_4 + R_5 \end{vmatrix}}$$

Two observations are made from the example above.

1. For simple networks, the choice of loops is by inspection. An analysis to determine the number of independent loops is unnecessary for the type of problems contemplated.
2. The loop currents are the unknowns in a loop analysis. The directions of the loop currents are assigned arbitrarily. If the direction of

a current is "incorrectly" assigned, its calculated value will be negative.

Example A-3. A bridge network in Fig. A-7 is driven by a constant-current source I. Find the current I_5.

Solution: Using node d as the reference (ground), the network has three independent nodes. The node equations are

Node a: $I_1 + I_5 - I_4 = 0$

Node b: $I - I_1 - I_2 = 0$

Node c: $I_2 - I_5 - I_3 = 0$

The voltages V_a, V_b, and V_c are measured relative to V_d (ground). The current through the resistor R_1 is $(V_b - V_a)/R_1$. Considering all three nodes, we obtain

Node a: $\dfrac{V_b - V_a}{R_1} + \dfrac{V_c - V_a}{R_5} - \dfrac{V_a - 0}{R_4} = 0$

Node b: $\dfrac{V_b - V_a}{R_1} + \dfrac{V_b - V_c}{R_2} \qquad = I$

Node c: $\dfrac{V_b - V_c}{R_2} - \dfrac{V_c - V_a}{R_5} - \dfrac{V_c - 0}{R_3} = 0$

The equations can be rearranged to solve for the unknown voltages. It can be shown that $I_5 = (V_c - V_a)/R_5$.

Two observations are made from the node analysis above.

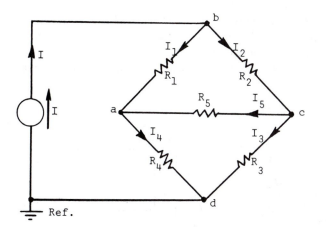

FIGURE A-7 Resistance bridge with a constant-current source.

1. For simple networks, the choice of nodes is by inspection.
2. The node voltages are the unknowns in a node analysis. The direction of node currents can be assigned arbitrarily. The calculated value will be negative, if the direction of a node current is "incorrectly" assigned.

C. Equivalent Networks

Equivalent network is an extremely useful concept for examining the performance of instruments. Thévenin's and Norton's theorem are described in this section.

Two networks are *equivalent* if one cannot be distinguished from the other by measurements at their terminals. For example, assume that a network A is connected to a network B to yield certain voltage–current relation in B. Network A and a network C are equivalent if C is substituted for A and the voltage–current relation in B remains unchanged. The equivalence is defined by means of the external performance of the network. Internally, the A and C may have different components.

Thévenin's theorem replaces the network in Fig. A-8a by means of its equivalent in Fig. A8-b. The equivalent voltage V_{eq} is the open-circuit voltage V_{oc}, and the output impedance R_o is the impedance of the network, measured across terminals a and b with all the energy sources in the network removed. A voltage source in a network is removed by replacing with a short circuit, and a current source is removed by replacing with an open circuit. In other words, R_o can be measured with an ohmmeter after all the energy sources in the network are removed.

Example A-4. (a) Find the Thévenin's equivalent of the circuit in Fig. A-9a. (b) If a resistor R_L is connected to the circuit, find the voltage and current of R_L. (c) Repeat part (b) without using Thévenin's theorem. Solution: (a) When R_L is not connected to the circuit, R_1 and R_2 form a voltage divider for V. The voltage across R_2 is the open-circuit voltage V_{oc}.

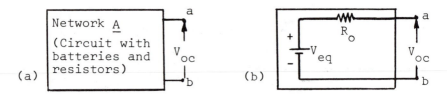

FIGURE A-8 (a) Electrical network. (b) Thévenin's equivalent.

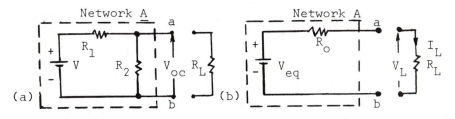

FIGURE A-9 (a) Electrical network. (b) Thévenin's equivalent.

$$V_{oc} = \frac{R_2}{R_1 + R_2} V$$

$V_{eq} = V_{oc}$ by Thévenin's theorem. To find the output impedance R_o, the battery V is removed by replacing with a short circuit. Thus R_1 and R_2 are in parallel, and

$$R_o = \frac{R_1 R_2}{R_1 + R_2}$$

The equivalent network in Fig. A-9b consists of V_{eq} and R_o.

(b) When R_L is connected to the equivalent network in Fig. A-9b, we get

$$V_{eq} = (R_o + R_L)I_L \quad \text{and} \quad V_L = R_L I_L$$

Substituting R_o and $V_{eq} = V_{oc}$ from above, we obtain

$$I_L = \frac{R_2}{R_1 R_2 + R_L(R_1 + R_2)} V \quad \text{and} \quad V_L = \frac{R_2 R_L}{R_1 R_2 + R_L(R_1 + R_2)} V$$

(c) When R_L is connected to a and b in Fig. A-9a, R_L and R_2 are in parallel. Their equivalent is $R_{eq} = R_2 R_L/(R_2 + R_L)$. Since R_1 and R_{eq} form a voltage divider for V, the voltage across R_{eq} is V_L.

$$V_L = \frac{R_{eq}}{R_1 + R_{eq}} V = \frac{R_2 R_L}{R_1 R_2 + R_L(R_1 + R_2)} V$$

$$I_L = \frac{V_L}{R_L} = \frac{R_2}{R_1 R_2 + R_L(R_1 + R_2)} V$$

The answers in parts (b) and (c) are identical, and the circuits in Fig. A-9a and b are equivalent.

Let us give a heuristic proof of Thévenin's theorem. Network A in Fig. A-10a consists of resistors and batteries, and network B consists of resistors only. The steps are as follows:

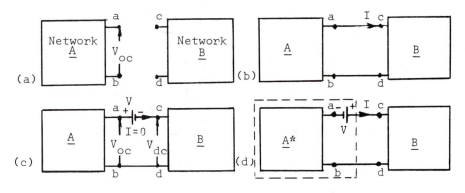

FIGURE A-10 Steps to establish Thévenin's equivalent network.

1. When network A is connected to B in Fig. A-10b, a current I flows from A to B.

2. Insert a battery V between A and B in Fig. A-10c to render $I = 0$. With zero current flow, there is no voltage drop across network B, and the voltage $V_{dc} = 0$. For the loop b–a–c–d, we have $V_{ba} - V + 0 = 0$, or $V_{ba} = V$. Furthermore, when $I = 0$, V_{ba} is the open-circuit voltage of network A; that is, $V_{ba} = V_{oc} = V = V_{eq}$.

3. Now, replace network A by A^*, as shown in Fig. A-10d. Network A^* is derived from A by removing all the energy sources in A. In other words, A^* consists of only the resistors in A. This is R_o of the network A by Thévenin's theorem. The combination of R_o in series with V in *reversed polarity* is enclosed in dashed lines in Fig. A-10d. Since V stops the current flow in Fig. A-10c, reversing its polarity and removing the energy sources in A will restore the current I in Fig. A-10b. Thus the Thévenin's equivalent of A consists of R_o and $V_{eq} = V$.

The *Norton's equivalent* in Fig. A-11a consists of a current source I_{eq} in parallel with R_o. The resistor R_o is the output resistance by Thévenin's theorem. The current source I_{eq} is the short-circuit current I_{sc} across output terminals a and b. Let us derive Norton's equivalent from the Thévenin's equivalent. If terminals a and b in Fig. A-11b are shorted, the current flowing in the short circuit is

$$I_{eq} = \frac{V_{eq}}{R_o} \tag{A-14}$$

If terminals a and b in Fig. A-11a are shorted, the entire current I_{eq} will flow through the short circuit. Thus $I_{sc} = I_{eq}$ by Norton's theorem, and the two networks in Fig. A-11 are equivalent.

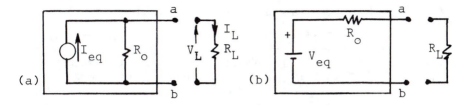

FIGURE A-11 Comparison of Norton's and Thévenin's theorem. (a) Norton's equivalent. (b) Thévenin's equivalent.

For example, let R_L be connected across terminals a and b in Fig. A-11a. The current I_L and the voltage V_L are

$$I_L = \frac{R_o}{R_L + R_o} I_{eq} \quad \text{and} \quad V_L = R_L I_L = \frac{R_L R_o}{R_L + R_o} I_{eq}$$

Substituting $I_{eq} = V_{eq}/R_o$ from Eq. A-14, we get

$$I_L = \frac{1}{R_L + R_o} V_{eq} \quad \text{and} \quad V_L = \frac{R_L}{R_L + R_o} V_{eq}$$

It can be shown readily that the values of I_L and V_L are identical if R_L is connected across terminals a and b in Fig. A-11b for the Thévenin's equivalent.

A-2. CIRCUIT ELEMENTS

The capacitor C and the inductor L are introduced as additional circuit elements in this section for ac circuit analysis. Circuit elements are either active or passive, as shown in Table A-1. *Active elements* are sources of energy and are idealized as voltage and current sources. A voltage source $v(t)$ maintains its voltage across the terminals, independent of the current flow. The $(\pm)$ signs indicate the instantaneous polarity of $v(t)$. A current source $i(t)$ maintains its current through the element independent of the voltage across its terminals. The arrow indicates the direction of the current flow at the instant when $i(t)$ is positive. The analogy of a voltage source is a constant-pressure pump, and that of a current source is a constant-delivery pump. For convenience, the voltage and current sources are indicated by $v(t)$ and $i(t)$, and the symbols v and i are used for instantaneous values in the circuit.

Passive elements are the resistor R in ohms Ω, capacitor C in farads F, and inductors L in henrys H. The resistor R is an energy-dissipating element. The voltage drop across R in the direction of current flow is

TABLE A-1 Circuit Elements and Voltage–Current Relations

Element	Symbol	Voltage	Current
Voltage source	$-\ \bigcirc\ +$ $\ v(t) \rightarrow$	Voltage independent of current flow	Current depends on network
Current source	$\bigcirc$ $i(t)$	Voltage depends on network	Current independent of voltage across terminals
Resistor	$+\ i\ R\ -$ $\overset{}{\longrightarrow}\!\!\text{W}\!\!\longrightarrow$ v	$v = Ri$	$i = \dfrac{v}{R}$
Inductor	$+\ i\ L\ -$ v	$v = L\,\dfrac{di}{dt}$	$i = \dfrac{1}{L}\displaystyle\int v\,dt$
Capacitor	$+\ i\ C\ -$ v	$v = \dfrac{1}{C}\displaystyle\int i\,dt$	$i = C\,\dfrac{dv}{dt}$

$$v = Ri \quad \text{or} \quad i = \frac{v}{R} \tag{A-15}$$

The power dissipation P in a resistor R in J/s or watts is

$$P = vi = i^2R = \frac{v^2}{R} \tag{A-16}$$

The capacitor C is an energy storage element. The voltage across C is proportional to the accumulation of electric charge. Since current i is the rate of flow of electric charge, the voltage or potential across C is

$$v = \frac{1}{C}\int i\,dt \quad \text{or} \quad i = C\frac{dv}{dt} \tag{A-17}$$

By analogy, the head or the hydraulic potential in a water tank is proportional to the accumulation of water in the tank. The energy stored in C in joules is

$$J = \int vi\,dt = \int v\left(C\frac{dv}{dt}\right)dt = \frac{C}{2}v^2 \tag{A-18}$$

Equation A-18 shows that the energy stored in C depends on the potential level v, and C stores potential energy.

The inductor L is also an energy storage element. The voltage drop across L in the direction of current flow is

$$v = L \frac{di}{dt} \quad \text{or} \quad i = \frac{1}{L} \int v \, dt \tag{A-19}$$

where i is the rate of flow of electric charge. By analogy, L is analogous to a mass, i to a velocity, di/dt to an acceleration, and v to a force. The energy stored in an inductor L in joules is

$$J = \int vi \, dt = L \int \frac{di}{dt} i \, dt = \frac{L}{2} i^2 \tag{A-20}$$

The equation shows that the energy stored in L depends on a rate, and L stores kinetic energy. C and L are called *reactive elements,* since energies are only stored in the elements and not dissipated.

Mutual inductance M appears as a parameter in networks, but M is not a circuit element. When two coils are adjacent to one another, they are coupled by flux linkages. The change in magnetic flux, due to a change in the current i in one coil, will induce a voltage v in the other coil, and vice versa, that is,

$$v_2 = M \frac{di_1}{dt} \quad \text{and} \quad v_1 = M \frac{di_2}{dt} \tag{A-21}$$

The transformer in Fig. A-12 is an application of mutual inductance. A transformer consists of a primary winding and a secondary winding on an iron core. In an ideal transformer, all the magnetic flux from the primary winding is intercepted by the secondary. The voltage and current relations of the coils are

$$\frac{v_1}{v_2} = \frac{n_1}{n_2} \quad \text{and} \quad \frac{i_1}{i_2} = \frac{n_2}{n_1} \tag{A-22}$$

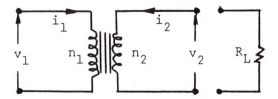

FIGURE A-12 Schematic of a transformer.

where n_1 and n_2 are the number of turns of the respective coils. If a resistor R_L is connected across the terminals of the secondary winding, the secondary current is $i_2 = v_2/R_L$. Substituting this in Eq. A-22 and simplifying, we get

$$i_1 = \left(\frac{n_2}{n_1}\right)^2 \frac{v_1}{R_L} = \frac{v_1}{R_{eq}} \qquad \text{or} \qquad R_{eq} = \left(\frac{n_1}{n_2}\right)^2 R_L \qquad (A\text{-}23)$$

A-3. GENERAL CIRCUIT ANALYSIS

Applications of Kirchhoff's voltage and current laws for general circuit analysis are illustrated briefly in this section. The objective is to give a background for ac circuit analysis. The equations for the analysis are differential equations with time as the independent variable. Since the transient solutions from the differential equations in circuit analysis are of secondary interest in this brief review, the examples to follow will show only the formulation of the problem.

Example A-5.   (a) Write the loop equation for the RLC circuit shown in Fig. A-13. (b) Modify the equation to yield v_o as the dependent variable.

Solution: (a) Applying Kirchhoff's voltage law from Eq. A-12 for the voltages across each of the circuit elements, we get

$$L \frac{di}{dt} + Ri + \frac{1}{C} \int i \, dt = v(t) \qquad (A\text{-}24)$$

(b) To use v_o as the dependent variable, we substitute

$$v_o = \frac{1}{C} \int i \, dt \qquad i = C \frac{dv_o}{dt} \qquad \frac{di}{dt} = C \frac{d^2 v_o}{dt^2}$$

into Eq. A-24 and obtain

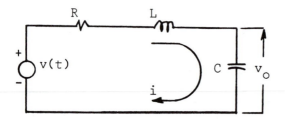

FIGURE A-13   One-loop ac network.

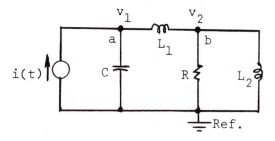

FIGURE A-14   Two-pole pair network.

$$LC \frac{d^2 v_o}{dt^2} + RC \frac{dv_o}{dt} + v_o = v(t)$$

Example A-6.   Write the node equations for the network shown in Fig. A-14.

Solution: Let the voltages v_1 and v_2 be measured relative to the reference node (ground). Applying Kirchhoff's current law from Eq. A-13 to each of the nodes and rearranging, we obtain

Node a: $C \dfrac{dv_1}{dt} + \dfrac{1}{L_1} \displaystyle\int (v_1 - v_2)\, dt = i(t)$

Node b: $\dfrac{1}{R} v_2 + \dfrac{1}{L_1} \displaystyle\int (v_2 - v_1)\, dt + \dfrac{1}{L_2} \displaystyle\int v_2\, dt = 0$

A-4.   ALTERNATING-CURRENT CIRCUITS

Ohm's law and Kirchhoff's voltage and current laws from dc analysis are extended for the study of alternating current circuit in this section. Since ac is a steady-state sinusoidal analysis, sinusoidal functions are reviewed before the discussion of ac circuits.

A.   Sinusoidal Functions

A sinusoidal function is a sine or a cosine or their combinations. Consider an ac voltage

$$v = V_p \sin \omega t \tag{A-25}$$

where V_p is the peak value or the amplitude of the sinusoidal voltage v, ω the circular frequency in rad/s, and t the time in seconds. The instantaneous value of v is denoted by the length OP along a straight line shown in Fig. A-15a. To a stationary observer, v simply changes in value between

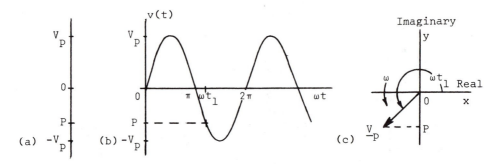

FIGURE A-15 Representations of $v(t) = V_p \sin \omega t$.

V_p and $-V_p$. Although v is a time function, the observer cannot see the past or the future values of v. Hence a sinusoidal function is in fact a "reciprocating" time function.

For the purpose of visualization, v is plotted versus time t or ωt shown in Fig. A-15b. The voltage can be displayed as a trace on the screen of an oscilloscope. The trace observed is due to the repetitive display, the persistency of vision, and the nature of the phosphorescent screen of the scope. The plot is an artifice for purpose of visualization only.

For purpose of manipulation, a sinusoidal function is presented as a rotating *vector* of constant magnitude and circular frequency ω, as shown in Fig. A-15c. The vectorial representation of an ac voltage is

$$v = \mathbf{V}_p = \bar{V}_p e^{j\omega t} = \underbrace{(V_p e^{j\alpha})}_{\text{phasor}} e^{j\omega t} \qquad (A\text{-}26)$$

where $\mathbf{V}_p$ is a rotating vector of magnitude $V_p = |\mathbf{V}_p|$, and $\bar{V}_p = V_p e^{j\alpha}$ is a *phasor*. Thus an ac voltage is denoted by means of the product of a phasor and $e^{j\omega t}$. A sine or a cosine function can be represented by the vector $e^{j\omega t}$ ($= \cos \omega t + j \sin \omega t$). For example, when $v = V_p \sin \omega t$, the value of v at any time is the projection of the vector $\mathbf{V}_p$ on the imaginary or the y axis. If $v = V_p \cos \omega t$, the value of v is the projection of $\mathbf{V}_p$ on the real or the x axis. Note that the representation in a complex plane in Fig. A-15c is for the convenience of manipulation only. In the real world, voltages, currents, and all physical quantities must be real.

The phasor $V_p e^{j\alpha}$ is a complex quantity, where α is its phase angle relative to a reference vector. In phasor notation, a quantity is often written without the subscript p and the bar over the letter. For example, the phasor $\bar{V}_p$ is often written as V.

A *phase angle* indicates the relative time between two events having the same frequency. For example, the phase angle ϕ between the ac voltages v_1 and v_2 of the same frequency ω is illustrated in Fig. A-16a. Let

$$v_1 = V_1 \sin \omega t$$
$$v_2 = V_2 \sin (\omega t - \phi) = V_2 \sin \omega(t - t_\phi) \tag{A-27}$$

where $t_\phi = \phi/\omega$ has the dimension of seconds. The phase angle ϕ in Eq. A-27 shows the *time lead* or *time lag* between the voltages v_1 and v_2. In this case v_1 leads v_2 by the time t_ϕ, or v_2 lags v_1 by the same amount of time. Due to the relative time between the two voltages, their traces do not cross the zero axis at the same time, or their maxima do not occur at the same time. The phase angle can also be conveniently shown in Fig. A-16b as an angle between two vectors $\mathbf{V}_1$ and $\mathbf{V}_2$. Again, this representation of phase angle is only a convenience. The physical interpretation of a phase angle is a time lead or lag. Note that a phase angle exists only between two periodic functions of the same frequency, and the phase angle between two periodic functions of different frequencies is meaningless.

Let us illustrate the ease of manipulating with vectors.

$$v_1 = V_1 \sin (\omega t + \alpha) \triangleq \mathbf{V}_1 = \bar{V}_1 e^{j\omega t} = (V_1 e^{j\alpha}) e^{j\omega t}$$
$$v_2 = V_2 \sin (\omega t + \beta) \triangleq \mathbf{V}_2 = \bar{V}_2 e^{j\omega t} = (V_2 e^{j\beta}) e^{j\omega t}$$

where $\mathbf{V}_1$ and $\mathbf{V}_2$ are vectors, α and β are their phase angles, and $\bar{V}_1$ and $\bar{V}_2$ are the phasors. The operations that follow are self-evident.

$$\frac{d}{dt} (\bar{V}_1 e^{j\omega t}) = j\omega(\bar{V}_1 e^{j\omega t}) \tag{A-28}$$

$$\int (\bar{V}_1 e^{j\omega t}) \, dt = \frac{1}{j\omega} (\bar{V}_1 e^{j\omega t}) \tag{A-29}$$

$$\bar{V}_1 \times \bar{V}_2 = (V_1 e^{j\alpha})(V_2 e^{j\beta}) = (V_1 \times V_2) e^{j(\alpha+\beta)} \tag{A-30}$$

$$\frac{\bar{V}_1}{\bar{V}_2} = \frac{V_1 e^{j\alpha}}{V_2 e^{j\beta}} = \frac{V_1}{V_2} e^{j(\alpha-\beta)} \tag{A-31}$$

The time derivative of a vector is the product of $j\omega$ and the vector itself. The integration is the vector divided by $j\omega$. Equations A-30 and A-31 are statements of the product and quotient of complex numbers.

For convenience, certain terms are often used to describe the magnitude of sinusoidal functions. If $v = V_p \sin \omega t$, the *peak value* is V_p. If the trace of a sine wave is displayed on an oscilloscope, it is convenient to read the *peak-to-peak* (V_{p-p}) *value,* as shown in Fig. A-17a.

An ac voltmeter generally indicates the nominal *root-mean-square* (rms) *value,* which is defined as

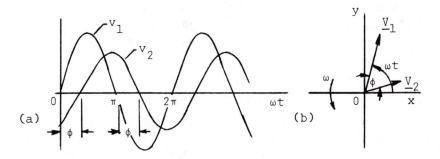

FIGURE A-16 Phase angle between two ac voltages.

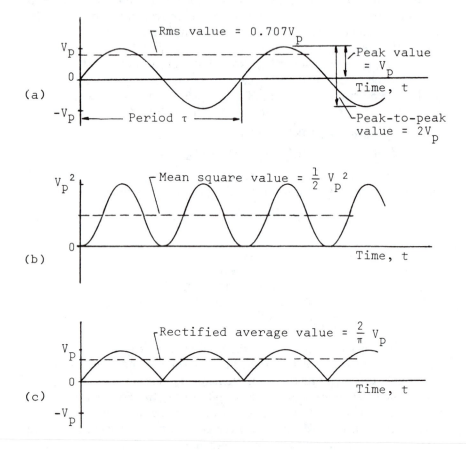

FIGURE A-17 Waveforms in ac analysis. (a) Sine wave (with no static value). (b) Sine square wave. (c) Rectified sine wave.

$$V_{rms} = \left[\frac{1}{\tau} \int_0^\tau (V_p \sin \omega t)^2 \, dt\right]^{1/2} = \left(\frac{1}{2} V_p^2\right)^{1/2} = \frac{1}{\sqrt{2}} V_p \qquad \text{(A-32)}$$

where $\tau = 2\pi/\omega$ is the period of the sinusoidal voltage. The square of a sine wave is shown in Fig. A-17b. The rms value is the square root of the instantaneous values that have been squared and averaged over one cycle, as illustrated in Eq. A-32. It can readily be shown that

$$v^2 = (V_p \sin \omega t)^2 = \frac{V_p^2}{2} - \frac{V_p^2}{2} \cos 2\omega t \qquad \text{(A-33)}$$

Hence the mean-square value is $V_p^2/2$. The rms value is $V_p/\sqrt{2} = 0.707 \, V_p$ and is denoted by the symbol V_{rms} or V. The rms value relates the power dissipation in a resistor by an ac voltage or current. For example, the instantaneous power P due to a voltage $v = V_p \sin \omega t$ through a resistor R is $P = v^2/R$. The mean power from Eq. A-33 is

$$\text{mean power} = \frac{I_p^2}{2} R = \frac{V_p^2}{2} \frac{1}{R} \qquad \text{(A-34)}$$

It can be shown that the *average value* of a rectified sine wave is $2V_p/\pi$. A rectified wave is obtained by converting the negative values to their corresponding positive values, while leaving the positive values unchanged, as shown in Fig. A-17c. The average value can be measured with a dc meter, and an ordinary ac voltmeter is basically a dc meter. The dial of the dc meter is adjusted to read ac rms values. It indicates the rms value only when the input waveform is sinusoidal.

B. Ohm's Law and Impedances

An impedance Z for ac analysis is defined from the generalization of Ohm's law in Eq. A-1. Impedances of circuit elements are described in this section. Impedances are complex numbers, and rms values are used only to denote voltages and currents.

Ohm's law is a component equation, since it describes the voltage/current relation of a component. The current I due to a voltage V applied across a component of impedance Z is

$$V = ZI \qquad \text{(A-35)}$$

where V and I are phasors, and Z is a complex number.

Let us derive the impedances for the circuit elements R, L, and C. When an ac voltage $v(t)$ is applied to the RLC circuit in Fig. A-13, the loop equation from Eq. A-24 is

$$L \frac{di}{dt} + Ri + \frac{1}{C} \int i \, dt = v(t)$$

Since the voltages and currents are sinusoidal, we substitute $(j\omega)$ from Eqs. A-28 and A-29 and obtain

$$\left(j\omega L + R + \frac{1}{j\omega C}\right)Ie^{j\omega t} = Ve^{j\omega t} \qquad (A\text{-}36)$$

Eliminating $e^{j\omega t}$ and defining $V = ZI$, we get

$$\left(j\omega L + R + \frac{1}{j\omega C}\right)I = V \qquad \text{or} \qquad Z = \left(j\omega L + R + \frac{1}{j\omega C}\right) \qquad (A\text{-}37)$$

where $j\omega L + R + 1/j\omega C$ is the impedance of R, L, and C in series, looking in from the driving voltage V.

The impedances of the R, L, and C are shown in Table A-2. The impedance of a resistor is R. Since R is real, the applied voltage and the current through R are in phase.

$$V = RI \qquad \text{or} \qquad I = \frac{V}{R} \qquad (A\text{-}38)$$

The impedance of an inductor L is $j\omega L$. The voltage drop across an inductor L is

TABLE A-2   Equations for AC Analysis

Circuit element	Impedance, Z	Reactance, X	Ohm's law: $V = ZI$	
Resistor, R	R	R	$V = R \times I$	$I = \dfrac{V}{R}$
Inductor, L	$j\omega L$	ωL	$V = j\omega L \times I$	$I = \dfrac{V}{j\omega L}$
Capacitor, C	$\dfrac{1}{j\omega C}$	$\dfrac{1}{\omega C}$	$V = \dfrac{I}{j\omega C}$	$I = j\omega C \times V$
Impedances in series:	$Z_{eq} = \sum\limits_{k} Z_k$			
Impedances in parallel:	$\dfrac{1}{Z_{eq}} = \sum\limits_{k} \dfrac{1}{Z_k}$			
Kirchhoff's voltage law:	$\sum\limits_{k} V_k = 0$	around a closed loop		
Kirchhoff's current law:	$\sum\limits_{k} I_k = 0$	at a common node		

$$V = ZI = (j\omega L)I \quad \text{or} \quad I = \frac{V}{j\omega L} \tag{A-39}$$

The current I flowing through L lags the applied voltage V by 90°. The impedance of a capacitor C is $1/j\omega C$. From Ohm's law, we get

$$V = ZI = \frac{1}{j\omega C}I \quad \text{or} \quad I = (j\omega C)V \tag{A-40}$$

The current I leads V by 90° for the capacitor. An ac current does not literally flow "conductively" through the dielectric of a capacitor. An ac current is due to the flow of electric charge alternately into and out of a capacitor. The plates of a capacitor are alternately charged and discharged accordingly.

Impedances in series and in parallel are analogous to resistors in series and in parallel. From Eqs. A-6 and A-10 we get

$$Z_{eq} = \sum_k Z_k \qquad \text{impedances in series} \tag{A-41}$$

$$\frac{1}{Z_{eq}} = \sum_k \frac{1}{Z_k} \qquad \text{impedances in parallel} \tag{A-42}$$

C. Kirchhoff's Laws and AC Circuits

The basic equations for ac circuit analysis are shown in Table A-2. Ohm's law is $V = ZI$ and Kirchhoff's voltage and current laws are

$$\sum_k V_k = 0 \qquad \text{around a closed loop} \tag{A-43}$$

$$\sum_k I_k = 0 \qquad \text{at a common node} \tag{A-44}$$

where V_k is a voltage drop (or rise) across a circuit element, and I_k a current flowing into (or away from) a common node. Both V_k and I_k are phasors and their rms values are commonly used for computations.

Example A-7.   An ac voltage $v(t) = V_p \sin \omega t$ is applied to the circuit in Fig. A-18. (a) Using vector notations, find the magnitude and phase angle of the ac current. (b) Repeat by using impedances.
Solution: (a) From Kirchhoff's voltage law, we get

$$L\frac{di}{dt} + Ri = V_p \sin \omega t \triangleq V_p e^{j\omega t}$$

where $V_p e^{j\omega t}$ is a voltage vector. Substituting the current vector $i = (I_p e^{-j\alpha})e^{j\omega t}$ into the equation and simplifying, we get

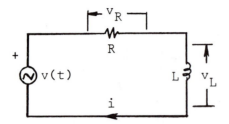

FIGURE A-18 Series circuit.

$$(j\omega L + R)(I_p e^{-j\phi}) = V_p \quad \text{or} \quad I_p e^{-j\phi} = \frac{1}{j\omega L + R} V_p$$

$$I_p = \frac{V_p}{|j\omega L + R|} = \frac{V_p}{[(j\omega L)^2 + R^2]^{1/2}} \quad \text{and} \quad \phi = \tan^{-1}\frac{\omega L}{R}$$

Since the voltage is a sine function, the current is

$$i = I_p \sin(\omega t - \phi)$$

(b) From Table A-2, the impedance of R and L in series is $Z = (R + j\omega L)$. From Ohm's law, we get

$$I = \frac{V}{Z} = \frac{V}{R + j\omega L} = \frac{V}{[R^2 + (\omega L)^2]^{1/2}} e^{-j\phi}$$

where $\phi = \tan^{-1}(\omega L/R)$, and $V/[R^2 + (\omega L)^2]^{1/2}$ is the magnitude of the current. The current lags the applied voltage by the phase angle ϕ. Under steady-state conditions, only the relative magnitude and phase angle of V and I are of interest.

Example A-8. Find the impedance Z_{ab} of the circuit shown in Fig. A-19.
Solution: The values of Z_1, Z_2, and Z_3 are

$$Z_1 = 3 + j0.5\omega \qquad Z_2 = 0 + j2\omega \qquad Z_3 = 4 - \frac{j4}{\omega}$$

Since Z_2 and Z_3 are in parallel, Z_{ab} is

$$Z_{ab} = Z_1 + \frac{Z_2 Z_3}{Z_2 + Z_3}$$

$$= (3 + j0.5\omega) + \frac{(j2\omega)(4 - j4/\omega)}{4 + j(2\omega - 4/\omega)}$$

Example A-9. Let $R = 6\,\Omega$, $X_L = 8\,\Omega$, $X_C = 16\,\Omega$, and $v(t) = 100$ V at 60

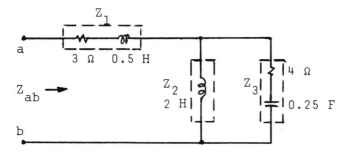

FIGURE A-19 Equivalent impedance Z_{ab}.

Hz, as shown in Fig. A-13. Find (a) the impedance of the circuit, and (b) the current associated with each of the circuit elements.

Solution: (a) The impedance of the series circuit is

$$Z = R + j(X_L - X_C) = 6 + j(8 - 16) = 10\underline{/-53.1°}$$

(b) From Ohm's law, the loop current is

$$I = \frac{V}{Z} = \frac{100}{10\underline{/-53.1°}} = 10\underline{/53.1°}$$

The rms value of the current is 10 A and the current leads the applied voltage by 53.1°.

The voltage across the resistor is $RI = 6(10) = 60$ V, and the voltage across R and the current I are in phase. The voltage V_L across L is $X_L I$ = 8(10) = 80 V, and V_L leads I by 90°. The voltage V_C across C is $X_C I = 16(10) = 160$ V, and V_C lags I by 90°. The same current I flows through all the circuit elements in a series circuit. The voltages across the elements are shown in Fig. A-20.

A-5. RESONANT CIRCUITS

Resonance occurs when a circuit is excited at one of its natural frequencies. If a system is undamped, no energy input is required to maintain the oscillation at a natural frequency. Hence the energy input at resonance is used to build up the amplitude of the oscillation. If a system possesses damping, the amplitude at resonance is finite. A steady state is reached when, over one cycle, the energy input is equal to the energy dissipation in the damper. Thus the amplitude at resonance is determined by the amount of damping in the system.

Consider the *series resonance* of the *RLC* circuit in Fig. A-13. Assume

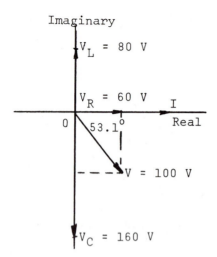

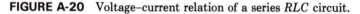

FIGURE A-20 Voltage–current relation of a series RLC circuit.

that the frequency ω of the applied voltage V is varied until a resonance occurs. The corresponding current I can be very large at resonance. Hence the impedance Z of a series circuit is a minimum at resonance. The impedance is

$$Z = R + j\left(\omega L - \frac{1}{\omega C}\right)$$

Since R is constant, a minimal Z requires that

$$\omega L = \frac{1}{\omega C} \quad \text{or} \quad \omega_0^2 = \frac{1}{LC} \tag{A-45}$$

where ω_0 rad/s is the resonant frequency. $Z = R$ at resonance, and the current is limited only by R or the energy-dissipating element in the circuit.

Example A-10. Let $R = 10\ \Omega$, $L = 250$ mH, $C = 0.1\ \mu$F, and $v(t) = 20$ V_{rms} for the series circuit shown in Fig. A-13. Find the voltage across each of the circuit elements at resonance.

Solution: Since $Z = R = 10\ \Omega$ at resonance, the current I is

$$I = \frac{V}{R} = \frac{20}{10} = 2\ \text{A}$$

The natural frequency from Eq. A-45 is

$$\omega_0 = (LC)^{-1/2} = [(250 \times 10^{-3})(0.1 \times 10^{-6})]^{-1/2} = 6.32 \times 10^3 \text{ rad/s}$$

The same current flows through all the circuit elements, and the magnitudes of the voltages across the elements are

$$V_R = RI = 10 \times 2 = 20 \text{ V}$$

$$V_L = X_L I = \omega L \times I = (6.32 \times 10^3)(250 \times 10^{-3})(2) = 3.16 \times 10^3 \text{ V}$$

$$V_C = X_C I = \frac{I}{\omega C} = \frac{2}{(6.32 \times 10^3)(0.1 \times 10^{-6})} = 3.16 \times 10^3 \text{ V}$$

Note the magnitude of the voltages V_L and V_C for the applied voltage of only 20 V. Kirchhoff's law is still obeyed. The voltage V_L and V_C are 180° out of phase and the overall voltage drop of the circuit is equal to the source voltage of 20 V.

An ideal *parallel resonant circuit* is shown in Fig. A-21a. Since a natural frequency is due to the rate of energy exchange between L and C in the circuit, the resonant frequency for the parallel circuit is the same as that for the series circuit. From Eq. A-42, the impedance Z_{ab} due to L and C in parallel is

$$Z_{ab} = \frac{j\omega L}{1 - \omega^2 LC} \tag{A-46}$$

Since $\omega^2 = \omega_0^2 = 1/LC$, the impedance at resonance is infinite for an ideal parallel circuit. In other words, the energy required to maintain the circuit at resonance is minimal. The circuit is called a "tank" circuit and is used in a radio for tuning to the radio frequencies of broadcasting stations. The resistance of the inductor must be included in the actual circuits shown in Fig. A-21b and c.

The sharpness of the resonance curve of a circuit is described by a quality factor Q:

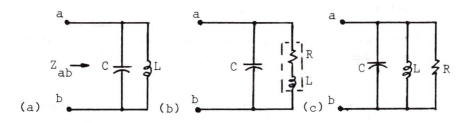

FIGURE A-21 Parallel resonant circuits. (a) Ideal. (b), (c) Actual.

$$Q = \frac{2\pi(\text{total energy stored})}{\text{energy dissipation per cycle}} \tag{A-47}$$

The total energy stored in a circuit at resonance is also the maximum energy. The total energy is stored either in the inductor L or the capacitor C. From Eqs. A-18 and A-20, we get

$$\text{total energy stored} = \tfrac{1}{2}LI_p^2 = \tfrac{1}{2}CV_p^2 \tag{A-48}$$

where I_p is the maximum current through L, and V_p the maximum voltage across C. From Eq. A-33, the energy dissipation per cycle in a series resistor R is

$$(\text{mean power})(\text{period}) = \frac{I_p^2 R}{2}\frac{2\pi}{\omega_0} = \frac{V_p^2}{2R}\frac{2\pi}{\omega_0} \tag{A-49}$$

For a series resonant RLC circuit, the same current flows through all the circuit elements. Using the terms involving I_p in the equations above, we get

$$\frac{\text{total energy stored}}{\text{energy dissipation per cycle}} = \frac{\tfrac{1}{2}LI_p^2}{\tfrac{1}{2}I_p^2 R(2\pi/\omega_0)} = \frac{\omega_0 L}{2\pi R}$$

Hence the Q of the circuit is

$$Q(\text{series}) = \frac{L}{R}\,\omega_0 = \frac{1}{R}\left(\frac{L}{C}\right)^{1/2} \tag{A-50}$$

For a high-Q circuit, the Q for a series circuit can also be found from the resonant curve in Fig. A-22. The maximum current at $\omega = \omega_0$ is $I_{\max} = V/R$. The half-power points occur at ω_1 and ω_2, when $I = 0.707I_{\max}$. The bandwidth is $\omega_2 - \omega_1$. It can be shown that

$$Q = \frac{\omega_0}{\omega_2 - \omega_1} \tag{A-51}$$

Evidently, the higher the Q, the narrower the bandwidth and the sharper the resonance curve. This is obvious from Eq. A-47, since a high-Q circuit is one with little energy dissipation.

For the parallel circuit in Fig. A-21c, the same voltage is applied across the circuit elements. Using the terms involving V_p in Eqs. A-48 and A-49, we get

$$Q(\text{parallel}) = \omega_0 RC = R\left(\frac{C}{L}\right)^{1/2} \tag{A-52}$$

Discussions of resonant circuits are found in the literature [see, e.g., B.

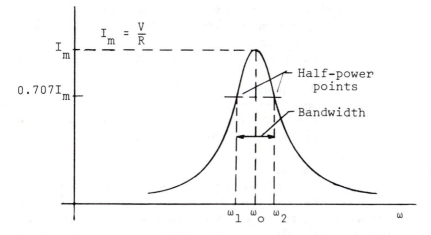

FIGURE A-22 Amplitude versus frequency of a series RLC resonant circuit.

Friedland, O. Wing, and R. Ash, *Principles of Linear Networks* (New York: McGraw-Hill Book Company, 1961), p. 225].

PROBLEMS

A-1. Find the currents I_1, I_2, and I_3 in the circuit shown in Fig. PA-1 by means of (a) the series–parallel technique, (b) loop equations, and (c) node equations.

A-2. Repeat Prob. A-1 for the circuit shown in Fig. PA-2.

A-3. Find the currents I_1, I_2, and I_3 in the circuit shown in Fig. PA-3 by means of (a) loop equations, and (b) node equations.

A-4. Write the loop equations for the circuit shown in Fig. PA-4.

A-5. Find the current I in Fig. PA-5.

A-6. The circuit in Fig. PA-6 is driven by a constant-current source. Deduce the current I_1 and I_2 and the voltages V_1 and V_2 in the circuit.

A-7. (a) Find the voltage V_o for the circuit shown in Fig. PA-7a. (b) Repeat for the circuit in Fig. PA-7b.

A-8. Find the input resistance R_{ab} of the circuit in Fig. PA-8.

A-9. Reduce the circuit to the left of the terminals a and b shown in Fig. PA-9 to its Thévenin's equivalent. Find the current through R_L when it is connected across a and b.

A-10. Repeat Prob. A-9 using Norton's theorem.

A-11. Repeat Prob. A-9 for the circuit shown in Fig. PA-11.

A-12. Repeat Prob. A-9 for the circuit shown in Fig. PA-11.

A-13. (a) Find the Thévenin's equivalent of the circuit shown in Fig. PA-12a. (b) Repeat for the circuit shown in Fig. PA-12b.

A-14. Write the node equation for the circuit shown in Fig. PA-13 and find V_a and V_b.

A-15. The voltage V_{ba} shown in Fig. PA-14 is measured with a multimeter. The specification is 5 kΩ/V_{ac}. Using the range 0 to 50 V, the measured voltage is 8.5 V. The meter is dialed to the range 0 to 10 V for a more accurate reading, and it reads 7 V. What is the true voltage?

A-16. An ammeter is used for measuring the current as shown in Fig. PA-15a. The meter consists of a 50-μA meter and a number of resistors, as shown in Fig. PA-15b. Using the range 0 to 150 mA, the measured current is $I = 14$ mA. Changing to the range 0 to 15 mA for a better reading, the measured current is 12 mA. What is the value of the true current?

A-17. Deduce the Thévenin's equivalent for the circuit shown in Fig. PA-16.

A-18. The circuit shown in Fig. PA-17a is called the Y network and that in Fig. PA-17b the Δ network. Derive the equations for the Y–Δ transformation.

A-19. Find the R_{eq} for the circuit shown in Fig. PA-18.

A-20. Repeat Prob. A-19 for the circuit shown in Fig. PA-19.

A-21. The circuit shown in Fig. PA-20 is the model of an ac amplifier, where v_{kg} is the voltage rise from K to G. Find the ac voltage ratio v_0/v_g.

A-22. Express the following in the polar form.

    (a) $(3 + j4)(3 - j4)$ (b) $\dfrac{1}{3 + j4}$

    (c) $3 + \dfrac{10}{1 + j}$ (d) $\dfrac{2 + j3}{3 + j2}$

A-23. Find the sum of the phasors.

    (a) $1\underline{/0°} + 1\underline{/90°}$ (b) $1\underline{/0°} + 1.732\underline{/90°} + 0°$

    (c) $3\underline{/-225°} + 3\underline{/45°}$ (d) $-3\underline{/215°} + 6\underline{/35°}$

    (e) $3\underline{/70°} + 4\underline{/160°}$ (f) $6\underline{/135°} + 1\underline{/270°}$

A-24. Find Z_{ab} for the circuits shown in Fig. PA-21.

A-25. Find the input voltage V for the circuit shown in Fig. PA-22. Assume that the frequency of V is $\omega = 3$ rad/s and the current is $I = 3\underline{/45°}$.

A-26. A voltage of 10 V at $\omega = 10^3$ rad/s is applied to the circuit shown in Fig. PA-23. (a) Find the voltages across each of the circuit elements. (b) Draw a vector diagram of the current and voltages.

A-27. (a) Find the currents I, I_R, and I_C for the circuit shown in Fig. PA-24. (b) Draw a vector diagram of the voltage and currents.

A-28. Find the Thévenin's equivalent of the circuit shown in Fig. PA-25. Assume that the excitation frequency is $\omega = 4$ rad/s.

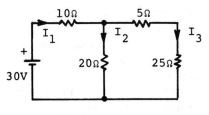

FIGURE PA-1

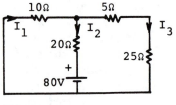

FIGURE PA-2

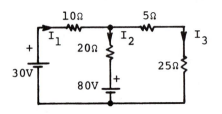

FIGURE PA-3

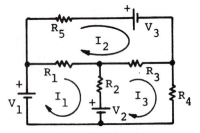

FIGURE PA-4

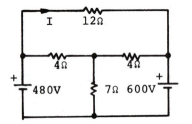

FIGURE PA-5

FIGURE PA-6

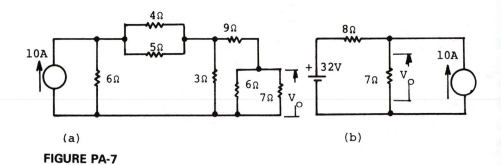

(a)

(b)

FIGURE PA-7

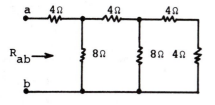

FIGURE PA-8

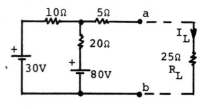

FIGURE PA-9

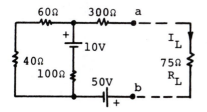

FIGURE PA-10

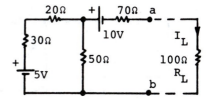

FIGURE PA-11

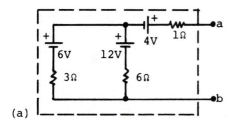

(a)

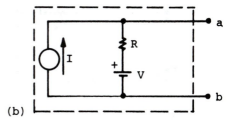

(b)

FIGURE PA-12

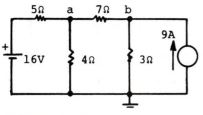

FIGURE PA-13

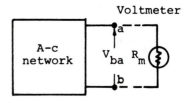

FIGURE PA-14

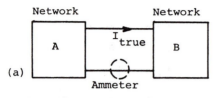

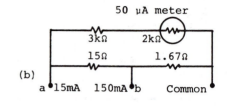

FIGURE PA-15

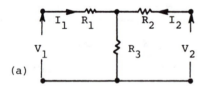

FIGURE PA-16

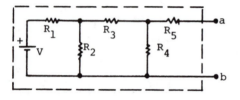

FIGURE PA-17

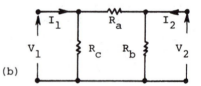

FIGURE PA-18

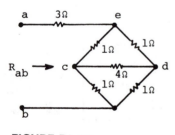

FIGURE PA-19

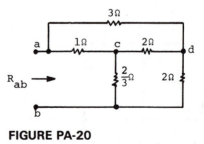

FIGURE PA-20

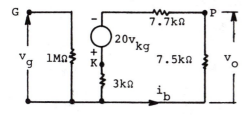

FIGURE PA-21

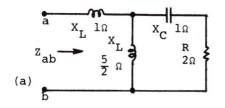

(a)

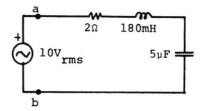

(b)

FIGURE PA-22

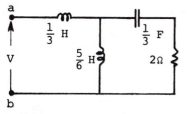

FIGURE PA-23

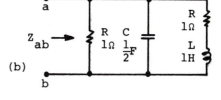

FIGURE PA-24

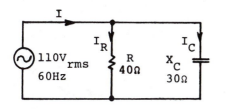

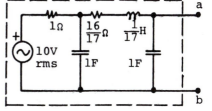

FIGURE PA-25

Appendix B
Linear Ordinary Differential Equations with Constant Coefficients

Solution of linear ordinary differential equations with constant coefficients by the "classical" method is reviewed briefly in this appendix.

B-1. PRELIMINARY

Some definitions are given in this preliminary section. A differential equation is an equation relating the variables and their derivatives or differentials. An nth-order ordinary differential equation with constant coefficients is

$$(a_n D^n + a_{n-1} D^{n-1} + \cdots + a_1 D + a_0)y = x(t) \tag{B-1}$$

where the a's are constants, n is the order of the differential equation, the time t is the independent variable, $D\ (= d/dt)$ is an operator, y is the unknown and the dependent variable, and $x(t)$ is a known time function.

The *solution* of a differential equation is the functional relation of the dependent and independent variables and no derivatives are involved in the relation. For example, a formal solution of Eq. B-1 is

$$y = f(t) \tag{B-2}$$

as shown in Fig. B-1. In other words, a value of y can be found for any given time at $t = t_1$, as illustrated in Fig. B-1.

It is convenient to use the D and the L linear differential operators for the discussions to follow. The familiar D *operator* is defined by the operations

$$Dy = \frac{d}{dt} y = \frac{dy}{dt} \qquad D^2 y = \frac{d}{dt} \frac{dy}{dt} = \frac{d^2 y}{dt^2} \qquad D^n y = \frac{d^n y}{dt^n} \tag{B-3}$$

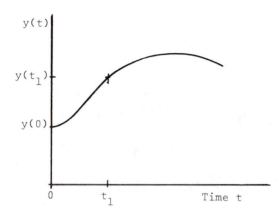

FIGURE B-1 Solution of $Ly = x(t)$.

The *L operator* is defined as

$$L = a_n D^n + a_{n-1} D^{n-1} + \cdots + a_1 D + a_0 \qquad (B\text{-}4)$$

Equation B-1 is written more precisely as

$$Ly = x(t) \qquad (B\text{-}5)$$

The superposition principle applies to the L operator.

$$L(C_1 y_1 + C_2 y_2) = C_1 Ly_1 + C_2 Ly_2 \qquad (B\text{-}6)$$

Equation B-1 or B-5 is an nth-order nonhomogeneous differential equation. The particular integral y_p is a solution of

$$Ly_p = x(t) \qquad (B\text{-}7)$$

The corresponding nth-order homogeneous differential equation is

$$Ly = 0 \qquad (B\text{-}8)$$

The solution of $Ly = 0$ is the complementary function $y_c(t)$. Let y_{c1}, $y_{c2}, \ldots, y_{cn}$ be the solutions of Eq. B-8. It can be shown that an nth-order homogeneous differential equation has n linearly independent solutions. Let y_c be a linear combination of the individual solutions.

$$y_c = C_1 y_{c1} + C_2 y_{c2} + \cdots + C_n y_{cn} \qquad (B\text{-}9)$$

where the C's are constants. Since L is a linear operator, a linear combination of the solutions of $Ly = 0$ is also a solution of the equation.

$$\begin{aligned} Ly_c &= L(C_1 y_{c1} + C_2 y_{c2} + \cdots + C_n y_{cn}) \\ &= C_1 Ly_{c1} + C_2 Ly_{c2} + \cdots + C_n Ly_{cn} = 0 \end{aligned} \qquad (B\text{-}10)$$

The *general solution* of $Ly = x(t)$ consists of the complementary function y_c and the particular integral y_p.

$$y = y_c + y_p \tag{B-11}$$

This is verified by substituting into Eq. B-7 to yield

$$Ly = L(y_c + y_p) = 0 + x(t) \tag{B-12}$$

B-2.   COMPLEMENTARY FUNCTION

The *complementary function* is the solution of the *homogeneous equation* in Eq. B-8.

$$(a_n D^n + a_{n-1}D^{n-1} + \cdots + a_1 D + a_0)y = 0 \tag{B-13}$$

Let $y_c = e^{st}$ be a solution, where s is a constant. Substituting y_c in Eq. B-13 and factoring out the e^{st} term, we get

$$a_n s^n + a_{n-1}s^{n-1} + \cdots + a_1 s + a_0 = 0 \tag{B-14}$$

This is the *characteristic equation*. It gives the constraint that must be satisfied in order that $y_c = e^{st}$ be a solution. By the fundamental theorem of algebra, Eq. B-14 has n number of roots.

If the n roots are distinct, the solution of Eq. B-13 is y_c.

$$y_c = C_1 e^{s_1 t} + C_1 e^{s_2 t} + \cdots + C_n e^{s_n t} \tag{B-15}$$

The roots of Eq. B-14 may be real, complex, imaginary, distinct, or repeating. All these cases can be examined by means of a second-order differential equation without loss of generality.

Case 1.   *Real and Distinct Roots*

Consider the second-order differential equation

$$(mD^2 + cD + k)y = 0 \tag{B-16}$$

The corresponding characteristic equation is

$$ms^2 + cs + k = 0 \tag{B-17}$$

The roots $s_{1,2}$ of Eq. B-17 and the solutions of Eq. B-16 are

$$s_{1,2} = \frac{1}{2m}[-c \pm (c^2 - 4mk)^{1/2}] \tag{B-18}$$

$$y_c = C_1 e^{s_1 t} + C_2 e^{s_2 t} \tag{B-19}$$

Since $(c^2 - 4mk) > 0$ for real roots, $s_{1,2}$ are real and negative.

Case 2.   Real and Repeating Roots

If the roots are real and repeating, Eq. B-18 gives

$$s = s_1 = s_2 = -\frac{c}{2m} \tag{B-20}$$

and there is only one solution from Eq. B-19. It can be shown that the solution y_c below will satisfy Eq. B-16.

$$y_c = (C_1 + C_2 t)e^{st} \tag{B-21}$$

Case 3.   Complex Roots

The roots are complex when $(c^2 - 4mk) < 0$. Complex roots occur in conjugate pairs. Let the complex conjugate roots be

$$s_{1,2} = -a \pm jb \tag{B-22}$$

where $j = \sqrt{-1}$, $a = c/2m$, $b = (4mk - c^2)^{1/2}/2m$, and a and b are real and positive. Substituting $s_{1,2}$ into Eq. B-19 and using Euler's formula, $e^{\pm j\theta} = \cos\theta \pm j\sin\theta$, the solution is

$$y_c = e^{-at}(C_1 e^{+jbt} + C_2 e^{-jbt}) \tag{B-23}$$

$$y_c = e^{-at}[(C_1 + C_2)\cos bt + j(C_1 - C_2)\sin bt] \tag{B-24}$$

$$y_c = e^{-at}(A \cos bt + B \sin bt) \tag{B-25}$$

Since y_c is a real physical quantity, A and B above are real coefficients. Similarly, the constants (C's) in Eq. B-23 are complex conjugates.

Case 4.   Imaginary Roots

The roots are imaginary when $c = 0$, and $s_{1,2} = \pm j\sqrt{4mk} = \pm j\omega_n$. This is a special case of complex roots when $a = 0$ and $b = \omega_n$. Thus

$$y_c = A \cos \omega_n t + B \sin \omega_n t \tag{B-26}$$

B-3.   PARTICULAR INTEGRAL

The *particular integral* y_p is a solution of the *nonhomogeneous equation* in Eq. B-1 or B-5.

$$Ly_p = x(t) \tag{B-27}$$

Since $x(t)$ is known, y_p does not contain arbitrary constants. If $x(t) = x_1(t) + x_2(t)$, the solution is obtained by superposition. From $Ly_{p1} = x_1(t)$ and $Ly_{p2} = x_2(t)$, we get

$$Ly_p = L(y_{p1} + y_{p2}) = x_1(t) + x_2(t) \tag{B-28}$$

For many practical problems, $x(t)$ is some combination of elementary functions, such as a sine function, a polynomial, or an exponential. The *method of undetermined coefficients* gives a simple procedure for finding y_p for these forcing functions.

The method is stated in two rules.

Rule 1. The trial function y_p in Eq. B-27 is a linear combination of $x(t)$ and all its independent derivatives.

Rule 2. When a trial function y_p from rule 1 has a term proportional to a term in y_c, a new trial function is substituted. The new trial function is the product of the initial trial function and the lowest integral power of t such that none of the terms in y_p is proportional to the terms in y_c.

Example B-1. Find the particular integral y_p of the equation

$$\tau \dot{y} + y = 5t \tag{B-29}$$

Solution: Let $y_p = At + B$, where A and B are the undetermined coefficients. Note that y_p is a linear combination of $5t$ and all its derivatives. Substituting y_p in Eq. B-29 and equating coefficients of t, we get $At = 5t$ and $A + B = 0$. The solution is

$$y_p = 5(t - \tau)$$

Example B-2. Find the particular integral of the equation

$$\ddot{y} + \omega_n^2 y = X \sin \omega t \tag{B-30}$$

Solution: The successive derivatives of $\sin \omega t$ consist of sine and cosine functions only. Hence the trial function is

$$y_p = A \sin \omega t + B \cos \omega t$$

Substituting this in Eq. B-30 and collecting the sine and cosine term, we obtain

$$(\omega_n^2 - \omega^2)A \sin \omega t + (\omega_n^2 - \omega^2)B \cos \omega t = X \sin \omega t$$

Hence the values of the undetermined coefficients A and B are

$$A = \frac{X}{\omega_n^2 - \omega^2} \quad \text{and} \quad B = 0$$

Example B-3. Find the particular integral of the equation

$$\ddot{y} + \omega_n^2 y = X \sin \omega_n t \tag{B-31}$$

Solution: The complementary function y_c of Eq. B-31 is

$$y_c = A \sin \omega_n t + B \sin \omega_n t$$

which is also the trial function y_p by rule 1. Hence the new trial function by rule 2 is

$$y_p = (A \sin \omega_n t + B \cos \omega_n t)t \tag{B-32}$$

Substituting Eq. B-32 in Eq. B-31, collecting the sine and cosine terms, and simplifying, we get

$$2\omega_n A \cos \omega_n t - 2\omega_n B \sin \omega_n t = X \sin \omega_n t$$

The coefficients in Eq. B-32 are $A = 0$ and $B = -X/(2\omega_n)$.

B-4. PARTICULAR SOLUTION

The general solution of the nonhomogeneous differential equation in Eq. B-1 is the sum of the complementary function y_c and the particular integral y_p, as shown in Eq. B-11. The *particular solution* is obtained from the general solution by giving particular values to these constants.

The constants in y_c are generally evaluated by the given initial conditions for the problem. There are as many constants in y_c as the order of the differential equation. For example, a third-order differential equation requires three initial conditions. Although the constants appear only in y_c, they are evaluated by applying the initial conditions to the general solution. In others words, the initial conditions should not be applied to y_c only, because it is the entire solution that must satisfied the conditions.

Example B-4.   Find the particular solution of the equation

$$20\ddot{y} + 30\dot{y} + 17,000y = 80 \cos 35t$$

for the initial conditions $y(0) = 0.025$ and $\dot{y}(0) = 0.300$.
Solution: It can be shown that

$$y_p = 0.0062 \cos (35t - 126°)$$
$$y_c = e^{-7.5t}(A \cos 28.2t + B \sin 28.2t)$$

The general solution is $y = y_c + y_p$. Applying the initial conditions to the general solution, we have

$$y(0) = 0.025 = A - 0.0036 \quad \text{or} \quad A = 0.0286$$
$$\dot{y}(0) = 0.300 = (28.2B - 7.5A) + 0.177 \quad \text{or} \quad B = 0.012$$

Thus the particular solution is

$$y = e^{-7.5t}(0.029 \cos 28.2t + 0.012 \sin 28.2t)$$
$$+ 0.0062 \cos (35t - 126°)$$

Appendix C
Fourier Transforms

The Fourier transform for transient studies in Chapter 4 are described in this appendix. The presentation includes the derivation of the Fourier transform from the exponential form of the Fourier series, the inverse transformation, and an illustration for obtaining the Fourier transform.

C-1. EXPONENTIAL FORM OF FOURIER SERIES

A periodic function $f(t)$ can be expressed in a *Fourier series:*

$$f(t) = \frac{a_0}{2} + \sum_{n=1}^{\infty} (a_n \cos n\omega_0 t + b_n \sin n\omega_0 t) \tag{C-1}$$

where τ is the period of $f(t)$ and $\omega_0 = 2\pi/\tau$ is the fundamental frequency in rad/s. The coefficients of the series are

$$a_n = \frac{2}{\tau} \int_{-\tau/2}^{\tau/2} f(t) \cos n\omega_0 t \, dt \qquad \text{for } n = 0, 1, 2, \ldots$$
$$b_n = \frac{2}{\tau} \int_{-\tau/2}^{\tau/2} f(t) \sin n\omega_0 t \, dt \qquad \text{for } n = 1, 2, 3, \ldots \tag{C-2}$$

The series is expressed in the *exponential form* using the identities

$$\cos n\omega_0 t = \frac{1}{2}(e^{jn\omega_0 t} + e^{-jn\omega_0 t})$$
$$\sin n\omega_0 t = \frac{1}{2j}(e^{jn\omega_0 t} - e^{-jn\omega_0 t}) \tag{C-3}$$

Substituting Eq. C-3 into Eq. C-1 and simplifying, we get

$$f(t) = c_0 + \sum_{n=1}^{\infty} (c_n e^{jn\omega_0 t} + c_{-n} e^{-jn\omega_0 t}) \tag{C-4}$$

where the new coefficients are identified as

$$c_0 = \tfrac{1}{2}a_0 \qquad c_n = \tfrac{1}{2}(a_n - jb_n) \qquad c_{-n} = \tfrac{1}{2}(a_n + jb_n) \tag{C-5}$$

The second term in the summation in Eq. C-4 is modified by substituting n for $-n$ and changing the limits of the summation:

$$\sum_{n=1}^{\infty} c_{-n}e^{-jn\omega_0 t} = \sum_{n=-1}^{\infty} c_n e^{jn\omega_0 t} \tag{C-6}$$

The exponential form of the Fourier series is obtained by substituting Eq. C-6 into Eq. C-4.

$$f(t) = \sum_{n=-\infty}^{\infty} c_n e^{jn\omega_0 t} \tag{C-7}$$

The complex coefficients c_n are obtained by substituting Eq. C-2 into Eq. C-5. It can be shown that

$$c_n = \frac{1}{\tau}\int_{-\tau/2}^{\tau/2} f(t)e^{-jn\omega_0 t}\, dt \tag{C-8}$$

Note that the c_n's are complex quantities and Eq. C-7 has negative frequencies when n is negative. Since $f(t)$ is a real physical quantity, the complex coefficients and negative frequencies are from the mathematical manipulations. The complex quantities must occur in conjugates to yield a real function $f(t)$.

Example C-1.　Find the exponential form of the Fourier series for the pulse train shown in Fig. C-1. Plot the frequency spectrum. Assume that the period is $\tau = \tfrac{1}{4}$ and the pulse duration is $T = \tau/6$.
Solution: The periodic pulse train $f(t)$ is

$$f(t) = \begin{cases} 0 \\ A \\ 0 \end{cases} \quad \text{for} \quad \begin{cases} -\tau/2 < t < 0 \\ 0 < t < T \\ T < t < \tau/2 \end{cases} \tag{C-9}$$

A coefficient c_n of the frequency spectrum is obtained by substituting Eq. C-9 into Eq. C-8.

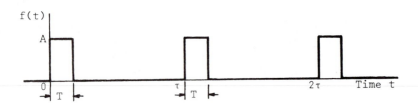

FIGURE C-1　Rectangular pulse train.

$$c_n = \frac{1}{\tau} \int_{-\tau/2}^{\tau/2} f(t) e^{-jn\omega_0 t}\, dt = \frac{1}{\tau} \int_0^T A e^{-jn(2\pi/\tau)t}\, dt$$

$$= \frac{A}{n\pi}\left(\sin \frac{n\pi T}{\tau}\right) e^{-jn\pi T/\tau} \tag{C-10}$$

$$|c_n| = \frac{A}{n\pi}\left|\sin \frac{n\pi T}{\tau}\right| \quad\text{or}\quad |c_n| = \frac{AT}{\tau}\left|\frac{\sin(n\pi T/\tau)}{n\pi T/\tau}\right| \tag{C-11}$$

The exponential form of the Fourier series is formed by substituting Eq. C-10 into Eq. C-7.

$$f(t) = \frac{A}{\pi} \sum_{n=-\infty}^{\infty} \frac{1}{n}\left(\sin \frac{n\pi T}{\tau}\right) e^{jn(2\pi/\tau)(t-T/2)} \tag{C-12}$$

where $\omega_0 = 2\pi/\tau$. Note that the magnitude d_n for the frequency spectrum from Eq. 4-78 (Chapter 4) is two times that of $|c_n|$ from Eq. C-5, and d_n is used for the plot in Fig. C-2. The value of c_0 is deduced from Eq. C-10 by l'Hospital's rule.

$$c_0 = \lim_{n\to 0} \frac{A}{n\pi} \sin \frac{n\pi T}{\tau} = A\frac{T}{\tau}$$

C-2. FOURIER INTEGRAL AND TRANSFORM PAIR

A transient is nonrepeating. It can be regarded as a periodic function with an extremely long period. A heuristic derivation of the Fourier integral is given in this section.

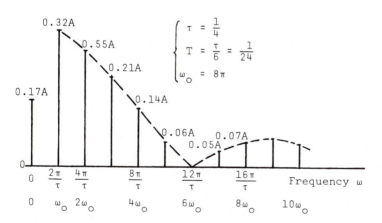

FIGURE C-2 Frequency spectrum of rectangular pulse train shown in Fig. C-1.

A periodic function has a discrete spectrum as illustrated in Fig. C-2. The interval between adjacent components in a spectrum is $\Delta\omega = \omega_0$, where ω_0 is the fundamental frequency in rad/s. The frequencies of the kth and the $(k + 1)$st components are

$$k\omega_0 = k\frac{2\pi}{\tau} \triangleq \omega_k \qquad \text{and} \qquad (k + 1)\omega_0 = (k + 1)\frac{2\pi}{\tau} \triangleq \omega_{k+1} \qquad \text{(C-13)}$$

$$\Delta\omega = (k + 1)\frac{2\pi}{\tau} - k\frac{2\pi}{\tau} = \frac{2\pi}{\tau} \qquad \text{(C-14)}$$

A typical Fourier coefficient c_k in Eq. C-8 is at the frequency $k\omega_0$. Using the notation $c(\omega_k)$ for c_k, and ω_k for $k\omega_0$, and substituting $1/\tau = \Delta\omega/2\pi$ into Eqs. C-7 and C-8, we get

$$f(t) = \sum_{k=-\infty}^{\infty} c(\omega_k)e^{j\omega_k t}$$

$$c(\omega_k) = \frac{\Delta\omega}{2\pi}\int_{-\tau/2} f(t)e^{-j\omega_k t}\,dt \qquad \tau/2$$

$$f(t) = \sum_{k=-\infty}^{\infty} \left[\frac{\Delta\omega}{2\pi}\int_{-\tau/2} f(t)e^{-j\omega_k t}\,dt\right]e^{j\omega_k t} \qquad \tau/2 \qquad \text{(C-15)}$$

As the period τ approaches infinity, $\Delta\omega$ becomes $d\omega$ in Eq. C-14, and the discrete spectrum in Fig. C-2 becomes a continuous spectrum. At the same time, the discrete variable ω_k in Eq. C-15 becomes a continuous variable ω, and the summation becomes an integration. Thus Eq. C-15 becomes the Fourier integral of the transient $f(t)$.

$$f(t) = \frac{1}{2\pi}\int_{-\infty}^{\infty}\left[\int_{-\infty}^{\infty} f(t)e^{-j\omega t}\,dt\right]e^{j\omega t}\,d\omega \qquad \text{(C-16)}$$

Defining the quantity inside the brackets as $g(j\omega)$, we obtain the Fourier transform pair:

$$f(t) = \frac{1}{2\pi}\int_{-\infty}^{\infty} g(j\omega)e^{j\omega t}\,d\omega = \mathscr{F}^{-1}[g(j\omega)] \qquad \text{(C-17)}$$

$$g(j\omega) = \int_{-\infty}^{\infty} f(t)e^{-j\omega t}\,dt = \mathscr{F}[f(t)] \qquad \text{(C-18)}$$

where $g(j\omega)$ is the Fourier transform $\mathscr{F}[f(t)]$ and $f(t)$ is the inverse transform $\mathscr{F}^{-1}[g(j\omega)]$.

The transform pair describes a physical event in two equivalent domains, where $f(t)$ is in the time domain and $g(j\omega)$ in the frequency domain. A problem can be analyzed in either domain and the results are mutually convertable. Hence it is possible to speak of the frequency content of a transient signal in Sec. 4-8.

The Fourier transform $g(j\omega)$ is a complex function of ω. Expanding $e^{-j\omega t}$ in Eq. C-18 as a sine and cosine function gives

$$g(j\omega) = \int_{-\infty}^{\infty} f(t) \cos \omega t \, dt - j \int_{-\infty}^{\infty} f(t) \sin \omega t \, dt \tag{C-19}$$

$$g(j\omega) = \text{Re}[g(j\omega)] + j \, \text{Im}[g(j\omega)] \tag{C-20}$$

where $\text{Re}[g(j\omega)]$ is the real part and $\text{Im}[g(j\omega)]$ the imaginary part of the Fourier transformation $g(j\omega)$. Hence

$$|g(j\omega)| = |\text{Re}[g(j\omega)] + j \, \text{Im}[g(j\omega)]| \tag{C-21}$$

$$\underline{/g(j\omega)} = \tan^{-1} \frac{\text{Im}[g(j\omega)]}{\text{Re}[g(j\omega)]} \tag{C-22}$$

Example C-2. (a) Determine the Fourier transform $g(j\omega)$ of the nonperiodic rectangular pulse $f(t)$ in Fig. C-3, where

$$f(t) = \begin{cases} A \\ 0 \end{cases} \quad \text{for} \quad \begin{cases} 0 < t < T \\ \text{everywhere else} \end{cases}$$

(b) Plot the corresponding $\text{Re}[g(j\omega)]$ and $\text{Im}[g(j\omega)]$.

Solution: (a) Substituting $f(t)$ in Eq. C-18, we get

$$g(j\omega) = \int_0^T A e^{-j\omega t} \, dt = \frac{A}{j\omega} (1 - e^{-j\omega T})$$

$$= \frac{A}{j\omega} (e^{+j\omega T/2} - e^{-j\omega T/2}) e^{-j\omega T/2}$$

$$= \frac{2A}{\omega} \left(\sin \frac{\omega T}{2} \right) e^{-j\omega T/2}$$

$$= A T \frac{\sin \omega T/2}{\omega T/2} e^{-j\omega T/2} \tag{C-23}$$

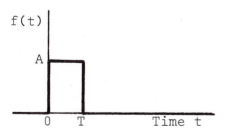

FIGURE C-3 Rectangular pulse.

(b) Substituting $e^{-j\omega T/2} = (\cos \omega T/2 - j \sin \omega T/2)$ in Eq. C-23 yields

$$g(j\omega) = \frac{2A}{\omega} \left(\sin \frac{\omega T}{2} \right) \left(\cos \frac{\omega T}{2} - j \sin \frac{\omega T}{2} \right)$$

$$= \frac{A}{\omega} \sin \omega T + j \frac{A}{\omega} (\cos \omega T - 1)$$

$$= \left(AT \frac{\sin \omega T}{\omega T} \right) + j \left(AT \frac{\cos \omega T - 1}{\omega T} \right)$$

$$= \text{Re}[g(j\omega)] + j \, \text{Im}[g(j\omega)] \tag{C-24}$$

The frequency spectrum $|g(j\omega)|$ versus frequency ω and the phase spectrum $/g(j\omega)$ versus ω plots are shown in Fig. C-4a. The $\text{Re}[g(j\omega)]$ and $\text{Im}[g(j\omega)]$ parts are shown in Fig. C-4b.

C-3.   INVERSE TRANSFORMATION

The transient $f(t)$ is obtained from the inverse Fourier transform of $g(j\omega)$ shown in Eq. C-17. The simplified inverse transformation and the properties of $g(j\omega)$ are described in this section.

   The integration in Eq. C-17 is simplified by assuming that $f(t)$ is a real-time function, and $f(t) = 0$ for $t < 0$. Expanding $g(j\omega)$ and $e^{j\omega t}$ in Eq. C-17 gives

$$f(t) = \frac{1}{2\pi} \int_{-\infty}^{\infty} \{\text{Re}[g(j\omega)] + j \, \text{Im}[g(j\omega)]\}(\cos \omega t + j \sin \omega t) \, d\omega$$

$$= \frac{1}{2\pi} \int_{-\infty}^{\infty} \{\text{Re}[g(j\omega)] \cos \omega t - \text{Im}[g(j\omega)] \sin \omega t\} \, d\omega$$

$$+ \frac{j}{2\pi} \int_{-\infty}^{\infty} \{\text{Re}[g(j\omega)] \sin \omega t + \text{Im}[g(j\omega)] \cos \omega t\} \, d\omega \tag{C-25}$$

From the assumption that $f(t)$ is a real-time function, the imaginary part of $f(t)$ in Eq. C-25 is zero. Note that the quantities in Eq. C-25 are functions of ω. The imaginary part has two terms in the integrand, and each must be an odd function of ω for the integral to be zero for $-\infty < \omega < \infty$. Since $\sin \omega t$ is an odd function, $\text{Re}[g(j\omega)]$ must be an even function of ω. Similarly, $\cos \omega t$ is an even function and $\text{Im}[g(j\omega)]$ is an odd function of ω.

   From the assumption that $f(t) = 0$ for $t < 0$, the real part of $f(t)$ in Eq. C-25 must be zero for $t < 0$. The real part has two terms in the integrand. The $\{\text{Re}[g(j\omega)] \cos \omega t\}$ term is an even function of t and $\{\text{Im}[g(j\omega)] \sin \omega t\}$ is an odd function of t. Their integrals for $t < 0$ must be equal and opposite to yield $f(t) = 0$ for $t < 0$. Furthermore, if these terms are even and odd

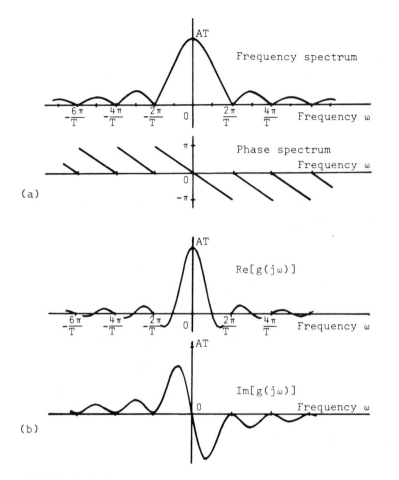

FIGURE C-4 Fourier transform $g(j\omega)$ of rectangular pulse shown in Fig. C-3.

functions of t, their integrals must be identical for $t > 0$. Finally, each term is an even function of ω, and the integration can be carried out over half of the range for $0 > \omega > \infty$.

By assuming that $f(t)$ is a real-time function and $f(t) = 0$ for $t < 0$, the simplified form of Eq. C-17 is

$$f(t) = \frac{2}{\pi} \int_0^\infty \text{Re}[g(j\omega)] \cos \omega t \, d\omega \qquad \text{for } t > 0 \qquad \text{(C-26)}$$

or

$$f(t) = -\frac{2}{\pi} \int_0^\infty \mathrm{Im}[g(j\omega)] \sin \omega t \; d\omega \qquad \text{for } t \quad 0 \tag{C-27}$$

Since the real and imaginary parts of $g(j\omega)$ are even and odd functions of ω, the following properties are self-evident:

$$\mathrm{Re}[g(-j\omega)] = \mathrm{Re}[g(j\omega)]$$
$$\mathrm{Im}[g(-j\omega)] = -\mathrm{Im}[g(j\omega)]$$
$$|g(-j\omega)| = |g(j\omega)| \tag{C-28}$$
$$/g(-j\omega) = -/g(j\omega)$$
$$g(-j\omega) = g^*(j\omega)$$

where $g^*(j\omega)$ is the complex conjugate of $g(j\omega)$. The first four properties above for a rectangular pulse are shown in Fig. C-4.

C-4. ILLUSTRATION FOR OBTAINING THE FOURIER TRANSFORM

Fourier transformation and the inverse are routinely performed by digital computes. An analog method for obtaining the transformation is described to give a better feeling for the subject.

The Fourier transform $g(j\omega)$ of a transient $f(t)$ is computed by means of Eq. C-19. Assuming that $f(t) = 0$ for $t < 0$, we get

$$g(j\omega) = \int_0^\infty f(t) \cos \omega t \; dt - j \int_0^\infty f(t) \sin \omega t \; dt \tag{C-29}$$

$$\mathrm{Re}[g(j\omega)] = \int_0^\infty f(t) \cos \omega t \; dt \tag{C-30}$$

$$-\mathrm{Im}[g(j\omega)] = \int_0^\infty f(t) \sin \omega t \; dt \tag{C-31}$$

Assume a transient $f(t)$ as shown in Fig. C-5a. $\mathrm{Re}[g(j\omega)]$ from Eq. C-30 is integrated with respect to time t with ω as a parameter. For a given $\omega = \omega_1$, the multiplication and integration are as shown in the figure. A value $\mathrm{Re}[g(j\omega_1)]$ is obtained when the integral reaches a constant value. The process is repeated for a range of ω to obtain $\mathrm{Re}[g(j\omega)]$ versus ω. Only positive values of ω need be considered, since $\mathrm{Re}[g(j\omega)]$ is an even function. Similarly, $\mathrm{Im}[g(j\omega)]$ can be computed as shown in Fig. C-5b.

The same procedure is used for finding the inverse transform for a given $g(j\omega)$, using the time t as a parameter. Either Eq. C-26 or C-27 can be used for the computation.

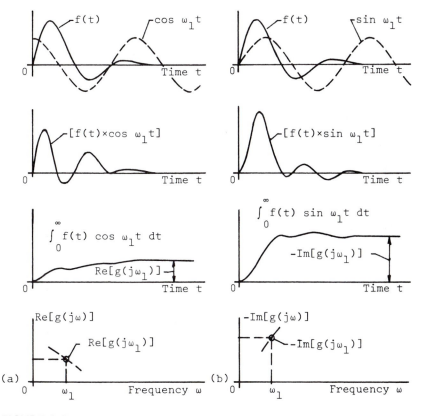

FIGURE C-5

SUGGESTED READINGS

Hsu, W. P., *Fourier Analysis,* Simon and Schuster, New York, 1967.

Papoulis, A., *The Fourier Integral and Its Applications,* McGraw-Hill Book Company, New York, 1962.

Sneddon, I. N., *Fourier Transforms,* McGraw-Hill Book Company, New York, 1951.

Stuart, R. D., *An Introduction to Fourier Analysis,* John Wiley & Sons, Inc., New York, 1966.

Index